Extremes in a Changing Climate

Water Science and Technology Library

VOLUME 65

For further volumes:
http://www.springer.com/series/6689

Amir AghaKouchak • David Easterling • Kuolin Hsu
Siegfried Schubert • Soroosh Sorooshian
Editors

Extremes in a Changing Climate

Detection, Analysis and Uncertainty

Springer

Editors
Amir AghaKouchak
Department of Civil and Environmental Engineering
University of California, Irvine
Irvine, CA
USA

Kuolin Hsu
Department of Civil and Environmental Engineering
University of California, Irvine
Irvine, CA
USA

Soroosh Sorooshian
Department of Civil and Environmental Engineering
University of California
Irvine, CA
USA

David Easterling
NOAA's National Climatic Data Center
Asheville, NC
USA

Siegfried Schubert
NASA/Goddard Space Flight Center
Greenbelt, MD
USA

ISSN 0921-092X
ISBN 978-94-007-4478-3 ISBN 978-94-007-4479-0 (eBook)
DOI 10.1007/978-94-007-4479-0
Springer Dordrecht Heidelberg New York London

Library of Congress Control Number: 2012951318

Printed on acid-free paper

Springer is part of Springer Science+Business Media (www.springer.com)

Foreword

The climate is changing owing to human activities. The predominant effect is from the changing atmospheric composition mainly from burning of fossil fuels that produces both visible pollution that blocks the sun and carbon dioxide. The particulates have a short lifetime as they are washed out of the atmosphere by rainfall, but carbon dioxide has a very long lifetime: typically over a 100 years and so it builds up and accumulates. Indeed carbon dioxide amounts have increased by over 40% since pre-industrial times and over half of this increase has occurred since 1970. Carbon dioxide is a greenhouse gas and produces a blanketing effect that results in a warming planet.

As well as increasing temperatures, a warming world is expected to alter precipitation in several ways. A warmer atmosphere can hold more water vapor, and will as long as there is a water supply nearby, as there always is over the oceans. The result is that more atmospheric moisture is pulled into storms resulting in heavier rains – or snows. But where it is not raining, the warmer atmosphere sucks moisture out of the soils and vegetation, promoting drought. Indeed, it is expected that a warming world alters the occurrence and magnitude of extremes such as droughts, heavy rainfalls and floods, as well as the geographic distribution of rain and snow. The winter snow season is likely to get shorter at each end and more precipitation events are apt to be rain. But snow events in winter can also be larger, as long as temperatures remain below about freezing. All of these changes are related to an acceleration of the hydrologic cycle and atmospheric circulation changes, and they include the direct impact of warmer conditions on atmospheric water vapor amounts, rainfall intensity, and snow-to-rain occurrence. However, extremes are inherently rare and difficult to observe reliably. But they are exceedingly important because of their nature and the tremendous damage that can ensue as conditions exceed those previously experienced. Key scientific issues relate to determining the statistics of extremes and how they are changing, and whether those changes are indeed caused by the human-induced changes in climate. These are very much the topic of this book.

A key issue then is what will happen in the future? For this we use climate models along with statistical approaches. But there are questions of how well models are able to handle extremes and how we can improve their capabilities. New improved and updated data sets at high frequency (e.g., hourly) are needed to properly characterize many of these facets of our climate and to allow for assessment against comparable model data sets. Then analyses are required that quantify which changes are consistent with our expectations and how we can best contribute to improving their prediction in a future climate. Confronting models with observationally-based products can lead to new metrics of performance and highlight shortcomings and developmental needs that may focus field programs, process studies, numerical experimentation, and model development. New applications should be developed for improved tracking and warning systems, and assessing changes in risk of drought, floods, river flow, storms, coastal sea level surges, and ocean waves. This book provides a welcome entrée into all of these areas.

Addressing the topic of changes in hydrologic extremes has been and continues to be one of the priorities of the World Climate Research Programme and especially the Global Energy and Water cycle Experiment (GEWEX). Indeed the workshop that is the focus of the material in this book grew directly from these interests.

National Center for Atmospheric Research Kevin E. Trenberth

Preface

The observed increase in weather and hydroclimatological extremes, particularly in the past decade, has brought much needed attention to the subject. In 2011, Texas experienced the driest year on record resulting in economic losses exceeding $7 billion. The Texas drought was by far the costliest drought on record with an approximately 90% increase from the previous global record set in 2006. In the past few years, major drought events have been recorded in Ethiopia, Australia, United States, Eurasia, Middle East, and southern Europe. At the other end of extreme, in 2010, Pakistan experienced the worst flooding in 80 years resulting in over 2,000 casualties, four million people displaced, and 20 million people affected directly or indirectly. The number of people suffering from this extreme event exceeds the combined total of the Indian Ocean tsunami (2004), Kashmir earthquake (2005), and Haiti earthquake (2010). Similar to droughts, every continent, in the past few years, has experienced major floods. For example, Australia, United States, Thailand, and China have set local records in flooding and the resulting economic losses. New records in terms of number of tornadoes in the United States in 2004 (1817) and 2011 (1691) and heat waves in 2003 (Europe) and 2010 (Russia) suggest that we should get used to climate extremes of all kinds and get more serious about developing better mitigation and adaptation strategies.

A question that has unfortunately become political and controversial is whether the observed changes are due to anthropogenic causes or whether our observations are just within the expected climate variability. Perhaps more important is the question of how climate extremes might change in the future. The first step to address these questions is detecting extremes and their variability. Predicting how weather and climate extremes might change in the future requires an understanding of their changes and behavior in the past. In fact, mitigating climate extremes requires a comprehensive and reliable study of statistics of extremes.

The main motivation for this book stems from the demand for more extensive and reliable methods for analyzing climate extremes in a nonstationary world. Given the limitations in the spatial and temporal resolutions of global climate data records, substantial progress in understanding extremes will largely rely on improvements in stochastic methods suitable for analyzing extremes. This book provides a collection of the state-of-the-art methodologies and approaches suggested for detecting extremes, trend analysis, accounting for nonstationarities, and uncertainties associated with extreme value analysis in a changing climate. This volume is designed so that it can be used as the primary reference on the available methodologies for studying climate extremes.

The first chapter introduces several statistical indices designed for detecting and diagnosing changes in climate extremes. Most statistical methods currently being used in engineering design are based on the stationary assumption (i.e., an unchanging climate in a statistical sense). Chapter 2 describes how extremal distributions can be retained under the nonstationary assumption. Chapter 3 extends the discussion by introducing a Bayesian framework for nonstationary analysis of extremes and their associated uncertainties. The chapter also provides a discussion of the so-called regional parameter concept using Bayesian hierarchical modeling. Chapter 4 is devoted to return-periods and return-levels estimation under climate change. The chapter examines two different definitions of return-period under the nonstationary assumption. Chapter 5 illustrates the application of Copula functions to multivariate extreme value analysis. Chapter 6 presents several parametric and nonparametric methods for tail dependence analysis. Chapter 7 discusses the theoretical framework, observational evidence, and related developments in stochastic modeling of weather and climate extremes. Chapter 8 surveys methods of projecting changes in extreme weather and climate statistics. Chapter 9 examines the simulated and observed short-term climate variability and weather extremes that have occurred over the last three decades with a focus on the winter hemispheres. Chapter 10 explores uncertainties in observed changes in climate extremes. Chapter 11 assesses uncertainties in the projection of future changes in precipitation extremes. Chapter 12 presents various data sets that are suitable for examining changes in extremes in the observed record. Finally, Chap. 13 is devoted to the concept of nonstationarity in extremes and engineering design.

Acknowledgment

Completing this book could not have been accomplished without the assistance and support of many individuals. First and foremost, the Editors wish to express sincere appreciation to the authors who shared their knowledge and expertise in climate extremes. We thank Kevin Trenberth for his thoughtful views on Extremes in a Changing Climate. We convey our gratitude to Professor Vijay Singh, Editor-in-Chief of the Springer's *Water Science and Technology Library*, and the members of the Editorial Advisory Board. Without their encouragement and support this work would not have been realized. We offer special thanks to the many reviewers of the book for their thoughtful comments and suggestions which led to improvements in the quality of the book. Appreciation is expressed to Scott Sellars and Andrea Thorstensen for reviewing some of the manuscripts and offering suggestions for improvement. We praise the professional service provided by the publication team of Springer Hermine Vloemans and Petra van Steenbergen throughout the preparation of this book. We wish to acknowledge that the idea for the book was motivated by discussions from a workshop on precipitation sponsored by the Army Research Office held in 2010. We thank Russell Harmon who has been instrumental in the planning and organization of the ARO workshop. Finally, the first author (Amir AghaKouchak) acknowledges the support from the US Bureau of Reclamation (Award No. R11AP81451).

Amir AghaKouchak
David Easterling
Kuolin Hsu
Siegfried Schubert
Soroosh Sorooshian

Contents

Contributors

Amir AghaKouchak Department of Civil and Environmental Engineering, University of California Irvine, Irvine, CA, USA

Joseph J. Barsugli Cooperative Institute for Research in Environmental Sciences, University of Colorado at Boulder, Boulder, CO, USA

Levi D. Brekke Bureau of Reclamation (U.S. Department of the Interior), Research and Development Office, Denver, CO, USA

Daniel Cooley Department of Statistics, Colorado State University, Fort Collins, CO, USA

Carlo De Michele Department of Hydraulic, Environmental, Roads and Surveying Engineering, Politecnico di Milano, Milano, Italy

David R. Easterling National Climatic Data Center, National Oceanic and Atmospheric Administration (NOAA), Asheville, NC, USA

Kuolin Hsu Department of Civil and Environmental Engineering, University of California Irvine, Irvine, CA, USA

Dörte Jakob School of Earth Sciences, Melbourne, Australia and Bureau of Meteorology, Climate and Water Division, University of Melbourne, Melbourne, Australia

Richard W. Katz Institute for Mathematics Applied to Geosciences, National Center for Atmospheric Research, Boulder, CO, USA

Kenneth E. Kunkel Department of Marine, Earth, and Atmospheric Sciences, North Carolina State University, Raleigh, NC, USA

Cooperative Institute for Climate and Satellites, National Climatic Data Center, National Oceanic and Atmospheric Administration (NOAA), Asheville, NC, USA

Michel Lang Irstea, UR HHLY, Hydrology-Hydraulics, Lyon, France

Young-Kwon Lim Global Modeling and Assimilation Office, NASA Goddard Space Flight Center (GSFC), Greenbelt, MD, USA

Goddard Earth Sciences Technology and Research (GESTAR), Greenbelt, MD, USA

Benjamin Renard Irstea, UR HHLY, Hydrology-Hydraulics, Lyon, France

Gianfausto Salvadori Dipartimento di Matematica e Fisica "E. De Giorgi", Università del Salento, Lecce, Italy

Siegfried D. Schubert Global Modeling and Assimilation Office, NASA Goddard Space Flight Center (GSFC), Greenbelt, MD, USA

Scott Sellars Department of Civil and Environmental Engineering, University of California Irvine, Irvine, CA, USA

Soroosh Sorooshian Department of Civil and Environmental Engineering, University of California Irvine, Irvine, CA, USA

Xun Sun Irstea, UR HHLY, Hydrology-Hydraulics, Lyon, France

Philip Sura Department of Earth, Ocean and Atmospheric Science, The Florida State University, Tallahassee, FL, USA

Kevin E. Trenberth National Center for Atmospheric Research, Boulder, CO, USA

Michael Wehner Lawrence Berkeley National Laboratory, Berkeley, CA, USA

Xuebin Zhang Climate Research Division, Environment Canada, Toronto, ON, Canada

Francis W. Zwiers Pacific Climate Impacts Consortium (PCIC), University of Victoria, Victoria, Canada

Chapter 1
Statistical Indices for the Diagnosing and Detecting Changes in Extremes

Xuebin Zhang and Francis W. Zwiers

Abstract This chapter introduces statistical indices that have been used to quantify the past changes in weather and climate extremes. These indices can also be used to assess changes in future extremes as projected by climate models. We also present examples in which the influence of anthropogenic climate change has been identified on extreme daily temperature, extreme daily precipitation, and the probability of occurrence for a specific extreme event.

1.1 Introduction

Weather and climate extremes have always played an important role in shaping the natural environment and pose significant challenges to society. For example, extremely cold winter temperatures strongly regulate over-winter survival of the spruce beetle in the Yukon and the mountain pine beetle in British Columbia, Canada. The insect freezes and dies below a threshold temperature of about −40°C, therefore, the occurrence of such very cold temperature events in winter affects the abundance of the beetle population in the following spring. Another example is the availability of water: too much or too little water can both pose strong challenges to society. As climate changes, weather and climate extremes will also change. Given their importance and the prospect of changes in the future, it is very important to understand how and why weather and climate extremes have changed in the past,

X. Zhang (✉)
Climate Research Division, Environment Canada, Toronto, Canada
e-mail: Xuebin.Zhang@ec.gc.ca

F.W. Zwiers
PCIC, University of Victoria, Victoria, Canada
e-mail: fwzwiers@uvic.ca

A. AghaKouchak et al. (eds.), *Extremes in a Changing Climate*, Water Science and Technology Library 65, DOI 10.1007/978-94-007-4479-0_1,

and how they will change in the future. One approach towards such an understanding would be to develop metrics with which weather and climate extremes can be characterized, quantified and monitored. However, there are no universally accepted metrics for such a purpose.

The word "extreme" can refer to many different things in the climate literature, and consequently, there is no unique climatological definition for extreme (Stephenson 2008). This occurs in part because the word "extreme" can be used to describe either a characteristic of a climate variable or that of an impact. In the case of a climate variable, such as surface air temperature or precipitation, an extreme can be reasonably well defined referring to values in the tails of the variable's distribution that would be expected to occur infrequently. In the case of an impact, an extreme may be less well defined since there may not be a unique way to quantify the impact. The linkage between an extreme event in the sense of a climate variable and an extreme event in the sense of the impact of a climate or weather event is not straightforward. A rare weather or climate event may not necessarily cause damages. For example, a strong wind associated with a tropical cyclone over the ocean may not result in any damage if there are no ships nearby. Similarly, not all damages are caused by rare weather and climate events. For instance, while the 2011 Thailand flood caused more than eight billion US dollars in insured damages, the amount of rain that fell in the region was not very unusual (van Oldenborgh et al. 2012).

In this chapter, we define some indices that can be derived from daily weather data. These indices allow the characterization of historical and future changes in extremes of weather variables. We will also provide some examples in which anthropogenic influence on weather and climate extremes can be detected and attributed. Our discussion will focus on weather and climate variables rather than impacts resulting from weather or climate events.

1.2 Indices of Extremes for Weather and Climate Variables

The Intergovernmental Panel on Climate Change in its 4th Assessment (Hegerl et al. 2007) defines an extreme climatic event as one that is rare at a particular place and time. The IPCC Special Report on Managing the Risks of Extreme Events and Disasters to Advance Climate Change Adaptation refines this definition, stating that "an extreme (weather of climate) event is generally defined as the occurrence of a value of a weather or climate variable above (or below) a threshold value near the upper (or lower) ends of the range of observed values of the variable" (Seneviratne et al. 2012). The idea of defining extremes as events in the tail(s) of a probability distribution is illustrated in Fig. 1.1. Figure 1.1 also shows, schematically, how warm or cold extremes are affected by changes in the mean or standard deviation of daily temperature and how extreme precipitation is affected by an increase in precipitation intensity. Exactly what constitutes an extreme event will depend on the context of usage.

When designing infrastructure, the engineering community accounts for climate extremes that occur only infrequently and are generally not expected to recur each year. Examples include the estimates of long period return values of annual maximum amount of rainfall within 5, 20 min, or within 1, 2, 6, 12, and 24 h to derive design rainfalls for sewage systems, or of annual maximum wind gust speed or annual maximum (accumulated) snow depth to derive codes for building design. In this case, the concept of extremes corresponds well to that used in statistical science, and thus powerful statistical tools based on extreme value theory are available to aid in the analysis of historical and future extremes (e.g. Coles 2001; Katz et al. 2002). Such tools were developed to infer extreme values that might occur beyond the range of the observed sample, such as the problem of estimating the 100 year return value on the basis of a 30-year sample. Increasingly, these tools are being used in the evaluation of extreme events simulated in climate models (e.g. Kharin et al. 2007; Wehner et al. 2010), in the characterization of the influence of large scale atmospheric circulation variations on extreme rainfall (Zhang et al. 2010), and in the detection of anthropogenic influence on temperature extremes (Zwiers et al. 2011).

To address the needs of various aspects of climate research on extremes as illustrated above, and to facilitate the monitoring of extremes, the Joint CCl[1]/CLIVAR[2]/JCOMM[3] Expert Team on Climate Change Detection and Indices (ETCCDI) defined a set of descriptive indices of extremes (Frich et al. 2002; Alexander et al. 2006; Klein Tank et al. 2009; Zhang et al. 2011). The indices were based on the European Climate Assessment indices (Klein Tank and Können 2003). They were chosen to sample a wide variety of climates and included indicators such as the total number of days annually with frost and the maximum number of consecutive dry days in a year (see Table 1.1). They can be derived from daily values of maximum and minimum temperatures, and daily precipitation amounts, and are designed to be easily updatable as more data become available, so as to facilitate monitoring. When developing those indices, it was realized they need to be as comparable as possible across different regions. Some of these indices have been used for the detection and attribution of changes in extremes.

The ETCCDI indices generally fall into three different types. One type of index measures the monthly and/or annual maxima or minima of daily temperature or annual maximum daily precipitation amounts. These types of extreme indices, like annual maximum 12 or 24 h precipitation amounts mentioned earlier, have been widely used in engineering applications to infer design values for engineered structures. Another type of index involves the calculation of the number of days

[1]The World Meteorological Organization (WMO) Commission for Climatology (CCl).

[2]The World Climate Research Program (WCRP) Climate Variability and Predictability Project (CLIVAR).

[3]The WMO and the UNESCO's Intergovernmental Oceanographic Commission (IOC) Joint Technical Commission for Oceanography and Marine Meteorology (JCOMM).

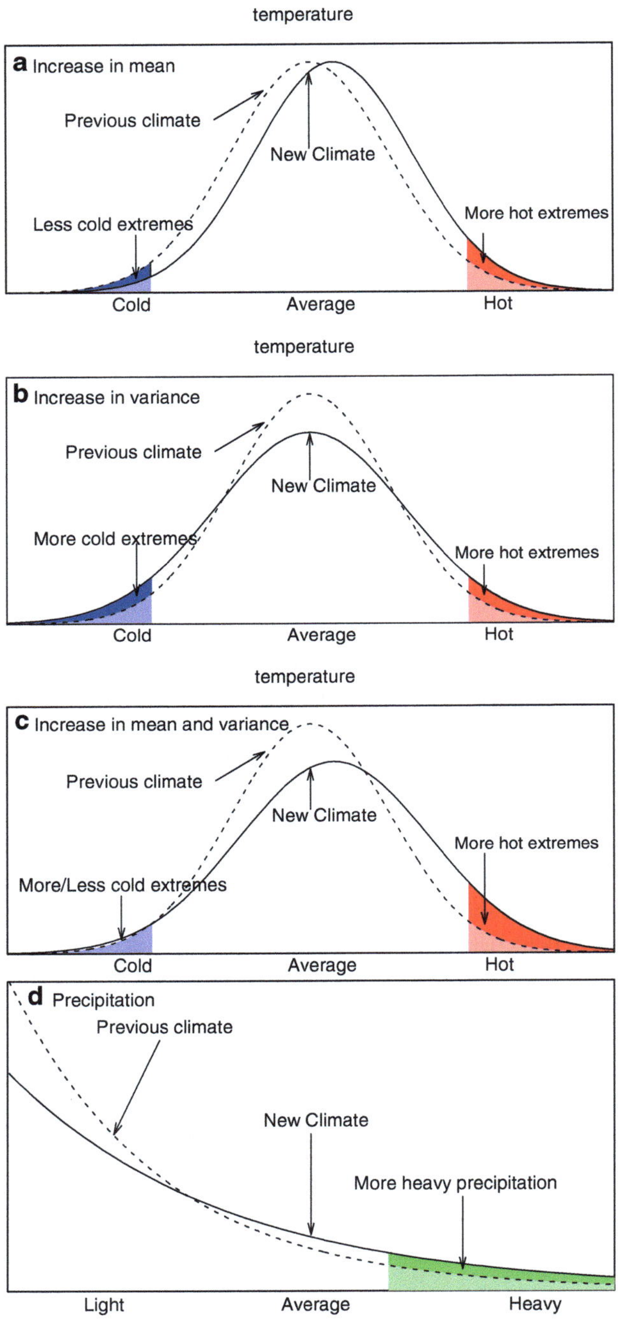

Fig. 1.1 (continued)

in a year exceeding specific thresholds that are relative to a fixed base period climatology. An example of this type of index is the number of days with daily minimum temperature below the long-term 10th percentile in the 1961–1990 base period for the respective calendar days. As the 10th percentile differs from one calendar day to another, extremes defined this way are a relative characteristic. Therefore, a minimum temperature of −10°C in Toronto in winter would not be considered as an extreme, but the same event would be very extreme in other seasons. Applying such a definition to data makes it possible to compare the indices from different places with different climates as the same part of the probability distribution of daily temperature is sampled at each location. A third type of index involves the calculation of the number of days in a year exceeding specific thresholds that are fixed across the space. The thresholds are typically impacts related. An example is the number of frost days per year (i.e., minimum temperature below 0°C). The phenomena may in itself not be an extreme in either the absolute or relative sense. The number of frost days and the number of growing season days are such examples. While such indices may not be applicable everywhere on the earth, and the phenomena may not occur in some places on the earth (e.g. a frost day would never occur in the tropics except perhaps in high mountains), they have a long history in many applications such as agriculture in many places.

1.3 Detection and Attribution of Changes in Climate Extremes

The identification of past changes in climate extremes is important. Understanding the possible causes of those changes is even more important. For example, understanding whether external forcing of the climate system, such as that due to human induced greenhouse gas emissions, had caused an observed change in extremes would affect confidence in the projection of future changes that are obtained from climate models that are driven with future emissions scenarios. One would have a

Fig. 1.1 Schematic representations of the probability density function of daily temperature, which tends to be approximately Gaussian (exceptions can be caused by soil freezing, energy balance constraints, and other factors such as snow or soil moisture feedbacks, e.g. Fischer and Schär 2009; Hirschi et al. 2011), and daily precipitation, which has a skewed distribution. The probability of occurrence, or frequency, of extremes is denoted by the shaded areas. In the case of temperature, changes in the frequencies of extremes are affected by changes in the mean, in the variance, and in both the mean and the variance. In a skewed distribution such as that of precipitation, a change in the mean of the distribution generally affects its variability or spread, and thus an increase in mean precipitation would also likely imply an increase in heavy precipitation extremes, and vice-versa. In addition, the shape of the right hand tail could also change, affecting extremes. Furthermore, climate change may alter the frequency of precipitation and the duration of dry spells between precipitation events. Figure 1.1a, b, c modified from Folland et al. (2001) and Fig. 1.1d modified from Peterson et al. (2008)

Table 1.1 The extreme temperature and precipitation indices recommended by the ETCCDI (some user defined indices are not shown). Precise definitions are given at http://cccma.seos.uvic.ca/ETCCDI/list_27_indices.html

ID	Indicator name	Indicator definitions	Units
TXx	Max Tmax	Monthly maximum value of daily max temperature	°C
TNx	Max Tmin	Monthly maximum value of daily min temperature	°C
TXn	Min Tmax	Monthly minimum value of daily max temperature	°C
TNn	Min Tmin	Monthly minimum value of daily min temperature	°C
TN10p	Cool nights	Percentage of time when daily min temperature <10th percentile	%
TX10p	Cool days	Percentage of time when daily max temperature <10th percentile	%
TN90p	Warm nights	Percentage of time when daily min temperature >90th percentile	%
TX90p	Warm days	Percentage of time when daily max temperature >90th percentile	%
DTR	Diurnal temperature range	Monthly mean difference between daily max and min temperature	°C
GSL	Growing season length	Annual (1st Jan to 31st Dec in NH, 1st July to 30th June in SH) count between first span of at least 6 days with TG > 5°C and first span after July 1 (January 1 in SH) of 6 days with TG < 5°C	days
FD0	Frost days	Annual count when daily minimum temperature <0°C	days
SU25	Summer days	Annual count when daily max temperature >25°C	days
TR20	Tropical nights	Annual count when daily min temperature >20°C	days
WSDI	Warm spell duration indicator	Annual count when at least 6 consecutive days of max temperature >90th percentile	days
CSDI	Cold spell duration indicator	Annual count when at least 6 consecutive days of min temperature <10th percentile	days

(continued)

Table 1.1 (continued)

ID	Indicator name	Indicator definitions	Units
CSDI	Cold spell duration indicator	Annual count when at least 6 consecutive days of min temperature <10th percentile	days
RX1day	Max 1-day precipitation amount	Monthly maximum 1-day precipitation	mm
RX5day	Max 5-day precipitation amount	Monthly maximum consecutive 5-day precipitation	mm
SDII	Simple daily intensity index	The ratio of annual total precipitation to the number of wet days (≥ 1 mm)	mm/day
R10	Number of heavy precipitation days	Annual count when precipitation ≥10 mm	days
R20	Number of very heavy precipitation days	Annual count when precipitation ≥20 mm	days
CDD	Consecutive dry days	Maximum number of consecutive days when precipitation <1 mm	days
CWD	Consecutive wet days	Maximum number of consecutive days when precipitation ≥1 mm	days
R95p	Very wet days	Annual total precipitation from days >95th percentile	mm
R99p	Extremely wet days	Annual total precipitation from days >99th percentile	mm
PRCPTOT	Annual total wet-day precipitation	Annual total precipitation from days ≥1 mm	mm

good reason to expect a change in extremes to continue in the future if the underlying cause of the change is well understood and is expected to continue into the future. In the climate literature, the identification of the changes and attribution to possible causes is usually termed detection and attribution. The good practice guidance paper on detection and attribution that was developed for the use by lead authors for the Intergovernmental Panel on Climate Change assessments (Hegerl et al. 2010) provides detailed definitions and outlines standard procedures for detection and attribution.

Detection is defined as the process of demonstrating that the climate or a system affected by climate has changed in some defined statistical sense without providing a reason for that change. An identified change is detected in observations if its likelihood of occurrence by chance due to variability generated by the climate system itself alone is determined to be small, for example, less than 10%. Attribution is defined as the process of evaluating the relative contributions of multiple causal factors to a change or event with an assignment of statistical confidence. Identification of a statistically significant trend in a time series of climate observations is a simple example of detection. Attribution is usually more complicated and can be conducted in several ways, though confidence in attribution will be differ from one approach to anther.

The observed climate variations, such as those that are seen by monitoring indices of extremes, generally reflect both natural variations internal to the climate system and the responses to external forcing such as changes in solar radiation and human induced changes in greenhouse gases. As the observed climate is only one realization of the climate system, it is difficult to separate internal variation from the response to external forcing using observations alone (Hegerl and Zwiers 2011). However, a climate model can be used to generate many realizations of the climate under the same forcing, enabling the estimation of both the response to external forcing, or signals, and the spatial and temporal structure of natural internal variation of the climate as represented by the climate model. As a result, a typical detection and attribution analysis involves the comparison of signals simulated by multiple climate models with the observations. A standard statistical approach to the detection and attribution problem is the optimal fingerprinting method (e.g., Allen and Stott 2003). This method is essentially a generalized regression, in which observed climate variations are regressed onto model simulated signals with the residual compared with model simulated variability.

Detection and attribution methods were first developed to consider changes in the mean of the distribution of climate variability, and because extremes are events that occur in the tails of such distribution, the detection and attribution of changes in extremes creates new challenges. Approaches that have been used either transfer the data into a form that allows the standard optimal fingerprinting method to be used, or explicitly considers the distributional properties of the extremes being analyzed. In the following, we illustrate, through examples, some detection/attribution approaches that have been employed in the context of extremes.

1.3.1 Changes in Extreme Temperatures

Zwiers et al. (2011) evaluated whether there has been an anthropogenic influence on long return period daily temperature extremes. In this study, they used observed 1961–2000 annual extreme temperatures, including annual maximum values of daily maximum (TXx) and daily minimum (TNx) temperatures, and annual minimum values of daily maximum (TXn) and daily minimum (TNn) temperatures, to characterize past changes in extreme temperatures. Recognizing that the real response to external forcing may have a different magnitude but similar spatial-temporal patterns to that simulated by climate models and that the distribution of annual maxima or minima temperatures is reasonably well approximated by the Generalized Extreme Value (GEV) distribution, they fit non-stationary GEV distributions (see Chap. 2 for details) for extreme temperatures with location parameters that vary in time according to climate model simulated responses. They found that the climate model simulated pattern of warming response to historical anthropogenic forcing in cold extremes fits observations best when its amplitude is scaled up by a factor greater than one, and that in warm extremes the fit to observations is best when the amplitude is scaled down. They quantified that globally, waiting times for extreme annual minimum daily minimum and daily maximum temperature events that were expected to recur once every 20 years in the 1960s are now estimated to exceed 35 and 30 years, respectively. In contrast, waiting times for circa 1960s 20-years extremes of annual maximum daily minimum and daily maximum temperatures are estimated to have decreased to fewer than 10 and 15 years, respectively. Figure 1.2 displays estimated return periods and their 5 and 95% uncertainty limits for circa 1960s 20-year return values of annual extreme daily temperatures in the 1990s climate for many sub-continental regions and the global land area in total.

Morak et al. (2012) analyzed global and regional long-term trends in the frequency of hot and cold temperature extremes defined as the number of days exceeding the 90th percentile or not reaching the 10th percentile of daily minimum (TN90, TN10, see Table 1.1 for definitions) and daily maximum (TX90 and TX10) temperatures. They compared the observed trends with those simulated by a climate model using the optimal fingerprinting method. The signals (or fingerprints) were obtained from an ensemble of historical forcing simulations conducted with a climate model developed at the UK Met Service Hadley Centre (HadGEM1). Both observations and the means of ensemble simulations under combined natural and anthropogenic forcings show an increase in the frequency of warm extreme and a decrease in cold extremes in both boreal warm and cold seasons, though there are regional differences. They also found that anthropogenic influence is detectable in the frequency of warm and cold temperatures.

1.3.2 Anthropogenic Influence on Annual Maximum 1- or 5-Day Precipitation

Detection and attribution of possible anthropogenic influence on extreme precipitation has proven to be a substantially bigger challenge. Observed daily precipitation is highly variable from place to place, and exhibits spatial scales of variations that are smaller than climate models can represent. This means that large numbers of densely spaced precipitation observing stations are required to be able to isolate the component of observed precipitation that is directly comparable to simulations. Unfortunately, there are few regions globally with a sufficiently dense observing network to enable this kind of comparison. Min et al. (2011) therefore tried to circumnavigate this scale issue by transforming the observed and simulated annual maximum 1- or 5-day precipitation amounts into probability based indices (PI) before applying the standard optimal detection method. To do this they used separate transformations based on the GEV distribution for observations and model output, thereby bringing both onto a dimensionless scale on the unit interval.

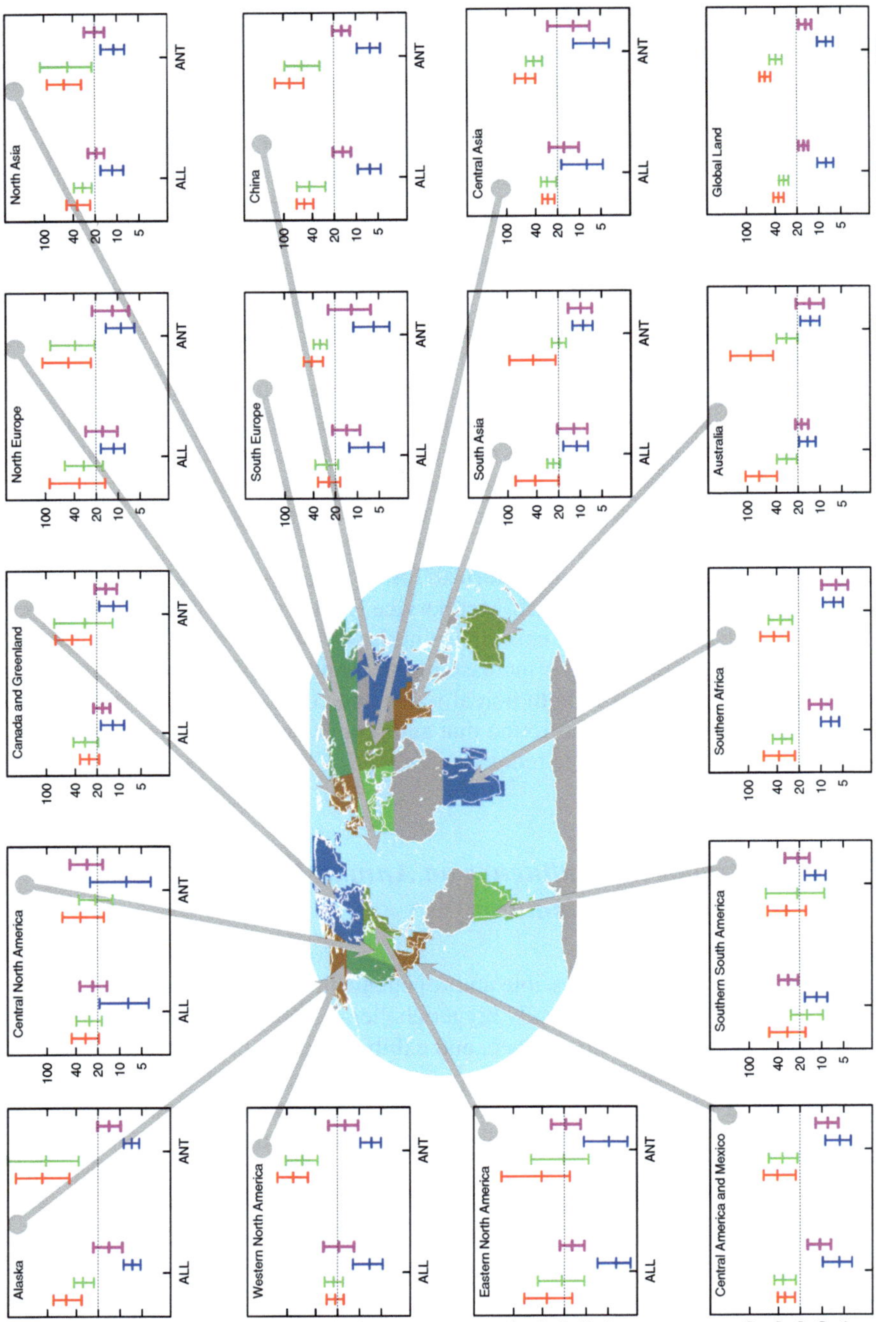

Fig. 1.2 (continued)

Figure 1.3 shows trends of extreme precipitation indices for annual maximum 1-day precipitation amounts during 1951–1999 from observations, and from model simulations under anthropogenic forcing or under combined effect of anthropogenic and natural forcings. They found evidence of human influence in observed changes in precipitation extremes during the latter half of the twentieth century .It was found that a best fit with observations required that the magnitude of the climate model simulated responses to external forcing be increased by a large factor, which limits confidence in the attribution of observed changes.

1.3.3 Event Attribution

When a rare and catastrophic meteorological extreme event occurs, a question that is often posed is whether such an event is due to anthropogenic influence. Because it is very difficult to rule out the occurrence of low probability events in an unchanged climate, and because the occurrence of such events usually involves multiple factors, it is very difficult to attribute an individual event to specific causes (Hegerl et al. 2007). However, in this case, it may nevertheless be possible to estimate the influence of external forcing on the likelihood of occurrence of such an event. Such estimates of how the likelihood of events has changed are especially useful from risk management point of view. The infamous 2003 European heat wave caused huge impacts in Europe, with an estimated 40,000 heat related deaths García-Herrera et al. 2010). To address the question whether anthropogenic contributed to the severity/occurrence of the 2004 European heat wave, Stott et al. (2004) used an event attribution method. With this method, they first detected anthropogenic influence on mean summer temperature in southern Europe; they then estimated the effect of anthropogenic forcing on the likelihood of a warm summer, and finally inferred an anthropogenic influence on the likelihood of the 2003 European heat wave. Note that the attribution result for one event may not necessarily extend to another event. Using another method, Dole et al. (2011) suggest that the 2010 Russian heatwave could have occurred without anthropogenic influence. This finding is not necessarily inconsistent with the possibility that human influence may have increased the odds of this occurrence of the event.

Fig. 1.2 Estimated return periods (years) and their 5 and 95% uncertainty limits for 1960s 20-years return values of annual extreme daily temperatures in the 1990s climate (see text for more details). *ANT* refers to model simulated responses with only anthropogenic forcing and *ALL* is both natural and anthropogenic forcing. Error bars are for annual minimum daily minimum temperature (*red*: TNn), annual minimum daily maximum temperature (*green*: TXn), annual maximum daily minimum temperature (*blue*: TNx), and annual maximum daily maximum temperature (*pink*: TXx), respectively. *Grey* areas have insufficient data

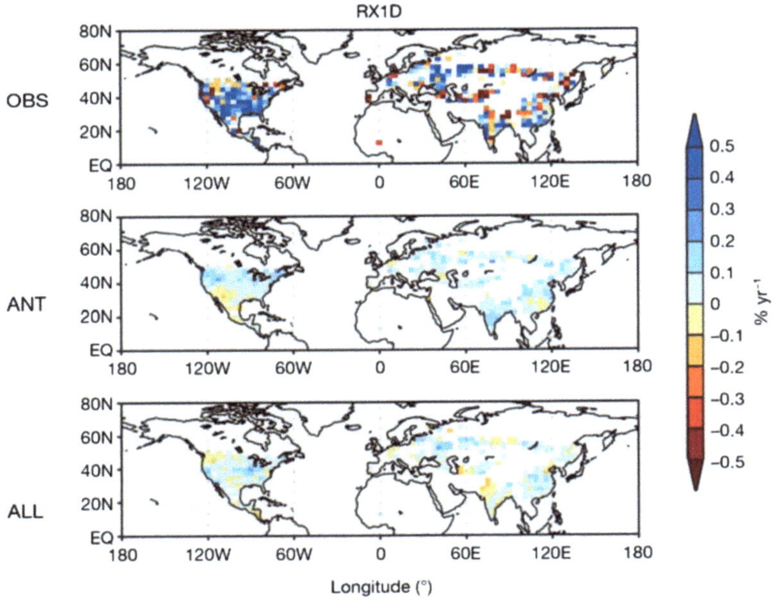

Fig. 1.3 Geographical distribution of trends of extreme precipitation indices (PI) for annual maximum daily precipitation amounts (*RX1D*) during 1951–99. Observations (*OBS*); model simulations with anthropogenic (*ANT*) forcing; model simulations with anthropogenic plus natural (*ALL*) forcing. For models, ensemble means of trends from individual simulations are displayed. Units: per cent probability per year (From Min et al. (2011; see paper for details))

1.4 Summary

In this chapter, we introduced some concepts for definition of statistical indices with which past changes in weather and climate extremes may be quantified. These indices can also be used to monitor ongoing changes in extremes, and they can be used to assess changes in future extremes as projected by climate models. We also present a few examples in which influence of anthropogenic climate changes on extremes has been identified. It should be noted that there appears to be no unique method to define what constitutes a weather or a climate extreme. While it is vastly important to understand how and why weather and climate extremes may have changed in the past and how they will behave in the future, there are significant challenges towards a full understanding.

References

Alexander LV et al (2006) Global observed changes in daily climate extremes of temperature and precipitation. J Geophys Res 111:D05109. doi:10.1029/2005JD006290

Allen MR, Stott PA (2003) Estimating signal amplitudes in optimal fingerprinting. Clim Dyn 15:419–434

Coles S (2001) An introduction to statistical modeling of extreme values. Springer, London, 208pp
Dole R, Hoerling M, Perlwitz J, Eischeid J, Pegion P, Zhang T, Quan X-W, Xu T, Murray D (2011) Was there a basis for anticipating the 2010 Russian heat wave? Geophys Res Lett 38:L06702. doi:10.1029/2010GL046582
Fischer EM, Schär C (2009) Future changes in daily summer temperature variability: driving processes and role for temperature extremes, Clim Dyn. doi:10.1007/s00382–008–0473–8
Folland CK et al (2001) Observed climate variability and change. In: Houghton JT et al (eds) Climate change 2001: the scientific basis. Contribution of Working Group I to the third assessment report of the Intergovernmental Panel on Climate Change. Cambridge University Press, Cambridge, United Kingdom and New York, NY, pp 99–181
Frich P, Alexander LV, Della-Marta P, Gleason B, Haylock M, Klein Tank AMG, Peterson T (2002) Observed coherent changes in climatic extremes during the second half of the twentieth century. Clim Res 19:193–212
García-Herrera R, Díaz J, Trigo RM, Luterbacher J, Fischer EM (2010) A review of the European summer heatwave of 2003. Crit Rev Environ Sci Technol 40:267–306
Hegerl G, Zwiers F (2011) Use of models in detection and attribution of climate change. WIREs Clim Change 2:570–591. doi:10.1002/wcc.121
Hegerl GC, Zwiers FW, Braconnot P, Gillett NP, Luo Y et al (2007) Understanding and attributing climate change. In: Solomon S, Qin D, Manning M, Chen Z, Marquis M et al (eds) Climate change 2007: the physical science basis. Contribution of Working Group I to the fourth assessment report of the Intergovernmental Panel on Climate Change. Cambridge University Press, Cambridge
Hegerl GC, Hoegh-Guldberg O, Casassa G, Hoerling MP, Kovats RS, Parmesan C, Pierce DW, Stott PA (2010) Good practice guidance paper on detection and attribution related to anthropogenic climate change. In: Stocker TF, Field CB, Qin D, Barros V, Plattner G-K, Tignor M, Midgley PM, Ebi KL (eds) Meeting report of the intergovernmental panel on climate change expert meeting on detection and attribution of anthropogenic climate change. IPCC Working Group I Technical Support Unit, University of Bern, Bern
Hirschi M, Seneviratne SI, Alexandrov V, Boberg F, Boroneant C, Christensen OB, Formayer H, Orlowsky B, Stepanek P (2011) Observational evidence for soil-moisture impact on hot extremes in southeastern Europe. Nat Geosci. doi:10.1038/NGEO1032
Katz RW, Parlange MB, Naveau P (2002) Statistics of extremes in hydrology. Adv Water Resour 25:1287–1304
Kharin VV, Zwiers FW, Zhang X, Hegerl GC (2007) Changes in temperature and precipitation extremes in the IPCC ensemble of global coupled model simulations. J Climate 20:1419–1444
Klein Tank AMG, Können GP (2003) Trends in indices of daily temperature and precipitation extremes in Europe, 1946–99. J Climate 16:3665–3680
Klein Tank AMG, Zwiers FW, Zhang X (2009) Guidelines on analysis of extremes in a changing climate in support of informed decisions for adaptation. WMO, Geneva, p 56 (WCDMP-72, WMO-TD/No.1500)
Min S-K, Zhang X, Zwiers FW, Hegerl GC (2011) Human contribution to more intense precipitation extremes. Nature 470:378–381
Morak S, Hegerl GC, Christidis N (2012) Detectable changes in temperature extremes. J Climate, accepted
Peterson TC, Anderson D, Cohen SJ, Cortez M, Murname R, Parmesan C, Phillips D, Pulwarty R, Stone J (2008) Why weather and climate extremes matter. In: Karl TR et al (eds) Weather and climate extremes in a changing climate. Regions of focus: North America, Hawaii, Caribbean, and U.S. Pacific Islands. U.S. Climate Change Science Program and the Subcommittee on Global Change Research, Washington, DC, pp 11–33
Seneviratne SI, Nicholls N, Eastering D, Goodness CM, Kanae S, Kossin J, Luo Y, Marengo J, McInnes K, Rahimi M, Reichstein M, Sorteberg A, Vera C, Zhang X (2012) Changes in climate extremes and their impacts on the natural physical environment. In: Field CB, Barros V, Stocker TF, Qin D, Dokken DJ, Ebi KL, Mastrandrea MD, Mach KJ, Plattner G-K, Allen SK, Tignor M, Midgley PM (eds) Managing the risks of extreme events and disasters to advance climate

change adaptation, A Special Report of Working Groups I and II of the Intergovernmental Panel on Climate Change (IPCC). Cambridge University Press, Cambridge/New York, pp 109–230
Stephenson DB (2008) Definition, diagnosis, and origin of extreme weather and climate events. In: Diaz HF, Murnane RJ (eds) Climate extremes and society. Cambridge University Press, Cambridge
Stott PA, Stone DA, Allen MR (2004) Human contribution to the European heat wave of 2003. Nature 432:610–614. doi:10.1038/nature0308
Van Oldenborgh GJ, van Urk A, Allens M (2012) The absence of a role of climate change in the 2011 Thailand floods, in Explaining extreme events of 2011 from a climate perspective. In: Peterson TC et al (ed) BAMS, 93:1041–1067. doi:10.1175/BAMS-D-12-00021.1
Wehner MF, Smith RL, Bala G, Duffy P (2010) The effect of horizontal resolution on simulation of very extreme precipitation events in a global atmospheric model. Clim Dyn 34:241–247. doi:10.1007/s00382-009-0656-y
Zhang X, Wang J, Zwiers FW, Ya Groisman P (2010) The influence of large scale climate variability on winter maximum daily precipitation over North America. J Climate 23:2902–2915
Zhang X, Alexander L, Hegerl GC, Jones P, Tank AK, Peterson TC, Trewin B, Zwiers FW (2011) Indices for monitoring changes in extremes based on daily temperature and precipitation data. WIREs Clim Change 2:851–870. doi:10.1002/wcc.147
Zwiers FW, Zhang X, Feng Y (2011) Anthropogenic influence on long return period daily temperature extremes at regional scales. J Climate. doi:10.1175/2010JCLI3908.1

Chapter 2
Statistical Methods for Nonstationary Extremes

Richard W. Katz

Abstract There is a long tradition of the use of methods based on the statistical theory of extreme values in hydrology, particular for engineering design (e.g., for the proverbial "100-yr flood"). For the most part, these methods are based on the assumption of stationarity (i.e., an unchanging climate in a statistical sense). The focus of this chapter is on how the familiar distributions that arise in extreme value theory, namely the generalized extreme value (GEV) and generalized Pareto (GP) distributions, can be retained under nonstationarity. But now the extremal distribution is allowed to gradually shift by introducing time as a covariate; that is, expressing one or more of the parameters of the distribution as a function of time. At least for the parameter estimation technique of maximum likelihood, it is straightforward to fit such statistical models. Some detailed examples are provided of how the proposed methods can be applied to the detection and statistical modeling of trends in hydrologic extremes, such as for stream flow and precipitation.

2.1 Introduction

There is a long tradition of the use of methods based on the statistical theory of extreme values in hydrology, particular for engineering design (e.g., for the proverbial "100-yr flood"). For the most part, these methods are based on the assumption of stationarity (i.e., an unchanging climate in a statistical sense). In recent years, the specter of global climate change has been raised in conjunction with the enhanced greenhouse effect. Physical considerations suggest an intensification of the hydrologic cycle, with increases in the frequency and intensity of extreme high

R.W. Katz (✉)
Institute for Mathematics Applied to Geosciences, National Center
for Atmospheric Research, Boulder, CO, USA
e-mail: rwk@ucar.edu

A. AghaKouchak et al. (eds.), *Extremes in a Changing Climate*, Water Science and Technology Library 65, DOI 10.1007/978-94-007-4479-0_2,

precipitation events (Allen and Ingram 2002; Emori and Brown 2005; Trenberth et al. 2003) and the possibility of more severe floods (Milly et al. 2008). So the question naturally arises of how the statistical theory of extreme values can be extended to the case of nonstationarity, the subject of the present chapter.

Although the influence of global climate change on hydrologic extremes such as high stream flow can be quite difficult to detect (Villarini et al. 2009), it is straightforward to find hydrologic examples in which nonstationarity is present because of more local, direct human influence. In particular, Fig. 2.1 shows the annual time series of mean and peak flow at Mercer Creek, WA, a relatively small drainage basin experiencing a period of rapid urbanization starting about 1970. Although no trend is evident in mean flow, a marked increase in peak flow occurred during the 1970s and early 1980s (Gilleland and Katz 2011). Such behavior concurs with our understanding of land use effects on stream flow, with development resulting in more rapid runoff (i.e., higher peak flows) even without any change in total flow (Konrad 2003).

When confronted with nonstationarity in extremes, hydrologists have tended to abandon methods based on extreme value theory (an inconsistency pointed out, for instance, by Clarke 2002). Instead, they have resorted to statistical methods either inappropriate or, at least, inefficient for extremes. While nonparametric techniques, such as a Mann-Kendall test for trend or a Pettitt test for a change point (e.g., Helsel and Hirsch 1993), have a long tradition of use in hydrology and have been applied in the case of extremes (e.g., Villarini et al. 2009), they were not designed for dealing with extremes per se. Given the necessarily quite limited observations available for extremes, it can be persuasively argued that making use of a theory, albeit approximate, would be at least potentially more efficient. Further, such a method can produce information about trends in a form desirable for water resource management (i.e., in terms of shifts in return levels).

An alternative approach common in hydrology has been to apply regression analysis with a trend variable to hydrologic time series transformed to an approximate normal distribution. For instance, Vogel et al. (2011) applied the logarithmic transformation to stream flow, as consistent with a lognormal distribution. From the perspective of extreme value theory, a fundamental limitation of this approach is the failure to permit a heavy tail, despite much evidence indicating that variables such as stream flow and precipitation can be heavy-tailed (e.g., Katz et al. 2002).

In the present chapter, we focus on how the familiar distributions that arise in extreme value theory, namely the generalized extreme value (GEV) and generalized Pareto (GP) distributions, can be retained under nonstationarity. But now the extremal distribution is allowed to gradually shift by introducing time as a covariate; that is, expressing one or more of the parameters of the distribution as a function of time. At least for the parameter estimation technique of maximum likelihood, it is straightforward to fit such statistical models.

The chapter is organized as follows. Section 2.2 deals with statistical methods for modeling nonstationary extremes, namely by means of the block maxima and point process approaches. Each technique is first reviewed under stationarity, before showing how it can be generalized to handle nonstationarity as well. Section 2.3 then provides some detailed examples of how the proposed methods can be applied

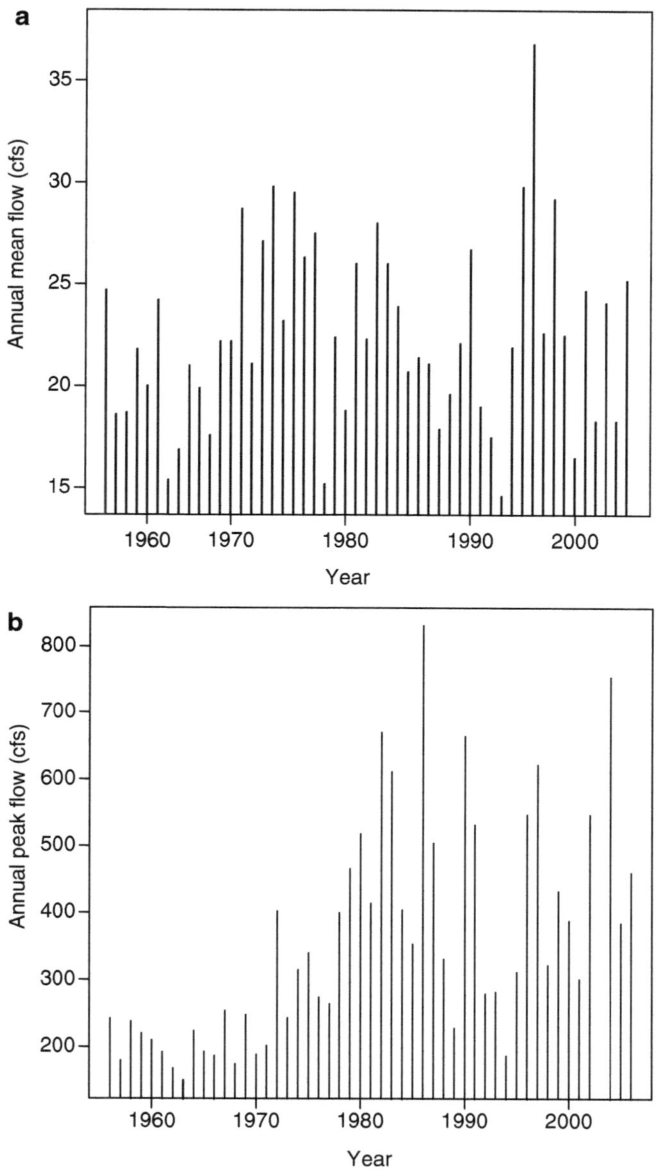

Fig. 2.1 Annual time series of mean (*top*) and peak (*bottom*) flow (cfs) at Mercer Creek, WA during 1956–2006 (*Note*: peak flow is missing for year 2003)

to the detection and statistical modeling of trends in hydrologic extremes. Finally, Sect. 2.4 consists of a discussion of outstanding issues in the statistical modeling of nonstationary extremes, including the challenging problem of extending regional analysis to allow for a common or smoothly varying trend across space.

2.2 Statistical Methods

Under the assumption of stationarity, the GEV and GP distributions arise as approximations for block maxima (e.g., annual peak stream flow) and for excesses over a high threshold, respectively. The GP distribution is utilized as part of the peaks over threshold (POT) or partial duration series (PDS) approach (Coles 2001), which can be formulated in terms of a two-dimensional, nonhomogeneous Poisson process (Davison and Smith 1990; Smith 1989).

The stationarity assumption does not necessarily require temporal independence, with the GEV approximation for block maxima still holding under a wide range of dependence conditions (Leadbetter et al. 1983), including clustering at high levels (a characteristic clearly exhibited by stream flow) and long-range dependence (a characteristic for which there is some evidence that hydrologic variables possess; Hurst 1951). With an adjustment such as declustering, the POT technique can still be applied under temporal dependence as well (Coles 2001).

Except under quite limited circumstances, it does not appear possible to directly extend extreme value theory to nonstationary time series (Leadbetter et al. 1983). Rather, we adopt a more pragmatic approach, with one or more of the parameters of the extremal distributions being expressed as functions of time (Coles 2001). In other words, at any given time, an extreme value distribution would still be used, but the distribution itself would shift over time. At least in this limited sense, the statistical model remains interpretable in terms of extreme value theory.

2.2.1 *Block Maxima*

The block maxima approach is predicated upon the Extremal Types Theorem. Let the time series, denoted by $\{X_1, X_2, \ldots, X_n\}$, be independent and identically distributed (iid) with common cumulative distribution function (cdf) F. Let $M_n = \max\{X_1, X_2, \ldots, X_n\}$ and suppose there exist normalizing constants, $a_n > 0$ and b_n, such that

$$\Pr\{(M_n - b_n)/a_n \leq x\} \rightarrow G(x) \text{ as } n \rightarrow \infty. \tag{2.1}$$

Then the cdf G must be the GEV; that is,

$$G(x; \mu, \sigma, \xi) = \exp\left\{-[1 + \xi(x - \mu)/\sigma]^{-1/\xi}\right\}, \quad 1 + \xi(x - \mu)/\sigma > 0 \tag{2.2}$$

(Coles 2001). Here μ, $\sigma > 0$, and ξ denote the location, scale, and shape parameters, respectively. With this notational convention (not always followed in hydrology), the GEV has three types: (i) $\xi > 0$ corresponds to a heavy-tailed or Fréchet type; (ii) $\xi < 0$ to a bounded tail or Weibull type; and (iii) $\xi = 0$ [obtained by taking the limit as $\xi \rightarrow 0$ in (2.2)] to the light-tailed or Gumbel type. The location parameter μ is not the mean, but does represent the "center" of the distribution, and the scale parameter σ is not the standard deviation, but does govern the "size" of the deviations about μ.

Under temporal clustering at high levels (i.e., dependence among extremes close together in time), (2.1) still holds with an adjustment to the location and scale parameters depending on the mean cluster length (i.e., the shape parameter is not affected; Coles 2001). Under long-range temporal dependence (i.e., dependence among extremes far apart in time), no adjustment necessarily need be made. Nevertheless, the rate of convergence might well be affected by either temporal clustering or long memory. In practice, one fits the GEV distribution directly to block maxima, so the parameters are automatically adjusted for any temporal clustering (Coles 2001).

Nonstationarity is introduced by expressing one or more of the parameters of the GEV as a function of time, say $\mu(t)$, $\sigma(t)$, and $\xi(t)$, $t = 1, 2, \ldots$. Here t would typically denote the year (or season) over which the maximum is taken. For instance, one might consider as a candidate model the nonstationary GEV distribution with

$$\mu(t) = \mu_0 + \mu_1 t, \quad \ln \sigma(t) = \sigma_0 + \sigma_1 t, \quad \xi(t) = \xi. \tag{2.3}$$

That is, the model consists of a GEV distribution with linear trends in the location and log-transformed scale parameters [to constrain $\sigma(t) > 0$], but no trend in the shape parameter. So at any point in time t, the distribution of the maximum remains the GEV consistent with the Extremal Types Theorem. The parameter μ_1 can be interpreted as the slope of a linear trend in the center of the distribution, and the transformed parameter $\exp(\sigma_1)$ as the proportionate rate of change in the scale (or size) of the distribution.

Such trends can be more readily interpreted in terms of the corresponding time varying quantile (or "effective" return level), which would reduce to a conventional return level (with return period $1/p$) if it did not vary with time. The $(1-p)$th quantile of the GEV distribution, as a function of time t, can be obtained by inverting (2.2):

$$\begin{aligned} G^{-1}[1-p;\mu(t),\sigma(t),\xi(t)] &= \mu(t) + [\sigma(t)/\xi(t)] \\ &\times \left\{[-\ln(1-p)]^{-\xi(t)} - 1\right\}, \quad 0 < p < 1. \end{aligned} \tag{2.4}$$

In particular, if the location and/or scale parameters have linear time trends, then the effective return level would also change linearly.

2.2.2 *Excesses Over High Threshold*

The analogue to the GEV distribution for block maxima is the GP distribution for excesses over a high threshold. For now, we make the same assumption that the time series $\{X_1, X_2, \ldots, X_n\}$ is iid with common cdf F. For sufficiently high threshold u, the distribution of the excess $Y_i = X_i - u$, conditional on $X_i > u$, has an approximate GP distribution with cdf

$$H[y;\sigma(u),\xi] = 1 - \{1 + \xi\,[y/\sigma(u)]\}^{-1/\xi},\quad 1 + \xi\,[y/\sigma(u)] > 0, y > 0 \qquad (2.5)$$

(Coles 2001). Here $\sigma(u) > 0$ and ξ denote the scale and shape parameters, respectively. The shape parameter has the same interpretation as for the GEV, with the GP having three types: (i) $\xi > 0$ corresponding to the heavy-tailed or Pareto type; (ii) $\xi < 0$ to the bounded tail or beta type; and (iii) $\xi = 0$ [obtained by taking the limit as $\xi \to 0$ in (2.5)] to the light-tailed or exponential type. The scale parameter $\sigma(u)$ is not the standard deviation, but does govern the "size" of the excesses.

The notation $\sigma(u)$ is intended to emphasize the dependence of the scale parameter on the threshold. That is, if the distribution of the excess Y_i has an exact GP distribution (rather than only approximate), then increasing the threshold from u to $u*$ would result in another GP distribution with the same shape parameter ξ, but an adjusted scale parameter given by

$$\sigma(u^*) = \sigma(u) + \xi\left(u^* - u\right),\quad u^* > u \qquad (2.6)$$

(Coles 2001). In particular, the scale parameter would increase [i.e., $\sigma(u*) > \sigma(u)$] if $\xi > 0$, decrease if $\xi < 0$. Consistent with the "memoryless" property of the exponential distribution, there would be no change in the scale parameter if $\xi = 0$.

The selection of the threshold u involves a delicate trade-off, high enough for the GP approximation to be fairly accurate, but not so high to make the number of exceedances of the threshold too small for parameter estimation to be reliable. Accounting for temporal dependence is a further complication, with one approach being to decluster the data (e.g., identifying flood peaks in stream flow) and fitting the GP distribution only to cluster maxima (Coles 2001).

2.2.3 *Point Process Approach*

To fully model extremes, the GP distribution for the excess over a high threshold u needs to be combined with the rate at which the threshold is exceeded. By what is sometimes called the Law of Small Numbers (i.e., the Poisson approximation to the binomial distribution for rare events), it is natural to approximate the time series of exceedances by a one-dimensional Poisson process with rate parameter $\lambda > 0$. In particular, the number of exceedances of u, $N_u(T)$ say, within a time interval of length T would be approximately Poisson with mean $E[N_u(T)] = \lambda T$. The POT technique combines these two components, the GP and Poisson, also being termed a Poisson-GP model, with parameters λ, $\sigma(u)$, and ξ.

A closely related approach consists of modeling both of these processes, the exceedances and the excesses, simultaneously as a two-dimensional point process, with the horizontal dimension corresponding to time and the vertical dimension to the excess over the threshold. For sufficiently high threshold u, the points in this two-dimensional space would be approximately distributed as a nonhomogenous Poisson process. Even though the process might well be homogenous in the time dimension, it would be inherently nonhomogeneous in the vertical dimension because larger excesses should be less likely than smaller ones (Coles 2001; Smith 1989).

For notational convenience, suppose that the sequence of observations $\{X_1, X_2, \ldots, X_n\}$ over which the maximum is taken corresponds to a time interval of length T. Because $N_u(T) = 0$ (i.e., no exceedances of u) if and only if $M_n \leq u$ (i.e., the maximum of the sequence of observations does not exceed u), the two-dimensional point process can be parameterized in terms of the GEV distribution for block maxima. Specifically, the parameters of the Poisson-GP model, λ, $\sigma(u)$, and ξ, can be expressed as functions of the parameters of the GEV distribution, μ, σ, and ξ, as follows:

$$\ln \lambda = -(1/\xi) \ln\left[1 + \xi\,(u - \mu)/\sigma\right], \quad \sigma(u) = \sigma + \xi\,(u - \mu), \tag{2.7}$$

with the shape parameter ξ being identical. The expression for λ closely resembles the exceedance probability for a GP distribution, that for $\sigma(u)$ the effect on the scale parameter when the threshold of a GP distribution is changed. Because the GEV distribution corresponds to the approximate distribution of block maxima (with block size n), strictly speaking the Poisson rate parameter λ depends on the block size as well (e.g., Katz et al. 2002). This parameterization has the advantage that the scale parameter σ of the GEV, unlike that of the GP $\sigma(u)$, does not depend on u.

If the parameterization in terms of the GEV distribution is used, then nonstationarity can be introduced into the point process model in the same way as for the block maxima approach, again with parameters, $\mu(t)$, $\sigma(t)$, and $\xi(t)$, dependent on time t [e.g., as in (2.3)]. But now the unit of time could be days or hours, rather than years. In particular, the interpretation of the parameters remains the same as for those in (2.3). If the nonstationarity were substantial enough in magnitude, then the threshold could be a function of time as well, $u(t)$ say, to ensure that the extreme value approximation remains valid over the entire time span.

Note that the point process approach, at least in terms of the GEV parameterization, is still not commonly used to model trends in climate or hydrologic extremes; rather, nonstationarity is typically introduced separately into the two components of the Poisson-GP model (one exception is Brown et al. 2008). Any such trends over time in these parameters, $\lambda(t)$, $\sigma(u, t)$, and $\xi(t)$, can be converted into the corresponding trends in the GEV parameters through the inversion of (2.7). But one disadvantage is that the direct interpretation of any trends in the parameters of the Poisson-GP model is not nearly as straightforward.

2.2.4 *Parameter Estimation*

2.2.4.1 Maximum Likelihood

We adopt the technique of maximum likelihood for estimating the parameters of extremal distributions (e.g., Coles 2001). This technique is first illustrated for the block maxima approach under stationarity, further assuming that the distribution of

M_n is GEV with parameters μ, σ, and ξ (i.e., that the block size n is sufficiently large). Let $G'(x;\ \mu,\ \sigma,\ \xi)$ denote probability density function (pdf) corresponding to cdf G given by (2.2) [i.e., obtained by differentiating the expression (2.2) for G with respect to x]. Suppose a sample of size T of block maxima (i.e., effectively an original sample of size nT before taking block maxima), denoted by $M_n(1) = m_1$, $M_n(2) = m_2, \ldots, M_n(T) = m_T$, is available. It need not be assumed that the original observations are temporally independent, only that the resultant block maxima are approximately so.

The likelihood function, denoted by $L(m_1, m_2, \ldots, m_T;\ \mu,\ \sigma,\ \xi)$, can be thought of as a measure of how likely these T observed block maxima, $m_1, m_2, \ldots, m_T$, are as a function of the unknown GEV parameters μ, σ, and ξ. The maximum likelihood estimates (MLEs) are those values of the parameters, μ, σ, and ξ, that maximize the likelihood function. In practice, it convenient to work in terms of the log likelihood function, with the logarithmic transformation converting a product to a sum (equivalent, but more tractable for the purpose of optimization), maximizing the expression

$$\ln L(m_1, m_2, ..., m_T; \mu, \sigma, \xi) = \ln G'(m_1; \mu, \sigma, \xi) + \cdots + \ln G'(m_T; \mu, \sigma, \xi), \tag{2.8}$$

with respect to μ, σ, and ξ. Instead of maximization, it is more convenient to minimize the negative log likelihood (nllh) function $-\ln L(m_1, m_2, \ldots, m_T;\ \mu,\ \sigma,\ \xi)$.

With this formulation, the extension of the optimization problem to the nonstationary case, in which the parameters of the GEV distribution depend on time t, $\mu(t)$, $\sigma(t)$, and $\xi(t)$, $t = 1, 2, \ldots, T$, is immediate. For instance, with the nonstationary model (2.3), the likelihood function, denoted by $L[m_1, m_2, \ldots., m_T;\ \mu(t),\ \sigma(t),\ \xi(t)]$, can be expressed as a function of five parameters; namely, μ_0, μ_1, σ_0, σ_1, and ξ, with the optimization problem now entailing minimizing the negative log likelihood function with respect to these five parameters.

Although the details are not provided here, the MLEs for the parameters in the point process model, or in the Poisson-GP model, can be obtained in the same manner (for an expression for the log likelihood function, see Katz et al. 2002). It should be noted that another estimation technique, namely probability-weighted or L-moments (Hosking and Wallis 1997), remains more popular than maximum likelihood in hydrologic applications. But it is not as straightforward to apply under nonstationarity.

2.2.4.2 Model Selection

Maximum likelihood also has the advantage of facilitating approximate tests of significance in the comparison of "nested" models. For instance, consider a possible linear trend in the location parameter of the GEV distribution; that is,

$$\mu(t) = \mu_0 + \mu_1 t, \quad \sigma(t) = \sigma, \quad \xi(t) = \xi. \tag{2.9}$$

The alternative model with no trend can be obtained as a special case of (2.9) by setting the slope $\mu_1 = 0$. For such a comparison of models, the likelihood ratio test provides an approximate procedure for large sample size (e.g., Coles 2001). In particular, the test of the null hypothesis of no trend (i.e., $\mu_1 = 0$) can be performed by comparing the minimized negative log likelihood function for the two competing models (one with μ_1 constrained to be zero, the other with μ_1 unconstrained). That is, a comparison of

$$\begin{aligned} \text{nllh}(3) &\equiv -\ln L(m_1, m_2, \ldots, m_T; \hat{\mu}, \hat{\sigma}, \hat{\xi}) \text{ vs.} \\ \text{nllh}(4) &\equiv -\ln L(m_1, m_2, \ldots, m_T; \hat{\mu}_0, \hat{\mu}_1, \hat{\sigma}, \hat{\xi}). \end{aligned} \tag{2.10}$$

Here the simplified notation nllh(k) denotes the minimized negative log likelihood for a model requiring the estimation of k parameters, and a caret over a parameter denotes the corresponding MLE.

Under null hypothesis of $\mu_1 = 0$, the likelihood ratio test statistic, based on twice the difference between nllh(3) and nllh(4), has an approximate chi squared distribution [with one degree of freedom (df), denoted by $\chi^2(1)$] for large sample size T. That is, the test is based on

$$2\,[\text{nllh}\,(3)\; - \;\text{nllh}\,(4)] \;\sim\; \chi^2\,(1) \tag{2.11}$$

(Coles 2001). This test statistic has one df, because the more complex model requires the estimation of one additional parameter (namely, μ_1).

When a number of candidate models are involved (still all nested), a model selection criteria would be more appropriate than repeated tests of significance whose outcomes lose their interpretibility. Two popular criteria are Akaike's information criteria (AIC) and the Bayesian information criterion (BIC) (e.g., Venables and Ripley 2002), both penalizing the minimized negative log likelihood function for the number of parameters estimated. From a collection of nested candidate models, AIC (or BIC) selects the model that minimizes the quantity

$$\text{AIC}(k) = 2\ \text{nllh}(k) + 2\ k, \quad \text{BIC}(k) = 2\ \text{nllh}(k) + k\ \ln\ T, \tag{2.12}$$

respectively.

2.2.4.3 Diagnostics

Quantile-Quantile (Q-Q) plots are a popular technique for checking the fit of stationary extremal distributions, as they are particularly effective at highlighting apparent discrepancies in the extreme upper tail. To apply this technique to a fitted nonstationary model, it is first necessary to transform the data to stationarity. For example, if the random variable $M_n(t)$ is distributed as a nonstationary GEV with parameters $\mu(t)$, $\sigma(t)$, and $\xi(t)$, then the transformed variable

$$\varepsilon_t = [1/\xi(t)] \ln\{1 + \xi(t)\,[M_n(t) - \mu(t)]\,/\sigma(t)\} \tag{2.13}$$

has a standard Gumbel distribution (i.e., $\mu = 0$, $\sigma = 1$) (Coles 2001). Substituting the MLEs into (2.13), a conventional Q-Q plot for block maxima can be generated in terms of the transformed variable ε_t as compared to a standard Gumbel distribution. Because this approach ignores the fact that the GEV parameters in (2.13) are actually unknown, such a Q-Q plot should only be viewed as a rough diagnostic.

For the nonstationary point process model with the GEV parameterization, the corresponding scale parameter $\sigma(u, t)$ of the GP distribution for the excess over a high threshold can be obtained using (2.7). If a random variable Y_t has a nonstationary GP distribution with parameters $\sigma(u, t)$ and $\xi(t)$, then the transformed variable

$$\varepsilon_t = [1/\xi(t)] \ln\{1 + \xi(t)\,[Y_t/\sigma(u,t)]\} \tag{2.14}$$

has a standard exponential distribution [i.e., $\sigma(u) = 1$] (Coles 2001). So, analogous to (2.13), a conventional Q-Q plot for excesses can be generated in terms of the transformed variable ε_t, defined by (2.14), as compared to a standard exponential distribution.

2.3 Examples

To obtain MLEs of the parameters of extremal distributions such as the GEV, minimizing the negative log likelihood function requires iterative numerical procedures. In the following examples, we use the R packages "extRemes" and "ismev" (Gilleland and Katz 2011; R Development Core Team 2011) that rely on "brute force" optimization. More efficient numerical optimization could be implemented using, for instance, vector generalized linear models (VGLM) (Yee and Stephenson 2007). For a review of statistical software to analyze extremes, see Stephenson and Gilleland (2006).

2.3.1 Trend in Block Maxima

2.3.1.1 Annual Peak Flow at Mercer Creek, WA

We now return to the example of annual peak flow at Mercer Creek, briefly mentioned in the Introduction (recall Fig. 2.1, lower panel). Motivated by information about the rapid urbanization in this drainage basin during the time period of roughly 1970–1985, the nonstationarity in peak flow is modeled in terms of piecewise linear trends in the location and log-transformed scale parameters of the GEV distribution. That is, a model of the following form is fitted:

$$\mu(t) = \mu_0 + \mu_1\, z(t), \quad \ln\sigma(t) = \sigma_0 + \sigma_1\, z(t), \quad \xi(t) = \xi, \quad t = 1, 2, \ldots, T. \tag{2.15}$$

Table 2.1 Parameter estimates and standard errors for two forms of nonstationary GEV distribution, along with those for stationary GEV, fitted to annual time series of peak flow (cfs) at Mercer Creek, WA during 1956–2006

Model	Parameter estimate	Standard error	nllh	*P*-value
Stationary			318.67	
Location μ	264.20	18.00		
Scale σ	101.51	15.51		
Shape ξ	0.311	0.185		
Nonstationary (μ)			309.78	$<10^{-4}$
Location μ_0	219.17	22.28		
μ_1	8.91	2.76		
Scale σ	97.33	12.70		
Shape ξ	0.064	0.141		
Nonstationary (μ, σ)			294.43	$<10^{-7}$
Location μ_0	194.05	8.01		
μ_1	14.67	2.15		
Scale (log) σ_0	3.419	0.193		
σ_1	0.109	0.018		
Shape ξ	−0.034	0.112		

Minimized negative log likelihood (nllh) and *P*-value for likelihood ratio test (i.e., comparing a model with the one immediately above) also included

Here the piecewise linear time covariate $z(t)$ is defined by

$$z(t) = 1, t \leq t_1; \;\; z(t) = t - t_1, t_1 + 1 \leq t \leq t_2; \;\; z(t) = t_2 - t_1, t_2 + 1 \leq t \leq T; \tag{2.16}$$

with $t = 1$ corresponding to the year 1956, $t_1 = 15$ to the year 1970, $t_2 = 30$ to the year 1985, and $T = 51$ to the year 2006. Rather than fill in the missing value for year 2003, it can be appropriately handled in the statistical software through simply omitting the record for that year from the data file.

Table 2.1 gives parameter estimates, along with standard errors, for three different forms of GEV distribution:

(i) Stationary GEV in which none of the parameters depend on time [i.e., $\mu_1 = \sigma_1 = 0$ in (2.15)];
(ii) Nonstationary GEV in which the location parameter, but not the scale parameter, depends linearly on time during 1971–1985 [i.e., $\sigma_1 = 0$ in (2.15)];
(iii) Nonstationary GEV in which both the location and log-transformed scale parameters depend linearly on time during 1971–1985 [i.e., general form of (2.15)].

Table 2.1 also includes the minimized negative log likelihood for these three candidate models. The likelihood ratio test (2.11) for comparing model (i) versus model (ii) (i.e., whether $\mu_1 = 0$) gives a *P*-value smaller than 0.0001, whereas the likelihood ratio test for comparing model (ii) versus model (iii) (i.e., whether $\sigma_1 = 0$, given $\mu_1 \neq 0$) yields an even smaller *P*-value. In other words, there is strong evidence that model (iii) fits the data better than model (i) or (ii).

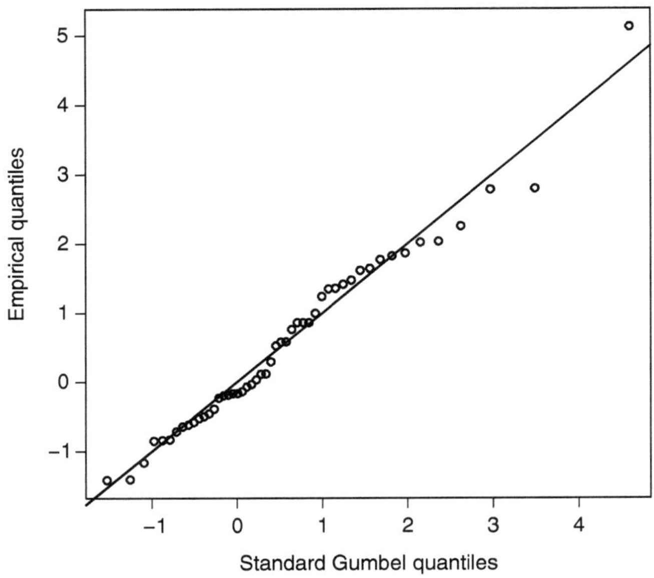

Fig. 2.2 Q-Q plot (based on transformation to standard Gumbel distribution) for nonstationary GEV distribution with piecewise linear trends in location and log-transformed scale parameters fitted to annual time series of peak flow (cfs) at Mercer Creek, WA during 1956–2006

Figure 2.2 shows a Q-Q plot for model (iii) based on the transformation to the standard Gumbel distribution (2.13), with the fit appearing reasonable. Figure 2.3 shows the corresponding trends in a few example time varying quantiles for the fitted nonstationary GEV of model (iii) [i.e., substituting the estimated parameters into (2.4)], along with the observed time series of annual peak flow. The quantile curves increase nonlinearly during 1971–1985 because of the scale parameter being modeled as an exponential function of time (2.15), with the increase being more rapid the higher the quantile. Figure 2.4 shows the fitted GEV pdfs for annual peak flow during the time periods of 1956–1970 and 1985–2006, with substantial increases in both the center and the spread of the distribution being associated with the changes in land use.

2.3.1.2 Winter Maximum Daily Precipitation at Manjimup, Western Australia

This example concerns the statistical modeling of trends in daily precipitation extremes at Manjimup, a site in southwest Western Australia. This region has experienced an overall drying trend in recent decades (Bates et al. 2010), particularly during the winter season and including a decrease in extreme high daily precipitation amounts (Li et al. 2005). Figure 2.5 shows the winter (May–October) time series of maximum daily precipitation amounts, with an extended period of markedly lower

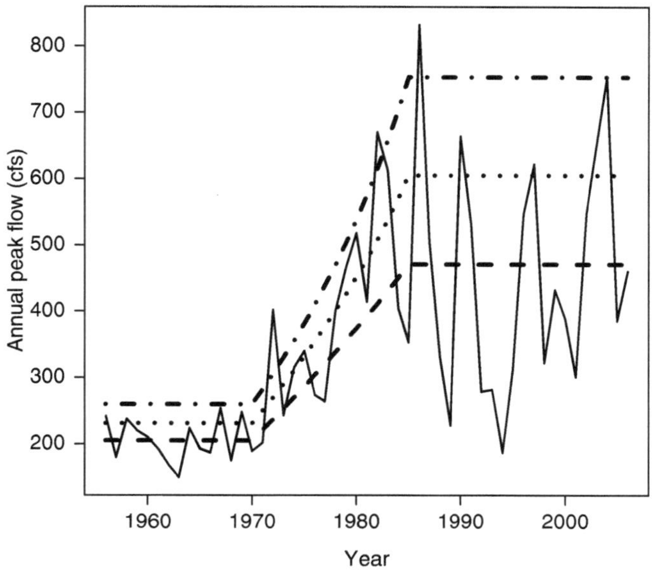

Fig. 2.3 Annual time series of peak flow (cfs) at Mercer Creek, WA during 1956–2006, along with selected quantiles [0.5 (*dashed curve*), 0.75 (*dotted curve*), 0.9 (*dot-dashed curve*)] for fitted nonstationary GEV distribution with piecewise linear trends in location and log-transformed scale parameters

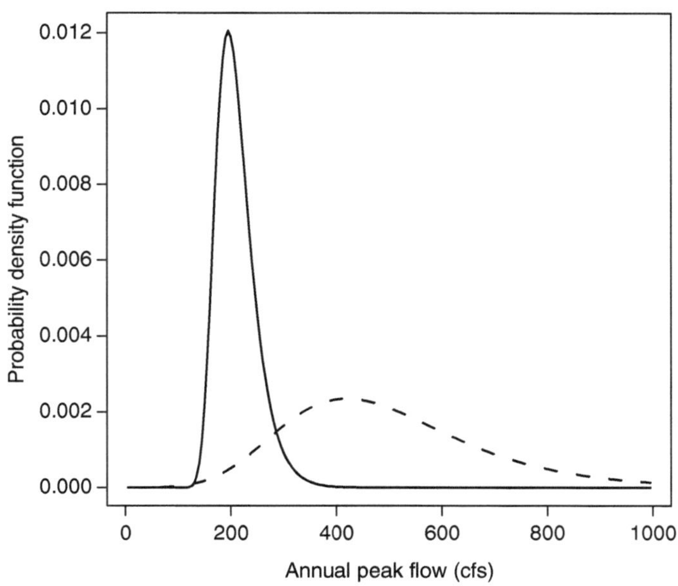

Fig. 2.4 Fitted GEV pdfs for annual peak flow at Mercer Creek, WA during time periods 1956–1970 (*solid curve*) and 1985–2006 (*dashed curve*)

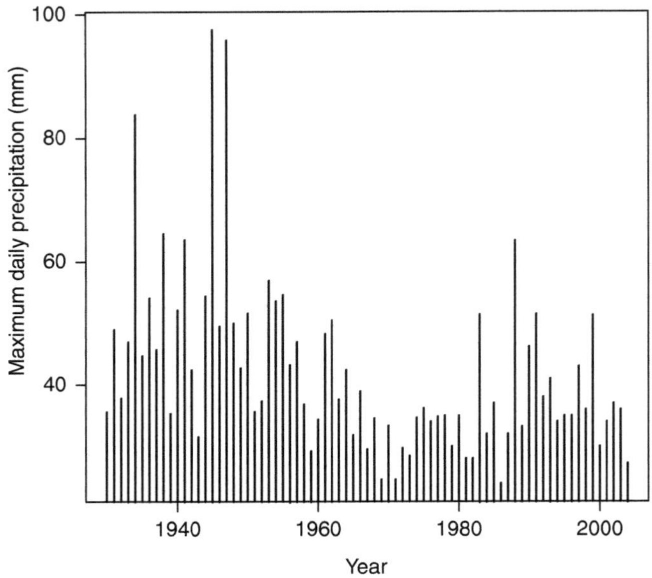

Fig. 2.5 Winter (May–October) time series of maximum daily precipitation amount (mm) at Manjimup, Western Australia during 1930–2004

extremes starting in the mid or late 1960s. Bates et al. (2010) found a gradual decline in variables such as mean stream flow, whereas Li et al. (2005) argued for an abrupt shift (or "change point") in extreme precipitation. Although some long-term changes in atmospheric/oceanic circulation patterns have been detected (Bates et al. 2010), attempts to relate the decreases in extreme high precipitation to, for instance, an apparent trend an index of the Antarctic Oscillation (AAO), remain inconclusive (Li et al. 2005).

A nonstationary GEV distribution of the general form

$$\mu(t) = \mu_0 + \mu_1\, t + \mu_2\, t^2, \quad \ln\sigma(t) = \sigma_0 + \sigma_1\, t + \sigma_2\, t^2, \quad \xi(t) = \xi, \qquad (2.17)$$

is fitted; that is, with possible quadratic trends in the location and log-transformed scale parameters. To avoid numerical problems, time has been rescaled to the interval $[-1, 1]$, with $t = -1$ corresponding to the year 1930, $t = 0$ to the year 1967, and $t = 1$ to the year 2004. This form of nonstationarity would be consistent with a gradual decline in precipitation extremes at Manjimup.

Starting with the stationary case, trends of increasing complexity are fitted as follows:

(i) Stationary GEV distribution [i.e., $\mu_1 = \mu_2 = \sigma_1 = \sigma_2 = 0$ in (2.17)];
(ii) Linear trend in location parameter [i.e., $\mu_2 = \sigma_1 = \sigma_2 = 0$ in (2.17)];
(iii) Quadratic trend in location parameter [i.e., $\sigma_1 = \sigma_2 = 0$ in (2.17)];

Table 2.2 Model selection for nonstationary GEV distributions fitted to winter (May–October) time series of maximum daily precipitation amount (mm) at Manjimup, Western Australia during 1930–2004

Model	Number of parameters	nllh	AIC	BIC
Stationary	3	285.93	577.87	584.82
Linear trend in μ	4	280.55	569.10	578.37
Quadratic trend in μ	5	278.23	566.45	578.04
Quadratic trend in μ, linear trend in $\ln \sigma$	6	275.79	563.57[a]	577.48[a]
Quadratic trends in μ and $\ln \sigma$	7	274.80	563.61	579.83

Minimized negative log likelihood (nllh), along with AIC and BIC values, are listed for each candidate model

[a]Denotes minimum value

Table 2.3 Parameter estimates and standard errors for nonstationary GEV distribution, with quadratic trend in location parameter and linear trend in log-transformed scale parameter, fitted to winter (May–October) time series of maximum daily precipitation amount (mm) at Manjimup, Western Australia during 1930–2004

	Parameter estimate	Standard error	P-value
Location parameter			
Constant μ_0	35.241	1.292	
Linear trend μ_1	−7.688	1.787	0.001
Quadratic trend μ_2	4.582	2.513	0.031
Scale parameter (log)			
Constant σ_0	2.050	0.102	
Linear trend σ_1	−0.414	0.170	0.027
Shape parameter			
Constant ξ	0.107	0.092	

P-values for likelihood ratio tests also included (i.e., comparing a model with the one immediately above)

(iv) Quadratic trend in location parameter and linear trend in log-transformed scale parameter [i.e., $\sigma_2 = 0$ in (2.17)];

(v) Quadratic trends in location and in log-transformed scale parameters [i.e., general form of (2.17)].

Because of so many candidate models being fitted, the AIC and BIC (2.12) are first applied instead of the likelihood ratio test. For each candidate model, Table 2.2 lists the minimized negative log likelihood, along with the AIC and BIC values. Model (iv), with a quadratic trend in the location parameter and a linear trend in the log-transformed scale parameter, is selected by both the AIC and BIC. A likelihood ratio test comparing this model to model (v), with quadratic trends in both the location and log-transformed scale parameters, yields a P-value of 0.161, in agreement with the AIC and BIC. Table 2.3 gives parameter estimates and standard errors for model (iv), along with P-values for likelihood ratio tests again consistent with the outcomes of the AIC and BIC given in Table 2.2. The fit appears reasonable according to a Q-Q plot (not shown) based on the same method as Fig. 2.2.

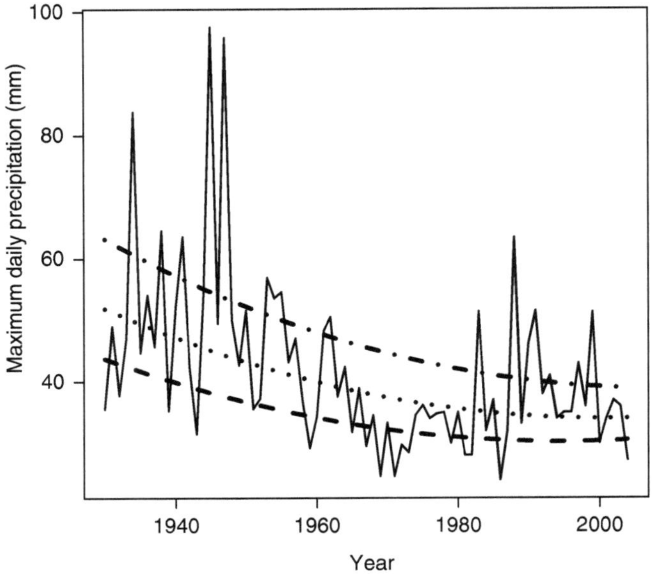

Fig. 2.6 Winter (May–October) time series of maximum daily precipitation amount (mm) at Manjimup, Western Australia during 1930–2004, along with selected quantiles [0.25 (*dashed curve*), 0.5 (*dotted curve*), 0.75 (*dot-dashed curve*)] for fitted nonstationary GEV distribution with quadratic trend in location parameter and linear trend in log-transformed scale parameter

Figure 2.6 shows the corresponding trends in a few example time varying quantiles for the fitted nonstationary GEV of model (iv) [i.e., substituting the estimated parameters into (2.4)], along with the observed time series of winter maximum daily precipitation. The quantile curves decrease nonlinearly, along with decreasing spread in recent decades. Still the actual time series exhibits more abrupt shifts on a decadal time scale, especially one starting in the mid or late 1960s, not well captured by the fitted model.

2.3.2 *Trend in Point Process*

2.3.2.1 Poisson-GP Model for Manjimup Winter Daily Precipitation

We continue with the example of trends in precipitation extremes at Manjimup, Western Australia, but now modeling trends in the daily extremes instead of only block maxima. First, the Poisson-GP model is fitted to the time series of daily precipitation amount in winter at Manjimup, including possible linear trends in the log-transformed rate parameter of the Poisson component for the exceedance of a threshold and in the log-transformed scale parameter of the GP component for the excess over the threshold. That is, models of the form

$$\ln \lambda(t) = \lambda_0 + \lambda_1\, t, \quad \ln \sigma(u,t) = \sigma_0 + \sigma_1\, t, \xi(t) = \xi, \tag{2.18}$$

Table 2.4 Parameter estimates and standard errors for Poisson-GP model, with linear trends in log-transformed Poisson rate parameter and in log-transformed GP scale parameter using a threshold of 30 mm, fitted to winter (May–October) time series of daily precipitation amount (mm) at Manjimup, Western Australia during 1930–2004

	Parameter estimate	Standard error	P-value
Poisson component			
Rate parameter (log):			
Constant λ_0	−3.564	0.123	
Linear trend λ_1	−0.021	0.004	$<10^{-9}$
GP component			
Scale parameter (log):			
Constant σ_0	2.265	0.155	
Linear trend σ_1	−0.0086	0.0038	0.029
Shape parameter ξ	0.111	0.079	

P-values for likelihood ratio tests for trends also included

are fitted, where the time variable $t = 1$ corresponds to the year 1930 and $t = 75$ to the year 2004. Note that the index of time is only allowed to shift from one winter to the next, rather than gradually within a given winter, consistent with the approach taken above for block maxima. It turns out that this form of nonstationary Poisson-GP model produces a gradual decline in extremes of approximately the same shape as that of (2.17) in terms of block maxima.

For a threshold of $u = 30$ mm as used by Li et al. (2005), Table 2.4 give parameter estimates, standard errors, and P-values for the two trend models in (2.18). There is strong evidence of a trend in the rate of threshold exceedance, with the fitted nonstationary mean of the Poisson distribution being shown in Fig. 2.7. The trend in the GP scale parameter is statistically significant at the 5% level, with a few example quantiles of the fitted nonstationary GP distribution being shown in Fig. 2.8. The fit appears reasonable according to a Q-Q plot (not shown), based on the transformation from nonstationary GP to standard exponential distribution (2.14).

The GEV parameters, $\mu(t)$ and $\sigma(t)$, can be expressed as functions of the Poisson-GP parameters, $\lambda(t)$ and $\sigma(u, t)$, by solving the system of equations (2.7). Derived from the fitted nonstationary Poisson-GP model, the corresponding trends in quantiles of the GEV are remarkably similar to those shown in Fig. 2.6, notwithstanding the different approaches (i.e., block maxima versus Poisson-GP) and different functional forms (i.e., quadratic versus exponential trends).

2.3.2.2 Point Process Applied to Manjimup Winter Daily Precipitation

As a closely related alternative to the Poisson-GP model, a two-dimensional point process is fitted to the same time series of daily precipitation extremes during winter at Manjimup. Consistent with Li et al. (2005) who detected a change point in precipitation extremes at Manjimup, the nonstationarity is modeled as an abrupt shift in the parameters of the point process, starting in 1966, from one stationary

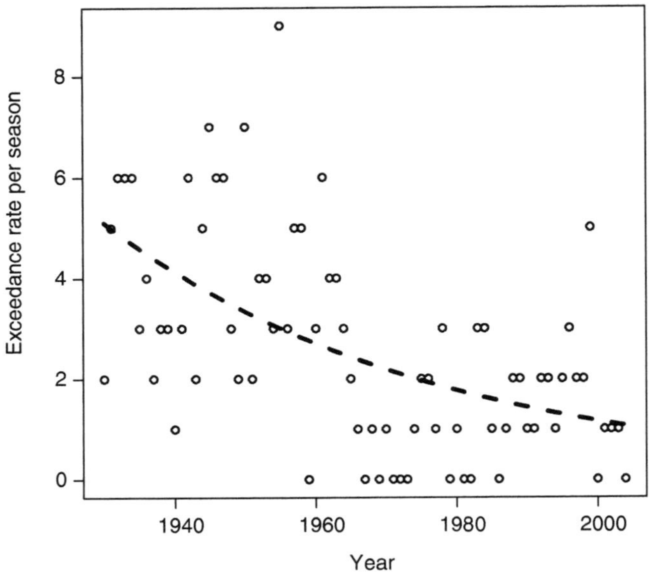

Fig. 2.7 Winter (May–October) time series of frequency of daily precipitation amount greater than a threshold of 30 mm at Manjimup, Western Australia during 1930–2004, along with mean of fitted nonstationary Poisson distribution (*dashed curve*) with linear trend in log-transformed rate parameter

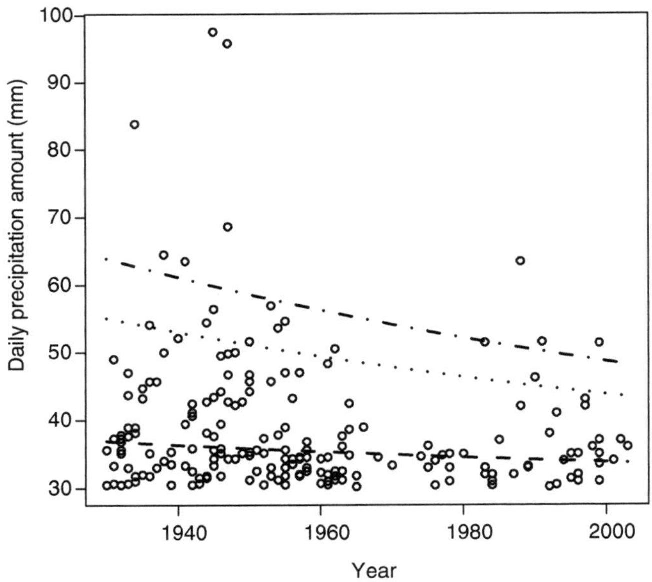

Fig. 2.8 Winter (May–October) time series of daily precipitation amount over a threshold of 30 mm at Manjimup, Western Australia during 1930–2004, along with selected quantiles [0.5 (*dashed curve*), 0.9 (*dotted curve*), 0.95 (*dot-dashed curve*)] for fitted nonstationary GP distribution with linear trend in log-transformed scale parameter

Table 2.5 Parameter estimates and standard errors for point process model, with different parameters for time period 1930–1965 (using a threshold of 30 mm) and 1966–2004 (using a threshold of 22 mm), fitted to winter (May–October) time series of daily precipitation amount (mm) at Manjimup, Western Australia

	Parameter estimate	Standard error	*P*-value
Time period: 1930–1965			
Location parameter μ	42.419	1.437	
Scale parameter σ	9.829	1.050	
Shape parameter ξ	0.143	0.094	
Time period: 1966–2004			
Location parameter μ	32.297	0.961	$< 10^{-13}$
Scale parameter σ	6.893	0.505	0.040
Shape parameter ξ	−0.093	0.058	0.027

P-values also included for likelihood ratio tests of equality of parameters in two time periods

model to another. With the time covariate $z(t)$ now being defined by $z(t) = 0$ if $t \leq 36$; $z(t) = 1$ if $t \geq 37$ instead of (2.16) and the time index t remaining as in (2.18), this shift in parameters can be expressed in the same way as in (2.15). If all the parameters change, then the point process could be fitted separately to the data for the two time periods, 1930–1965 and 1966–2004. The advantage of fitting a nonstationary model, via the time covariate, to the entire data set is that constraints can be readily imposed (e.g., allowing the location and scale parameters to shift, but not the shape parameter).

Table 2.5 gives the parameter estimates and standard errors for the point process with shifts in all three GEV parameters between the two time periods. To keep the rate of exceeding the threshold roughly the same for the two time periods, the results are based on a threshold of 30 mm during the first time period, 22 mm during the second. As explained in Sect. 2.2.3, the use of a time varying threshold does not create any difficulties in interpretation, provided the point process is parameterized in terms of the GEV. Also included in Table 2.5 are *P*-values for the likelihood ratio tests of equality of parameters in the two time periods. The strongest evidence is for a shift in the location parameter, as compared to a fully stationary model, with a *P*-value of virtually zero. The evidence for a shift in the scale parameter, in addition to the location parameter, or for a shift in the shape parameter, in addition to the location and scale parameters, is weaker with *P*-values of about 0.04 and 0.03, respectively. A physical explanation for this apparent shift in the shape parameter, from a heavy to a bounded tail, would be desirable.

The fit appears reasonable according to a Q-Q plot (not shown), again based on the transformation from nonstationary GP to standard exponential distribution (2.14). Figure 2.9 shows a few example quantiles of the fitted point process, as compared to the time series of observed winter daily maxima. As in the case of quadratics being used to model nonstationarity in the block maxima approach (see Fig. 2.6), the present model still does not necessarily capture all of the apparent variations on a roughly decadal time scale in precipitation extremes during winter at Manjimup.

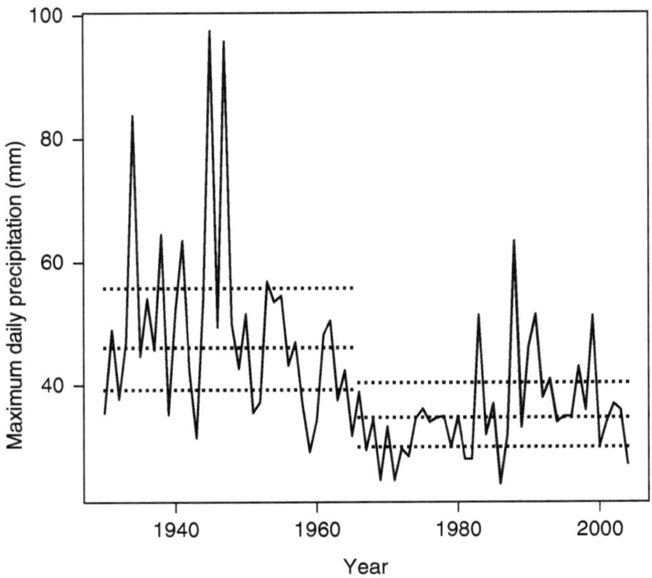

Fig. 2.9 Winter (May–October) time series of maximum daily precipitation amount (mm) at Manjimup, Western Australia, along with selected quantiles (0.25, 0.5, and 0.75 indicated by dotted lines) for fitted nonstationary point process with different parameters for time periods 1930–1965 (threshold = 30 mm) and 1966–2004 (threshold = 22 mm)

2.4 Discussion

In the examples presented, only relatively simple forms of nonstationarity have been considered which can be implemented through linear models. For examples of fitting more flexible forms of nonstationarity to climate or hydrologic extremes, see Chavez-Demoulin and Davison (2005), Davison and Ramesh (2000), and Ramesh and Davison (2002). Techniques for fitting such possibly nonlinear models include vector generalized additive models (VGAM; Yee and Stephenson 2007) and generalized additive models for location, scale, and shape (GAMLSS; Rigby and Stasinopoulos 2005).

Instead of, or in addition to, time trends, physically based covariates could be introduced into extremal models as well. For instance, annual peak flow could be fitted by a GEV distribution conditional on the state of an index of the El Niño-Southern Oscillation (ENSO) phenomenon (e.g., Katz et al. 2002). Similarly to the trend examples presented here, the location and log-transformed scale parameters of the GEV might be varied with the ENSO index, keeping the shape parameter constant. Among other things, the unconditional distribution of annual peak flow would be a mixture of GEV distributions, providing a chance mechanism by which more complex forms of extremal distributions could naturally arise.

Our applications have only dealt with data for one site at a time. Trend detection would be more powerful if applied to multiple sites within a relatively homogenous region. The rationale being that nonstationarity driven by global climate change should be of a similar magnitude at adjacent sites. The idea of "borrowing strength" across space in trend analysis is an natural extension of the technique of regional analysis commonly applied in hydrology under stationary to estimate high flood quantiles (Hosking and Wallis 1997). For instance, the location and scale parameters of the GEV distribution could be site-dependent, with the shape parameter site-independent, and a linear trend in the location parameter having the same slope across the region. Hanel et al. (2009) constitutes one example of such an approach applied to the output of a regional climate model.

Quantile regression is an alternative approach, not based on extreme value theory, with the quantiles of a distribution being directly modeled as functions of covariates. As such, it would produce trends in quantiles that could be directly compared to those obtained in the examples presented here. Nevertheless, the question remains of how well quantile regression performs in practice for high quantiles. See Yu et al. (2003) for a general review of quantile regression and Friederichs and Hense (2007) for a hydrologic-related application.

In many climate change applications, information is needed not only about extremes, but about the low to moderate levels of the hydrologic variable as well. As clear in the example shown in Fig. 2.1, it would not necessarily be reasonable to assume the same form of trend for extremes as for the remainder of the process. Attempts to unify the modeling of high levels of a hydrologic variable with lower levels include Furrer and Katz (2008) and Vrac and Naveau (2007). But these attempts have not yet focused directly on the issue of trends.

Lastly, an important issue concerns how to convey information about the risk of extreme events under nonstationary. The familiar concepts of return level and return period are predicated upon the assumption of stationarity. One way to treat nonstationarity is to evaluate the model in terms of time varying quantiles. In the present chapter, the graphical displays of the fitted nonstationary extremal models have been constructed using these quantities. Yet, for some applications, it would be unrealistic to provide a measure of risk that shifts from one year to the next. For example, it would not necessarily be feasible to change the width of the flood plain on an annual basis. So an alternative measure of risk would be needed; for instance, based on the expected waiting time until the next event (Olsen et al. 1998). This topic is the focus of the next chapter in this book.

Acknowledgments I gratefully acknowledge the comments of one anonymous reviewer. The Mercer Creek stream flow time series was obtained from the U.S. Geological Survey (USGS) web site (http://nwis.waterdata.usgs.gov/usa/nwis/peak). I thank Yun Li (CSIRO, Perth, Australia) for providing the daily precipitation data for southwest Western Australia. This research was partially supported by the National Center for Atmospheric Research (NCAR) Weather and Climate Impacts Assessment Science Program. NCAR is sponsored by the National Science Foundation.

References

Allen MR, Ingram WJ (2002) Constraints on future changes in climate and the hydrologic cycle. Nature 419:224–232
Bates BC, Chandler RE, Charles SP, Campbell EP (2010) Assessment of apparent nonstationarity in time series of annual inflow, daily precipitation, and atmospheric circulation indices: a case study from southwest Western Australia. Water Resour Res 46:W00H02. doi:10.1029/2010WR009509
Brown SJ, Caesar J, Ferro CAT (2008) Global changes in extreme daily temperature since 1950. J Geophys Res 113:D05115. doi:10.1029/2006JD008091
Chavez-Demoulin V, Davison AC (2005) Generalized additive modeling of sample extremes. Appl Stat 54:207–222
Clarke RT (2002) Fitting and testing the significance of linear trends in Gumbel-distributed data. Hydrol Earth Syst Sci 6:17–24
Coles S (2001) An introduction to statistical modeling of extreme values. Springer, London
Davison AC, Ramesh NL (2000) Local likelihood smoothing of sample extremes. J R Stat Soc Ser B 62:191–208
Davison AC, Smith RL (1990) Models for exceedances over high thresholds. J R Stat Soc Ser B 52:393–442
Emori S, Brown SJ (2005) Dynamic and thermodynamic changes in mean and extreme precipitation under changed climate. Geophys Res Lett 32:L17706. doi:10.1029/2005GL023272
Friederichs P, Hense A (2007) Statistical downscaling of extreme precipitation events using censored quantile regression. Mon Weather Rev 135:2365–2378
Furrer EM, Katz RW (2008) Improving the simulation of extreme precipitation events by stochastic weather generators. Water Resour Res 44:W12439. doi:10.1029/2008WR007316
Gilleland E, Katz RW (2011) A new software to analyze how extremes change over time. Eos 92:13–14
Hanel M, Buishand TA, Ferro CAT (2009) A nonstationary index flood model for precipitation extremes in transient regional climate model simulations. J Geophys Res 114:D15107. doi:10.1029/2009JD011712
Helsel DR, Hirsch RM (1993) Statistical methods in water resources. Elsevier, Amsterdam
Hosking JRM, Wallis JR (1997) Regional frequency analysis: an approach based on L-moments. Cambridge University Press, Cambridge
Hurst HE (1951) Long-term storage capacity of reservoirs. Trans Am Soc Civil Eng 116:770–799
Katz RW, Parlange MB, Naveau P (2002) Statistics of extremes in hydrology. Adv Water Resour 25:1287–1304
Konrad CP, (2003) Effects of urban development on floods. USGS Fact Sheet FS-076-03. Available at http://pubs.usgs.gov/fs/fs07603/
Leadbetter MR, Lindgren G, Rootén H (1983) Extremes and related properties of random sequences and processes. Springer, New York
Li Y, Cai W, Campbell EP (2005) Statistical modeling of extreme rainfall in southwest Western Australia. J Climate 18:852–863
Milly PCD, Betancourt J, Falkenmark M, Hirsch RM, Kundzewicz ZW, Lettenmaier DP, Stouffer RJ (2008) Stationarity is dead: Whither water management? Science 319:573–574
Olsen JR, Lambert JH, Haimes YY (1998) Risk of extreme events under nonstationary conditions. Risk Anal 18:497–510
R Development Core Team (2011) R: a language and environment for statistical computing. R Foundation for Statistical Computing, Vienna. http://www.R-project.org
Ramesh NL, Davison AC (2002) Local models for exploratory analysis of extremes. J Hydrol 256:106–119
Rigby RA, Stasinopoulos DM (2005) Generalized additive models for location, scale and shape. Appl Stat 54:507–554

Smith RL (1989) Extreme value analysis of environmental time series: an application to trend detection in ground-level ozone. Stat Sci 4:367–393

Stephenson A, Gilleland E (2006) Software for the analysis of extreme events: the current state and future directions. Extremes 8:87–103

Trenberth KE, Dai A, Rasmussen RM, Parsons DB (2003) The changing character of precipitation. Bull Am Meteorological Soc 84:1205–1217

Venables WN, Ripley BD (2002) Modern applied statistics with S, 4th edn. Springer, New York

Villarini G, Serinaldi F, Smith JA, Krajewski WF (2009) On the stationarity of annual flood peaks in the continental United States during the 20th century. Water Resour Res 45:W08417. doi:10.1029/2008WR007645

Vogel RM, Yaindl C, Walter M (2011) Nonstationarity: flood magnification and recurrence reduction factors in the United States. J Am Water Resour Assoc 47:464–474

Vrac M, Naveau P (2007) Stochastic downscaling of precipitation: from dry events to heavy rainfalls. Water Resour Res 43:W07402. doi:10.1029/2006WR005308

Yee TW, Stephenson AG (2007) Vector generalized linear and additive extreme value models. Extremes 10:1–19

Yu K, Lu Z, Stander J (2003) Quantile regression: applications and current research areas. J R Stat Soc Ser D 52:331–350

Chapter 3
Bayesian Methods for Non-stationary Extreme Value Analysis

Benjamin Renard, Xun Sun, and Michel Lang

Abstract Non-stationary models for extremes have attracted significant attention in recent years. These models require adapted estimation methods. Bayesian inference offers an attractive framework to estimate non-stationary models and, importantly, to quantify estimation and predictive uncertainties.

This chapter therefore focuses on the application of Bayesian inference to non-stationary extreme models. It is organized as a step-by-step building of non-stationary models of increasing generality. The principles of Bayesian inference are introduced using the simple case of a univariate and stationary distribution. The construction of at-site non-stationary models, using regression functions linking parameter values with time-varying covariates, is then presented. The difficulty of identifying non-stationary components based on the sole use of at-site data is also discussed, and motivates the construction of regional non-stationary models. Such models are based on the concept of "regional parameters", i.e. parameters being assumed identical for all sites within a homogeneous region. The inference of regional models poses an additional difficulty compared to the at-site approach: the existence of spatial dependences makes the derivation of the inference equations challenging. A practical solution, based on the use of spatial copulas, is briefly presented. Lastly, a generalization of the "regional parameter" paradigm, based on Bayesian hierarchical modeling, is discussed.

3.1 Introduction

Non-stationary models for extremes have attracted significant attention in recent years. The most common approach has been to introduce non-stationarity through the parameters of the extreme distribution, by using regression models linking

B. Renard (✉) • X. Sun • M. Lang
Irstea, UR HHLY, Hydrology-Hydraulics, 3 bis quai Chauveau, CP 220 F-69336 Lyon, France
e-mail: Benjamin.renard@irstea.fr; xun.sun@irstea.fr; michel.lang@irstea.fr

A. AghaKouchak et al. (eds.), *Extremes in a Changing Climate*, Water Science and Technology Library 65, DOI 10.1007/978-94-007-4479-0_3,

parameter values with some time-varying covariate. This approach requires developing adapted estimation methods. In most cases, inference based on the likelihood has been favored over e.g. moment-based methods, due to its generality and flexibility to introduce various forms of non-stationarity (e.g., Coles et al. 2003). In particular, Bayesian inference offers an attractive framework to estimate non-stationary models and, importantly, to quantify estimation and predictive uncertainties.

This chapter therefore focuses on the application of Bayesian inference to non-stationary extreme models. It is organized as a step-by-step building of non-stationary models of increasing generality (and, as a consequence, of increasing complexity). We start in Sect. 3.2 by a short introduction to Bayesian inference and related tools. This introduction is illustrated using the simplest possible case of estimating the parameters of an univariate and stationary distribution. In particular, Sect. 3.2 will define the basic pieces forming Bayesian inference (likelihood function, prior, posterior and predictive distributions) and will briefly describe Monte-Carlo Markov Chain samplers, which are in practice the inseparable companions of the Bayesian Hydrologist. Section 3.3 will then illustrate the construction of at-site non-stationary models, using regression functions linking parameter values with time-varying covariates. The generality of this approach will be illustrated using several examples of non-stationary models. The difficulty of identifying non-stationary components based on the sole use of at-site data will also be discussed, and will motivate the construction of regional non-stationary models. Such models are presented in Sect. 3.4, and are based on the concept of "regional parameters", i.e. parameters being assumed identical for all sites within an homogeneous region. The idea behind this assumption is to enable using data from several sites simultaneously in order to "share information" between sites and make the identification of non-stationary components more robust. The inference of regional models poses an additional difficulty compared to the at-site approach: the existence of spatial dependences makes the derivation of the inference equations challenging. A practical solution, based on the use of spatial copulas, will be briefly presented. Lastly, Sect. 3.5 discusses the generalization of such regional models, in order to make a less stringent assumption than the regional parameter assumption. This generalization is based on Bayesian hierarchical modeling, which has been recently applied in hydrological extreme analyses, and whose use is likely to grow in the future in the authors' opinion.

3.2 What Is Bayesian Inference?

3.2.1 Basics of Bayesian Inference

This section presents the general principles of Bayesian inference. We consider the simple case of estimating the distribution of a one-dimensional random variable, based on an independent and identically distributed (*iid*) sample of observations. Further refinements (e.g. non-stationarity, multivariate case, non-independence, hierarchical modeling, etc.) will be explored throughout the chapter to address the issues posed by the modeling of non-stationary data. Moreover, readers interested in a more in-depth discussion of the Bayesian paradigm may refer to Gelman et al. (2004), amongst many other textbooks on this subject.

3.2.1.1 Notation

Let Y_t denote the random variable of interest (e.g. annual maximum of year t recorded at some site). Observations are denoted by y_t and are considered as realizations from Y_t. Bold letters are used to denote the observation vector $\boldsymbol{y} = (y_t)_{t=1:N_t}$ and the corresponding random vector $\boldsymbol{Y} = (Y_t)_{t=1:N_t}$. The following assumptions are made in the remainder of this section:

- ***Independence***: Random variables $(Y_t)_{t=1:N_t}$ are mutually independent. In particular, this implies that there is no serial correlation between consecutive values.
- ***Identical distribution***: All random variables $(Y_t)_{t=1:N_t}$ have the same distribution, parameterized by some vector $\boldsymbol{\theta}$. The probability density function (pdf) of this distribution, evaluated at some value y, is noted $p(y|\boldsymbol{\theta})$.

Note that the second assumption implies stationarity, since the distribution of Y_t does not evolve with time.

3.2.1.2 Likelihood

The likelihood is a central concept in several inference paradigms, including the Bayesian one. It quantifies the information brought by the data. Loosely speaking, the likelihood corresponds to the probability of observing what was actually observed, according to the probabilistic model. More correctly, the likelihood of observations $\boldsymbol{y}$ is equal to the joint pdf of the random vector $\boldsymbol{Y} = (Y_1, ..., Y_{N_t})$, evaluated at observations $\boldsymbol{y}$. In standard Bayesian notation, the likelihood is therefore noted as $p(\boldsymbol{y}|\boldsymbol{\theta})$. Given the independence assumption, the joint pdf of the random vector $\boldsymbol{Y}$ is equal to the product of its marginal pdfs, yielding:

$$p(\boldsymbol{y}|\boldsymbol{\theta}) = \prod_{t=1}^{N_t} p(y_t|\boldsymbol{\theta}) \tag{3.1}$$

The likelihood in Eq. (3.1) may suffice to infer the parameters: for instance, the maximum likelihood estimate is derived by maximizing the likelihood viewed as a function of $\boldsymbol{\theta}$. However, Bayesian inference requires the specification of an additional piece of information: the prior distribution.

3.2.1.3 Prior Distribution

This distribution aims at encoding any prior knowledge on parameters $\boldsymbol{\theta}$ the modeler may have. Its pdf is denoted by $p(\boldsymbol{\theta})$. As suggested by this notation, the prior distribution does not depend on the observations $\boldsymbol{y}$, and should therefore be specified without using observations, but rather using any external source of knowledge. Note that in general, the parameter vector $\boldsymbol{\theta}$ is multi-dimensional, which implies that the prior distribution is also a multi-dimensional distribution.

The following Sect. 3.2.1.4 will describe how this prior distribution is used together with the likelihood to produce the inference. Independently of this technical derivation, the existence and the utilization of a prior distribution makes an important difference between Bayesian and frequentist (e.g. maximum likelihood) inferences. It can be viewed as both a strength and a weakness of the Bayesian paradigm. On the one hand, it can be argued that the prior distribution introduces a form of subjectivity in the inference: modelers using different prior assumptions will make different estimations and predictions. On the other hand, it can also be argued that prior knowledge frequently exists, especially in hydrology, and that the Bayesian framework offers an opportunity to openly and rigorously include it to the inference. This topic has been the subject of extensive debates in the statistical community (see e.g. Clark 2005; Efron 2005; Gelman 2008 for recent discussions). However, Bayesian inference is now pragmatically considered as a valuable tool to solve practical problems by most statisticians and hydrologists.

The specification of the prior distribution can be made in numerous ways. In hydrology, the following approaches have been used (this list is not exhaustive):

- **Using "expert knowledge"**: the "geophysical prior" of Martins and Stedinger (2000) is a prior distribution restricting the range of possible values for the shape parameter of a GEV distribution, based on past experience in hydrological case studies. Note that in many cases, expert knowledge may be difficult to express in terms of parameters, but might be easier to elicit once cast in terms of quantiles. Coles and Powell (1996) therefore proposed to use rough quantile estimates made by expert hydrologists, and to transfer this knowledge on quantiles into a prior distribution for parameters.
- **Using regional information**: data from sites similar to the target site may be used to derive a prior distribution. This has been implemented by e.g. Ribatet et al. (2006).
- **Non-informative priors**: perhaps surprisingly, the definition of a prior distribution that reflects a complete lack of knowledge is not straightforward. Jeffreys (1946) proposed a general rule to construct such prior for a given inference problem. Note that it is also a current practice to use "flat" distributions to denote a lack of knowledge, e.g. a uniform distribution with wide support or a Gaussian distribution with large variance.
- **Conjugate priors**: in some cases, the choice of a particular family of prior distribution facilitates computations, by avoiding the use of MCMC sampling (see subsequent Sect. 3.2.2). Such priors are known as "conjugate priors".

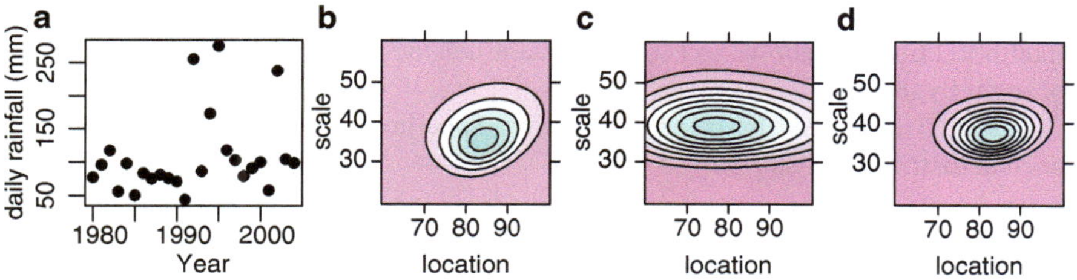

Fig. 3.1 Illustration of a simple Bayesian inference, based on annual maxima of daily rainfall (**a**) with a Gumbel distributional assumption. Contour plots of the likelihood function (**b**), the prior pdf (**c**) and the posterior pdf (**d**) are shown

However, such families are rarely tractable for extreme value distributions (but see Parent and Bernier 2003 for an exception), and will therefore not be further considered in this chapter.

3.2.1.4 Posterior Distribution

The Bayes theorem combines the knowledge brought by the prior distribution and the data (through the likelihood) into the posterior distribution of parameters $\boldsymbol{\theta}$, whose pdf is noted $p(\boldsymbol{\theta}\,|\,\boldsymbol{y})$:

$$p(\boldsymbol{\theta}\,|\,\boldsymbol{y}) = \frac{p(\boldsymbol{y}\,|\,\boldsymbol{\theta})p(\boldsymbol{\theta})}{\int p(\boldsymbol{y}\,|\,\boldsymbol{\theta})p(\boldsymbol{\theta})d\boldsymbol{\theta}} \tag{3.2}$$

Note that the denominator in Bayes theorem is a normalizing constant since it only depends on $\boldsymbol{y}$, but not on $\boldsymbol{\theta}$, which is integrated out. As we shall see later on, computing this multi-dimensional integral is in general not required. Consequently, Bayes theorem is often written in the following simple form:

$$p(\boldsymbol{\theta}\,|\,\boldsymbol{y}) \propto p(\boldsymbol{y}\,|\,\boldsymbol{\theta})p(\boldsymbol{\theta}) \tag{3.3}$$

where the symbol '$\propto$' denotes proportionality.

Equation (3.3) shows how Bayes' theorem combines prior and data knowledge through a disarmingly simple multiplication. The resulting posterior distribution summarizes what is known about parameters $\boldsymbol{\theta}$ whose inference is sought.

As an illustration, consider annual maxima of daily rainfall represented in Fig. 3.1a. Let us assume that these data are realizations from a Gumbel distribution with unknown location and scale parameters $\boldsymbol{\theta} = (\mu, \lambda)$. This example will be used throughout Sect. 3.2 to illustrate the basics of Bayesian inference.

Figure 3.1b shows the likelihood function, viewed as a function of parameters, and derived as:

$$p(\boldsymbol{y}\,|\,\mu, \lambda) = \prod_{t=1}^{N_t} p_{Gumbel}(y_t\,|\,\mu, \lambda) \tag{3.4}$$

where $p_{Gumbel}(y_t|\mu, \lambda)$ is the pdf of the Gumbel distribution evaluated at y_t. The likelihood function shown in Fig. 3.1b quantifies the information on the parameters brought by the data.

Figure 3.1c shows the prior distribution, specified using two independent log-normal distributions as follows:

$$p(\mu, \lambda) = p_{LogN}(\mu|\log(80), 0.2)\, p_{LogN}(\lambda|\log(40), 0.15) \tag{3.5}$$

where $p_{LogN}(z|a, b)$ is the pdf of a Log-Normal distribution with parameters a and b, and evaluated at value z. Note that we use the parameterization where a and b are the mean and the standard deviation of the log-transformed variable, respectively. This prior distribution corresponds to assuming that the location parameter is about 80, with a standard error of about 20%, and that the scale parameter is about 40, with a standard error of about 15%.

Lastly, Fig. 3.1d shows the (unnormalized) posterior pdf resulting from the application of Bayes' theorem in Eq (3.3). The posterior distribution combines the knowledge brought by the data (through the likelihood) and prior information (through the prior distribution).

Subsequent Sect. 3.2.3 will describe in details several ways of using the posterior distribution for parameter estimation and prediction. For instance, the posterior mode (i.e. parameters maximizing the posterior distribution), or marginal means/medians yield point-estimates of the parameters. However, this would be a somewhat restrictive use of the posterior distribution: indeed, the posterior also quantifies the uncertainty in the estimation of $\boldsymbol{\theta}$. Consequently, Bayesians favor a more thorough exploration of the posterior distribution (i.e. not restricted to optimal values or moment characteristics). However, this becomes challenging when the dimension of the inference increases: for instance, graphical representations like Fig. 3.1c, d are impossible in dimension 3 and above. Consequently, this exploration is achieved in practice by generating a large number of realizations from the posterior distribution, using a Markov Chain Monte Carlo (MCMC) sampler.

3.2.2 MCMC Samplers

3.2.2.1 General Principles

The basic aim of MCMC sampling is to generate values from a target distribution. In particular, it is useful in the following cases:

(i) The target distribution is multi-dimensional, making it impossible to use standard cdf-inversion techniques.
(ii) The target distribution does not belong to any standard family (e.g. Gaussian, Exponential, Elliptical, etc.), leaving little hope to find an existing simulation algorithm.
(iii) The target distribution is only known up to a constant of proportionality.

Typically, posterior distributions are characterized by all three points (i)–(iii) above, which make MCMC samplers a natural tool for their exploration.

A detailed description of the general theory of MCMC sampling lies well beyond the scope of this chapter. Textbooks are entirely devoted to this topic, for instance Robert and Casella (2004). We therefore restrict ourselves to a description of general principles, and briefly present the most widely used sampler, the Metropolis-Hastings algorithm (Metropolis and Ulam 1949; Metropolis et al. 1953; Hastings 1970). The interested reader may refer to existing textbooks for more information, and use existing software and packages implementing MCMC samplers, for instance packages mcmc (http://www.stat.umn.edu/geyer/mcmc/) and MCMCpack (Martin et al. 2011) in R, or the WinBugs (Lunn et al. 2000) and JAGS (http://www-ice.iarc.fr/~martyn/software/jags/) software.

3.2.2.2 A General-Purpose Sampler: The Metropolis-Hastings Algorithm

The idea behind MCMC sampling is to generate a random walk whose asymptotic distribution is the target distribution (e.g. the posterior distribution). More accurately, this random walk is constructed as a Markov Chain with the target distribution as equilibrium distribution (see Robert and Casella 2004 for more details).

Let $f(\boldsymbol{x})$ denote the (possibly unnormalized) pdf of the target distribution. In most cases, this distribution is multidimensional and $\boldsymbol{x}$ is hence a vector. Since it is in general not possible to generate values directly from f, the basic idea behind MCMC samplers is to use another distribution, called the proposal or the jump distribution, to propose candidate values. Those candidate values will then be accepted or rejected, according to a given acceptance rule. A well-chosen acceptance rule ensures that the asymptotic distribution of generated values is the target distribution.

More accurately, let $J(z|\boldsymbol{x})$ denote the pdf of the jump distribution (evaluated at z), which is conditional on $\boldsymbol{x}$ and is used to generate a candidate $\boldsymbol{x}^*$. This distribution should be chosen so that one can easily generate a value from it. A common choice is to use a Gaussian distribution centered on $\boldsymbol{x}$ and with covariance matrix $\boldsymbol{\Sigma}$ (whose choice will be discussed later on). In this case, $J(z|\boldsymbol{x}) = p_N(z|\boldsymbol{x}, \boldsymbol{\Sigma})$.

With this notation, the Metropolis-Hastings algorithm can be described as follows:

Algorithm 1. Metropolis-Hastings

Initialization: choose a starting value $\boldsymbol{x}^{(0)}$
For $i = 1: N_{\text{iter}}$

step 1 Generate a candidate value $\boldsymbol{x}^*$ from the jump distribution with pdf $J(z|\boldsymbol{x}^{(i-1)})$

step 2 Compute the acceptance ratio $\tau = \frac{f(\boldsymbol{x}*)}{f(\boldsymbol{x}^{(i-1)})}\frac{J(\boldsymbol{x}^{(i-1)}|\boldsymbol{x}*)}{J(\boldsymbol{x}*|\boldsymbol{x}^{(i-1)})}$

step 3 Accept the candidate ($\boldsymbol{x}^{(i)} = \boldsymbol{x}^*$) with probability max(1;τ); otherwise, reject the candidate ($\boldsymbol{x}^{(i)} = \boldsymbol{x}^{(i-1)}$).

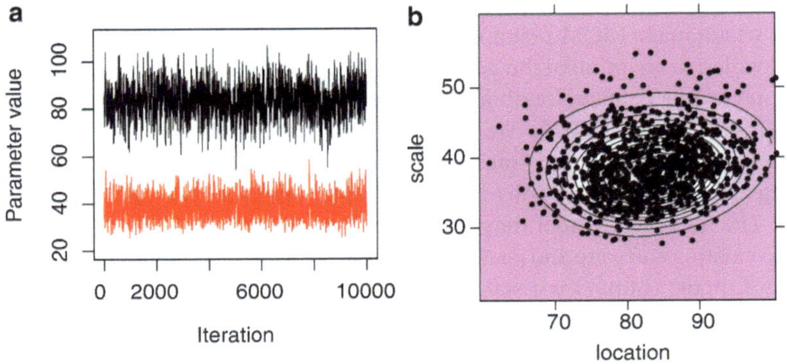

Fig. 3.2 MCMC-sampling from the posterior distribution: (**a**) time series of generated values for the location (*black*) and scale (*red*) parameters; (**b**) scatterplot of generated (location,scale) pairs, overlaying a contour plot of the posterior pdf

The generated values $(\boldsymbol{x}^{(i)})_{i=1:N_{iter}}$ form a Markov Chain: the distribution of $\boldsymbol{x}^{(i)}$ solely depends on previous value $\boldsymbol{x}^{(i-1)}$ given the definition of the jump distribution. Moreover, the Metropolis-Hastings acceptance rule ensures that its equilibrium distribution is the target distribution (Hastings 1970).

Before discussing some properties of this sampler, a simple illustration is given in Fig. 3.2, based on the same setup as in previous Sect. 3.2.1 (see Fig. 3.1). Algorithm 1 is applied with the unnormalized posterior (3.3) as target distribution. The jump distribution is a Gaussian distribution centered on previous value $\boldsymbol{x}^{(i-1)} = (\mu^{(i-1)}, \lambda^{(i-1)})$ and with covariance matrix equal to identity. Figure 3.2a shows the chains generated for the location and scale parameters, while Fig. 3.2b shows the scatterplot of $(\mu^{(i)}, \lambda^{(i)})$ pairs. A comparison with the contour plot illustrates that those values are realizations from the target posterior distribution.

3.2.2.3 Monitoring Convergence

As explained in the previous section, MCMC-generated values can be considered as realizations from the target distribution only asymptotically. This raises the question of monitoring the convergence of the simulated chain toward the target distribution. In particular, a poor choice of starting points may result in generating values that are not realizations from the equilibrium distribution during the first iterations. This is illustrated in Fig. 3.3, which shows the evolution of three parallel chains sampled from the same posterior distribution but with different starting values. The poorly-chosen starting values result in the chains not reaching convergence before several hundreds iterations. However, all three chains converge after a few hundreds iterations. Consequently, the influence of starting values is in general minimized by using a burn-in period, i.e. by discarding a sufficient number of iterations at the beginning of the chain.

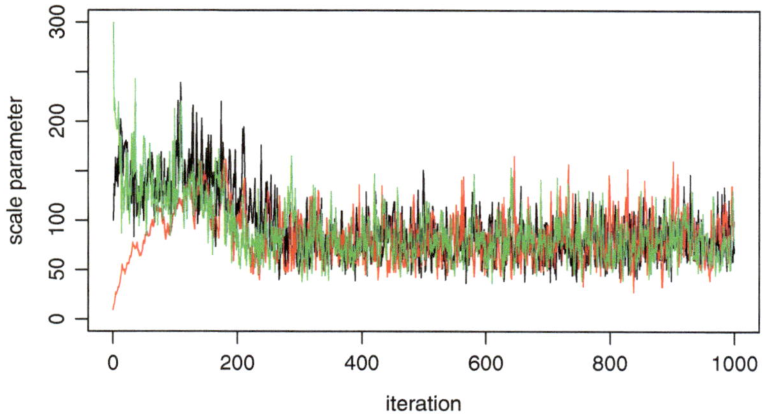

Fig. 3.3 Illustration of the influence of starting values on MCMC convergence

Several methods have been proposed to monitor MCMC convergence (see El Adlouni et al. 2006 for a review and comparison). They are based on running several chains in parallel (e.g. Gelman et al. 2004) or using subsequences of a single chain (e.g. Geweke 1992), and verifying that they generate values from a similar distribution. Note that the R package CODA (Plummer et al. 2006) implements numerous diagnostics for MCMC convergence.

3.2.2.4 Building Efficient Samplers

As most numerical methods, MCMC samplers require some tuning that strongly affects numerical efficiency. It is difficult to provide universal guidelines for efficiently tuning MCMC samplers, but the following points are important:

- **Choice of starting values**: choosing starting values located in a high-density area of the posterior space will limit the length of the burn-in period. If informative priors are used, the prior mean or mode can be chosen as a first approximation. Applying an optimization algorithm to the posterior distribution is another possibility.
- **Choice of the jump distribution**: this is the most important point, since a poorly-chosen jump distribution may result in a prohibitively slow algorithm. As a general rule of thumb, the jump distribution should be “similar” to the target distribution in terms of size and orientation. When a Gaussian jump distribution is used, these are controlled by the choice of the covariance matrix:
  - The diagonal terms (jump variances) should be neither too small nor too large: large jumps will result in candidate values being generated far from the high density region, thus leading to a very low acceptance rate and a high

number of iterations to visit the whole parameters space. Alternatively, small jumps will result in a too high acceptance rate, with the parameter space being visited too slowly.

  - The off-diagonal terms (covariances) should be chosen so that the correlations of the jump distribution are similar to the posterior correlations. This will ensure that jumps are made preferentially in a direction consistent with the orientation of the posterior distribution.
  - Let us also mention that non-Gaussian jump distributions may be interesting, in particular fat-tailed distributions (e.g. Student). Indeed, such jump distributions will occasionally generate a candidate far from the current location of the chain, thus limiting the risk of the chain being stuck in a local mode in the case of multi-modal posterior distributions.

- **Adaptive algorithms**: The specification of an adequate covariance matrix for the jump distribution is in general difficult, since little is known about the posterior distribution before sampling from it. A possibility to overcome this issue is to change the characteristics of the jump distribution during iterations, so that the algorithm can "learn" from previous iterations to adapt the jump size and orientation. Such algorithms are known as adaptive samplers (see e.g. Haario et al. 2001, 2005; Renard et al. 2006).
- **Block Sampling**: In multi-dimensional problems, it is possible to split the inference space into several blocks, and to perform a Metropolis update on a block-by-block basis, rather than attempting to update the current vector in one single step. This strategy increases the complexity of the algorithm (since several jumps and acceptance/rejection steps are performed at each iteration), but may facilitate the specification of "good" jump distributions and fasten convergence. Marshall et al. (2004) provide a detailed description of such samplers.

Finally, note that there exist other Monte-Carlo methods that may be useful to the Bayesian Hydrologist (e.g. importance sampling, acceptation-rejection sampling, see Robert and Casella 2004). However, MCMC samplers, when properly tuned, are in general the most efficient way to generate values from a posterior distribution.

3.2.3 Using the Posterior Distribution for Inference and Prediction

For simplicity, we will assume in this section that the posterior distribution has been explored using a MCMC sampler, yielding a large sample of values $(\boldsymbol{\theta}^{(i)})_{i=1:N_{sim}}$ that can be considered as realizations from the posterior distribution $p(\boldsymbol{\theta}|\mathrm{y})$. This is not a restrictive assumption given that cases where results can be derived analytically are the exception rather than the rule, especially in the context of extreme value distributions.

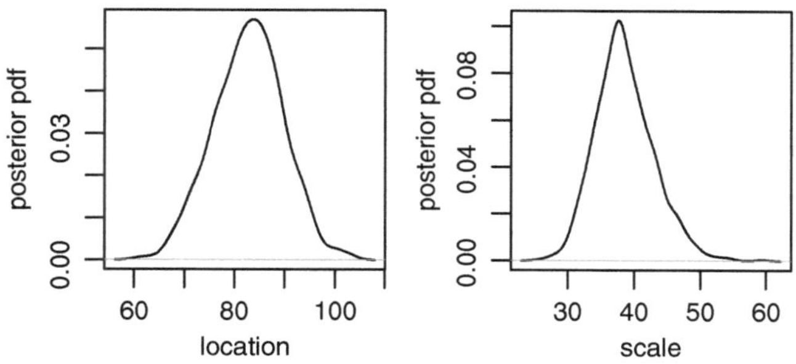

Fig. 3.4 Marginal posterior pdf of the location and scale parameters, estimated from the MCMC samples using a kernel density. The data used for inference are shown in Fig. 3.1a

3.2.3.1 Posterior-Based Inference

In a first step, MCMC samples are usually used to derive parameter estimates. The notation $\boldsymbol{\theta} = (\theta_1, ..., \theta_{N_D})$ is used to denote each component of the parameter vector. Marginal estimates (i.e. estimates of each individual parameter θ_k) are simply obtained by considering values $\left(\theta_k^{(i)}\right)_{i=1:N_{sim}}$, which are realizations from the marginal posterior distribution of parameter θ_k. As an illustration, consider the values plotted in Fig. 3.2a: a histogram of those values will approximate the marginal posterior pdf of each parameter. This is implemented in Fig. 3.4 (smoothing has been performed using a Gaussian kernel), and yields an estimation of each inferred parameter. Importantly, this estimation is not a single value, but a whole distribution. This is an important characteristic of Bayesian inference: it directly provides the uncertainty in parameter estimates. By contrast, consider standard estimation approaches (e.g. moment-based, maximum likelihood ML): in a first step, point-estimates are sought using some "optimality" criterion (e.g. maximizing the likelihood or recovering empirical moments). In a second step, the sampling distributions of these point-estimates are derived, either using some resampling technique (e.g. bootstrapping) or some asymptotic approximation (e.g. asymptotic normality of ML estimates). Note that the latter option can be problematic when short samples are used (which is frequent when analyzing hydrological extremes).

Note that it is still possible to obtain point-estimates from the posterior distribution. For instance, computing the marginal mean or median from the samples $(\boldsymbol{\theta}^{(i)})_{i=1:N_{sim}}$ yields point-estimates for each parameter. Alternatively, a modal estimate (i.e. a "maximum posterior estimate", mimicking the idea behind maximum likelihood estimation) can be derived by selecting the parameter vector $\boldsymbol{\theta}^{(i_{\max})}$ corresponding to the largest posterior value. Note that since MCMC methods are not optimization procedures, such modal estimate might be refined by using an optimization method to get closer to the posterior mode.

Estimates for quantities that can be expressed as functions of the parameters can easily be derived in a similar way. This is in particular the case for quantiles,

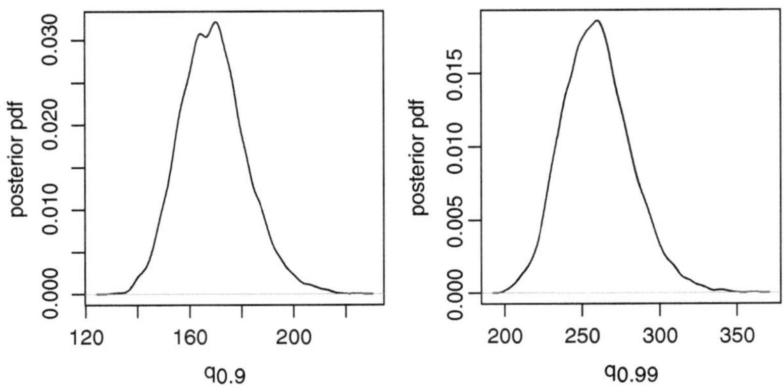

Fig. 3.5 Posterior pdf of 0.9- and 0.99-quantiles estimated from the MCMC samples using a kernel density

which are of primary interest in extreme value analyses. Indeed, quantiles (whose probability of exceedance is often prescribed by regulation) are used to design civil engineering structures (e.g. reservoirs, dam spillways, etc.) or to map flood-prone areas where building restrictions may be enforced.

The p-quantile q_p is a function of the parameters, $q_p = Q(p, \boldsymbol{\theta})$ (e.g. for a Gumbel distribution, $q_p = \mu - \lambda \log(-\log(p))$). Consequently, the posterior distribution of q_p can be approximated by computing the p-quantile corresponding to each MCMC-generated parameter, yielding $\left(q_p^{(i)}\right)_{i=1:N_{sim}} = \left(Q(p, \boldsymbol{\theta}^{(i)})\right)_{i=1:N_{sim}}$. Figure 3.5 shows the corresponding posterior pdf for 0.9- and 0.99-quantiles obtained by applying a kernel density estimation to $\left(q_p^{(i)}\right)_{i=1:N_{sim}}$. As for parameter estimates, Bayesian inference directly yields a full distribution – i.e., it directly yields the uncertainty in quantile estimates.

The quantile curve is a standard representation used in extreme values analyses. Figure 3.6a shows the quantile curve estimation resulting from the Bayesian inference. In this figure, quantiles (also termed return levels) are expressed as a function of the return period T instead of the non-exceedance probability p ($T = 1/(1\text{-}p)$). Each light gray line is a particular quantile curve corresponding to a MCMC-generated parameter vector. A "median quantile curve" (dashed line) can be plotted by computing, for each return period T, the median of the posterior distribution of q_T. Note that the resulting curve joins posterior medians but does not correspond to any of the light gray lines. Alternatively, the "modal quantile curve" (thick solid line) corresponds to the modal parameter estimate. As seen from Fig. 3.6a, both estimates are in general similar. Estimation uncertainties can be represented using posterior 90% intervals (dotted lines). Those are obtained by computing 5 and 95% quantiles from the posterior distribution of q_T. Note that the terminology "confidence interval" is in general not used in a Bayesian setting, because the interpretation of such interval is different in Bayesian and frequentist

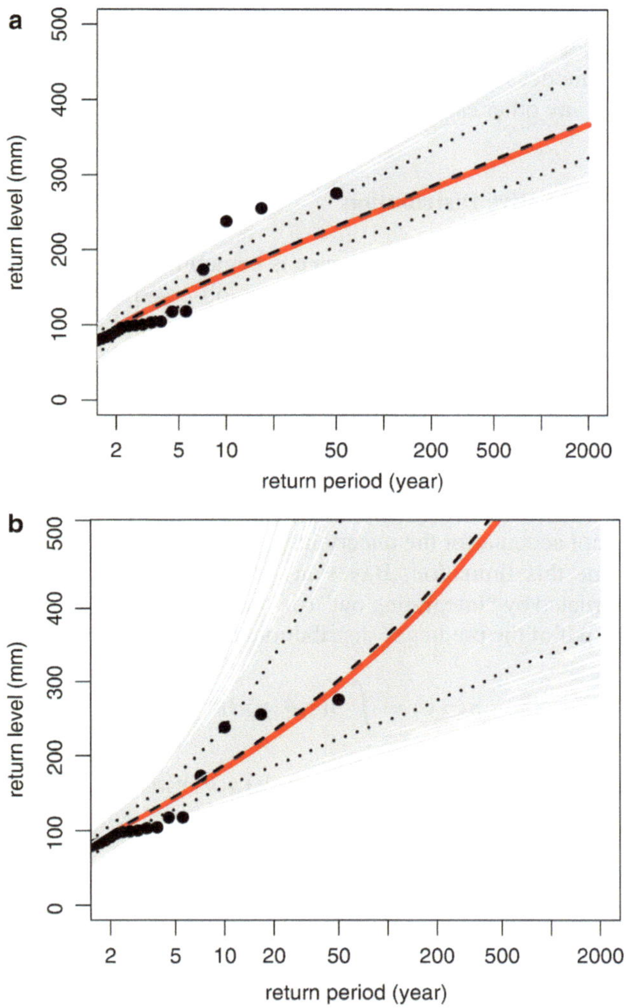

Fig. 3.6 Quantile curves obtained with Gumbel (**a**) and GEV (**b**) distributional assumption. Each *light gray line* corresponds to a MCMC-generated parameter vector. *Dotted lines* represent a 90% posterior interval for the return level, while the *dashed line* is the median of the return level posterior distribution. The *thick red line* corresponds to the modal parameter estimates

contexts (see e.g. Gelman et al. 2004 for additional discussion on this topic). Bayesians favor the terminology "posterior interval", or "credibility interval".

It is important to keep in mind that the posterior distribution only quantifies estimation uncertainties. However, the posterior distribution remains conditional on the modeling assumptions (here, assumption of a Gumbel distribution). To illustrate this point, consider Fig. 3.6b, which shows the quantile curve derived with a

Generalized Extreme Value (GEV) distributional assumption. Overall, using a GEV assumption yields higher quantiles and, even more markedly, higher uncertainties. Consequently, model checking and validation remains a vital task in Bayesian inference, as in any other estimation approach.

3.2.3.2 The Predictive Distribution

The ultimate aim of a statistical inference is sometimes (even frequently) not parameter estimation, but rather prediction of an unobserved value. As an illustration, when statistical estimates are used for designing some civil engineering structures or for hazard mapping, the operational objective is to predict upcoming events, as opposed to describe past events or provide an "optimal" parameter value.

The standard approach to estimate the distribution of an unobserved outcome is to use the assumed distribution with some "good" value of the parameters, i.e. the pdf of a future outcome would be $p(z|\hat{\boldsymbol{\theta}})$, where $\hat{\boldsymbol{\theta}}$ corresponds for instance to the modal estimate or to the vector of marginal means/medians. The drawback of this approach is that it does not account for the uncertainty in estimating the parameters $\boldsymbol{\theta}$.

To overcome this limitation, Bayesians often use the predictive distribution, which is constructed by "integrating out" (or "marginalizing") the parameters. More accurately, the pdf of the predictive distribution, noted $p(z|\mathbf{y})$, is defined as follows:

$$\begin{aligned} p(z|\mathbf{y}) &= \int p(z, \boldsymbol{\theta}|\mathbf{y}) d\boldsymbol{\theta} \\ &= \int p(z|\boldsymbol{\theta}, \mathbf{y}) p(\boldsymbol{\theta}|\mathbf{y}) d\boldsymbol{\theta} \\ &= \int p(z|\boldsymbol{\theta}) p(\boldsymbol{\theta}|\mathbf{y}) d\boldsymbol{\theta} \end{aligned} \tag{3.6}$$

In the equation above, $p(z|\boldsymbol{\theta})$ corresponds to the pdf of the assumed distribution, while $p(\boldsymbol{\theta}|\mathbf{y})$ corresponds to the pdf of the posterior distribution. Consequently, the predictive distribution can be interpreted as follows: while the standard approach for prediction is to use the assumed distribution $p(z|\boldsymbol{\theta})$ with a value of $\boldsymbol{\theta}$ that (for instance) *maximizes* the posterior pdf, the predictive distribution uses the assumed distribution $p(z|\boldsymbol{\theta})$ *integrated* over possible posterior realizations of $\boldsymbol{\theta}$. Since the posterior distribution describes the estimation uncertainty, the predictive distribution directly accounts for uncertainties in prediction mode. If the uncertainty if very small, the predictive distribution will be very close to the standard "modal" predictor.

In practice, the integration in Eq. (3.6) is rarely performed analytically. Instead, it can be approximated using the MCMC samples $(\boldsymbol{\theta}^{(i)})_{i=1:N_{sim}}$ from the posterior distribution. Algorithm 2 shows how this approximation can be performed. In some cases, it may be more practical to directly generate values from the predictive distribution. This can be achieved using Algorithm 3.

Algorithm 2. Estimation of the Predictive pdf

- Initialization: choose a grid $z_1, \ldots, z_m$ where the predictive pdf will be evaluated
- For $i = 1$: N_{sim}
- Compute $g_k^{(i)} = p\left(z_k|\boldsymbol{\theta}^{(i)}\right)$ for $k = 1{:}m$
- The predictive pdf at value z_k can be approximated by $\hat{p}(z_k|\boldsymbol{y}) = \frac{1}{N_{sim}} \sum_{i=1}^{N_{sim}} g_k^{(i)}$ for $k = 1{:}m$

Algorithm 3. Generating Values from the Predictive Distribution

- For $i = 1$: N_{sim}
- Generate a sample of size m from the assumed distribution with parameters $\boldsymbol{\theta}^{(i)}: \left(z_1^{(i)}, \ldots, z_m^{(i)}\right) \underset{iid}{\leftarrow} p\left(z|\boldsymbol{\theta}^{(i)}\right)$
- The sample $\left(z_k^{(i)}\right)_{k=1:m, i=1:N_{sim}}$ is a realization from the predictive distribution

Figure 3.7 provides an illustration of the predictive distribution and compares it to more standard predictions using an "optimal" predictor augmented with an uncertainty interval. In the Gumbel case (Fig. 3.7a), both estimates are similar: this is because uncertainties are relatively small in this case, and integrating them do not markedly change the "optimal" prediction. However, the GEV predictive distribution yields markedly higher quantiles than the "optimal" prediction (Fig. 3.7b), which is a consequence of the large estimation uncertainties.

Although the use of predictive distributions has been proposed by several authors in the research literature (in particular, Stedinger 1983; Kuczera 1999; Coles 2001, chapter 9; Cox et al. 2002; Meylan et al. 2008, chapter 7), it remains scarcely used (if at all) in operational applications of extreme value analyses. This is surprising given the current emphasis on uncertainty analysis and the need to communicate uncertainty. We claim that the benefits of predictive distributions should be thoroughly evaluated and, if proven beneficial, the use of predictive distributions in operational applications should be encouraged.

Finally, the principles underlying the derivation of predictive distributions have strong link with decision theory and the concepts of loss (or cost) functions. The reader interested in these additional developments may refer to the books by Bernier et al. (2000) or Robert (2001) for additional details.

3.2.3.3 Model Comparison and Bayesian Model Averaging

An interesting property of Bayesian inference is the possibility to associate probabilities to competing models. Let us assume that $\{M_1, \ldots, M_q\}$ is a collection of

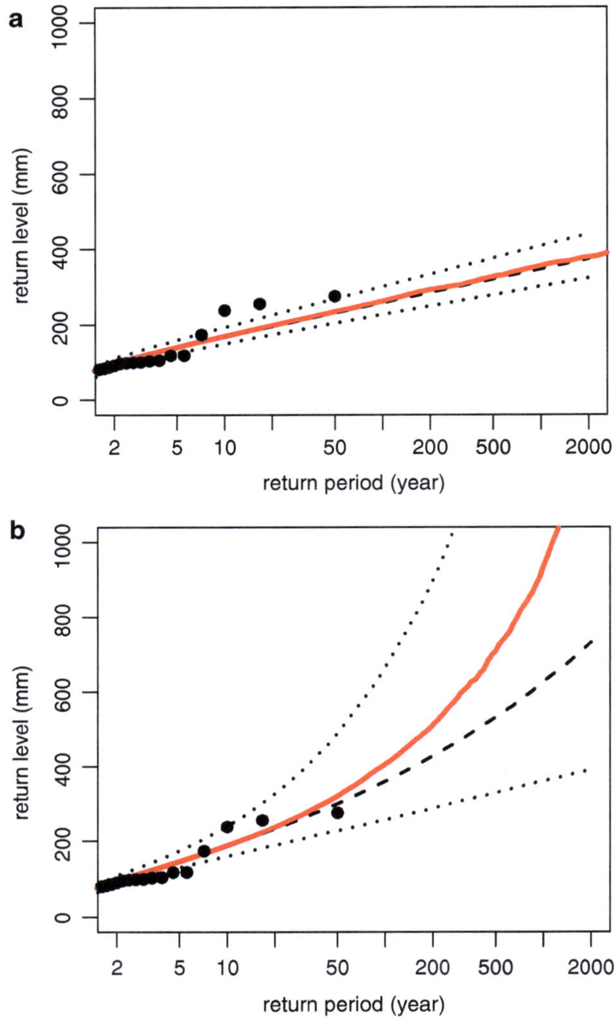

Fig. 3.7 Comparison of modal (*dashed black lines*) and predictive (*solid red lines*) distributions. *Dotted lines* represent a 90% posterior interval. (**a**) Gumbel assumption; (**b**) GEV assumption

candidate models for describing the data $\boldsymbol{y}$. Each model M_j uses a parameter vector $\boldsymbol{\theta}^{M_j}$ (whose dimension may differ from model to model). Moreover, let $p(M_1), \ldots, p(M_q)$ denote prior probabilities assigned to each model, so that $\sum_{i=1}^{q} p(M_i) = 1$. Note that the latter condition implies that the model collection is considered as exhaustive.

Bayes theorem can then be applied in its discrete form to define the posterior probability of each model M_j given observations $\boldsymbol{y}$ as follows:

$$p(M_j|\boldsymbol{y}) = \frac{p(\boldsymbol{y}|M_j)p(M_j)}{\sum_{i=1}^{q} p(\boldsymbol{y}|M_i)p(M_i)} \tag{3.7}$$

The term $p(\boldsymbol{y}|M_j)$ in Eq. (3.7) is the marginal likelihood of observations, and can be computed using standard conditional probability algebra as follows:

$$p(\boldsymbol{y}|M_j) = \int p\left(\boldsymbol{y}|\boldsymbol{\theta}^{M_j}, M_j\right) p\left(\boldsymbol{\theta}^{M_j}|M_j\right) d\boldsymbol{\theta}^{M_j} \tag{3.8}$$

One can recognize in Eq. (3.8) the denominator in the Bayes theorem (3.2), i.e. the normalizing constant of the posterior distribution. Although computing this quantity is not necessary for posterior sampling and prediction (see Sects. 3.2.2, 3.2.3.1 and 3.2.3.2), it is required if one wishes to compute model posterior probabilities.

Computing the marginal likelihood is in general difficult, and strongly suffers from the curse of dimensionality. However, practical solutions exist and give good results, at least when the size of the parameter vector remains moderate. Bos (2002) compares various methods for computing marginal likelihoods. One of the most efficient methods was proposed by Chib (1995) and is described in an Appendix to this chapter. We refer to Bos (2002) and references therein for alternative methods.

Once model posterior probabilities have been computed, they can be used both for model comparison and for prediction. One possible approach to model comparison is to use Bayes factors (Kass and Raftery 1995; see Perreault et al. 2000a, b for an hydrological application), which provide a measurement of the relevance of one model compared to another. The Bayes factor between models M_i and M_j is defined as follows:

$$B_{i,j} = \frac{p(M_i|\boldsymbol{y})}{p(M_j|\boldsymbol{y})} \Big/ \frac{p(M_i)}{p(M_j)} = \frac{p(\boldsymbol{y}|M_i)}{p(\boldsymbol{y}|M_j)} \tag{3.9}$$

The Bayes factor $B_{i,j}$ can be compared to 1, with stronger confidence in model M_i for high values of $B_{i,j}$. Kass and Raftery (1995) provide guidelines for interpreting Bayes factors, together with additional developments. In particular, it is possible to compute composite Bayes factors between two sets of models (e.g., several stationary models vs. several non-stationary models). Let S_1 and S_2 denote two disjoint sets of models, the composite Bayes factor between sets S_1 and S_2 is defined as:

$$B_{S_1,S_2} = \frac{\sum_{M \in S_1} p(M|\boldsymbol{y})}{\sum_{M \in S_2} p(M|\boldsymbol{y})} \Big/ \frac{\sum_{M \in S_1} p(M)}{\sum_{M \in S_2} p(M)} \tag{3.10}$$

Posterior probabilities of models can also be used to perform multi-model predictions. This can be achieved by computing a weighted average of individual model predictions, with the posterior model probabilities used as weights. As an illustration, let us consider the case of predicting the value of some p-quantile q_p. Each model M_i provides the posterior pdf of the p-quantile, noted as $p(q_p|\mathbf{y}, M_i)$. The marginal posterior distribution of q_p (i.e., unconditional on the model) can be derived as the following weighted average (which corresponds to the total probability law):

$$p(q_p|\mathbf{y}) = \sum_{i=1}^{q} p(q_p|\mathbf{y}, M_i)p(M_i|\mathbf{y}) \tag{3.11}$$

Notice that other predictive quantities (in particular, predictive distributions) can also be averaged in a similar way.

Bayesian model averaging allows accounting, at least partly, for model uncertainty. It is particularly useful when distinct models provide a comparably acceptable description of the data, but yield markedly different predictions in extrapolation. However, it makes the assumption that the model collection is exhaustive, since prior model probabilities sum up to one. Consequently, averaging models does not exempt from evaluating the performance of individual models. In particular, it would be unreasonable to expect that averaging a set of poor models may result in anything else than poor predictions.

3.3 Local Inference of Non-stationary Models

In this section, the standard Bayesian framework described in Sect. 3.2 is generalized in order to account for non-stationarity. A general non-stationary modeling framework is first described, before presenting several illustrations using distinct non-stationary models. The limitations of describing non-stationarity using at-site data only are also illustrated.

3.3.1 Introducing Non-stationarity Using Covariate Modeling

In a non-stationary context, the assumption of identical distribution made in Sect. 3.2.1.1 does not hold any more: by definition, non-stationarity refers to the fact that the distribution of the variable of interest is time-dependent. A pragmatic solution to account for this time-dependence has been proposed and explored by many authors (Coles 2001; Strupczewski and Kaczmarek 2001; Strupczewski et al. 2001; Katz et al. 2002; Khaliq et al. 2006): data are still assumed to be realizations from a given distribution family (e.g. Gumbel, GEV, etc.), but whose parameters may vary with time.

More formally, the random variable Y_t is assumed to follow a distribution with pdf $p(y|\boldsymbol{\theta}(t))$, which will be noted as follows in the sequel of this chapter:

$$Y_t \sim p(y|\boldsymbol{\theta}(t)) \tag{3.12}$$

In a nutshell, Eq. (3.12) simply means that the parameter values are different at each time step. However, Eq. (3.12) cannot be used in this basic form in practice: indeed, estimating one different parameter vector at each time step based on a single series of observations is impossible, simply due to a lack of degrees of freedom. Consequently, additional hypotheses on the form of the non-stationarity are required.

This can be achieved by assuming a regression model linking the parameter value at time t with the value of some time-varying covariate(s). More formally, denoting by $\boldsymbol{\theta}(t) = (\theta_1(t), ..., \theta_{N_D}(t))$ the N_D-dimensional parameter vector at time t, a regression model for each component $\theta_k(t)$ can be written as follows:

$$g_k(\theta_k(t)) = h_k(\boldsymbol{x}_k(t); \boldsymbol{\beta}_k) \tag{3.13}$$

Equation (3.13) is constructed with the following components:

- The one-to-one function g_k is a simple parameter transformation, termed the "link function" by analogy with generalized linear models (Dobson 2001). The identity function may be used in cases were no transformation is desired. However, non-trivial functions may be useful for some parameters, e.g. a logarithm function to ensure the positivity of a scale parameter, or a logit function for the parameter of a Binomial distribution which should be comprised between zero and one.
- The vector $\boldsymbol{x}_k(t)$ contains the values of a set of covariates varying in time, and used to describe the temporal evolution of the distribution of the variable Y_t. Such covariate could simply be the time itself, $\boldsymbol{x}_k(t) = t$. It may also be augmented with additional information, for instance the value of some climate index $CI(t)$ (e.g. the Southern Oscillation Index, see El Adlouni et al. 2007). In this case, $\boldsymbol{x}_k(t) = (t, CI(t))$.
- The function $h_k(.; \boldsymbol{\beta}_k)$ is the regression function, parametrized by a vector $\boldsymbol{\beta}_k$, and linking the values of parameter $\theta_k(t)$ with the values taken by the covariates $\mathbf{x}_k(t)$. The most common regression function is the linear function, $h_k(\boldsymbol{x}_k(t); \boldsymbol{\beta}_k) = \boldsymbol{x}_k(t).\boldsymbol{\beta}_k^T$, but other functions may be used.

The regression model in Eq. (3.13) is a very general tool to introduce different forms of non-stationarity, e.g. linear trend, step-change, seasonal cycle, or some combination of them (see e.g. Maraun et al. 2010). This will be illustrated through several examples in following Sect. 3.3.3.

3.3.2 Inference

Parameter estimation follows the steps outlined in Sect. 3.2.1. The main difference is that instead of directly inferring the parameters $\boldsymbol{\theta}(t)$, estimates are sought for the regression parameters $\boldsymbol{\beta} = (\boldsymbol{\beta}_1, \ldots, \boldsymbol{\beta}_{N_D})$ in Eq. (3.13). In order to avoid confusion, we introduce the following naming: D-parameters (Distribution parameters $\boldsymbol{\theta}(t)$) correspond to the parameter of the distribution in its original parameterization, while R-parameters correspond to the Regression parameters $\boldsymbol{\beta}$. Inference is performed for R-parameters, and D-parameters can then be retrieved by applying the regression model in Eq. (3.13), i.e. $\theta_k(t) = g_k^{-1}\left(h_k\left(\boldsymbol{x}_k(t); \boldsymbol{\beta}_k\right)\right)$ for $k = 1{:}N_D$.

The shorthand notation $\boldsymbol{x} = (\boldsymbol{x}_k(t))_{k=1:N_D, t=1:N_t}$ is used to denote the set of all covariates values. The posterior distribution of R-parameters is given by the Bayes' theorem as follows:

$$p(\boldsymbol{\beta}|\boldsymbol{y}, \boldsymbol{x}) \propto p(\boldsymbol{y}|\boldsymbol{\beta}, \boldsymbol{x})\, p(\boldsymbol{\beta}|\boldsymbol{x}) \tag{3.14}$$

In Eq. (3.14), the term $p(\boldsymbol{y}|\boldsymbol{\beta}, \boldsymbol{x})$ is the likelihood of observations $\boldsymbol{y} = (y_t)_{t=1:N_t}$. To compute this term, let us first consider the pdf of the variable Y_t at a given time step t. This pdf is derived as follows:

$$\begin{aligned} p(y|\boldsymbol{\beta}, \boldsymbol{x}(t)) &= p\left(y|\theta_1(t), \ldots, \theta_{N_D}(t)\right) \\ \text{where } \theta_k(t) &= g_k^{-1}\left(h_k\left(\boldsymbol{x}_k(t); \boldsymbol{\beta}_k\right)\right) \text{ for } k = 1:N_D \end{aligned} \tag{3.15}$$

Assuming independence between observations $\boldsymbol{y} = (y_t)_{t=1:N_t}$, the likelihood function is then equal to:

$$p(\boldsymbol{y}|\boldsymbol{\beta}, \boldsymbol{x}) = \prod_{t=1}^{N_t} p(y_t|\boldsymbol{\beta}, \boldsymbol{x}(t)) \tag{3.16}$$

The second term in Eq. (3.14), $p(\boldsymbol{\beta}|\boldsymbol{x})$, is the prior distribution of R-parameters $\boldsymbol{\beta} = (\boldsymbol{\beta}_1, \ldots, \boldsymbol{\beta}_{N_D})$. As suggested by the notation, the covariates $\boldsymbol{x}$ may be used for prior specification, but not observations $\boldsymbol{y}$.

3.3.3 Example: Extreme Rainfalls

An illustration of the "covariate modeling" approach presented in Sects. 3.3.1 and 3.3.2 is provided using a series of extreme rainfall. Annual maxima are extracted from the series of daily rainfall at the raingauge of Marignane, in South-East France. Data are available over the period 1946–2004 and are shown in Fig. 3.8 (source: Meteo France). In all three subsequent examples, a GEV distribution is used to describe observations. For the sake of simplicity, time is the sole covariate, but the general principle would remain identical with other, possibly more meaningful, covariates.

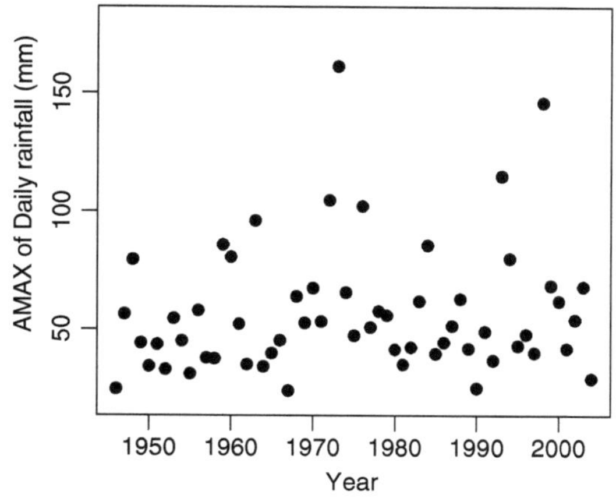

Fig. 3.8 Annual maxima of daily rainfall at the Marignane raingauge

3.3.3.1 Trend Model

In this first example, non-stationarity is introduced through a trend on the location parameter, with the scale and shape parameters remaining constant:

$$\begin{aligned} \mu(t) &= \mu_0 (1 + \mu_1 t) \\ \sigma(t) &= \sigma_0 \\ \xi(t) &= \xi_0 \end{aligned} \tag{3.17}$$

Compared with the general Eq. (3.13), the trend model for location parameter in Eq. (3.17) corresponds to:

- Using the identity as link function, $g(u) = u$
- Using time as the sole covariate, $\boldsymbol{x}(t) = t$
- Using the regression function $h(t;\boldsymbol{\beta}) = \beta_0(1 + \beta_1 t)$

Inference is hence sought for the four R-parameters (μ_0, μ_1, σ_0, ξ_0), through the posterior distribution in Eq. (3.14).

Prior distributions are specified using expert knowledge as follows:

$$\begin{aligned} \mu_0 &\overset{prior}{\sim} LogN(\log(40); 0.25) \\ \mu_1 &\overset{prior}{\sim} N(0; 0.02) \\ \sigma_0 &\overset{prior}{\sim} LogN(\log(20); 0.25) \\ \xi_0 &\overset{prior}{\sim} N(0; 0.25) \end{aligned} \tag{3.18}$$

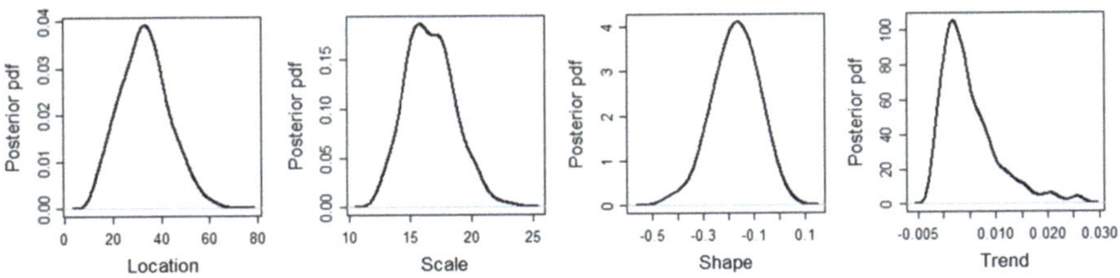

Fig. 3.9 Posterior pdf of R-parameters: location (μ_0), scale (σ_0), shape (ξ_0) and trend (μ_1)

Fig. 3.10 Annual maxima of daily rainfall and evolution of quantile estimates in time. The *blue lines* refer to the estimated median, the *red lines* to the estimated 0.99-quantile. *Dotted lines* represent 90% posterior intervals

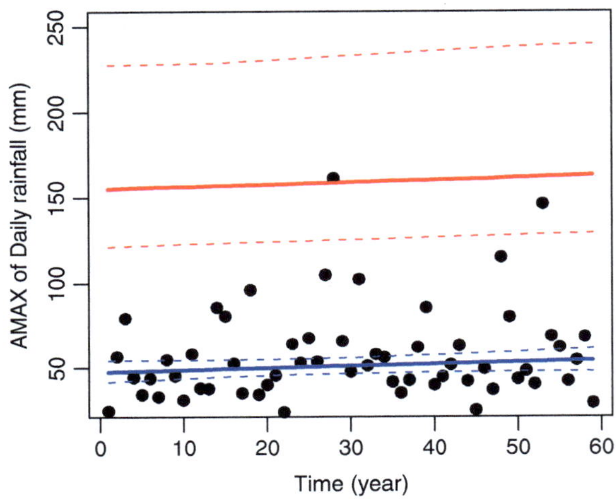

Priors for the location (μ_0) and scale (σ_0) parameters are based on the consideration of typical values found in this area of France. Similarly, the prior distribution for the shape parameter (ξ_0) encompasses typical shape values found in hydrology. This prior distribution can be compared with the "geophysical prior" proposed by Martins and Stedinger (2000). Lastly, the Gaussian prior distribution for the trend (μ_1) parameter is centered on zero and has a standard deviation of 0.02, which corresponds to a rather vague prior. Indeed, a trend with value 0.02 would correspond in the parameterization of Eq. (3.17) to a doubling of the location parameter only 50 years after the beginning of the series.

Figure 3.9 shows the posterior pdf of each R-parameter. The posterior pdf of the shape parameter mostly corresponds to negative values, suggesting a heavy-tailed distribution. Moreover, a positive trend may be suspected, with parameter μ_1 having its pdf mostly concentrated on positive values.

These estimates are easier to interpret once transformed into quantile estimates. Figure 3.10 hence shows the evolution of the median (blue line) and the 0.99-quantile (red line). The median is precisely estimated, and its value very slightly increases from about 47 mm to about 54 mm between the beginning and the end of the record. This trend is also discernible for the 0.99-quantile estimates. However,

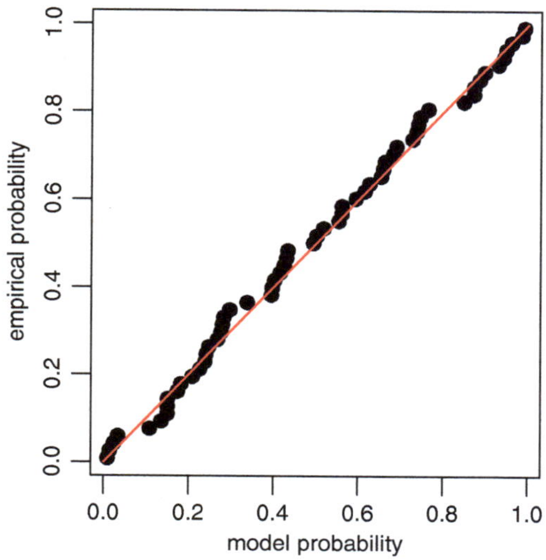

Fig. 3.11 pp-plot obtained with modal parameters

the magnitude of this trend has to be compared with the very large uncertainty in the 0.99-quantile: while the posterior median of the 0.99-quantile increases of about 10 mm, the width of 90% posterior intervals exceeds 100 mm! This observation illustrates that even with relatively long series (59 years), estimating extreme quantiles based on the sole use of at-site values yields considerable uncertainty, especially in a non-stationary context.

Note that model checking becomes more challenging in a non-stationary context. For instance, graphical diagnostics like Fig. 3.7 are not possible since all observations have distinct distributions. One possibility is to transform all observations into probability space by applying the corresponding cdf:

$$\pi_t = \hat{F}_{Y_t}(y_t) \tag{3.19}$$

In Eq. (3.19), $\hat{F}_{Y_t}$ corresponds to the estimated cdf of the variable Y_t. The cdf obtained with the modal parameter estimates may be used for example (but note that other choices are possible, for instance the cdf of the predictive distribution). Under the assumption that observations can be considered as realizations from the distribution with cdf $\hat{F}_{Y_t}$, the transformed values π_t should be realizations from a uniform distribution. This can be evaluated graphically by using a probability-probability plot (pp-plot), as illustrated in Fig. 3.11.

3.3.3.2 Step-Change Model

An alternative representation of non-stationarity can be obtained by assuming a sudden change at an unknown date in the location parameter. The corresponding

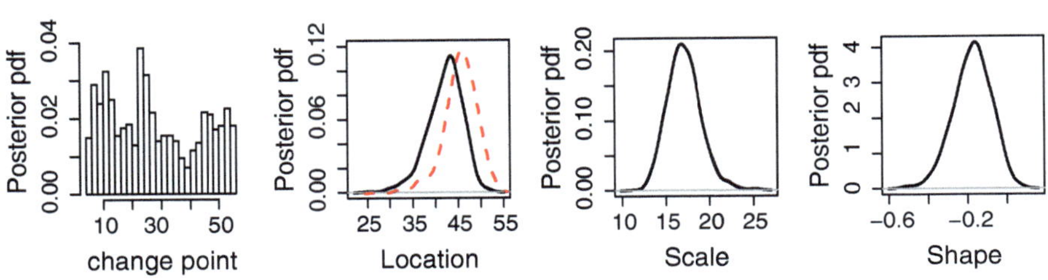

Fig. 3.12 Posterior pdf of parameters of the step-change model. For the location parameter, the *solid/dashed lines* represent the parameter values before/after change, respectively

model in Eq. (3.20) results in five unknown R-parameters: the change point τ_0, the location parameter before (μ_1) and after (μ_2) change, the scale parameter σ_0 and the shape parameter ξ_0. Priors similar to previous section are used, with the addition of a uniform prior for the change point parameter τ_0.

$$\mu(t) = \begin{cases} \mu_1 \text{ if } t \leq \tau_0 \\ \mu_2 \text{ otherwise} \end{cases}$$
$$\sigma(t) = \sigma_0$$
$$\xi(t) = \xi_0 \tag{3.20}$$

Figure 3.12 shows the posterior pdf of each R-parameter. The most likely change point is located around the 24th value, but this estimation is not precise, since the posterior pdf is not peaky and spans over all possible change point values. This suggests a difficulty to locate a marked change point in this series. This is further confirmed by considering the location parameter values before and after change, which are fairly similar. Lastly, the evolution of 0.5- and 0.99-quantiles shown in Fig. 3.13 is rather small. Note that the width of the 90% posterior interval for the 0.99-quantile now exceeds 110 mm, which is slightly higher than in previous Sect. 3.3.3.1. This might be due to the inclusion of a fifth parameter in the change-point model.

3.3.3.3 Model Comparison and Model Averaging

The trend and step-change models can be compared with a standard stationary GEV model using the approach described in Sect. 3.2.3.3. Table 3.1 shows the results of this model comparison exercise. The stationary model is assigned a prior probability of 0.5, while each non-stationary model (trend and step-change) is assigned an identical prior probability of 0.25. This prior elicitation aims at assigning an equal prior probability to the stationarity and non-stationarity assumptions.

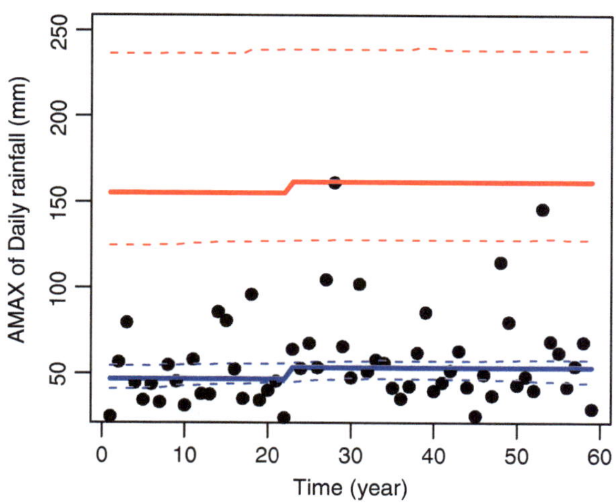

Fig. 3.13 Annual maxima of daily rainfall and evolution of quantile estimates in time. The *blue lines* refer to the estimated median, the *red lines* to the estimated 0.99-quantile. *Dotted lines* represent 90% posterior intervals

Table 3.1 Comparison between models M_0 to M_2

	M_0: stationary	M_1: trend	M_2: step-change
Prior: $p(M_i)$	0.50	0.25	0.25
Marginal Likelihood: $\log(p(\mathbf{y}\|M_i))$	−260.44	−261.06	−261.69
Bayes Factor: $B_{0,i}$	1.00	1.86	3.48
Posterior: $p(M_i\|\mathbf{y})$	0.71	0.19	0.10

The marginal likelihood of each model is then computed using Chib's approach (see Appendix). This allows computing pairwise Bayes Factors (BFs) between: (i) the stationary model in the one hand; (ii) The trend or the step-change model in the other hand. Pairwise BFs are equal to 1.86 and 3.48 for the trend and step-change models, respectively. The BF values larger than one suggest that the stationary model may be more relevant. According to Raftery's scale (1996), this evidence is "*not worth more than a bare mention*" for the trend model, and is "*positive*" for the step-change model. A composite BF {*stationary*} vs. {*trend, step-change*} can also be computed and yields a value of 2.43, which is "*not worth more than a bare mention*". Overall, these BFs values suggest that there is no strong evidence to favor or reject any of the three models. In particular, although the stationary model cannot be firmly rejected, it does not mean that there is no change: we may just be unable to detect it, due to the strong natural variability of those extreme data.

The posterior probabilities of each model shown in Table 3.1 support this interpretation, with no probability being close to zero or one, indicating the absence of strong evidence to reject or favor any particular model, respectively. This uncertainty in the choice of a particular model can be at least partly accounted for by deriving a composite prediction using Bayesian model averaging, rather than using only one particular model. Figure 3.14 shows an example of such prediction for the 0.99 quantile. This prediction shows a barely noticeable increase of the

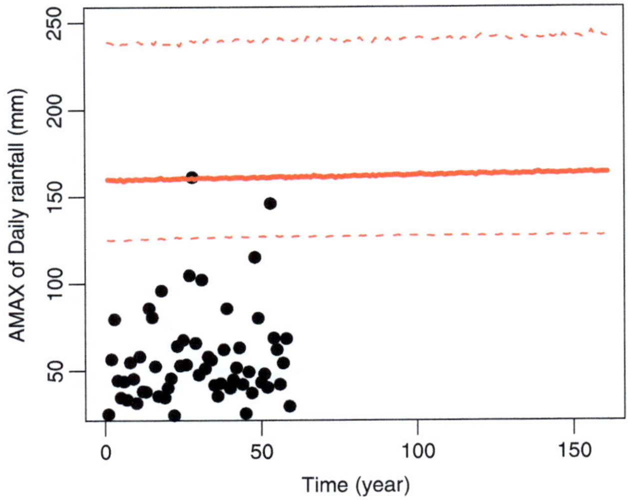

Fig. 3.14 Prediction of the 0.99-quantile using Bayesian model averaging. *Dashed lines* correspond to a 90% probability interval

0.99-quantile, even when extrapolated well beyond the period of availability of data. As in previous Sects. 3.3.3.1 and 3.3.3.2, the most noticeable characteristic of this prediction is its considerable uncertainty, with the width of the 90% interval exceeding 115 mm.

3.3.3.4 Identifiability Issues

Both previous examples introduced non-stationarity through the location parameter only. However, one may also wish to investigate trends in scale or shape characteristics. Although the general framework described in Sect. 3.3.1 enables such analysis in principle, its practical feasibility is questionable. Indeed, given the limited amount of data and the considerable uncertainties existing even with simple trend or step-change models like (3.17) and (3.20), there might be a limit to what can be identified from the data.

This issue is explored by considering a "trend on everything" model, where the location, scale and shape parameters are all affected by a linear trend:

$$\begin{aligned} \mu(t) &= \mu_0(1+\mu_1 t) \\ \log(\sigma(t)) &= \sigma_0(1+\sigma_1 t) \\ \xi(t) &= \xi_0(1+\xi_1 t) \end{aligned} \tag{3.21}$$

Vague priors are used for all six R-parameters in Eq. (3.21). The resulting posterior distribution is explored using the same MCMC sampler than in previous

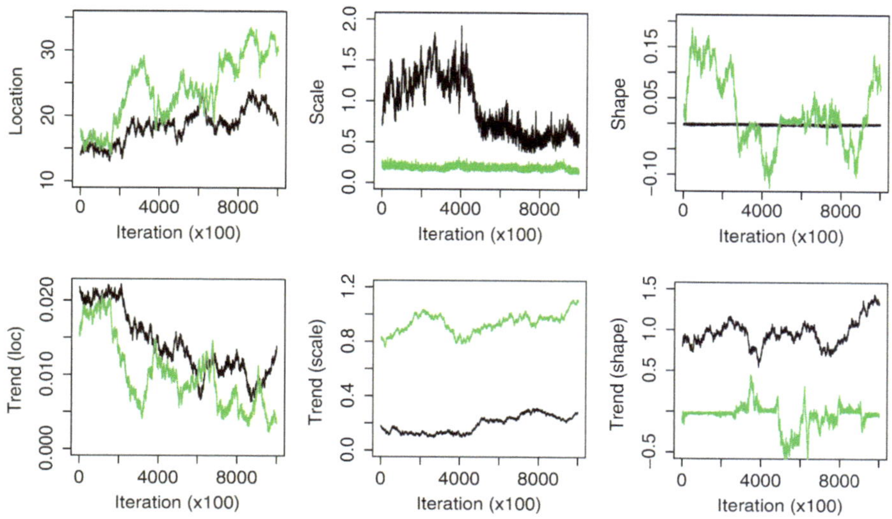

Fig. 3.15 Illustration of MCMC convergence failure due to a lack of identifiability. For each inferred parameter, two parallel chains obtained from the same posterior distribution using an identical MCMC sampler are shown

Sects. 3.3.3.1 and 3.3.3.2. However, the MCMC sampler was still far from convergence after 10^6 iterations (whereas convergence was readily achieved in previous examples). Figure 3.15 illustrates this convergence failure, using two parallel chains that are exploring distinct regions of the parameter space, and do not converge to an identical distribution.

Confronted to such convergence failures, the first reaction of most modelers is to blame the inefficiency of the MCMC sampler or some other numerical difficulty. For instance, one of the chains might be stuck in a secondary mode, and might therefore generate values from a low-density region that could be discarded.

However, results reported in Fig. 3.16 demonstrate that the two distinct regions explored by the two chains correspond to similar posterior density values: both regions are hence equally acceptable according to the posterior values. Moreover, running additional chains results in exploring other regions of the parameter space, but still with similar posterior density.

The lack of MCMC convergence is therefore not a mere numerical issue, but is rather symptomatic of a more profound problem affecting the inference we are trying to perform: the parameters are not identifiable from the data. In a nutshell, this means that the information content of the data is not sufficient to support the estimation of all parameters. More technically, non-identifiability manifests itself by the likelihood function being constant over infinite-size subspaces of the parameter space. Since non-informative priors are used in this example, the posterior distribution is also constant over such subspaces, yielding an ill-posed problem (see e.g. Gelfand and Sahu 1999; Renard et al. 2010 for further illustrations of non-identifiability and ill-posedness).

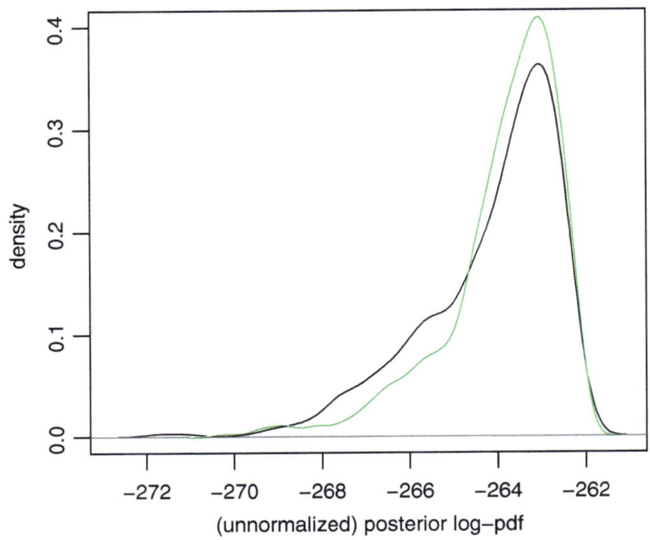

Fig. 3.16 Unnormalized posterior values corresponding to the two parallel chains in previous Fig. 3.15

Overcoming this ill-posedness issue can only be made in two ways. The first way is to use informative priors to constrain the inference and avoid exploring regions of the parameter space that are deemed unrealistic. Note that this strategy has the potential to make the inference well-posed, but does not make parameters identifiable *from the data*: instead, the inference becomes controlled by the prior assumptions. In this situation, it becomes of primary importance that the prior information represents genuine knowledge, rather than be considered as a mere numerical trick to ensure MCMC convergence.

The second strategy is to use more data. Unfortunately, this is easier said than done, since the length of hydrological series is limited. However, it might still be possible to include data from other "similar" sites through a regionalization procedure. This will be thoroughly discussed in the subsequent sections.

3.4 Regional Inference of Non-stationary Models

Previous Sect. 3.3 described a general Bayesian framework for building and inferring site-specific non-stationary models. Although several examples demonstrated the feasibility of this approach to describe non-stationarity, they also illustrated that resulting uncertainties are considerable. This section explores one possibility to reduce these uncertainties, through the derivation of regional non-stationary models that enable sharing information between sites.

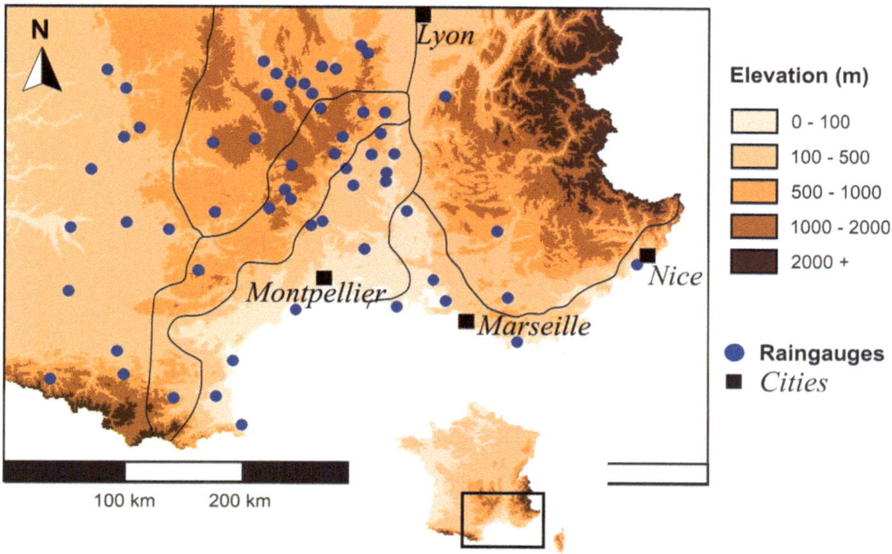

Fig. 3.17 Raingauges location. The *thin lines* represent six homogeneous regions defined by Pujol et al. (2007)

3.4.1 Motivation: Looking at Local Results at a Regional Scale

The interest of regional modeling is motivated using an example based on extreme rainfall in the French Mediterranean region (Fig. 3.17). Sixty raingauges, represented as blue dots, are used (source: Meteo France). The thin lines in Fig. 3.17 represent six homogeneous regions defined by Pujol et al. (2007). Annual maxima series are extracted over the period 1955–2004.

Each of the 60 series of annual maxima is studied with the trend model of Sect. 3.3.3.1. The posterior distributions of 0.9- and 0.99-quantiles are shown as boxplots in Fig. 3.18. Those quantiles are computed at the end of the period, i.e. at $t = 2004$. The spread of each boxplot confirms the considerable uncertainties in such estimates. Moreover, quantiles show significant spatial variability, with values varying by a factor of more than 2 from site to site. Figure 3.19 shows similar boxplots for the shape and trend parameters. Both parameters have a high level of uncertainty, which propagates to quantiles.

Such a considerable level of uncertainty is a fundamental limitation to the modeling of non-stationary extremes using local models. It makes the identification of non-stationary components a very challenging task. Similarly, it confuses the comparison between competing models (e.g. differing by the covariate used to introduce non-stationarity). Lastly, it is a strong limitation to the transfer of information from gauged to ungauged sites. In particular, the use of standard statistical methods like regression and/or kriging may be inadequate because uncertainties are not only

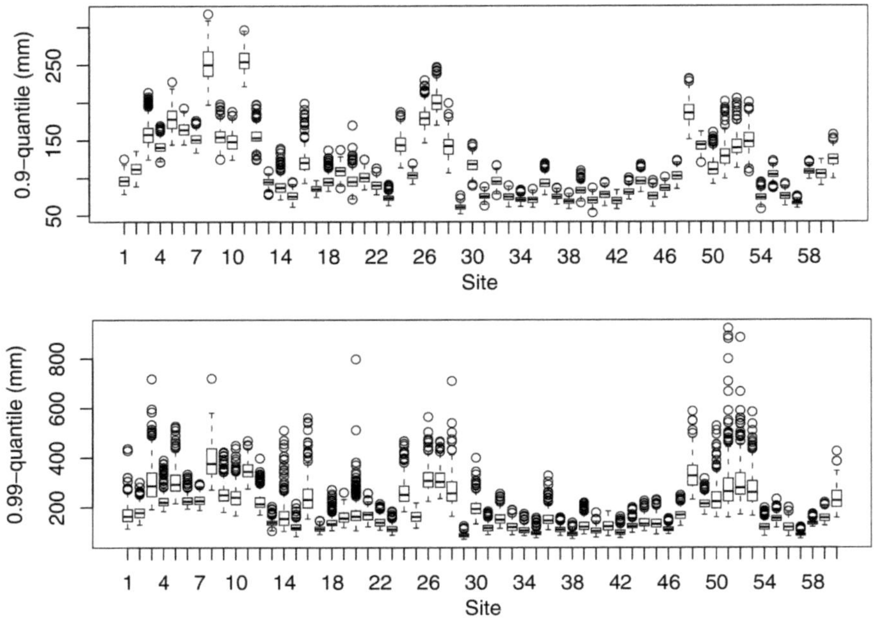

Fig. 3.18 Boxplots of the posterior pdf of 0.9- and 0.99-quantiles in 2004 for 60 raingauges

large, but they are also significantly varying from site to site (as an illustration, see the generalized least squares approach developed by Stedinger and Tasker 1985; Reis et al. 2005 to account for contrasted errors at different sites).

However, another indication from Fig. 3.19 is that the shape parameters are fairly similar across sites, which suggests that the spatial variability in quantiles does not result from the spatial variability in the shape parameters. This further suggests that some assumption constraining the spatial variability of this parameter is possible – this provides a basis for the construction of regional models.

3.4.2 Notation

The notation introduced in previous Sect. 3.2 is slightly modified to include a mention of the site. More precisely, *Y(s, t)* represents the random variable of interest at time *t* and site *s*. Observations are denoted by *y(s,t)* and are considered as realizations from *Y(s,t)*. $\boldsymbol{y} = (y(s,t))_{t=1:N_t, s=1:N_s}$ may for instance represent a set of annual maxima recorded at N_s sites during N_t years, and considered as realizations from the random variables $\boldsymbol{Y} = (Y(s,t))_{t=1:N_t, s=1:N_s}$.

The random variable *Y(s,t)* is assumed to follow a distribution with pdf $p(y|\boldsymbol{\theta}(s,t))$:

$$Y(s,t) \sim p\left(y|\boldsymbol{\theta}(s,t)\right) \tag{3.22}$$

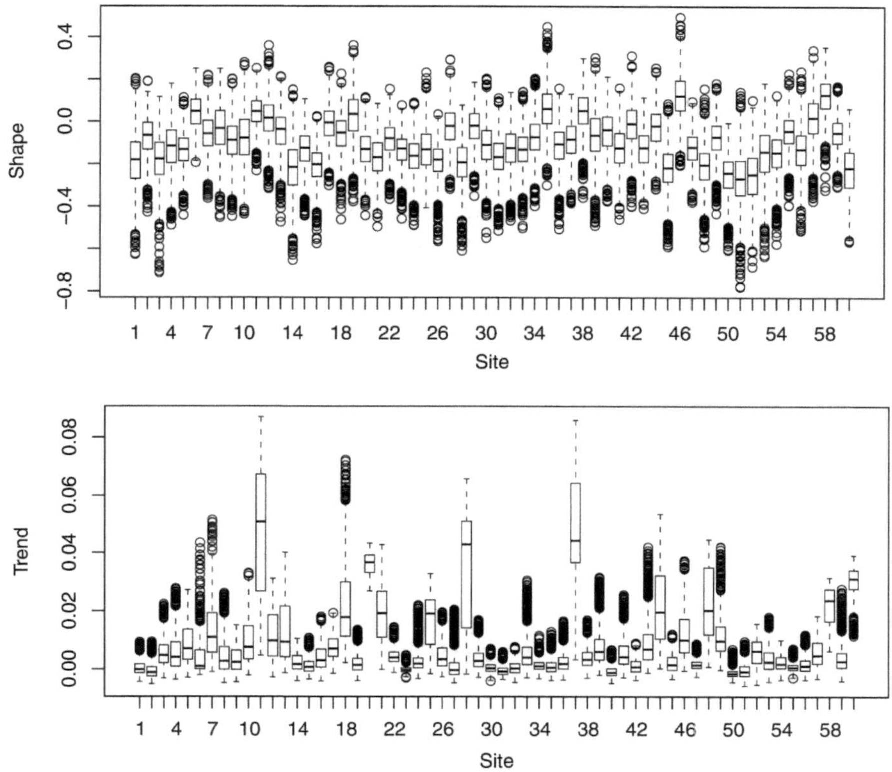

Fig. 3.19 Boxplots of the posterior pdf of shape and trend parameters for 60 raingauges

Note that compared with the at-site modeling framework presented in Sect. 3.3, the D-parameters $\boldsymbol{\theta}(s,t)$ now varies in both space and time.

3.4.3 The Notion of Regional Parameters

Following the general principle of previous Sect. 3.3, variations in time are described using a regression model for each D-parameter $\theta_k(s, t)$ $(k = 1{:}N_D)$ as follows:

$$g_k\left(\theta_k(s,t)\right) = h_k\left(\boldsymbol{x}_k(s,t); \boldsymbol{\beta}_k(s)\right) \tag{3.23}$$

Comparing Eq. (3.23) with the at-site regression in Eq. (3.13), the following comments can be made:

- Both the link function g_k and the regression function h_k are assumed identical for all sites.

- The covariate vector $\boldsymbol{x}_k(s,t)$ may now depend on both space and time. Indeed, while some covariates may be identical for all sites (e.g. the time or the value of some climate index), site-specific covariates may be useful (e.g. the "weather pattern" corresponding to the annual maximum at each site).
- The vector of R-parameters $\boldsymbol{\beta}_k(s)$ may also depend on site, allowing for site-specific relationships between D-parameters $\theta_k(s, t)$ and covariates $\boldsymbol{x}_k(s, t)$.

The construction of a regional model is then based on assumptions regarding the spatial variability of R-parameters $\boldsymbol{\beta}_k(s)$. More precisely, two different kinds of R-parameters can be considered:

- Local R-parameters denote site-specific parameters $\boldsymbol{\beta}_k(s)$. This is the most flexible assumption, but it does not allow sharing information between different sites.
- Regional R-parameters denote parameters that will be assumed identical for all sites within the region, i.e. $\boldsymbol{\beta}_k(s) = \boldsymbol{\beta}_k$. This assumption enables the sharing of information between sites. However, it also restricts the flexibility in describing spatial variability. As any assumption, its adequacy should be carefully assessed.

An example of such a regional model can be given by considering the trend model presented in Sect. 3.3.3.1:

$$\begin{aligned} \mu(s,t) &= \mu_0(s)(1+\mu_1(s)t) \\ \sigma(s,t) &= \sigma_0(s) \\ \xi(s,t) &= \xi_0(s) \end{aligned} \tag{3.24}$$

In Eq. (3.24), all R-parameters are local (i.e. site-specific). One may wish to make some of these R-parameters regional. For instance, Fig. 3.19 suggests that the shape parameter values are similar across sites. It might therefore be possible to assume a regional shape parameter, i.e. $\xi_0(s) = \xi_0$. Similarly, if the interest is to describe climate-related trends, one might expect some consistency across sites. The trend parameter could hence also be assumed regional, i.e. $\mu_1(s) = \mu_1$. Those two additional assumptions yield the following regional model:

$$\begin{aligned} \mu(s,t) &= \mu_0(s)(1+\mu_1 t) \\ \sigma(s,t) &= \sigma_0(s) \\ \xi(s,t) &= \xi_0 \end{aligned} \tag{3.25}$$

In model (3.25), all sites will contribute to the estimation of regional parameters ξ_0 and μ_1. Hopefully, this might reduce their uncertainty, at the cost of making a restrictive assumption on their spatial variability.

Finally, note that model (3.25) is just one amongst many other possible regional formulations. For instance, it would be possible to use a non-stationary model complying with the popular index flood formalism (Dalrymple 1960), by forcing the ratio between the scale and location parameters to remain constant is space. Such model could take the following form:

$$\begin{aligned}\mu(s,t) &= \mu_0(s)(1+\mu_1 t)\\ \sigma(s,t) &= \delta \times \mu_0(s)\\ \xi(s,t) &= \xi_0\end{aligned} \tag{3.26}$$

Note that in model (3.26), the ratio between the scale and location parameters is constant in space but not in time. An alternative model, forcing this ratio to remain constant in both space and time, could be:

$$\begin{aligned}\mu(s,t) &= \mu_0(s)(1+\mu_1 t)\\ \sigma(s,t) &= \delta \times \mu_0(s)(1+\mu_1 t)\\ \xi(s,t) &= \xi_0\end{aligned} \tag{3.27}$$

Several other examples of non-stationary regional models can be found in the literature (Perreault et al. 2000c; Cunderlik and Burn 2003; Renard et al. 2006; Hanel et al. 2009).

3.4.4 *Inference*

3.4.4.1 The Spatial Independence Case

Parameter inference is based on the posterior distribution:

$$p(\boldsymbol{\beta}|\boldsymbol{y},\boldsymbol{x}) \propto p(\boldsymbol{y}|\boldsymbol{\beta},\boldsymbol{x})p(\boldsymbol{\beta}|\boldsymbol{x}) \tag{3.28}$$

where we use the shorthand notation $\boldsymbol{x} = (\boldsymbol{x}_k(s,t))_{k=1:N_D,s=1:N_s,t=1:N_t}$ to denote the set of all covariates, and $\boldsymbol{\beta} = (\boldsymbol{\beta}_k(s))_{k=1:N_D,s=1:N_s}$ to denote the set of all R-parameters. The second term in Eq. (3.28), $p(\boldsymbol{\beta}|\boldsymbol{x})$, is the prior distribution of R-parameters. As in previous Sect. 3.3.2, the covariates $\boldsymbol{x}$ may be used for specifying the prior distribution, but not observations $\boldsymbol{y}$.

The derivation of the likelihood $p(\boldsymbol{y}|\boldsymbol{\beta},\boldsymbol{x})$ is more challenging than in previous Sect. 3.3.2, due to the spatial nature of the data. Let us start by considering the pdf of the spatial random vector $\boldsymbol{Y}(t) = (Y(s,t))_{s=1:N_s}$ at a given time step t. This pdf corresponds to a N_s-dimensional distribution, whose marginal (i.e. at-site) distributions are known from the regression model in Eq. (3.23):

$$\begin{aligned}&p(y|\boldsymbol{\beta}(s),\boldsymbol{x}(s,t)) = p\left(y|\theta_1(s,t),\ldots,\theta_{N_D}(s,t)\right)\\ &\text{where } \theta_k(s,t) = g_k^{-1}\left[h_k\left(\boldsymbol{x}_k(s,t);\boldsymbol{\beta}_k(s)\right)\right] \text{ for } k=1:N_D\end{aligned} \tag{3.29}$$

The following task is to combine those marginal distributions to derive the joint distribution of random vector $\boldsymbol{Y}(t)$. In the case where data can be assumed spatially independent, this combination is straightforwardly achieved by multiplying the marginal pdfs in Eq. (3.29):

$$p\left(y_1, \ldots, y_{N_s} | \boldsymbol{\beta}, \boldsymbol{x}(t)\right) = \prod_{s=1}^{N_s} p\left(y_s | \boldsymbol{\beta}(s), \boldsymbol{x}(s,t)\right) \tag{3.30}$$

Finally, the full likelihood of data $\boldsymbol{y}$ is derived by assuming temporal independence between successive time steps, yielding:

$$p(\boldsymbol{y}|\boldsymbol{\beta}, \boldsymbol{x}) = \prod_{t=1}^{N_t} p\left(y(1,t), \ldots, y(N_s,t) | \boldsymbol{\beta}, \boldsymbol{x}(t)\right) \tag{3.31}$$

3.4.4.2 Accounting for Spatial Dependence

When the spatial independence assumption does not hold (which is likely to happen in many spatial extreme analyses), another combination of marginal distributions needs to be implemented. One possible way is to use a copula for this purpose. Indeed, copulas are flexible tools to derive a N_S-dimensional joint distribution by combining a set of N_S marginal distributions.

An entire book could be devoted to this topic (see e.g. Salvadori et al. 2007). Moreover, given that this subject is discussed in details in Chap. 5 of this book, we restrict ourselves to a brief description of two particular copulas belonging to the family of elliptical copulas, which is well suited to the context of regional inference (see e.g. Bárdossy and Li 2008 for alternatives).

The elliptical copula family is described in details by Genest et al. (2007). It has the particularity of being parameterized by a N_SxN_S dependence matrix, describing the strength of dependence between pairs of variables (pairs of sites in a spatial context). This is of particular interest because it provides an interface with geostatistical tools, where pairwise dependences are expressed as a function of intersite distance.

We restrict ourselves to the description of two particular members of the elliptical copula family: the Gaussian and the Student copulas. Both copulas are derived in a similar way: let $F_1(y), \ldots, F_{N_S}(y)$ denote the marginal cumulative distribution functions (cdf). The cdf of the joint distribution built using a Gaussian/Student copula can be written as follows:

$$F\left(y_1, \ldots, y_{N_S}\right) = G\left(g^{-1}(F_1(y_1)), \ldots, g^{-1}\left(F_{N_S}(y_{N_S})\right)\right) \tag{3.32}$$

In Eq. (3.32), g is the cdf of a univariate Gaussian/Student distribution, while G is the cdf of a multivariate Gaussian/Student distribution. More precisely:

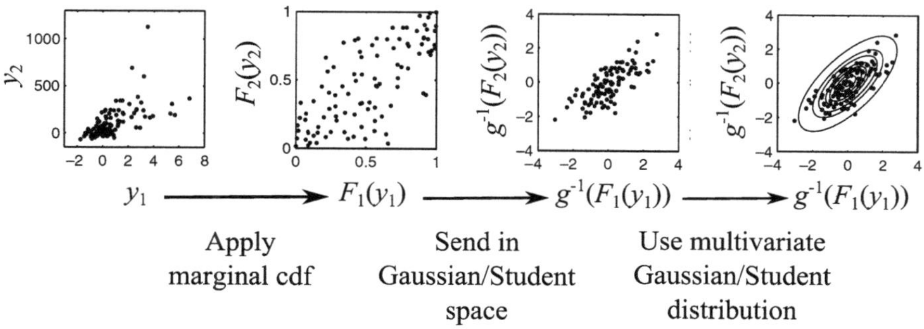

Fig. 3.20 Schematic of the Gaussian and Student copulas

$$\textbf{Gaussian Copula}: \begin{cases} g(u) = \phi(u), \text{ cdf of a standard Gaussian distribution} \\ G(y_1, ..., y_{N_S}) = \Phi_\Sigma(y_1, ..., y_{N_S}), \text{ joint cdf of a} \\ \quad \text{multivariate Gaussian} \\ \text{distribution with mean 0 and correlation matrix } \Sigma \end{cases}$$

$$\textbf{Student Copula}: \begin{cases} g(u) = \tau_\nu(u), \text{ cdf of the Student distribution with} \\ \nu \text{ degrees of freedom (d.o.f.)} \\ G(y_1, ..., y_{N_S}) = \Theta_{\Sigma,\nu}(y_1, ..., y_{N_S}), \text{ joint cdf of a multivariate} \\ \text{Student distribution with mean 0, covariance} \\ \text{matrix } \Sigma \text{ and } \nu \text{ d.o.f.} \end{cases}$$

Equation (3.32) shows that the Gaussian and Student copulas combine marginal distributions as a two-step procedure, illustrated in Fig. 3.20:

- Marginal values are transformed from original space (y_i) to Gaussian/Student space (u_i), using the transformation $u_i = g^{-1}(F_i(y_i))$
- Transformed values are then modeled using a multivariate Gaussian/Student distribution.

Note that the transformation $u_i = g^{-1}(F_i(y_i))$ ensures that the transformed variable u_i has a Gaussian/Student distribution. However, it does not imply that the *joint* distribution of $u_1, ..., u_{N_S}$ is a multivariate Gaussian/Student distribution (since marginal normality does not imply joint normality, and similarly for the Student case). Consequently, using a Gaussian/Student copula corresponds to making an assumption on the nature of dependence between data. As any assumption, it requires to be assessed based on all available evidence.

The joint pdf of a distribution constructed with a Gaussian/Student copula can easily be obtained by differentiating the cdf in Eq. (3.32), yielding:

$$\begin{cases} \textbf{Gaussiancopula:} f(y_1, ..., y_{N_S}) = \left(\dfrac{\prod_{i=1}^{N_S} f_i(y_i)}{\prod_{i=1}^{N_S} \psi(u_i)} \right) \Psi_{\Sigma}(u_1, ..., u_{N_S}) \\ \textbf{Studentcopula:} f(y_1, ..., y_{N_S}) = \left(\dfrac{\prod_{i=1}^{N_S} f_i(y_i)}{\prod_{i=1}^{N_S} t_\nu(u_i)} \right) T_{\Sigma,\nu}(u_1, ..., u_{N_S}) \end{cases}$$

where:

$f_i(y)$ marginal pdfs

$\psi(u)$ standard Gaussian pdf

$\Psi_{\Sigma}(u_1, ..., u_{N_S})$ multivariate Gaussian pdf with mean 0 and correlation matrix Σ

$t_\nu(u)$ pdf of a Student distribution with ν d.o.f.

$T_{\Sigma,\nu}(u_1, ..., u_{N_S})$ multivariate Student pdf with mean 0, covariance matrix Σ and ν d.o.f. (3.33)

Note that although the Gaussian and Student copulas are derived in a very similar way, their characteristics are different. In particular, the Gaussian copula is an asymptotically independent model (see e.g. Coles et al. 1999; Renard and Lang 2007 for discussions), while the Student copula can describe asymptotic dependence, whose strength is controlled by the d.o.f. parameter ν.

Both copulas are parameterized by a symmetric dependence matrix $\boldsymbol{\Sigma}$ describing pairwise dependences. The dimension of this matrix grows quadratically with the number of sites. In order to simplify the model, it is further assumed that the dependence between data from two sites solely depends on the intersite distance. An analogy can be drawn with the common treatment of stationary and isotropic spatial random fields in geostatistics (e.g. Chiles and Delfiner 1999). It follows that the elements of the pairwise dependence matrix $\boldsymbol{\Sigma}$ can be expressed as a function of the intersite distance, parameterized by some vector $\boldsymbol{\psi}$:

$$\Sigma(i, j) = \Psi\left(\left\|s_i - s_j\right\|; \boldsymbol{\psi}\right) \tag{3.34}$$

3.4.4.3 Inference with Spatially Dependent Data

Accounting for spatial dependences introduces additional parameters into the inference, namely the vector $\boldsymbol{\psi}$ used in the dependence-distance function (3.34) and the d.o.f. parameter ν in the case of the Student Copula. Consequently, the inference equations used in the spatial independence case (see previous Sect. 3.4.4.1) need to be modified. The posterior distribution of all inferred quantities becomes:

$$p(\boldsymbol{\beta}, \boldsymbol{\psi}, \nu | \boldsymbol{y}, \boldsymbol{x}) \propto p(\boldsymbol{y} | \boldsymbol{\beta}, \boldsymbol{\psi}, \nu, \boldsymbol{x}) p(\boldsymbol{\beta}, \boldsymbol{\psi}, \nu | \boldsymbol{x}) \quad (3.35)$$

In order to compute the likelihood $p(\boldsymbol{y} | \boldsymbol{\beta}, \boldsymbol{\psi}, \nu, \boldsymbol{x})$, the first task is to derive the marginal distributions, which remain identical to previous Eq. (3.29). Those marginal distribution are then combined using one of the joint pdf f in the copula Eq. (3.33):

$$p\left(y_1, \ldots, y_{N_s} | \boldsymbol{\beta}, \boldsymbol{\psi}, \nu, \boldsymbol{x}(t)\right) = f\left(y_1, \ldots, y_{N_s}\right) \quad (3.36)$$

Finally, the full likelihood of data $\boldsymbol{y}$ can be derived by assuming temporal independence, identically to previous Eq. (3.31).

3.4.5 Example: Extreme Rainfalls

This case study uses the same rainfall data as in previous Sect. 3.4.1. Annual maxima series are modeled with a GEV distribution:

$$Y(s,t) \sim GEV\left(\mu(s,t), \sigma(s,t), \xi(s,t)\right) \quad (3.37)$$

Let $r(s)$ denotes the region of site s. The following regional trend model is used:

$$\begin{aligned} \mu(s,t) &= \mu_0(s)\left(1 + \mu_1(r(s)) * t\right) \\ \sigma(s,t) &= \sigma_0(s) \\ \xi(s,t) &= \xi_0(r(s)) \end{aligned} \quad (3.38)$$

In Eq. (3.38), location and scale parameters are assumed local, while the trend and shape parameters are assumed constant within each homogeneous region (see Fig. 3.17 for the delineation of these regions).

Dependence between data is accounted for using a Gaussian copula, whose dependence matrix is parameterized as follows:

$$\Sigma(i,j) = \psi_0 \exp\left(-\psi_1 \left\|s_i - s_j\right\|\right) + (1 - \psi_0) \exp\left(-\psi_2 \left\|s_i - s_j\right\|\right) \quad (3.39)$$

Figure 3.21 shows the estimated dependence-distance relationship, and confirms that there exists a significant level of spatial dependence at least within a short distance range. Moreover, the 90% posterior interval is quite narrow, which indicates that the spatial dependence structure can be precisely identified. These observations indicate that elliptical copulas are a viable way to account for the existence of spatial dependence in regional models. However, note that although elliptical copulas are useful to describe dependence at observed levels, they might be inadequate

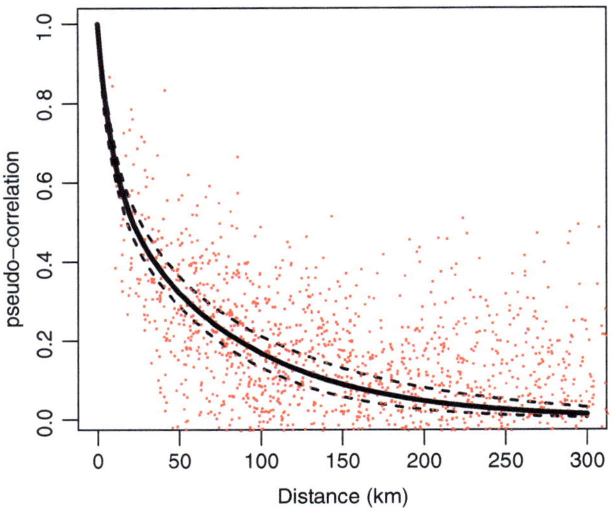

Fig. 3.21 Estimated dependence-distance relationship, with 90% posterior interval. *Dots* represent empirical pairwise correlations, computed between pairs of raingauges

when used in extrapolation (i.e. when the dependence structure is used to predict rare multivariate events, far beyond the observation range). This topic is further discussed in Sect. 3.6.2.

Figures 3.22 and 3.23 show the posterior distributions of 0.9- and 0.99-quantiles (at $t = 2004$). These posterior distributions can be compared with those obtained using purely local models estimated independently at each site (cf. previous Fig. 3.18). To ease this comparison, the latter posterior distributions are represented in light pink.

0.9-quantiles are fairly similar, both in terms of spatial pattern and uncertainty. This indicates that the regional model has no clear advantage over local models for the estimation of such a "low" quantile. However, stronger differences appear for the 0.99-quantiles. In particular, uncertainties are markedly reduced with the regional model (Fig. 3.23). This indicates that regional modeling may be beneficial in terms of uncertainty reduction for higher quantiles.

This reduction of uncertainty is primarily explained by the use of regional parameters. Indeed, Fig. 3.24 shows the posterior distributions of the regional shape parameter (black box). The box range is strongly reduced compared to the local estimates (light pink boxes). This reduction is even more noticeable for the trend parameter (Fig. 3.25), with its range being reduced by a factor of about 10 for some regions! This result indicates that a robust identification of non-stationary components may be difficult to achieve using local models, but that regional modeling has the potential to improve this identification.

Although these results are encouraging, it has to be kept in mind that the reduction of uncertainties has been obtained at the cost of making strong assumptions on

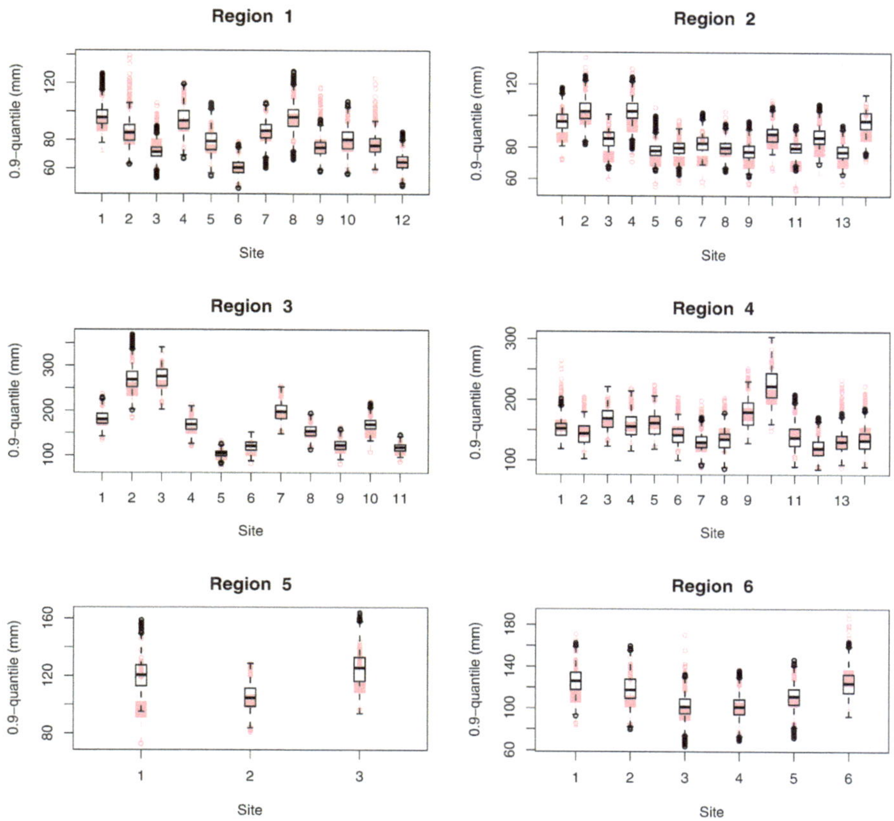

Fig. 3.22 Boxplots of the posterior pdf of 0.9-quantiles in 2004 for each region. *Light pink boxes* represent estimates using the local model, *black boxes* represent estimates using the regional model

the spatial variability of some parameters. If these assumptions are unrealistic, the reduction of uncertainties might be unduly optimistic – i.e. uncertainties could be under-estimated.

Consequently, it is important to verify that the regional model is supported by the data. To this aim, the pp-plot diagnostic presented in Sect. 3.3.3.1 is applied to all 60 sites. Overall, results shown in Fig. 3.26 indicate a good fit of the model to the data for most sites. However, a few systematic departures from the 1:1 line may be suspected, indicating that the regional model is not flexible enough to describe observed data for a few sites. In particular, the assumption of strict equality of regional parameters might be too restrictive. A possibility to remedy such departures is to allow for an additional source of variability in the regional model in order to cope with local peculiarities. The following Sect. 3.5 proposes a practical solution based on Hierarchical modeling.

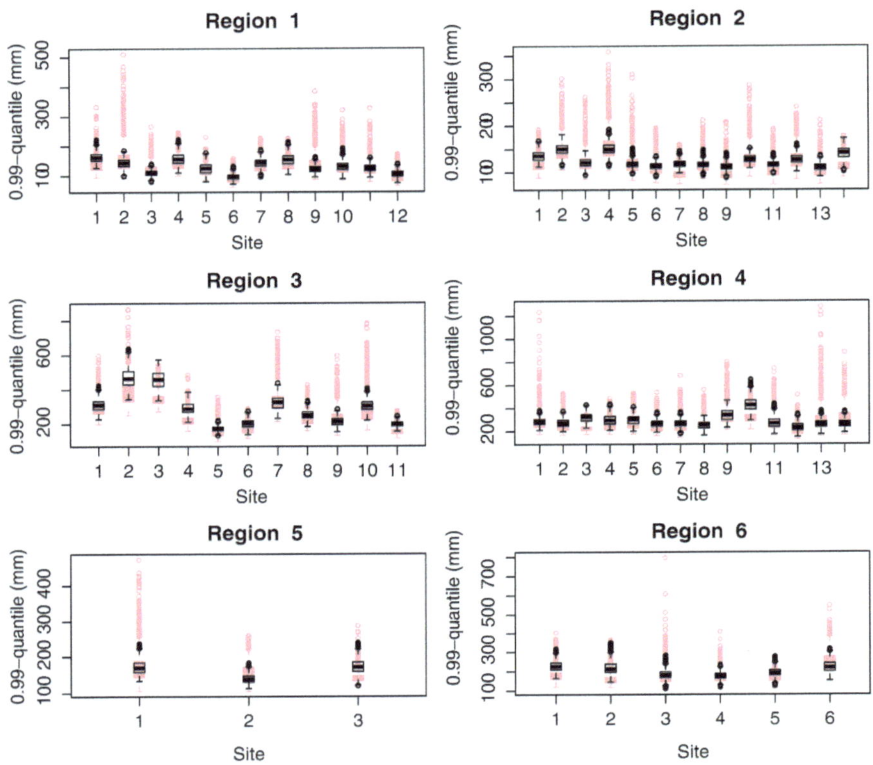

Fig. 3.23 Boxplots of the posterior pdf of 0.99-quantiles in 2004 for each region. *Light pink boxes* represent estimates using the local model, *black boxes* represent estimates using the regional model

3.5 Hierarchical Modeling

3.5.1 Principles of Hierarchical Modeling

3.5.1.1 Motivation

Previous Sect. 3.4 described a general framework for building regional non-stationary models. In this framework, inferred parameters are assumed to be of one of the following two types:

- "Regional parameters", i.e. parameters being assumed identical for all stations of a given homogeneous region. The advantages of this assumption are that (i) all sites will contribute to the estimation of the regional parameter, thus reducing estimation uncertainty; (ii) regional estimates can be used at an ungauged site within the region. However, these advantages come at a cost: the

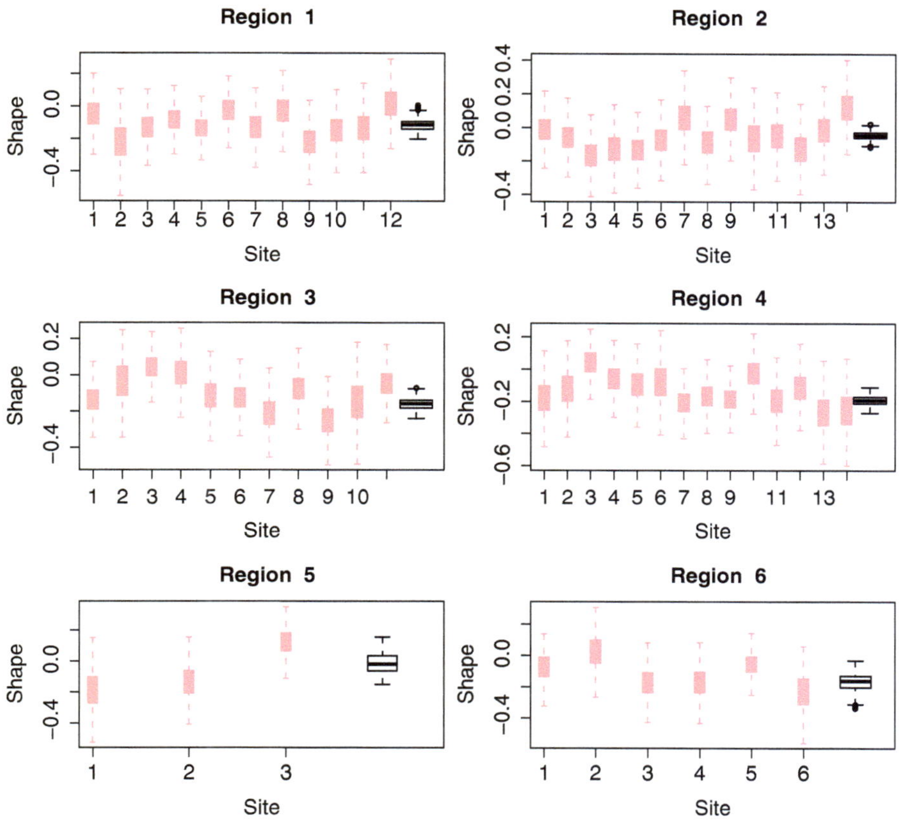

Fig. 3.24 Boxplots of the posterior pdf of shape parameters for each region. *Light pink boxes* represent estimates using the local model, the *black box* represent the regional estimate

regional parameter assumption is very strong since it reduces (indeed, annihilate) spatial variability. If such assumption proves unrealistic, the uncertainty may be significantly underestimated.

- "Local parameters" are site-specific parameters. Such parameters can obviously not be directly transferred to an ungauged site (it would requires additional regression and/or kriging steps, with the difficulties described in Sect. 3.4.1). Moreover, it does not enable sharing information between sites. The main advantage of local parameters is their ability to comply with the observed spatial variability. However, this flexibility comes at the cost of considerable uncertainties.

This description highlights that local and regional parameters have complementary advantages and drawbacks. It raises the following question: is it possible to define a third parameter type that would be somehow *in between* local and regional

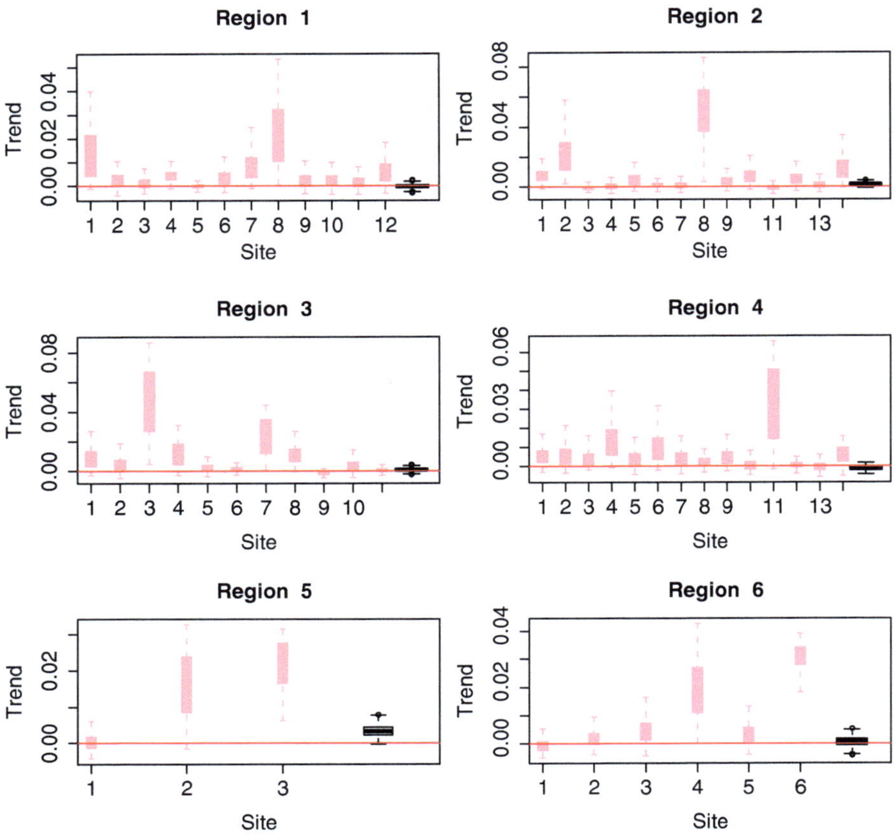

Fig. 3.25 Boxplots of the posterior pdf of trend parameters for each region. *Light pink boxes* represent estimates using the local model, the *black box* represent the regional estimate

parameters, and hence cumulate the advantages of both assumptions while limiting their drawbacks? For instance, would it be possible to assume that parameter values are indeed different at each site, but should still show some form of spatial consistency, with nearby sites having closer parameter values than sites located further away?

The objective of this section is to discuss a possible solution, using the notion of *stochastic parameters*. It is based on Bayesian Hierarchical modeling, which constitutes a very flexible framework to implement models where parameters can be either local, regional or stochastic.

Hierarchical modeling is a useful tool when the inference problem requires using several modeling layers. In general, the first layer describes the data, by assuming observations are realizations from a given distribution family, depending on some parameters. The second layer describes the variability of the parameters, by assuming they are themselves realizations from some stochastic process. For

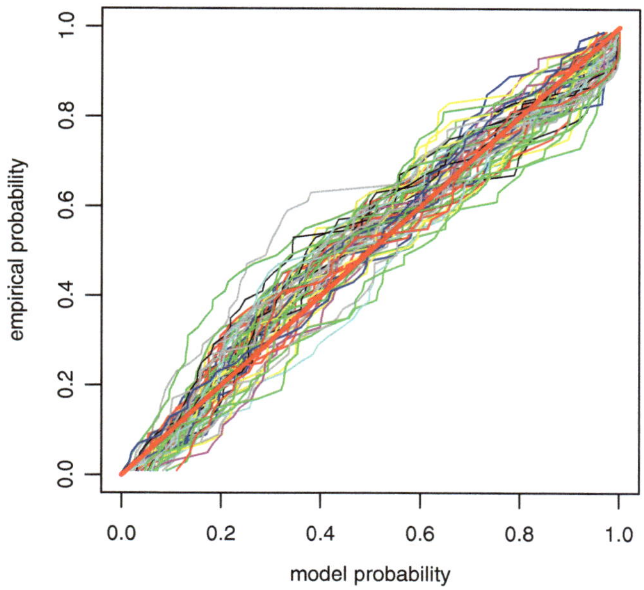

Fig. 3.26 pp-plots obtained with modal parameters for 60 sites

instance, in the regional problem described in the previous Sect. 3.4.5, the spatial variability in parameters may be described by assuming they are realizations from some Gaussian spatial field. This section elaborates on the derivation of hierarchical models in a regional context.

3.5.1.2 A Simple Example

Following the notation defined earlier, observations $y(s,t)$ are considered as realizations from $Y(s,t)$. For simplicity, we will first consider a single-parameter and stationary example, i.e. the random variable $Y(s,t)$ is assumed to follow a distribution with pdf $p(y|\theta(s))$. This represents the first layer *(data layer)* of the hierarchical model. The second layer *(process layer)* describes the spatial variability of parameter θ. We will assume here that θ values are realizations from a Gaussian spatial field:

$$(\theta(1), ..., \theta(N_s)) \sim N(\boldsymbol{\mu}, \boldsymbol{\Sigma}) \tag{3.40}$$

This Gaussian distribution is called the *hyper-distribution*, and its parameters $\boldsymbol{\mu}$ and $\boldsymbol{\Sigma}$ are called the *hyper-parameters*.

In this hierarchical setup, unknown quantities are the parameter values $(\theta(1), ..., \theta(N_s))$ and their hyper-parameters controlling spatial variability $\boldsymbol{\mu}$ and

$\boldsymbol{\Sigma}$. Inference of these quantities is based on the following posterior distribution, derived by applying Bayes' theorem:

$$p\left(\boldsymbol{\theta}(1:N_s), \boldsymbol{\mu}, \boldsymbol{\Sigma} \mid \boldsymbol{y}\right) \propto p\left(\boldsymbol{y} \mid \boldsymbol{\theta}(1:N_s), \boldsymbol{\mu}, \boldsymbol{\Sigma}\right) p\left(\boldsymbol{\theta}(1:N_s), \boldsymbol{\mu}, \boldsymbol{\Sigma}\right) \tag{3.41}$$

The first term in Eq. (3.41) is the likelihood of observations $\boldsymbol{y}$. Since observations are realizations from the distribution with pdf $p(y|\theta(s))$, the likelihood solely depends on the parameter values $\theta(1{:}\ N_s)$, but not on hyper-parameters $\boldsymbol{\mu}$ and $\boldsymbol{\Sigma}$. Consequently:

$$p(\boldsymbol{y} \mid \boldsymbol{\theta}(1:N_s), \boldsymbol{\mu}, \boldsymbol{\Sigma}) = p(\boldsymbol{y} \mid \boldsymbol{\theta}(1:N_s)) \tag{3.42}$$

The second term in Eq. (3.41) is the joint prior pdf of parameters and hyper-parameters. It can be decomposed by applying conditional probability algebra:

$$p(\boldsymbol{\theta}(1:N_s), \boldsymbol{\mu}, \boldsymbol{\Sigma}) = p(\boldsymbol{\theta}(1:N_s) | \boldsymbol{\mu}, \boldsymbol{\Sigma}) p(\boldsymbol{\mu}, \boldsymbol{\Sigma}) \tag{3.43}$$

Combining Eqs. (3.42) and (3.43), the posterior distribution can be written as follows:

$$p(\boldsymbol{\theta}(1:N_s), \boldsymbol{\mu}, \boldsymbol{\Sigma} \mid \boldsymbol{y}) \propto \underbrace{p(\boldsymbol{y} \mid \boldsymbol{\theta}(1:N_s))}_{Likelihood} \underbrace{p(\boldsymbol{\mu}, \boldsymbol{\Sigma})}_{prior} \underbrace{p(\boldsymbol{\theta}(1:N_s) | \boldsymbol{\mu}, \boldsymbol{\Sigma})}_{Hierarchical\ component} \tag{3.44}$$

Let us now compare the "hierarchical" posterior distribution (3.44) with the "standard" posterior in Eq. (3.3). In addition to the likelihood and prior terms, a new term appears in the hierarchical setup: the "hierarchical component" $p(\boldsymbol{\theta}(1{:}\ N_s)|\boldsymbol{\mu}, \boldsymbol{\Sigma})$, corresponding to the pdf of parameters according to their hyper-distribution. This term exerts an additional constraint on $\theta(s)$ values, compared to the standard setup where $\theta(s)$ are purely local parameters. It forces the spatial consistency of parameter values, by forcing them to comply with the spatial model expressed by the hyper-distribution. On the other hand, this constraint is far weaker that the constraint induced by assuming a regional parameter, since $\theta(s)$ values are not strictly equal on all sites. Consequently, this hierarchical setup fulfills the objective expressed in previous Sect. 3.5.1.1 by defining a third "stochastic parameter" type, which lies "in between" local and regional parameters in terms of inference constraints. Moreover, note that prediction at an ungauged site is still possible for stochastic parameters. Indeed, an estimate at an ungauged site can be obtained by using the estimated hyper-distribution as predictor.

3.5.2 Regional Hierarchical Modeling

This section illustrates the use of stochastic parameters to build regional hierarchical models.

3.5.2.1 Regional, Local and Stochastic Parameters

We still consider a stationary regional model, with the random variable $Y(s,t)$ following a distribution with pdf $p(y|\boldsymbol{\theta}(s))$. We assume that the parameter vector $\boldsymbol{\theta}(s)$ can be partitioned into $\{\boldsymbol{\theta}^{(\mathrm{I})}(s), \boldsymbol{\theta}^{(\mathrm{II})}(s), \boldsymbol{\theta}^{(\mathrm{III})}(s)\}$, representing the three possible parameter types:

- Type-I parameters are regional parameters, i.e. $\boldsymbol{\theta}^{(\mathrm{I})}(s) = \boldsymbol{\theta}^{(\mathrm{I})} \forall s$
- Type-II parameters are local parameters, i.e. $\boldsymbol{\theta}^{(\mathrm{II})}(s)$ values are different at each site.
- Type-III parameters are stochastic parameters. We will assume here that they are realizations from a Gaussian spatial field, i.e. $(\boldsymbol{\theta}^{(\mathrm{III})}(1), \ldots, \boldsymbol{\theta}^{(\mathrm{III})}(N_s)) \sim N(\boldsymbol{\mu}, \boldsymbol{\Sigma})$.

It is further assumed that the hyper-mean vector $\boldsymbol{\mu}$ is constant (i.e. all components are equal) and that the elements of the covariance matrix $\boldsymbol{\Sigma}$ can be expressed as a function of the intersite distance:

$$\Sigma(i,j) = \Psi(||s_i - s_j||; \boldsymbol{\psi}) \tag{3.45}$$

3.5.2.2 Inference

In this regional hierarchical model, unknown quantities requiring inference are the following:

- Regional parameters $\boldsymbol{\theta}^{(\mathrm{I})}$
- Local parameters $\boldsymbol{\theta}^{(\mathrm{II})}(1{:}N_s)$
- Stochastic parameters $\boldsymbol{\theta}^{(\mathrm{III})}(1{:}N_s)$
- Hyper-parameters $\boldsymbol{\mu}$ and $\boldsymbol{\Psi}$

The posterior distribution of these quantities is derived in a similar way than previous Sect. 3.5.1.2, yielding:

$$p\left(\boldsymbol{\theta}^{(\mathrm{I})}, \boldsymbol{\theta}^{(\mathrm{II})}(1{:}\,N_s), \boldsymbol{\theta}^{(\mathrm{III})}(1{:}N_s), \boldsymbol{\mu}, \boldsymbol{\psi}|\boldsymbol{y}\right) \propto \underbrace{p(\boldsymbol{y}|\boldsymbol{\theta}^{(\mathrm{I})}, \boldsymbol{\theta}^{(\mathrm{II})}(1{:}N_s), \boldsymbol{\theta}^{(\mathrm{III})}(1{:}N_s))}_{Likelihood}$$
$$\times \underbrace{p(\boldsymbol{\theta}^{(\mathrm{I})}, \boldsymbol{\theta}^{(\mathrm{II})}(1:N_s))p(\boldsymbol{\mu}, \boldsymbol{\psi})}_{priors} \times \underbrace{p(\boldsymbol{\theta}^{(\mathrm{III})}(1:N_s)|\boldsymbol{\mu}, \boldsymbol{\psi})}_{Hierarchical component} \tag{3.46}$$

3.5.3 *Case Study*

3.5.3.1 Model Specification

The same rainfall dataset as in Sect. 3.4 is used. In the first layer of the hierarchical model, the following assumption is used for at-site data:

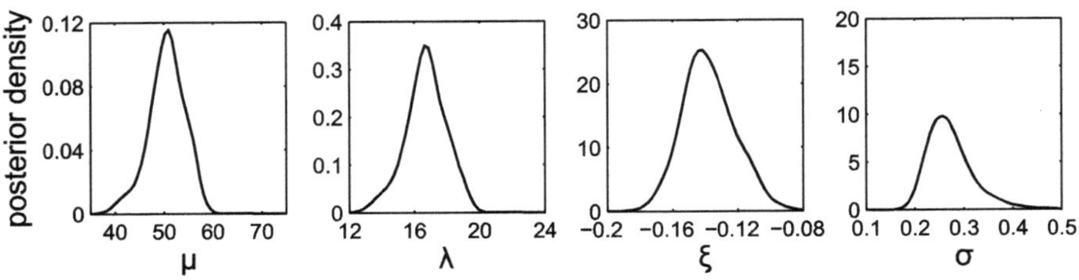

Fig. 3.27 Posterior pdfs of regional parameters and hyper-standard deviation

$$Y(s,t) \sim GEV(\delta(s)\mu, \delta(s)\lambda, \xi) \tag{3.47}$$

Equation (3.47) corresponds to a standard "index flood" model, with all sites having an identical shape parameter, and the ratio between location and scale parameters remaining constant (see e.g. *Ribatet et al.* 2006). The joint distribution of a spatial vector of annual maxima $\boldsymbol{Y}(t) = (Y(s,t))_{s=1:N_s}$ is derived using a Gaussian copula, with the same dependence-distance model as in previous Sect. 3.4.5:

$$\Sigma(i,j) = \psi_0 \exp(-\psi_1 ||s_i - s_j||) + (1 - \psi_0) \exp(-\psi_2 ||s_i - s_j||) \tag{3.48}$$

The second layer of the hierarchical model assumes that the "index flood" coefficients $\delta(s)$ are realizations from the following spatial process:

$$\log(\boldsymbol{\delta}(1 : N_s)) \sim N(\boldsymbol{0}, \boldsymbol{\Gamma}) \tag{3.49}$$

The covariance matrix $\boldsymbol{\Gamma}$ is parameterized as follows:

$$\Gamma(i,j) = \sigma^2 \left[\upsilon_0 \exp(-\upsilon_1 ||s_i - s_j||) + (1 - \upsilon_0) \exp(-\upsilon_2 ||s_i - s_j||) \right] \tag{3.50}$$

In this hierarchical model, three parameters are considered as regional (μ, λ and ξ) while a fourth parameter is considered as stochastic ($\delta(s)$), with a log-normal hyper-distribution.

3.5.3.2 Estimation

Figure 3.27 shows the posterior pdfs of regional parameters (location, scale and shape). The posterior pdf of the shape parameter has its mode around −0.14 and ranges between −0.18 and −0.08, which is consistent with the estimates obtained in local and regional models (see Figs. 3.19 and 3.24). The last plot in Fig. 3.27 shows the posterior pdf of the hyper-standard deviation (i.e. the standard deviation of $\log(\delta(s))$ values, see Eq. (3.50)). This term is of primary importance since it controls

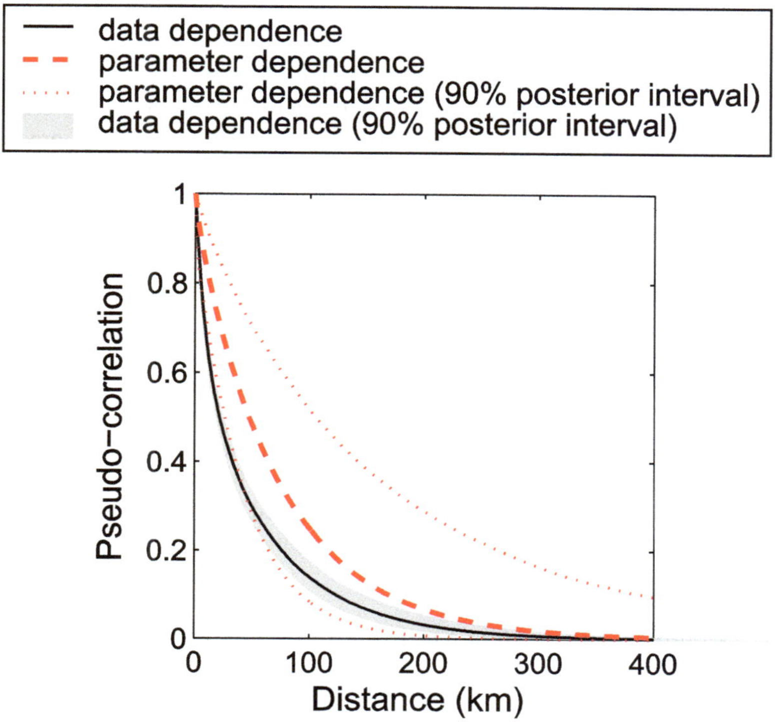

Fig. 3.28 Data and parameter dependences

the spatial variability in the stochastic parameter $\delta(s)$ and hence the uncertainty for prediction at ungauged sites. The mode of this posterior distribution roughly corresponds to a 25% coefficient of variation in $\delta(s)$ values. The regional location and scale parameters will be modulated by these $\delta(s)$ values to derive at-site location and scale values (see Eq. (3.47)).

Figure 3.28 shows the estimated dependence-distance relationships, for both the data dependence model (3.48) and the parameter dependence model (3.50). Note that both spatial dependence structures arise from distinct processes, as discussed by Cooley et al. (2007): in a nutshell, the former relates to weather spatial dependence, while the latter relates to climate spatial dependence. They also have contrasted impacts on regional modeling: while climate spatial dependence allows transferring information from gauged to ungauged sites, the main effect of weather dependence is to diminish the information content of data collected at nearby sites. Figure 3.28 shows that the dependence between data is precisely estimated, while the dependence between $\delta(s)$ values is far more uncertain (see 90% posterior intervals).

3.5.3.3 Prediction

As stated in Sect. 3.5.1.2, an attractive property of stochastic parameters is that they can be transferred from gauged to ungauged locations by using the hyper-distribution. As an illustration, the hierarchical model is used to predict the 0.99-quantile on a 40*50 grid. Note that high-elevation areas from the Alps and the Pyrenees (whose maximal elevations reach 4,810 and 3,404 m, respectively) were excluded from the prediction (white areas in Fig. 3.29). Figure 3.29a shows the contour plot of the 0.99-quantile estimated on the grid. An area with higher quantiles appears distinctly: it corresponds to the Cevennes mountain range, a region in France well known for its extreme rainfall events.

The Bayesian framework also enables a direct assessment of the uncertainties affecting predictions using the posterior distribution. Figure 3.29b shows the uncertainty in 0.99-quantiles, measured by the posterior coefficient of variation. The spatial pattern is governed by the location of gauged sites (dots). In particular, uncertainty is smaller nearby gauged sites, and increases in poorly-gauged areas. This is a consequence of the spatial dependence identified for the stochastic parameter $\delta(s)$: this dependence favors the transfer of information between nearby sites.

3.5.4 Towards a Complete Spatiotemporal Modeling of Extreme Values

The previous section aimed at illustrating the potential of Hierarchical models to study extremes at a regional scale, without necessarily making the strong assumption that regional parameters are strictly equal for all sites. However, the proposed example remains incomplete and would benefit from several additional developments to improve the description of both temporal and spatial variability:

- **Spatial variability**: The assumption that the stochastic parameter $\delta(s)$ is a realization from a homogeneous Gaussian process is unrealistic – elevation is likely to play an important role in the spatial variability of $\delta(s)$. This remark illustrates the need to include spatial covariates in the hierarchical model. For instance, this could be implemented by using a regression model to link the value of the hyper-mean μ in Eq. (3.40) with spatial covariates (e.g. elevation, distance to sea, etc.).
- **Temporal variability**: In principle, the introduction of non-stationarity into the hierarchical model could be made in a similar away, by using time-varying covariates. However, this task is not straightforward because temporal effects may vary in space. As an illustration, consider the simple trend model used throughout this chapter: the value of the trend parameter is likely to vary in space, and may also depend on spatial covariates like elevation. Note that variability in space and time do not play symmetric roles in this respect: while spatial effects will usually be considered as stationary in time, temporal effects are likely to vary in space.

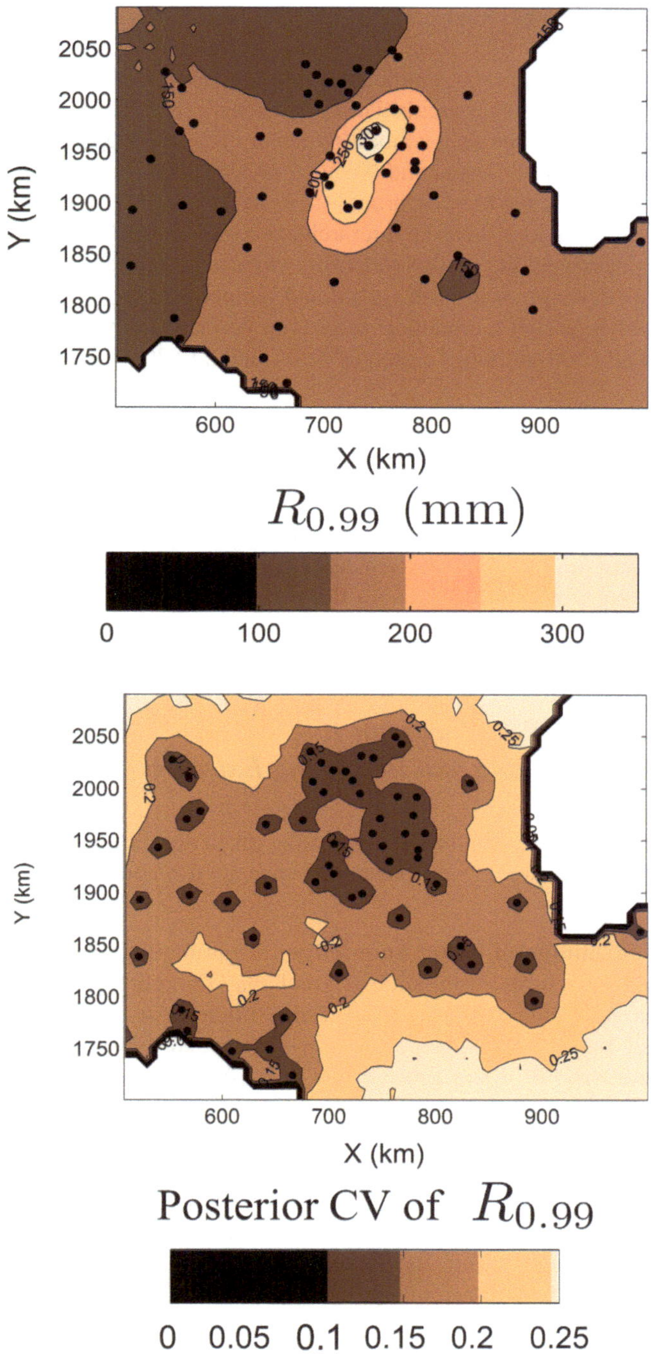

Fig. 3.29 Contour plots of: (**a**) estimated 100-year daily rainfall; (**b**) corresponding uncertainty (measured by the posterior coefficient of variation)

A general modeling framework to jointly describe the variability of hydrological extremes in space and time therefore remains to be defined. Note that the recent research literature suggests significant advances toward this goal, with several applications of Bayesian hierarchical models in a hydrological context. For instance, Perreault (2000) proposed such a model to detect a regional step-change into several series of annual runoff. Thyer and Kuczera (2000, 2003a, b) used a hierarchical model based on a hidden Markov process to describe temporal variability. Micevski (2007, 2006) proposed a Bayesian hierarchical regional flood model, while Cooley et al. (2007) described the spatial variability of extreme rainfall. Aryal et al. (2009) extended this description to both spatial and temporal variability. Similarly, Lima and Lall used Bayesian hierarchical models to describe daily rainfall occurrences (Lima and Lall 2009) or runoff extremes (Lima and Lall 2010) in a regional context.

3.6 Conclusion

This chapter proposed an overview of possible applications of Bayesian concepts to the modeling of non-stationary extremes. We conclude with a summary of the main benefits of Bayesian inference, and a discussion of important topics that require further developments.

3.6.1 Benefits of the Bayesian Inference for Describing, Understanding and Predicting Extremes

Bayesian inference presents several appealing properties, amongst which we consider the following are of particular relevance in the context of non-stationary extremes:

- **Taking advantage of prior knowledge**: the prior distribution offers an opportunity to include information beyond that brought by the data used for inference. We claim that such knowledge often exists in hydrology. This is illustrated by the recent discussions by Merz and Bloschl (2008a, b), who investigate ways of doing a better use of existing hydrological knowledge by using temporal, spatial or causal information beyond the sole flood sample. Note however that transforming such information into a proper prior distribution may be challenging.
- **Taking advantage of MCMC samplers**: the need to use MCMC samplers in most practical applications of Bayesian inference might lead to some initial reluctance to step into the Bayesian world. MCMC theory is indeed not straightforward. Moreover, understanding, implementing and optimizing MCMC samplers certainly requires time and efforts. However, once those initial difficulties are overcome, MCMC samplers turn out to be very robust and general

tools, even in relatively high-dimensional problems. Note that the existence of several packages and software (see Sect. 3.2.2) is of great help, even if specific problems may still motivate the implementation of specific samplers to gain more efficiency.

- **Focusing on the probabilistic model**: Bayesian inference offers a very general and flexible framework to the estimation problem. As an illustration, this chapter has shown that introducing non-stationarity was fairly easy, and required only minimal adaptation to the overall inference mechanism. While some estimation frameworks (in particular, maximum likelihood inference) also offer similar generality and flexibility, others are less suited to the introduction of non-stationary components. As an illustration, using moment-based estimators would require deriving theoretical moments for each new non-stationary model. In addition to the technical tediousness of such endeavor, it may distract from the arguably more important objectives of building realistic models for describing non-stationarity, and assessing them based on all available evidence.
- **A natural framework for uncertainty quantification and prediction**. The direct outcome of Bayesian inference is a (posterior) distribution. This makes the quantification of estimation uncertainties in parameters, or any function of the parameters (e.g. quantiles), straightforward. Moreover, Bayesian model averaging offers an opportunity to account for model uncertainty, at least to some extent (since a finite collection of models can hardly be exhaustive). Moreover, the Bayesian paradigm naturally handles uncertainty in predicting a future realization by means of the predictive distribution, which averages out the uncertainty in parameters.
- **Modeling complex systems using hierarchical models**. Hierarchical models appear naturally in a Bayesian framework: since parameters are considered as random variables, they can themselves be modeled using some probabilistic process. This is of particular interest because some problems encountered in Hydrology are intrinsically hierarchical. In particular, this is the case of regional estimation: data are assumed to be realizations from some distribution (level 1), whose parameters are themselves realizations from some spatial process (level 2). Bayesian hierarchical modeling offers an elegant and general way of estimating such models, without unnecessarily separating the inference process in several steps.

3.6.2 Challenges and Future Research

The context of non-stationary extremes raises a number of important challenges. Although the use of Bayesian inference is undoubtedly useful for implementing non-stationary models, it does not solve all issues. In particular, we would like to highlight three topics deserving further research in our opinion (this list being far from exhaustive): model comparison, model validation and spatial extremes.

Model comparison cannot be solely based on goodness-of-fit considerations. Indeed, increasing model complexity will in general result in a better fit to observations. However, model complexity may come at the cost of reduced predictive ability. Consequently, a trade-off between descriptive and predictive abilities has to be found. There exists no unique way of performing this trade-off: the approach presented in this chapter using Bayes factors is one possibility, but many others exist. As an illustration, numerous model comparison criteria exist in the literature, e.g. the AIC (Akaike 1973), BIC (Schwarz 1978), DIC (Spiegelhalter et al. 2002), etc. Such comparison is even more difficult when hierarchical models are used: indeed, the number of parameters is not a good description of the complexity of a hierarchical model, because stochastic parameters are constrained by the hyper-distribution and are therefore not independent parameters (see Spiegelhalter et al. 2002 for a thorough discussion on this topic). The difficulty in model comparison and selection is problematic in the context of non-stationary extremes, because distinct models may yield a similar fit to the data but result in dramatically different predictions when extrapolated.

The topic of model validation is also challenging in an extreme value context, all the more so if non-stationarity is considered. Note that the term *validation* refers to the assessment of the *predictive* performance of a model (as opposed to its *descriptive* performance as measured by goodness-of-fit), which can only be evaluated using data *not used* for estimation purposes (see discussion in Gunasekara and Cunnane 1992). The fundamental limitation to model validation is the scarcity of extreme values. However, this scarcity does not mean that validation is impossible: ambitious validation studies have been implemented using extensive datasets (see e.g. Interagency Advisory Committee on Water Data 1982, Appendix 14; Garavaglia et al. 2011). In addition, despite the growing emphasis on the importance of quantifying uncertainties, the issue of questioning the validity of such estimated uncertainties has received less attention in extreme value modeling. However, estimated uncertainties depend on the assumptions made in the model, and may therefore be unreliable if those assumptions are unrealistic. Further research is needed to develop tools for scrutinizing uncertainty estimates, and could build on progresses achieved in other fields (see e.g. Laio and Tamea 2007 in the field of probabilistic forecasting).

Lastly, modeling spatial extremes is a major challenge for the hydrologic community. Indeed, the relevant scale of many hydrological variables includes a spatial component (e.g. the areal rainfall over a catchment). Understanding the development of hydrological extreme events in space may therefore result in significant advances. However, one needs to make a distinction between this objective and the perhaps less ambitious objective of accounting for spatial dependence in a given spatial dataset to avoid biasing the inference. In the latter case, the dependence model is solely used to describe data dependence at observed levels – in particular, this dependence model is not extrapolated. The analyses carried out in this chapter belong to this second objective, and relatively simple elliptical copulas may be sufficient for this purpose. However, more advanced tools are required for a deeper understanding of spatial extremes. This area of research is rapidly evolving, both in Hydrology (Genest et al. 2007; Renard and Lang 2007; Bárdossy and Li 2008; Keef et al. 2009; AghaKouchak et al. 2010a, b, c, d, 2011) and in Statistics (Diggle et al. 1998; Schalther and Tawn 2003; Heffernan and Tawn 2004; Mikosch 2005; Naveau et al. 2005; De Haan and Pereira 2006; Padoan et al. 2010). Significant advances are therefore foreseeable in this topic in the next future.

Acknowledgments Part of this work is funded by the French Research Agency (ANR) through the project EXTRAFLO (https://extraflo.cemagref.fr/). Météo France is gratefully acknowledged for providing the data.

A.1 Appendix

A.1.1 The Chib Method for Computing Marginal Likelihoods

The marginal likelihood is the normalizing constant of the posterior distribution. For a given model M (omitted from the notation for simplicity), it can therefore be written as follows:

$$p(\boldsymbol{y}) = \frac{p(\boldsymbol{y}|\boldsymbol{\theta})p(\boldsymbol{\theta})}{p(\boldsymbol{\theta}|\boldsymbol{y})} = \frac{f(\boldsymbol{\theta})}{p(\boldsymbol{\theta}|\boldsymbol{y})} \tag{3.51}$$

In this equation, the numerator is the unnormalized posterior pdf and can therefore be computed for any $\boldsymbol{\theta}$. The difficulty is to compute the denominator, i.e. the normalized posterior pdf $p(\boldsymbol{\theta}|\boldsymbol{y})$.

Chib's approach (1995) to this problem is first based on the observation that relation (3.51) holds for any $\boldsymbol{\theta}$ value. For a given value $\boldsymbol{\theta}^*$, it is then possible to decompose the normalized posterior pdf $p(\boldsymbol{\theta}^*|\boldsymbol{y})$ evaluated at $\boldsymbol{\theta}^*$ as follows:

$$\begin{aligned} p(\boldsymbol{\theta}^*|\boldsymbol{y}) = {} & p(\theta_1^*|\boldsymbol{y})p(\theta_2^*|\theta_1^*,\boldsymbol{y}) \times ... \times p(\theta_k^*|\theta_1^*,...,\theta_{k-1}^*,\boldsymbol{y}) \times ... \\ & \times p(\theta_{N_D}^*|\theta_1^*,...,\theta_{N_D-1}^*,\boldsymbol{y}) \end{aligned} \tag{3.52}$$

Equation (3.52) decomposes the computation of a N_D-dimensional pdf into the multiplication of N_D one-dimensional pdfs. Each term $p(\theta_k^*|\theta_1^*,...,\theta_{k-1}^*,\boldsymbol{y})$ is the first marginal distribution of the distribution $p(\theta_k,\theta_{k+1}...,\theta_{N_D}|\theta_1^*,...,\theta_{k-1}^*,\boldsymbol{y})$, evaluated at θ_k^*. Consequently, Chib's proposal is to perform additional MCMC sampling from the conditional distributions $p(\theta_k,\theta_{k+1}...,\theta_{N_D}|\theta_1^*,...,\theta_{k-1}^*,\boldsymbol{y})$ for $k=2{:}N_D$, and to use the first marginal sample to compute each term $p(\theta_k^*|\theta_1^*,...,\theta_{k-1}^*,\boldsymbol{y})$. Since the latter is a one-dimensional distribution, estimating its normalized pdf based on MCMC samples poses no difficulty, either using a kernel density estimate or one-dimensional numerical integration.

Although Chib's approach is computationally demanding in high-dimensional problems, it remains one of the most stable approximations of the marginal likelihood (Bos 2002). Note that $\boldsymbol{\theta}^*$ should preferably be chosen in a high-density area of the posterior distribution (e.g. the posterior mode) to improve the efficiency of the approximation.

References

AghaKouchak A, Bardossy A, Habib E (2010a) Conditional simulation of remotely sensed rainfall data using a non-Gaussian v-transformed copula. Adv Water Resour 33(6):624–634. doi:10.1016/j.advwatres.2010.02.010

AghaKouchak A, Bardossy A, Habib E (2010b) Copula-based uncertainty modelling: application to multisensor precipitation estimates. Hydrol Process 24(15):2111–2124. doi:10.1002/hyp.7632

AghaKouchak A, Habib E, Bardossy A (2010c) A comparison of three remotely sensed rainfall ensemble generators. Atmos Res 98(2–4):387–399. doi:10.1016/j.atmosres.2010.07.016

AghaKouchak A, Ciach G, Habib E (2010d) Estimation of tail dependence coefficient in rainfall accumulation fields. Adv Water Resour 33(9):1142–1149. doi:10.1016/j.advwatres.2010.07.003

AghaKouchak A, Behrangi A, Sorooshian S, Hsu K, Amitai E (2011) Evaluation of satellite-retrieved extreme precipitation rates across the central United States. J Geophys Res Atmos 116:D02115. doi:10.1029/2010jd014741

Akaike H (1973) Information theory and an extension of the maximum likelihood principle. In: Petrov BN, Csaki F (eds) 2nd International symposium on information theory. Akademiai Kiado, Budapest

Aryal SK, Bates BC, Campbell EP, Li Y, Palmer MJ, Viney NR (2009) Characterizing and modeling temporal and spatial trends in rainfall extremes. J Hydrometeorol 10(1):241–253

Bárdossy A, Li J (2008) Geostatistical interpolation using copulas. Water Resour Res 44(7):W07412. doi:10.1029/2007wr006115

Bernier J, Parent E, Boreux J-J (2000) Statistique pour l'environnement: traitement bayésien de l'incertitude. Technique & Documentation, Paris, 364 pp

Bos CS (2002) A comparison of marginal likelihood computation methods. In: Hardle W, Ronz B (eds) Compstat2002. Physica-Verlag, Heidelberg

Chib S (1995) Marginal likelihood from the Gibbs output. J Am Stat Assoc 90(432):1313–1321

Chiles J-P, Delfiner P (1999) Geostatistics: modeling spatial uncertainty. Wiley, New York

Clark JS (2005) Why environmental scientists are becoming Bayesians. Ecol Lett 8:2–14

Coles S (2001) An introduction to statistical modeling of extreme values. Verlag, London, 210 pp

Coles SG, Powell EA (1996) Bayesian methods in extreme value modelling: a review and new developments. Int Stat Rev 64(1):119–136

Coles S, Heffernan JE, Tawn JA (1999) Dependence measures for extreme value analyses. Extremes 2:339–365

Coles S, Pericchi LR, Sisson S (2003) A fully probabilistic approach to extreme rainfall modelling. J Hydrol 273(1–4):35–50

Cooley D, Nychka D, Naveau P (2007) Bayesian spatial modeling of extreme precipitation return levels. J Am Stat Assoc 102(479):824–840

Cox DR, Isham VS, Northrop PJ (2002) Floods: some probabilistic and statistical approaches. Philos Trans R Soc Math Phys Eng Sci 360(1796):1389–1408

Cunderlik JM, Burn DH (2003) Non-stationary pooled frequency analysis. J Hydrol 276:210–223

Dalrymple T (1960) Flood frequency analyses. In: US Geological Survey (ed) Water-supply paper 1543-A. U.S. G.P.O, Washington, DC

De Haan L, Pereira TT (2006) Spatial extremes: models for the stationary case. Ann Stat 34(1):146–168
Diggle PJ, Tawn JA, Moyeed RA (1998) Model-based geostatistics. J R Stat Soc Ser C Appl Stat 47:299–326
Dobson AJ (2001) An introduction to generalised linear models. Chapman & Hall, London, 240 pp
Efron B (2005) Bayesians, frequentists, and scientists. J Am Stat Assoc 100(469):1–5. doi:10.1198/016214505000000033
El Adlouni S, Favre AC, Bobee B (2006) Comparison of methodologies to assess the convergence of Markov chain Monte Carlo methods. Comput Stat Data Anal 50(10):2685–2701
El Adlouni S, Ouarda TBMJ, Zhang X, Roy R, Bobée B (2007) Generalized maximum likelihood estimators for the nonstationary generalized extreme value model. Water Resour Res 43(3):W03410. doi:10.1029/2005wr004545
Garavaglia F, Lang M, Paquet E, Gailhard J, Garcon R, Renard B (2011) Reliability and robustness of a rainfall compound distribution model based on weather pattern sub-sampling. Hydrol Earth Syst Sci 15(2):519–532. doi:10.5194/hess-15-519-2011
Gelfand AE, Sahu SK (1999) Identifiability, improper priors, and Gibbs sampling for generalized linear models. J Am Stat Assoc 94:247–253
Gelman A (2008) Objections to Bayesian statistics. Bayesian Anal 3(3):445–450
Gelman A, Carlin JB, Stern HS, Rubin DB (2004) Bayesian data analysis, 2nd edn. Chapman & Hall, London, 696 pp
Genest C, Favre AC, Béliveau J, Jacques C (2007) Metaelliptical copulas and their use in frequency analysis of multivariate hydrological data. Water Resour Res 43:W09401. doi:10.1029/2006WR005275
Geweke J (1992) Evaluating the accuracy of sampling-based approaches to the calculation of posterior moments. In: Bernardo JM, Berger JO, Dawid AP, Smith AFM (eds) Bayesian statistics 4. Oxford University Press, Oxford, pp 169–193
Gunasekara TAG, Cunnane C (1992) Split sampling technique for selecting a flood frequency-analysis procedure. J Hydrol 130(1–4):189–200
Haario H, Saksman E, Tamminen J (2001) An adaptive metropolis algorithm. Bernoulli 7(2):223–242
Haario H, Saksman E, Tamminen J (2005) Componentwise adaptation for high dimensional MCMC. Comput Stat 20(2):265–273
Hanel M, Buishand TA, Ferro CAT (2009) A nonstationary index flood model for precipitation extremes in transient regional climate model simulations. J Geophys Res Atmos 114:D15107. doi:10.1029/2009jd011712
Hastings WK (1970) Monte Carlo sampling methods using Markov chains and their applications. Biometrika 57:97–109
Heffernan JE, Tawn JA (2004) A conditional approach for multivariate extreme values. J R Stat Soc 66:497–546
Interagency Advisory Committee on Water Data (1982) Guidelines for determining flood-flow frequency: Bulletin 17B of the Hydrology Subcommittee. U.S. Geological Survey, Reston
Jeffreys H (1946) An invariant form for the prior probability in estimation problems. Proc R Soc Lond A Math Phys Sci 186(1007):453–461
Kass RE, Raftery AE (1995) Bayes factors. J Am Stat Assoc 90(430):773–795
Katz RW, Parlange MB, Naveau P (2002) Statistics of extremes in hydrology. Adv Water Resour 25(8–12):1287–1304
Keef C, Svensson C, Tawn JA (2009) Spatial dependence in extreme river flows and precipitation for Great Britain. J Hydrol 378(3–4):240–252
Khaliq MN, Ouarda TBMJ, Ondo JC, Gachon P, Bobee B (2006) Frequency analysis of a sequence of dependent and/or non-stationary hydro-meteorological observations: a review. Water Resour Res 329(3–4):534–552
Kuczera G (1999) Comprehensive at-site flood frequency analysis using Monte Carlo Bayesian inference. Water Resour Res 35(5):1551–1557

Laio F, Tamea S (2007) Verification tools for probabilistic forecasts of continuous hydrological variables. Hydrol Earth Syst Sci 11(4):1267–1277
Lima CHR, Lall U (2009) Hierarchical Bayesian modeling of multisite daily rainfall occurrence: rainy season onset, peak, and end. Water Resour Res 45:W07422. doi:10.1029/2008WR007485
Lima CHR, Lall U (2010) Spatial scaling in a changing climate: a hierarchical Bayesian model for non-stationary multi-site annual maximum and monthly streamflow. J Hydrol 383(3–4): 307–318
Lunn DJ, Thomas A, Best N, Spiegelhalter D (2000) WinBUGS – a Bayesian modelling framework: concepts, structure, and extensibility. Stat Comput 10(4):325–337
Maraun D, Rust HW, Osborn TJ (2010) Synoptic airflow and UK daily precipitation extremes development and validation of a vector generalised linear model. Extremes 13(2):133–153. doi:10.1007/s10687-010-0102-x
Marshall L, Nott D, Sharma A (2004) A comparative study of Markov chain Monte Carlo methods for conceptual rainfall-runoff modeling. Water Resour Res 40:W02501. doi:10.1029/2003WR002378
Martin AD, Quinn KM, Park JH (2011) MCMCpack: Markov chain Monte Carlo in R. J Stat Softw 42(9)
Martins ES, Stedinger JR (2000) Generalized maximum-likelihood generalized extreme-value quantile estimators for hydrologic data. Water Resour Res 36(3):737–744
Merz R, Bloschl G (2008a) Flood frequency hydrology: 1. Temporal, spatial, and causal expansion of information. Water Resour Res 44:W08432. doi:10.1029/2007WR006744
Merz R, Bloschl G (2008b) Flood frequency hydrology: 2. Combining data evidence. Water Resour Res 44:W08433. doi:10.1029/2007WR006745
Metropolis N, Ulam S (1949) The Monte Carlo method. J Am Stat Assoc 44:335–341
Metropolis N, Rosenbluth AW, Rosenbluth MN, Teller AH, Teller E (1953) Equation of state calculations by fast computing machines. J Chem Phys 21:1087–1092
Meylan P, Favre A-C, Musy A (2008) Hydrologie fréquentielle: Une science prédictive. Presses polytechniques et universitaires romandes, Lausanne, 173 pp
Micevski T (2007) Nonhomogeneity in eastern Australian flood frequency data: identification and regionalisation. PhD thesis, University of Newcastle, Newcastle, Australia, 129 pp
Micevski T, Kuczera G, Franks SW (2006) A Bayesian hierarchical regional flood model. Paper presented at 30th hydrology and water resources symposium, Engineers Australia, Launceston, Tas, Australia, 4–7 December
Mikosch T (2005) How to model multivariate extremes if one must? Stat Neerl 59(3):324–338
Naveau P, Cooley D, Poncet P (2005) Spatial extremes analysis in climate studies. Paper presented at extreme value analysis, Gothenburg, Sweden
Padoan SA, Ribatet M, Sisson SA (2010) Likelihood-based inference for max-stable processes. J Am Stat Assoc 105(489):263–277
Parent E, Bernier J (2003) Encoding prior experts judgments to improve risk analysis of extreme hydrological events via POT modeling. J Hydrol 283(1–4):1–18
Perreault L (2000) Analyse bayésienne rétrospective d'une rupture dans les séquences de variables aléatoires hydrologiques. PhD thesis, ENGREF/INRS-Eau, 200 pp
Perreault L, Bernier J, Bobee B, Parent E (2000a) Bayesian change-point analysis in hydrometeorological time series. Part 2. Comparison of change-point models and forecasting. J Hydrol 235(3–4):242–263
Perreault L, Bernier J, Bobee B, Parent E (2000b) Bayesian change-point analysis in hydrometeorological time series. Part 1. The normal model revisited. J Hydrol 235(3–4):221–241
Perreault L, Parent E, Bernier J, Bobee B, Slivitzky M (2000c) Retrospective multivariate Bayesian change-point analysis: a simultaneous single change in the mean of several hydrological sequences. Stoch Environ Res Risk Assess 14(4–5):243–261
Plummer M, Best N, Cowles K, Vines K (2006) CODA: convergence diagnosis and output analysis for MCMC. R News 6(1):7–11
Pujol N, Neppel L, Sabatier R (2007) Regional tests for trend detection in maximum precipitation series in the French Mediterranean region. Hydrol Sci J J Sci Hydrol 52(5):956–973

Raftery AE (1996) Approximate Bayes factors and accounting for model uncertainty in generalized linear models. Biometrika 83(2):251–266

Reis DS, Stedinger JR, Martins ES (2005) Bayesian generalized least squares regression with application to log Pearson type 3 regional skew estimation. Water Resour Res 41(10)

Renard B, Lang M (2007) Use of a Gaussian copula for multivariate extreme value analysis: some case studies in hydrology. Adv Water Resour 30(4):897–912

Renard B, Garreta V, Lang M (2006) An application of Bayesian analysis and MCMC methods to the estimation of a regional trend in annual maxima. Water Resour Res 42(12)

Renard B, Kavetski D, Thyer M, Kuczera G, Franks SW (2010) Understanding predictive uncertainty in hydrologic modeling: the challenge of identifying input and structural errors. Water Resour Res 46:W05521. doi:10.1029/2009WR008328

Ribatet M, Sauquet E, Gresillon JM, Ouarda TBMJ (2006) A regional Bayesian POT model for flood frequency analysis. Stoch Environ Res Risk Assess 21(4):327–339

Robert CP (2001) The Bayesian choice: from decision-theoretic motivations to computational implementation. Springer, New York

Robert CP, Casella G (2004) Monte Carlo statistical methods. Springer, New York, 650 pp

Salvadori G, De Michele C, Kottegoda NT, Rosso R (2007) Extremes in nature: an approach using copulas. Springer, Dordrecht, 292 pp

Schalther M, Tawn JA (2003) A dependence measure for multivariate and spatial extreme values: properties and inference. Biometrika 90(1):139–156

Schwarz G (1978) Estimating the dimension of a model. Ann Stat 6(2):461–464

Spiegelhalter DJ, Best NG, Carlin BR, van der Linde A (2002) Bayesian measures of model complexity and fit. J R Stat Soc Ser B Stat Methodol 64:583–616

Stedinger JR (1983) Design-events with specified flood risk. Water Resour Res 19(2):511–522

Stedinger JR, Tasker GD (1985) Regional hydrologic analysis: 1. Ordinary, weighted and generalized least squares compared. Water Resour Res 21(9):1421–1432, [Correction, Water Resour Res 1422(1425): 1844, 1986.]

Strupczewski WG, Kaczmarek Z (2001) Non-stationary approach to at-site flood frequency modelling II. Weighted least squares estimation. J Hydrol 248(1–4):143–151

Strupczewski WG, Singh VP, Feluch W (2001) Non-stationary approach to at-site flood frequency modelling I. Maximum likelihood estimation. J Hydrol 248(1–4):123–142

Thyer M, Kuczera G (2000) Modeling long-term persistence in hydroclimatic time series using a hidden state Markov model. Water Resour Res 36(11):3301–3310

Thyer M, Kuczera G (2003a) A hidden Markov model for modelling long-term persistence in multi-site rainfall time series. 2. Real data analysis. J Hydrol 275(1–2):27–48

Thyer M, Kuczera G (2003b) A hidden Markov model for modelling long-term persistence in multi-site rainfall time series 1. Model calibration using a Bayesian approach. J Hydrol 275(1–2):12–26

Chapter 4
Return Periods and Return Levels Under Climate Change

Daniel Cooley

Abstract We investigate the notions of return period and return level for a nonstationary climate. We discuss two general methods for communicating risk. The first eschews the term return period and instead communicates yearly risk in terms of a probability of exceedance. The second extends the notion of return period to the non-stationary setting. We examine two different definitions of return period under non-stationarity. The first, which appears in Olsen et al. (Risk Anal 18:497–510, 1998), defines the m-year return level as the level for which the expected waiting time until the exceedance is m years. The second, which appears in Parey et al. (Clim chang 81(3):331–352, 2007) and Parey et al. (Environmetrics 21:698–718, 2010), defines the m-year return level as the level for which the expected number of events in an m year period is one. We illustrate the various risk communications with an application to annual peak flow measurements for the Red River of the North.

4.1 Introduction

4.1.1 Return Periods and Return Levels Under Stationarity

In many disciplines, return levels and return periods are used to describe and quantify risk. Classical work in probabilistic hydrology risk, precipitation frequency analysis, and other fields assumes a stationary climate. There is a general consensus in the scientific community that climate change has accelerated over the past few decades and that climate will continue to change in the coming decades primarily due to anthropogenic modifications of the Earth's atmosphere (IPCC 2007).

D. Cooley (✉)
Department of Statistics, Colorado State University, Fort Collins, CO, USA
e-mail: cooleyd@stat.colostate.edu

A. AghaKouchak et al. (eds.), *Extremes in a Changing Climate*, Water Science and Technology Library 65, DOI 10.1007/978-94-007-4479-0_4,

Consequently, there is growing interest to consider and account for non-stationarity when assessing risk. However, the assumption of stationarity has pervaded all areas of the above disciplines, including even the basic terminology. Two fundamental terms to these disciplines are return level and return period. These terms are relatively easily understood given stationarity, but become ambiguous in the non-stationary setting. The aim of this chapter is to explore these concepts under non-stationarity.

Because we will illustrate by analyzing an annual maximum series in Sect. 4.3, let us define the m-year return level as the high quantile for which the probability that the annual maximum exceeds this quantile is $1/m$. Under an assumption of stationarity, the return level is the same for all years, and this gives rise to the notion of the return period. Under stationarity, the return period of a particular event is the inverse of the probability that the event will be exceeded in any given year. Thus, the m-year return level is associated with a return period of m years. A return period can be *interpreted* in different ways, and we will explore different interpretations shortly. For now, let us simply acknowledge that in the stationary case there is a one-to-one relationship between a return level (the quantile) and a return period (the associated time interval).

The above definition in which return period is defined in terms of the *annual maximum's* probability of exceedance is not universal. Mays (2001, p. 317) equates the definition of return period with "average recurrence interval"; that is, the time between exceedance events. The difference between the two definitions arises because of the probability of having more than one exceedance in a given year. Because of the ambiguity of the definition of return period, NOAA's latest precipitation atlas effort (Bonnin et al. 2004) avoids using the term return period altogether. NOAA, like Mays, uses the terms average recurrence interval (ARI) and "annual exceedance probability" (AEP) to speak of return period as we defined it above. NOAA notes that the difference between the return levels associated with ARI and AEP is noticeable only for time intervals shorter 20 years, as the probability of multiple exceedances of very high thresholds in any year becomes negligible (Bonnin et al. 2004, Sect. 3.2). For the purposes of discussing the implications of non-stationarity in this chapter, the definitions of return period and return level in terms of the annual maximum are sufficient.

Due to the one-to-one relationship between return level and return period, in the stationary case it is straightforward to solve for either quantity given the value of the other. Design criteria may specify that a structure be built to withstand the (say) 100-year event, or regulations may require the designation of the 100-year floodplain, and the corresponding return level can be found for this period. A less common calculation is to find the return period associated with a particular level. An individual could wish to calculate the return period associated with his or her particular structure being flooded, or after a large event has occurred, one might wish to ascertain the event's associated return period.

4.1.2 Statistical Models for the Distribution's Tail

In practice, one does not know the exact distribution of the individual events or annual maximum, and one must estimate the upper tail of a distribution to assess risk. From the estimated distribution one can calculate the return period or return level of interest and additionally provide measures of uncertainty.

In the application in Sect. 4.3, we will rely on probabilistic results from extreme value theory to model the distribution of the annual maximum observation. There is a rich literature which describes the fundamental probability results of extreme value theory as well as the resulting statistical practice for modeling the tail derived from these results. Some recent references are de Haan and Ferreira (2006), Beirlant et al. (2004) and Coles (2001). Let $\{X_t\}$ denote a time series of our quantity of interest, for example, daily maximum streamflow or daily total precipitation. Let $M_n = \max_{t=1,\ldots,n} X_t$. The foundational result of extreme value theory states that if X_t are iid, and M_n can be linearly renormalized in such a way that its distribution converges as n grows, then it will converge to an extreme-value (equivalently, max-stable) distribution (Fisher and Tippett 1928; Gnedenko 1943). Further theoretical results state that the distribution of exceedances $P(X_t > x + u | X_t > u)$ should be well approximated by a generalized Pareto (GP) distribution above a sufficiently high threshold u (Balkema and De Haan 1974; Pickands 1975). If $\{X_t\}$ are not independent but are still identically distributed, the asymptotic distributions do not change, so long as certain relatively weak mixing conditions are met (Leadbetter et al. 1983).

In practice, an extreme value statistical analysis extracts a subset of data deemed extreme, and then fits a theoretically-justified distribution to this data subset. One approach is to construct a time series of block (e.g. annual) maxima, and another approach is to construct a subset of data which exceed a previously defined threshold (deemed a partial-duration series in hydrology). When the block-size n is fixed and large enough to assume that the asymptotic results provide a good approximation, the above theory suggests modeling M_n with a generalized extreme value (GEV) distribution. Threshold exceedance data can be modeled with GP distribution or an equivalent point process representation (Davison and Smith 1990). Traditional extreme value results assume the data are identically distributed, but methods have been developed enable construction of non-stationary models based on the GEV and GP distributions or the point process representation of threshold exceedances (Katz 2012; Smith 1989).

The GEV and GP distributions' advantage is their asymptotic justification which is particularly useful for extrapolating the tail beyond the range of the data. However, risk analyses are not always based on distributions derived from extreme value theory. The log-Pearson type III distribution is widely used in hydrology and its use is sometimes mandated by government agencies (Vogel and Wilson 1996). This chapter's discussion of return levels and periods is not inherently tied to the GEV distribution that is used in Sect. 4.3; any model for the distribution's tail can be used.

4.1.3 Interpretations of Return Periods Under Stationarity

In the stationary case there is a one-to-one relationship between the m-year return level and m-year return period which is defined implicitly as the reciprocal of the probability of an exceedance in any 1 year. Return periods were assumedly created for the purpose of interpretation: a 100-year event may be more interpretable by the general public than a 0.01 probability of occurrence in any particular year. But this implicit definition gives rise to at least two interpretations of "an m-year event". The first interpretation is that the expected waiting time until the next exceedance is m years. The second is that the expected number of events in m years is 1. We show below that both of these interpretations are correct under a stationary assumption.

Let M_y denote the random variable representing the annual maximum for year y. Note that we have omitted the notational dependence on block size n and now include a year index y. For now, assume $\{M_y\}$ are iid with distribution function F. Given a return period of interest m, we can solve the equation

$$F(r_m) = P(M_y \leq r_m) = 1 - 1/m$$

for r_m, the associated return level. The first interpretation of the m-year event is the expected waiting time until an exceedance occurs. Let T be the year of the first exceedance. One recognizes

$$\begin{aligned} P(T = t) &= P(M_1 \leq r_m, M_2 \leq r_m, \ldots, M_{t-1} \leq r_m, M_t > r_m) \\ &= P(M_1 \leq r_m)P(M_2 \leq r_m)\ldots P(M_{t-1} \leq r_m)P(M_t > r_m) \\ &= P(M_1 \leq r_m)^{t-1} P(M_1 > r_m) \\ &= F^{t-1}(r_m)(1 - F(r_m)) \\ &= (1 - 1/m)^{t-1}(1/m), \end{aligned}$$

where the second line follows from the independence assumption and the third from stationarity. T is a geometric random variable, and it is well known that its expected value is m. That is, the expected waiting time for an m-year event is m years.

An alternative interpretation of an m-year event is that the expected number of events in m years is 1. Let N be the random variable representing the number of exceedances in m years; that is $N = \sum_{y=1}^{m} I(M_y > r_m)$ where I is the indicator function. Each year can be viewed as a trial, and since we have assumed M_y are iid, N has a binomial distribution:

$$P(N = k) = \binom{m}{k}(1/m)^k(1 - 1/m)^{m-k}. \tag{4.1}$$

It is straightforward to show that the expected value of N is 1.

4.1.4 Outline

The outline for the remainder of this chapter is as follows. In the next section, we discuss several options for communicating risk in the non-stationary case. In Sect. 4.3, we fit a stationary model and some non-stationary models to a time series of annual maximum river flows and convey the different risk measures for these various models. We conclude with a discussion in Sect. 4.4 where we touch on both implications of non-stationary models and expressions of risk.

4.2 Communicating Risk Under Non-stationarity

In this section we outline two general ideas for conveying risk when the process is assumed to be non-stationary. We start by explicitly conveying the changing nature of the risk by expressing the yearly probability of exceedance given a fixed level, or conversely the yearly exceedance level for a fixed probability of exceedance. Later, we extend the notion of return period to the non-stationary setting via both the expected waiting time and the expected number of events interpretations of return period. Throughout Sect. 4.2, we keep the independence assumption, but drop the assumption that the sequence $\{M_y\}$ is identically distributed.

4.2.1 Communicating Changing Risk

The most straightforward approach to communicating risk in the non-stationary setting is to give yearly estimates of risk. However, the idea of a yearly return period seems illogical, and it makes more sense to communicate risk via probabilities. Let F_y denote the distribution function of M_y. In any particular year, there still exists a one-to-one correspondence between a probability of exceedance and a high quantile.

Given a particular level of interest r, it is straightforward to express yearly risk in terms of probability. Letting $p(y) = P(M_y > r) = 1 - F_y(r)$, once F_y is estimated it is simple to provide yearly point estimates of the probability of an exceedance $p(y)$. These point estimates could be supplemented by employing the delta method (Casella and Berger, 2002, Sect. 5.5.4) to obtain a confidence interval.

Generally risk calculations proceed in the opposite direction: one starts with a measure of risk (e.g., a return period in the stationary case) and finds the corresponding level. Simply inverting the procedure in the previous paragraph allows one to start with a probability of exceedance p and solve $F_y(r_p(y)) = 1 - p$. As the exceedance level $r_p(y)$ changes with year, it clearly conveys to the user the changing nature of risk. Again, there are existing methods to obtain confidence intervals to convey uncertainty associated with $r_p(y)$.

4.2.2 *Return Periods and Return Levels Under Non-stationarity*

Structures are designed for certain return periods based on the economic and human impacts of their failure. The dynamic risk valuations in Sect. 4.2.1 must be extended to make lifespan calculations in the non-stationary setting. Below we extend the expected waiting time and expected number of events interpretations of return period given in Sect. 4.1.3 to the non-stationary setting.

4.2.2.1 Return Period as Expected Waiting Time

One interpretation of return period given in Sect. 4.1.3 is the expected waiting time until an exceedance occurs. Olsen et al. (1998) use this definition to compute an m-year return level for non-stationary time series and compare this to the $1/m$ probability-of-exceedance level described above. Olsen et al. (1998) illustrate the difference in the definitions using simulated data. Here, we extend the basic definition set forth in Olsen et al. (1998) and discuss computational aspects. Specifically, we aim to find the level r_m for which the expected waiting time for an exceedance of this level is m years.

Let T be the waiting time (from $y = 0$) until an exceedance over a general level r occurs. Starting as before,

$$
\begin{aligned}
P(T = t) &= P(M_1 \leq r)P(M_2 \leq r)\dots P(M_{t-1} \leq r)P(M_t > r) \\
&= \prod_{y=1}^{t-1} F_y(r)(1 - F_t(r)) \\
\Rightarrow E[T] &= \sum_{t=1}^{\infty} t \prod_{y=1}^{t-1} F_y(r)(1 - F_t(r)) \qquad (4.2) \\
&= 1 + \sum_{i=1}^{\infty} \prod_{y=1}^{i} F_y(r),
\end{aligned}
$$

where (4.2) is the definition that appears in Olsen et al. (1998) and the last line results from expanding the sum in the previous line and then collecting terms (shown in the appendix).

Defining the m-year return level r_m as the level which the expected waiting time until an exceedance occurs is m years, then r_m is the solution to the equation

$$
m = 1 + \sum_{i=1}^{\infty} \prod_{y=1}^{i} F_y(r_m). \qquad (4.3)
$$

Because (4.3) cannot be written as a geometric series, solving for r_m is not straightforward. However, in the case that $F_y(r)$ is monotonically decreasing as $y \to \infty$ (that is, the extremes are getting more extreme), it is possible to bound the right-hand side of (4.3). For any positive integer L,

$$m = 1 + \sum_{i=1}^{L} \prod_{y=1}^{i} F_y(r_m) + \sum_{i=L+1}^{\infty} \prod_{y=1}^{i} F_y(r_m)$$

$$\Rightarrow m > 1 + \sum_{i=1}^{L} \prod_{y=1}^{i} F_y(r_m). \tag{4.4}$$

Furthermore,

$$m = 1 + \sum_{i=1}^{L} \prod_{y=1}^{i} F_y(r_m) + \prod_{y=1}^{L} F_y(r_m) \sum_{i=L+1}^{\infty} \prod_{y=L+1}^{i} F_y(r_m)$$

$$m \leq 1 + \sum_{i=1}^{L} \prod_{y=1}^{i} F_y(r_m) + \prod_{y=1}^{L} F_y(r_m) \sum_{i=L+1}^{\infty} (F_{L+1}(r_m))^{i-L}$$

$$= 1 + \sum_{i=1}^{L} \prod_{y=1}^{i} F_y(r_m) + \prod_{y=1}^{L} F_y(r_m) \frac{F_{L+1}(r_m)}{1 - F_{L+1}(r_m)}, \tag{4.5}$$

where the inequality follows from the fact that $F_{L+1} \geq F_y$ if $y > L + 1$. One can achieve bounds to any desired width by choosing L to be large enough. Solving for r_m must be done numerically, but this is done relatively easily since m is monotonically increasing with r_m.

4.2.2.2 Return Period as Expected Number of Events

The other interpretation of an m-year return period given in Sect. 4.1.3 is that the expected number of events in m years is one. Parey et al. (2007, 2010) extend this definition to the non-stationary case.

We aim to find the level r_m for which the expected number of exceedances in m years is one. Let N be the number of exceedances that occur in the m years beginning with year $y = 1$ and ending with year $y = m$. As the probability of an exceedance is no longer constant from year to year, the distribution of N is no longer binomial. Letting I be an indicator variable and r be a general threshold level, one obtains

$$
\begin{aligned}
N &= \sum_{y=1}^{m} I\left(M_y > r\right) \\
\Rightarrow E[N] &= \sum_{y=1}^{m} E[I\left(M_y > r\right)] \\
&= \sum_{y=1}^{m} P(M_y > r) \\
&= \sum_{y=1}^{m} (1 - F_y(r)). \qquad (4.6)
\end{aligned}
$$

Setting (4.6) equal to one and solving, we define the m-year return level r_m to be the solution to the equation

$$1 = \sum_{y=1}^{m} (1 - F_y(r_m)). \qquad (4.7)$$

4.3 Illustrative Example: Red River at Halstad

We examine annual peak flow measurements from the Red River of the North at Halstad, Minnesota, USA (station #05054500). Data were obtained from the US Geological Survey's National Water Information System.[1] The Halstad site has annual maximum data from 1942–2010. Additionally, we include the highest reading for 2011 (61,600 cfs, 7:00 a.m., April 12) from the site's real-time data archive.[2]

The Red River was selected because it has experienced recent newsworthy floods in 1997, 2009, and 2011. In Sect. 4.4, we further discuss possible implications of the selection of the Red River.

Figure 4.1 shows the annual maximum flow measurements at Halstad. There is some indication that the annual maximum flows have increased over this period. The highest measurements tend to occur in the latter part of the series, and the magnitude of the lowest annual maximum flows for the last 15 years seems to be higher than previous years.

As we have annual maximum data, we fit various stationary and non-stationary GEV distributions to the data. We assume

$$F_y(r) = P(M_y \le r) = \exp\left\{-\left[1 + \xi\left(\frac{r - \mu_y}{\sigma_y}\right)\right]^{-1/\xi}\right\}, \qquad (4.8)$$

[1] http://nwis.waterdata.usgs.gov/nwis

[2] The real-time data is marked as "Provisional and subject to revision"

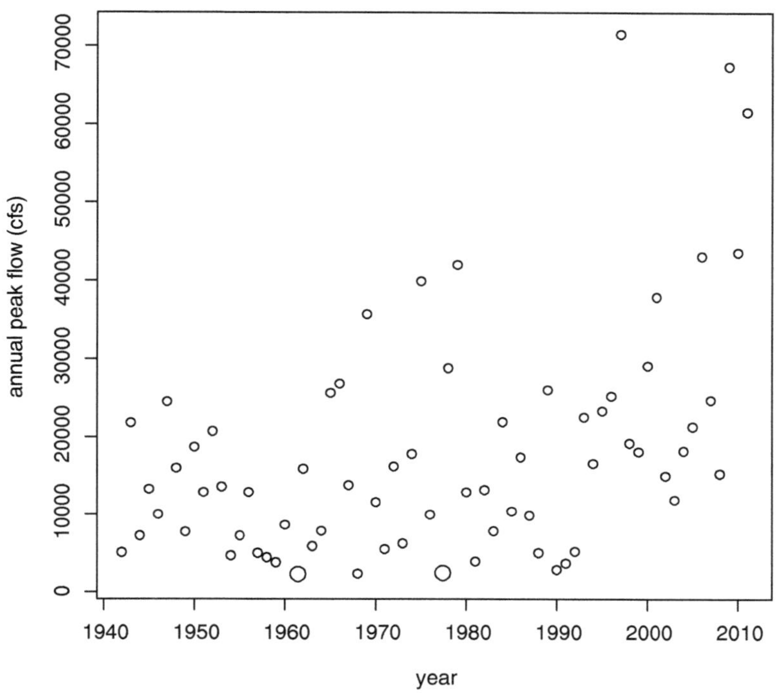

Fig. 4.1 Annual peak flow measurements for the Red River at Halstad, MN

where μ_y and σ_y are (possibly time-varying) location and scale parameters repectively, ξ controls the tail behavior, and $\left[1 + \xi\left(\frac{x-\mu_y}{\sigma_y}\right)\right] > 0$. A positive ξ indicates a heavy tail. We restrict ourselves to the case where ξ is not time-varying, although the model could be generalized to allow this parameter to vary with time as well.

4.3.1 The Stationary Model

We begin by fitting a stationary model to this annual maximum series. Via numerical maximum-likelihood, we fit the stationary GEV; that is, $\mu_y = \mu$ and $\sigma_y = \sigma$ in (4.8). Our estimates are $\hat{\mu} = 10,392$, $\hat{\sigma} = 7,924$ and $\hat{\xi} = 0.323$ with respective standard errors of 1,095, 944, and 0.137. The log-likelihood of the fitted model is -751.59. We note that the MLE for ξ seems a bit higher than tail parameter estimates typically associated with river flow data. Also note the large uncertainty associated with this difficult-to-estimate parameter.

Given a return period m, plugging $\hat{\mu}$, $\hat{\sigma}$, and $\hat{\xi}$ into (4.8), setting equal to $1-1/m$ and solving yields the estimated m-year return level. We will focus on the 50-year return period, however, analogous analyses could be performed any return period. Our point estimate for the 50-year return level is 72,306 cfs, and we note the largest observation in our 70-year record is 71,500 cfs.

Approximate confidence intervals for the return level can be obtained by the delta method (Casella and Berger, 2002, Sect. 5.5.4) which relies on the asymptotic normality of maximum-likelihood estimators and produces a symmetric confidence interval. Alternatively, profile likelihood methods (Coles 2001, Sect. 2.6.6) provide asymmetric confidence intervals, which better capture the skewness generally associated with return level estimates. For later comparison with the non-stationary cases we employ only the delta method procedure here, as profile likelihood methods for non-stationary measures of risk have not been developed. The delta method yields a 95% confidence interval of (33,324, 111,286).

4.3.2 A Nonstationary Model

We next fit a non-stationary model to the data. We allow the location parameter in (4.8) to be a linear function of time; that is, $\mu_y = \beta_0 + \beta_1(y - 1942)$, where y denotes year. Later we will consider other non-stationary models.

Again we fit the model using numerical maximum likelihood and use the maximum likelihood estimates from the stationary model and $\beta_1 = 0$ as the initial values for the numerical optimization procedure. Our estimates are $\hat{\beta}_0 = 7,976$, $\hat{\beta}_1 = 106$, $\hat{\sigma} = 8,630$, and $\hat{\xi} = 0.216$, with respective standard errors of 2,247, 65, 1,024, and 0.132. The log-likelihood of the model is -749.90. Model selection procedures such as the AIC (Akaike 1974) would select the model with time-varying μ as the better model (AIC scores of 1509.2 and 1507.8 respectively for the stationary and time-varying μ models, lower is better). Interestingly, the standard error associated with β_1 would fail to reject a test with the null hypothesis that the slope is 0, as a Gaussian-based 95 % confidence interval for this parameter estimate is $(-21.4, 233.4)$.

4.3.2.1 Communicating Changing Risk

Next, we find the level for each year which has a 0.02 probability of exceedance (that is, a 1-in-50 chance) according to our fitted non-stationary model. Substituting our estimates for $\hat{\mu}_y = \hat{\beta}_0 + \hat{\beta}_1 y, \hat{\sigma}$, and $\hat{\xi}$ into (4.8) and setting equal to $1 - 0.02$, we obtain

$$\hat{r}_{0.02}(y) = 106(y - 1942) + 60,821.$$

The only peak flow to exceed the 0.02 probability-of-exceedance level is that of 1997 where the measurement of 71,500 exceeds model's point estimate of $\hat{r}_{0.02}(1997) = 66,651$. The 2009 peak flow of 67,400 nearly exceeds $\hat{r}_{0.02}(2009) = 67,923$.

It is interesting that $\hat{r}_{0.02}(y)$ does not exceed the stationary model's 50 year return level estimate of 72,306 until the year 2050. The fact that the non-stationary model's estimated 0.02 probability-of-exceedance levels for the entire data record (1942–2011) are lower than the 50-year return level estimate of the stationary model can largely be attributed to the difference in the estimates of ξ.

Confidence intervals for the 0.02 probability-of-exceedance level for any particular year can be obtained via the delta method. The 95 % confidence interval is (28,989, 92,652) for the year 1942, is (39,183, 97,148) for 2011, and extrapolating the model into the future yields (44,996, 101,978) for 2061.

4.3.2.2 Return Period as Expected Waiting Time

As explained in Sect. 4.2.2, design criteria may require that a return level be estimated for a given return period. In the non-stationary setting, any definition of return period has to be associated with a specific time. The expected waiting time definition of return period has a specific starting time, as it is an infinite sum beginning in a particular year. The expected number of events definition has a specific time interval corresponding the years over which the sum is calculated.

We extrapolate the trend in our model and use the expected waiting time definition of return period to calculate the 50-year return level beginning in 2011. So that the equations in Sect. 4.2.2.1 make sense, we redefine y to be the number of years since 2011. We let $L = 200$, and get an estimated 50-year return level of 73,150 cfs. That is, beginning in 2011, the expected waiting time until we see an exceedance of 73,150 cfs is 50 years. With $L = 200$, the difference between the upper and lower bounds on our expected waiting time is less than 0.1 years.

Olsen et al. (1998) does not consider the uncertainty associated with the return level calculated from (4.3). We wish to employ the delta method to produce a confidence interval, but cannot do so directly since (4.3) does not yield an explicit expression for the m-year return level r_m. In the appendix, we show how the delta method can be used implicitly to obtain the variance of the return level. Using this method, we obtain a 95 % confidence interval of (44,383, 101,916) for the 50-year return level beginning in 2011.

4.3.2.3 Return Period as Expected Number of Events

We now change our definition of return period in the non-stationary case to be the amount of time for which the expected number of events is one. Again extrapolating the trend in our model into the future, we calculate the return level associated with the 50-year return period from 2012–2061. Using (4.7), setting $m = 50$ and numerically solving yields $\hat{r}_m = 70,950$ cfs. That is, in the 50-year period from 2012–2061, the expected number of exceedances of 70,950 cfs is one.

Parey et al. (2010) employ a bootstrap method to obtain confidence intervals for return periods defined in this manner. We will instead resort to the implicit delta method outlined in the appendix. The advantage of the bootstrap approach is that it will likely result in an asymmetric confidence interval, better capturing the skewness likely in the distribution of the return level estimate. The disadvantage of the bootstrap is that since we are dealing with only annual maximum data, our sample is rather small for bootstrapping. Bootstrapping heavy-tailed phenomena such as this river-flow data is particularly difficult, and Resnick (2007, Sect. 6.4) gives a thorough investigation. Our opinion is that any method for generating confidence intervals has drawbacks, but most methods provide the user a useful measure of uncertainty for the return level estimate. The implicit delta method yields a 95 % confidence interval of (42,458, 99,441) for the 50-year return level associated with the period from 2012–2061.

4.3.3 Other Possible Non-stationary Models

The model in Sect. 4.3.2 is not the only way to model non-stationary behavior in the annual maximum time series. We consider two additional models. In the first, we hold the location parameter μ_y constant and allow the scale parameter σ_y to vary linearly with year. In the second, we allow both μ_y and σ_y to vary linearly with year. One could also consider other parametric forms for the behavior of these parameters or could consider non-parametric representations such as the work done by Chavez-Demoulin and Davison (2005). However, extrapolating non-parametric approaches to consider future risk could prove challenging. Additionally, one could consider allowing the tail parameter ξ to vary with time, which would imply that the fundamental behavior of the tail is changing. Cooley and Sain (2010) and Schliep et al. (2010) construct models where ξ is allowed to vary spatially, but not in time. Knowing that ξ is difficult to estimate even when held constant, we choose to allow only μ_y and σ_y to vary in time.

Table 4.1 gives the parameter estimates for these new models as well as the stationary and non-stationary models fit in Sects. 4.3.1 and 4.3.2. All parameters estimated via numerical maximum likelihood. If the AIC is used as a strict model-selection criterion, each successive model is an improvement over the previous model. Figure 4.2 shows the QQ plot for each of the four models after transformation to standard Gumbel to account for non-stationarity (Katz 2012) as well as the occurrence year of the largest five empirical quantiles (after transformation). That the largest empirical quantiles for the non-stationary model (Fig. 4.2, upper-left) all occur after 1997 is some indication that the stationary model is failing to capture changing behavior, although the largest three quantiles for each of the non-stationary models are 1997, 2009, and 2011.

For each of the four models, we fit the 0.02 probability-of-exceedance level for the years 1942–2061. We also use the delta method to obtain 95 % confidence intervals for these levels. Figure 4.3 shows the results for each of the four models. Clearly, the different non-stationary models give very different risk estimates over this period.

Table 4.1 Maximum likelihood parameter estimates, log-likelihood values, $\ell(\boldsymbol{\theta}; \boldsymbol{m})$, and AIC values for the four models fit to the annual maximum series

Model	$\hat{\mu}_y$	$\hat{\sigma}_y$	$\hat{\xi}$	$\ell(\boldsymbol{\theta}; \boldsymbol{m})$	AIC
1	10,392	7,924	0.323	−751.59	1509.18
2	$7,975 + 106(y - 1942)$	8,630	0.216	−749.90	1507.80
3	9,681	$4,480 + 109(y - 1942)$	0.189	−748.70	1505.40
4	$8,854 + 78(y - 1942)$	$4,060 + 133(y - 1942)$	0.133	−746.53	1503.06

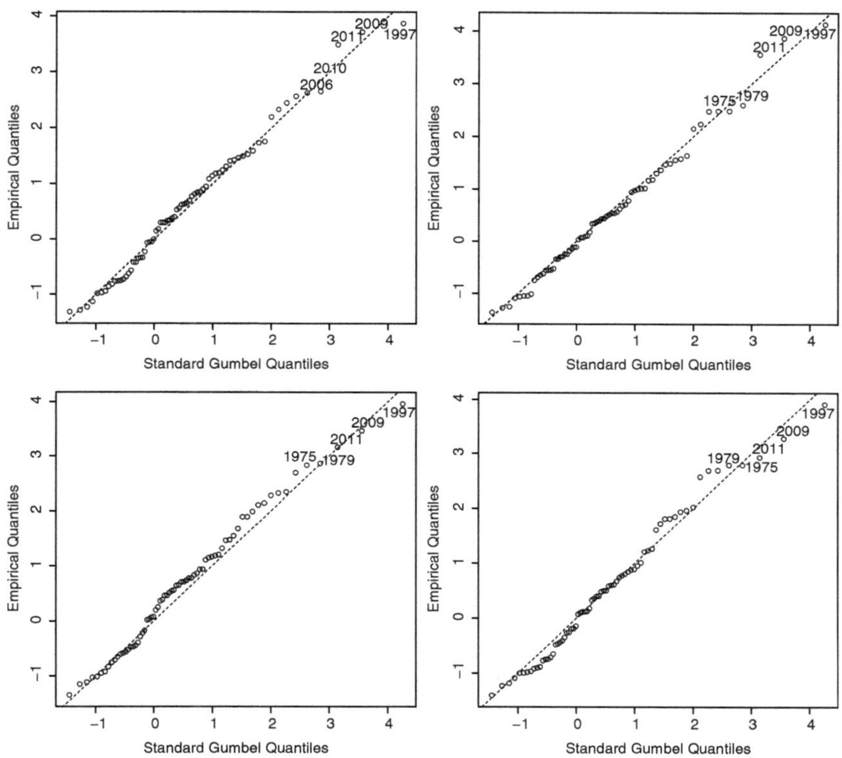

Fig. 4.2 QQ plots based on transformation to a standard Gumbel distribution for the stationary model (*top left*), time varying μ model (*top right*), time-varying σ model (*bottom left*), and time-varying μ and σ model (*bottom right*). The occurrence year of the largest five empirical quantiles (after transformation) are also shown

4.4 Discussion

The aim of this chapter has been to investigate the concept of return period in the context of non-stationarity due to climate change. Our example is drawn from hydrology, but the ideas for conveying risk under non-stationarity can be used for most any application, for instance precipitation or temperature data. Also, the

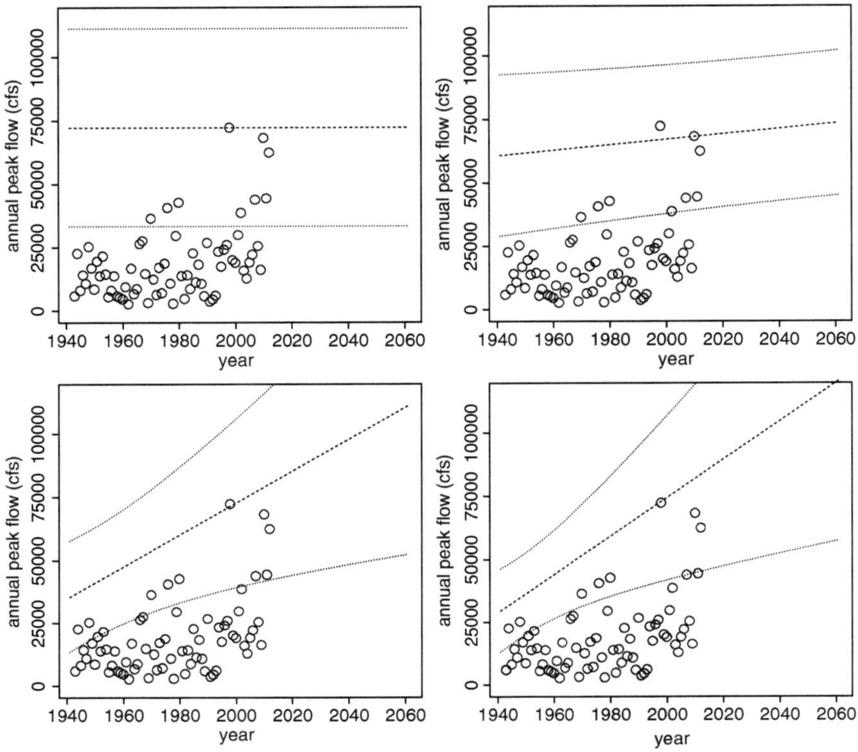

Fig. 4.3 Yearly 0.02 probability-of-exceedance levels for the stationary model (*top left*), time varying μ model (*top right*), time-varying σ model (*bottom left*), and time-varying μ and σ model (*bottom right*)

methods for conveying risk are not tied to fitting annual maximum data with a GEV distribution. Equations (4.3) and (4.7) require only estimates of $F_y(r_m)$, which could be obtained from fitting a GP or a log-Pearson type III distribution. Nor are the methods tied to the notion of return period being defined in terms of the annual maximum. Equations (4.3) and (4.7) could be adapted for use with the ARI rather than the AEP.

The Red River data example not only illustrates the different ways to communicate risk under non-stationarity, but also illustrates the questions that arise from fitting a non-stationary model and using it to convey future risk. A question that naturally arises regards extrapolating the fitted trend into the future. Figure 4.3 shows there is a significant effect on our risk measures based on which trend we choose. One of the basic lessons taught in an introductory regression course is the danger of extrapolating a model beyond the range of the data. Like it or not, estimating future risk requires extrapolation. All one can do is be clear in the assumptions made, and convey the uncertainty as best as one can.

Throughout, we have used a data-based approach. One can assess the fit of the four models by using tools such as the AIC, but we would be hesitant to select a model based *solely* on this or similar model selection criteria. One should always keep in mind that because extremes are rare, we are always data-poor when analyzing extremes, and thus it may be helpful to include outside expert knowledge into the model selection process.

The Red River example is interesting because, to our knowledge, the recent flood activity has not been directly linked to climate change. The Red River flows north, and floods most often occur in the spring when snow and ice begins melting in the south and the river remains frozen in the north. The recent floods do not seem to be caused by the exact same circumstances. The 1997 flood is largely attributed to snowmelt and extreme temperatures. The 2009 flood resulted from a combination of saturated and frozen ground, and snowmelt excerbated by rainfall. And the 2011 flood is at least partially blamed on high soil moisture in the previous year.

In part because we have used a data-driven approach, we have chosen our parameters to be linear functions of time. If extreme behavior is changing due to climate change, it is unlikely that they are behaving linearly. The recent mean temperature record, for example, does not appear to be increasing linearly but rather seems to have periods of rapid increase followed by plateaus. Still, linear trends may still be useful as models for long-term behavior and this is especially useful for extrapolation.

Another approach would be to use something other than time as the covariate in the model. For instance, one could imagine linking temperature data directly to CO_2 level rather than time. However, linking to a climatological covariate makes extrapolation into the future more difficult, as one would need to extrapolate the covariate as well. No obvious climatological covariate comes to mind for the Red River application.

There are other questions that arise with our particular application. The Red River was selected specifically because we were aware of recent flood activity. It is natural to ask if the non-stationarity we are seeing is due to selection bias and results from chance alone. Another question deals with the data itself. There is a well-documented flood in 1950, but the peak flow for this year is recorded as 18,700 cfs, a rather unremarkable level. If this data point is in error and the 1950 level is in fact much higher, this would drastically impact our non-stationary model.

Finally, one can ask which method of communicating risk is best? The yearly probability-of-exceedance level communicates most clearly the changing nature of risk, but is less useful for design criteria. The two return period definitions each have their advantages. The expected waiting time definition is more closely linked to a lifespan calculation than the expected number of events definition. However, the expected waiting time definition extrapolates the trend indefinitely, whereas the expected number of events definition only extrapolates over the m years used in the calculation.

Acknowledgements The author would like to thank Rick Katz, National Center for Atmospheric Research, for a helpful initial discussion, and for informing the author about important references. The author also thanks the editors for the invitation to write this chapter and an anonymous reviewer for helpful comments. The author acknowledges support from National Science Foundation grant DMS-0905315.

A Appendix

A.1 Expansion of (4.2)

Here we show how the expansion of (4.2) results in (4.3).

$$\begin{aligned}
E[T] &= \sum_{t=1}^{\infty} t \prod_{y=1}^{t-1} F_y(r)(1 - F_t(r)) \\
&= 1(1 - F_1(r)) + 2F_1(r)(1 - F_2(r)) + 3F_1(r)F_2(r)(1 - F_3(r)) + \\
&\qquad 4F_1(r)F_2(r)F_3(r)(1 - F_4(r)) \ldots \\
&= 1 - F_1(r) + 2F_1(r) - 2F_1(r)F_2(r) + 3F_1(r)F_2(r) - 3F_1(r)F_2(r)F_3(r) + \\
&\qquad 4F_1(r)F_2(r)F_3(r) - 4F_1(r)F_2(r)F_3(r)F_4(r) + \ldots \\
&= 1 + F_1(r) + F_1(r)F_2(r) + F_1(r)F_2(r)F_3(r) + F_1(r)F_2(r)F_3(r)F_4(r) + \ldots \\
&= 1 + \sum_{i=1}^{\infty} \prod_{y=1}^{i} F_y(r).
\end{aligned}$$

A.2 Implicit Delta Method

Let $\boldsymbol{\theta}$ be a d-dimensional parameter vector for a particular model, let $\hat{\boldsymbol{\theta}}$ be the maximum likelihood estimates, and let $V(\hat{\boldsymbol{\theta}})$ be the approximated covariance matrix of these estimates, obtained by inverting the hessian from the numerical optimization procedure.

The m-year return level is a function of the parameter vector: $r_m = f(\boldsymbol{\theta})$. Denote by ∇r_m the gradient of this function. The delta method says the variance of r_m is approximately

$$\nabla r_m^T V(\hat{\boldsymbol{\theta}}) \nabla r_m.$$

Unfortunately, we do not have explicit expressions for the function f for either of the m-year return level definitions in the non-stationary case. Rather, Eqs. (4.3) and (4.7) can be generalized as $m = g(\boldsymbol{\theta}, r_m)$. Let m_0 be our desired return period and let r_{m_0} denote the corresponding return level. Knowing g we are able to find $\partial m/\partial r_m$, and $\partial m/\partial \theta_i$ for $i = 1, \ldots, d$. Thus

$$\begin{aligned} \left.\frac{\partial r}{\partial \theta_i}\right|_{\theta=\hat{\theta},m=m_0} &= \left.\frac{\partial r}{\partial m}\right|_{\theta=\hat{\theta},m=m_0} \left.\frac{\partial m}{\partial \theta_i}\right|_{\theta=\hat{\theta},m=m_0} \\ &= \left(\left.\frac{\partial m}{\partial r}\right|_{\theta=\hat{\theta},r=r_{m_0}} \right)^{-1} \left.\frac{\partial m}{\partial \theta_i}\right|_{\theta=\hat{\theta},m=m_0}, \end{aligned}$$

which allows us to calculate the needed gradient.

References

Akaike H (1974) A new look at the statistical model identification. IEEE Trans Autom Control 19(6):716–723

Balkema A, De Haan L (1974) Residual life time at great age. Ann Probab 2(5):792–804

Beirlant J, Goegebeur Y, Segers J, Teugels J, Waal DD, Ferro C (2004) Statistics of extremes: theory and applications. Wiley, New York

Bonnin G, Todd D, Lin B, Parzybok T, Yekta M, Riley D (2004) NOAA Atlas 14, precipitation frequency Atlas of the United States, vol 1. U.S. Department of Commerce, National Oceanic and Atmospheric Administration, National Weather Service, Silver Spring

Casella G, Berger R (2002) Statistical inference, 2nd edn. Duxbury, Pacific Grove

Chavez-Demoulin V, Davison A (2005) Generalized additive models for sample extremes. J R Stat Soc Ser C (Appl Stat) 54(1):207–222

Coles SG (2001) An introduction to statistical modeling of extreme values. Springer series in statistics. Springer, London

Cooley D, Sain SR (2010) Spatial hierarchical modeling of precipiation extremes from a regional climate model. J Agric Biol Envrion Stat 15:381–402

Davison AC, Smith RL (1990) Models for exceedances over high thresholds. J R Stat Soc B 52:393–442

de Haan L, Ferreira A (2006) Extreme value theory. Springer series in operations research and financial engineering. Springer, New York

Fisher RA, Tippett LHC (1928) Limiting forms of the frequency distribution of the larges or smallest members of a sample. Proc Camb Philos Soc 24:180–190

Gnedenko B (1943) Sur la distribution limite du terme maximum d'une série aléatoire. Ann Math 44(3):423–453

IPCC (2007) In: Solomon S, Qin D, Manning M, Chen Z, Marquis M, Avery KB, Tignor M, Miller HL (eds) 2007a. Climate change 2007: the physical science basis. Contribution of working group I to the fourth assessment report of the intergovernmental panel on climate change. Cambridge, United Kingdom and New York, NY: Cambridge University Press, 996 p.

Katz R (2012) Statistical methods for nonstationary extremes. In: AghaKouchak A, Easterling D, Hsu K, Schubert S, Sorooshian S (eds) Extremes in a changing climate, Springer doi: 10.1007/978-94-007-4479-02

Leadbetter M, Lindgren G, Rootzén H (1983) Extremes and related properties of random sequences and processes. Springer, New York

Mays LW (2001) Water resources engineering. Wiley, New York

Olsen J, Lambert JH, Haimes Y (1998) Risk of extreme events under nonstationary conditions. Risk Anal 18:497–510

Parey S, Malek F, Laurent C, Dacunha-Castelle D (2007) Trends and climate evolution: statistical approach for very high temperatures in france. Clim chang 81(3):331–352

Parey S, Hoang TTH, Dacunha-Castelle D (2010) Different ways to compute temperature return levels in the climate change context. Environmetrics 21:698–718

Pickands J (1975) Statistical inference using extreme order statistics. Ann Stat 3:119–131

Resnick S (2007) Heavy-tail phenomena: probabilistic and statistical modeling. Springer series in operations research and financial engineering. Springer, New York

Schliep E, Cooley D, Sain SR, Hoeting JA (2010) A comparison study of extreme precipitation from six different regional climate models via spatial hierarchical modeling. Extremes 13:219–239

Smith RL (1989) Extreme value analysis of environmental time series: an application to trend detection in ground-level ozone. Stat Sci 4:367–393

Vogel R, Wilson I (1996) Probability distribution of annual maximum, mean, and minimum streamflows in the united states. J Hydrol Eng 1(2):69–76

Chapter 5
Multivariate Extreme Value Methods

Gianfausto Salvadori and Carlo De Michele

Abstract Multivariate extremes occur in several hydrologic and water resources problems. Despite their practical relevance, the real-life decision making as well as the number of designs based on an explicit treatment of multivariate variables is yet limited as compared to univariate analysis. A first problem arising when working in a multidimensional context is the lack of a "natural" definition of extreme values: essentially, this is due to the fact that different concepts of multivariate order and failure regions are possible. Also, in modeling multivariate extremes, central is the issue of dependence between the variables involved: again, several approaches are possible. A further practical problem is represented by the construction of multivariate Extreme Value models suitable for applications: the task is indeed difficult from a mathematical point of view. In addition, the calculation of multivariate Return Periods, quantiles, and design events, which represent quantities of utmost interest in applications, is rather tricky. In this Chapter we show how the use of Copulas may help in dealing with (and, possibly, solving) these problems.

5.1 Introduction

Multivariate extremes occur in several hydrologic and water resources problems including, among others: (1) flood and drought at different sites (Singh 1986; Pons 1992; Kim et al. 2003; Keef et al. 2009); (2) the precipitation dynamics (rain and

---

G. Salvadori (✉)
Dipartimento di Matematica e Fisica "E. De Giorgi", Università del Salento, Provinciale Lecce-Arnesano, P.O.Box 193, I-73100 Lecce, Italy
e-mail: gianfausto.salvadori@unisalento.it

C. De Michele
Department of Hydraulic, Environmental, Roads and Surveying Engineering, Politecnico di Milano, Piazza Leonardo da Vinci 32, I-20133 Milano, Italy
e-mail: carlo.demichele@polimi.it

A. AghaKouchak et al. (eds.), *Extremes in a Changing Climate*, Water Science and Technology Library 65, DOI 10.1007/978-94-007-4479-0_5,

snow) (Wilks 1998; Herr and Krzysztofowicz 2005); (3) the link between water quality and quantity in a river section (Grenney and Heyse 1985); (4) the hydraulic conductivity in porous media (Journel and Alabert 1988; Russo 2009). In recent years, the number of articles on multivariate methods in hydrology has exploded. However, despite the practical relevance of multivariate extremes, the real-life decision making, as well as the number of designs based on an explicit treatment of multivariate variables, are yet limited as compared to univariate analysis.

In applications, the notion of extreme event in a multidimensional context is closely related to that of *failure region* in structural design, as defined in Coles and Tawn (1994). Practically, the multivariate observation $\mathbf{x} = (x_1, \ldots, x_d)$ is extreme if it falls into some failure region $\mathscr{F} \subset \mathbf{R}^d$ having a "small" probability of being reached. For instance, $\mathscr{F}$ can be defined as

$$\mathscr{F} = \left\{ (x_1, \ldots, x_d) \in \mathbf{R}^d \colon \max_{1 \le i \le d} \{x_i\} > x^+ \right\}$$

or

$$\mathscr{F} = \left\{ (x_1, \ldots, x_d) \in \mathbf{R}^d \colon \min_{1 \le i \le d} \{x_i\} < x^- \right\},$$

for given large, or small, thresholds x^+ and x^-.

A first problem arising when working in a multidimensional context is the lack of a "natural" definition of extreme values: essentially, this is due to the fact that different concepts of multivariate order and failure regions are possible (see later). In addition, both model validation and computation may be difficult in multidimensional spaces, and models are less fully prescribed by the general theory. Furthermore, in two and more dimensions, there does not exist a simple distinction in three basic Extreme Values domains of attraction as in the univariate case (namely, Type I, II, and III laws): in fact, there is no reason for the univariate marginals of a multivariate distribution to share the same type of limiting Extreme Value probability law.

In modeling multivariate extremes, central is the issue of *association* between the variables involved. Actually, the degree of association may change by considering different "intensities" of the process under investigation: for instance, it may become weaker for more extreme events, to the extent that the most extreme ones are practically independent. In addition the notion of association can involve two or more dimensions: in the former case we refer to a pair-wise association, while in the latter one to a cluster association. We shall see later how to properly quantify the multivariate degree of association.

Another important issue is represented by the construction of multivariate Extreme Value models. In the past, the joint probability distributions were often assumed to be Multinormal (or its extensions, such as the multivariate Lognormal, t-Student's, and Fisher's distributions—see, e.g., Alexander (1954), Stedinger (1983), Hosking and Wallis (1988), Pons (1992) and Kottegoda and Natale (1994)), or multivariate logistic Gumbel (see, e.g., Raynal-Villasenor and Salas 1987; Bacchi et al. 1994; Yue 2000). Unfortunately, quite a few multivariate distributions present in literature, which are direct extensions of well known univariate ones, suffer from several limitations and constraints (see later). Only recently, new multivariate Extreme Value models were introduced to overcome some of these limitations (see below, and also Tawn 1990; Coles and Tawn 1991; Joe 1994; Rootz and Tajvidi 2006; Durante and Salvadori 2010; Salvadori et al. 2010).

A further fundamental question is represented by the estimate of the mean occurrence (viz., the Return Period) of a multivariate event. While the theory is clear in the univariate case, it requires some care in the multivariate one, where different definitions are possible (see below, and also Zhang (2005) and Singh et al. (2007), and references therein). Similarly, the absence of a "natural" order in multidimensional Euclidean spaces allows for the introduction of various definitions of multivariate quantile (see below, and also Serfling 2002; Chebana and Ouarda 2009a,b, 2011).

This chapter is articulated as follows. In Sect. 5.2 we introduce some basic notions concerning the concepts of Copulas, multivariate measures of dependence, asymptotic dependence, and the simulation of multivariate probability distributions via copulas (a tool of great importance in applications). In Sect. 5.3 we introduce the Extreme Value copulas, as well as special techniques for constructing multivariate Extreme Value models of practical utility. In Sect. 5.4 we address the problems regarding the calculation of the multivariate Return Periods, quantiles, and design events, which represent quantities of utmost interest in applications. Finally, in Sect. 5.5 we outline some open problems, and discuss the perspectives of the research in the area of multivariate extremes.

5.2 Copulas

The use of copulas in environmental sciences is recent and rapidly growing. For a thorough theoretical introduction to copulas see Joe (1997) and Nelsen (2006); for a practical approach see Salvadori et al. (2007) and Jaworski et al. (2010). In order to avoid troublesome situations, hereinafter we shall assume that the multivariate distribution $\mathbf{F}$ is continuous (but not necessarily absolutely continuous), and strictly increasing in each marginal: these regularity constraints are rather weak, and satisfied by the majority of the laws used in applications. Clearly, pathological cases can also be addressed, but they require suitable techniques that go beyond the scope of this short introduction.

For the sake of clarity and simplicity, hereinafter we shall mainly deal with the bivariate case: a thorough practical introduction can be found in Genest and Favre (2007). However, extensions to the general d-dimensional case ($d > 2$) are usually possible, and easy to derive from the point of view of probability theory, although the statistical side of things may become far more challenging as d gets larger and larger, and is still an open research area of mathematical statistics (this includes the construction of flexible parametric families of high-dimensional copulas, estimation procedures, simulation algorithms, etc.). The main result is represented by Sklar's Theorem (1959)—see Theorem 1 below, which splits any multivariate distribution into two different components: the univariate marginals, and a global dependence function (i.e., the Copula) ruling the joint behavior of the different variables involved.

An advantage of using copulas is that even complex marginal distributions, such as *finite mixtures* (Titterington et al. 1985), can be applied easily to the model of interest (for practical examples see, e.g., Favre et al. (2004) and De Michele et al. (2007)). Incidentally, we stress that all the multivariate distributions currently used in applications can be described (and generalized) in a straightforward manner in terms of suitable copulas. In addition, the copula approach does not suffer from several drawbacks of standard families of multivariate distributions (e.g., the multivariate Gaussian, Student, etc.), such as, for instance, the facts that the marginals may belong to the same probability family, or the parameters of the marginals also rule the dependence between the variables considered. For an exhaustive list of models see, e.g., Hutchinson and Lai (1990), Joe (1997), Nelsen (2006) and Salvadori et al. (2007), and references therein. Clearly, a note of caution is necessary here: using a copula corresponds to making a parametric assumption on the nature of dependence, and this may have important consequences in terms of, e.g., prediction and extrapolation (an issue that is sometimes overlooked in practical applications).

5.2.1 Basic Features

In this Section we outline briefly the mathematics of 2-copulas needed in the sequel. All the theoretical justifications can be found in Joe (1997), Nelsen (2006) and Salvadori et al. (2007). Hereinafter we shall write "i.i.d." for independent identically distributed, and "r.v." for random variable or vector.

Definition 1 (2-Copula). Let $\mathbf{I} = [0, 1]$. A 2-copula is a bivariate function $\mathbf{C}: \mathbf{I} \times \mathbf{I} \to \mathbf{I}$ such that:

1. (Uniform marginals) for all $u, v \in \mathbf{I}$,

$$\mathbf{C}(u, 0) = 0, \quad \mathbf{C}(u, 1) = u, \quad \mathbf{C}(0, v) = 0, \quad \mathbf{C}(1, v) = v; \tag{5.1}$$

2. (2-increasing) for all $u_1, u_2, v_1, v_2 \in \mathbf{I}$ such that $u_1 \le u_2$ and $v_1 \le v_2$,

$$\mathbf{C}(u_2, v_2) - \mathbf{C}(u_2, v_1) - \mathbf{C}(u_1, v_2) + \mathbf{C}(u_1, v_1) \ge 0. \tag{5.2}$$

A 2-copula $\mathbf{C}$ is uniformly continuous on its domain. In applications, it is often quite useful to consider 2-copulas as restrictions to $\mathbf{I}^2$ of joint distribution functions whose marginals are Uniform laws on $\mathbf{I}$. The following definition is a natural consequence of this fact.

Definition 2. A 2-copula $\mathbf{C}$ induces a probability measure $\mu_{\mathbf{C}}$ on $\mathbf{I}^2$, called $\mathbf{C}$-measure, given by

$$\mu_{\mathbf{C}}\{[0,u]\times[0,v]\} = \mathbf{C}(u,v)\,. \tag{5.3}$$

Often $\mathbf{C}$-measures are also called doubly stochastic measures.

Intuitively, the $\mathbf{C}$-measure $\mu_{\mathbf{C}}$ of a (measurable) subset A of $\mathbf{I}^2$ is the probability that two r.v.'s (U, V) Uniform on $\mathbf{I}$, and having joint distribution function $\mathbf{C}$, take values in A. Practically, $\mu_{\mathbf{C}}\{A\}$ is the probability that a simulation of the pair (U, V) belongs to A.

The link between 2-copulas and bivariate distributions is provided by the following fundamental result (Sklar 1959). Henceforth F_X, F_Y (respectively, F_U, F_V) will denote the marginal distribution functions of the r.v.'s X, Y (respectively, U, V), and Ran their Range.

Theorem 1 (Sklar's Theorem). *Let F_{XY} be a joint distribution function with marginals F_X and F_Y. Then there exists a 2-copula $\mathbf{C}$ such that*

$$F_{XY}(x,y) = \mathbf{C}\left(F_X(x), F_Y(y)\right) \tag{5.4}$$

for all reals x, y. If F_X, F_Y are continuous, then $\mathbf{C}$ is unique; otherwise, $\mathbf{C}$ is uniquely defined on $Ran(F_X)\times Ran(F_Y)$.

Conversely, if $\mathbf{C}$ is a 2-copula and F_X, F_Y are distribution functions, then the function F_{XY} given by Eq. (5.4) is a joint distribution with marginals F_X and F_Y.

Note 1. As anticipated above, although no theoretical constraints exist on the choice of F_X and F_Y, for the sake of simplicity we shall limit our investigation to *continuous strictly increasing* marginals: this is the case of most interest in applications. As a consequence, by virtue of Theorem 1, the copula representation will always be unique. However, only minor changes (involving the use of suitable *quasi-inverses*, written as $F_\bullet^{(-1)}$ or $F_\bullet^{[-1]}$—see Nelsen (2006)) are required in case F_X, F_Y do not satisfy such an assumption: for instance, this may be the case of daily rainfall, with a probability mass at zero, or integer-valued variables like occurrence (or number) of floods at several locations. In this Chapter the quasi-inverse will coincide with the ordinary inverse, i.e. the quantile function.

The following "inversion" result plays an important role in applications.

Corollary 1 (Sklar's Inversion). *Let $\mathbf{C}$, F_{XY}, and F_X, F_Y be as in Theorem 1, and suppose that F_X, F_Y are continuous. If $F_X^{(-1)}$ and $F_Y^{(-1)}$ denote, respectively, the quasi-inverses of F_X and F_Y, then*

$$\mathbf{C}(u,v) = F_{XY}\left(F_X^{(-1)}(u), F_Y^{(-1)}(v)\right) \tag{5.5}$$

for any $(u,v)\in\mathbf{I}^2$.

As a matter of fact, Theorem 1 (as well as Corollary 1) has a natural extension to a d-dimensional framework ($d > 2$) as follows:

$$F_{X_1,\dots,X_d}(x_1,\dots,x_d) = \mathbf{C}\left(F_{X_1}(x_1),\dots,F_{X_d}(x_d)\right) \tag{5.6}$$

for all reals $x_1,\dots,x_d$ (and similarly for Corollary 1). Here the function $\mathbf{C}$ (i.e., the d-copula) is only required to be a d-dimensional probability distribution on the unit hyper-cube $\mathbf{I}^d$, with Uniform marginals over $\mathbf{I}$. The following example shows the properties of three fundamental 2-copulas.

Example 1. Three special 2-copulas deserve a particular attention, since they provide the so-called *Fréchet-Hoeffding bounds*.

1. The *Fréchet-Hoeffding lower bound* $\mathbf{W}_2$ given by

$$\mathbf{W}_2(u,v) = \max\{u+v-1,0\}. \tag{5.7}$$

2. The *Fréchet-Hoeffding upper bound* $\mathbf{M}_2$ given by

$$\mathbf{M}_2(u,v) = \min\{u,v\}. \tag{5.8}$$

3. The *independence* (or *product*) 2-copula Π_2 given by

$$\Pi_2(u,v) = uv. \tag{5.9}$$

A family of 2-copulas which includes $\mathbf{W}_2$, $\mathbf{M}_2$ and Π_2 is called *comprehensive*. The copulas $\mathbf{W}_2$ and $\mathbf{M}_2$ provide general bounds, since for any 2-copula $\mathbf{C}$ and any pair $(u,v) \in \mathbf{I}^2$

$$\mathbf{W}_2(u,v) \le \mathbf{C}\,(u,v) \le \mathbf{M}_2(u,v). \tag{5.10}$$

When X and Y are continuous r.v.'s, the following characterization holds.

1. The variable Y is almost surely a strictly decreasing function of X if, and only if, $\mathbf{C}_{XY} = \mathbf{W}_2$. In particular, $\mathbf{W}_2$ is the distribution of the random vector $(U, 1-U)$, where U is Uniform on $\mathbf{I}$. Random variables with copula $\mathbf{W}_2$ are often called *counter-monotonic*.
2. The variable Y is almost surely a strictly increasing function of X if, and only if, $\mathbf{C}_{XY} = \mathbf{M}_2$. In particular, $\mathbf{M}_2$ is the distribution of the random vector (U, U), where U is Uniform on $\mathbf{I}$. Random variables with copula $\mathbf{M}_2$ are often called *co-monotonic*.
3. The 2-copula Π_2 describes the absence of dependence between X and Y: in fact, if $F_{XY} = \Pi_2(F_X, F_Y)$, then $F_{XY} = F_X F_Y$. Therefore, X and Y are independent if, and only if, their 2-copula is Π_2.

While both $\mathbf{M}_2$ and Π_2 can easily be extended to a multidimensional framework by considering the d-copulas ($d > 2$)

$$\mathbf{M}_d(u_1, \ldots, u_d) = \min\{u_1, \ldots, u_d\} \tag{5.11}$$

and

$$\Pi_d(u_1, \ldots, u_d) = u_1 \cdots u_d, \tag{5.12}$$

the function $\mathbf{W}_2$ does not have a multivariate copula extension: indeed, for a consistency problem, it is impossible to have more than two variables which are decreasing functions of the others.

A fundamental feature of copulas is that they are invariant under strictly increasing transformations, as stated by the following result in the bivariate case.

Proposition 1 (Invariance). *Let X and Y be continuous r.v.'s with 2-copula $\mathbf{C}_{XY}$. Also let α and β be strictly increasing functions on, respectively, Ran(X) and Ran(Y), and set $A = \alpha(X)$ and $B = \beta(Y)$. Then*

$$\mathbf{C}_{AB} = \mathbf{C}_{\alpha(X)\beta(Y)} = \mathbf{C}_{XY}. \tag{5.13}$$

The following important properties hold for 2-copulas.

Proposition 2. *Let X and Y be continuous r.v.'s with 2-copula $\mathbf{C}_{XY}$, and let α and β be strictly monotonic functions on, respectively, Ran(X) and Ran(Y).*

1. *If α is increasing and β decreasing then*

$$\mathbf{C}_{\alpha(X)\beta(Y)}(u, v) = u - \mathbf{C}_{XY}(u, 1 - v). \tag{5.14}$$

2. *If α is decreasing and β increasing then*

$$\mathbf{C}_{\alpha(X)\beta(Y)}(u, v) = v - \mathbf{C}_{XY}(1 - u, v). \tag{5.15}$$

3. *If α and β are both decreasing then*

$$\mathbf{C}_{\alpha(X)\beta(Y)}(u, v) = u + v - 1 + \mathbf{C}_{XY}(1 - u, 1 - v). \tag{5.16}$$

Given these invariance properties, and using the *Probability Integral Transform*, we may restrict our attention to the pair of r.v.'s (U, V) given by

$$(U = F_X(X), V = F_Y(Y)) \iff (X = F_X^{(-1)}(U), Y = F_Y^{(-1)}(V)) \tag{5.17}$$

where $F_X^{(-1)}, F_Y^{(-1)}$ are the *quasi-inverses* of the corresponding distribution functions. Clearly, U and V are Uniform on $\mathbf{I}$, i.e. $U \sim \mathrm{U}(0, 1)$ and $V \sim \mathrm{U}(0, 1)$, and (U, V) has the same 2-copula as the pair (X, Y), i.e. $(U, V) \sim \mathbf{C}_{UV} = \mathbf{C}_{XY}$. This may turn the original problem into a marginal-independent one, which may be less difficult to solve. The extension of Eq. (5.17) to the general d-dimensional case $(d > 2)$ is obvious.

Interestingly enough, any finite *convex linear combination* of 2-copulas $\mathbf{C}_i$'s is itself a 2-copula. In fact, for $k \in \mathbf{N}$, let $\mathbf{C}$ be given by

$$\mathbf{C}(u,v) = \sum_{i=1}^{k} \lambda_i \, \mathbf{C}_i(u,v), \tag{5.18}$$

where $\lambda_i \geq 0$ for all indices, and $\sum_{i=1}^{k} \lambda_i = 1$. Then $\mathbf{C}$ is a proper 2-copula.

This can be made more general, e.g. by considering d-dimensional copulas, and extended to the case of a continuous mixing parameter as follows. Let $\{\mathbf{C}_\lambda\}$ be an infinite collection of copulas indexed by a continuous parameter $\lambda \in \mathbf{R}$. Now, suppose that λ is the observation of a continuous r.v. with cumulative distribution function L. Then, setting

$$\mathbf{C}(u,v) = \int_{\mathbf{R}} \mathbf{C}_\lambda(u,v)\, \mathrm{d}L(\lambda), \tag{5.19}$$

it can be shown that $\mathbf{C}$ is a copula. Usually, L is referred to as the *mixing distribution* of the family $\{\mathbf{C}_\lambda\}$, and $\mathbf{C}$ is called the *convex sum* of $\{\mathbf{C}_\lambda\}$ with respect to L.

In the univariate case, the *empirical distribution function* represents a useful tool for investigating the behavior of the variables of interest. Similarly, the *empirical copula* (Deheuvels 1979; Genest and Favre 2007) defined below provides valuable information about the joint behavior of pairs of r.v.'s associated via a 2-copula $\mathbf{C}$.

Definition 3 (Empirical Copula). Let $\{(R_k, S_k)\}$ be the ranks associated with the sample $\{(X_k, Y_k)\}$, $k = 1, \dots, n$. The corresponding empirical copula $\mathbf{C}_n$ is defined as

$$\mathbf{C}_n(u,v) = \frac{1}{n} \sum_{k=1}^{n} \mathbf{1}\left(\frac{R_k}{n+1} \leq u, \frac{S_k}{n+1} \leq v \right), \tag{5.20}$$

where $u, v \in \mathbf{I}$ and $\mathbf{1}$ is an indicator function. The extension to the general d-dimensional case ($d > 2$) is obvious.

As in the univariate case, the empirical copula practically counts the number of pairs that satisfy given constraints, in order to provide an approximation of the copula linking the pair (X, Y)—incidentally, this may provide preliminary hints for performing visual fitting and model choice. Most importantly, the construction of the empirical copula is non-parametric, since only the ranks of the data are involved. Actually, the use of ranks in copula analysis is quite common (Genest and Favre 2007)—see also Sect. 5.2.2.

Example 2. As an illustration, we consider here the case study investigated in De Michele et al. (2005). The data are given by the pairs (Q, V), where the maximum annual flood peaks Q and volumes V are collected at the Ceppo Morelli dam (Northern Italy) from 1937 to 1994 (for a total of 49 observations). As a result of a thorough investigation, almost all of the occurrence dates of the Q's and the V's

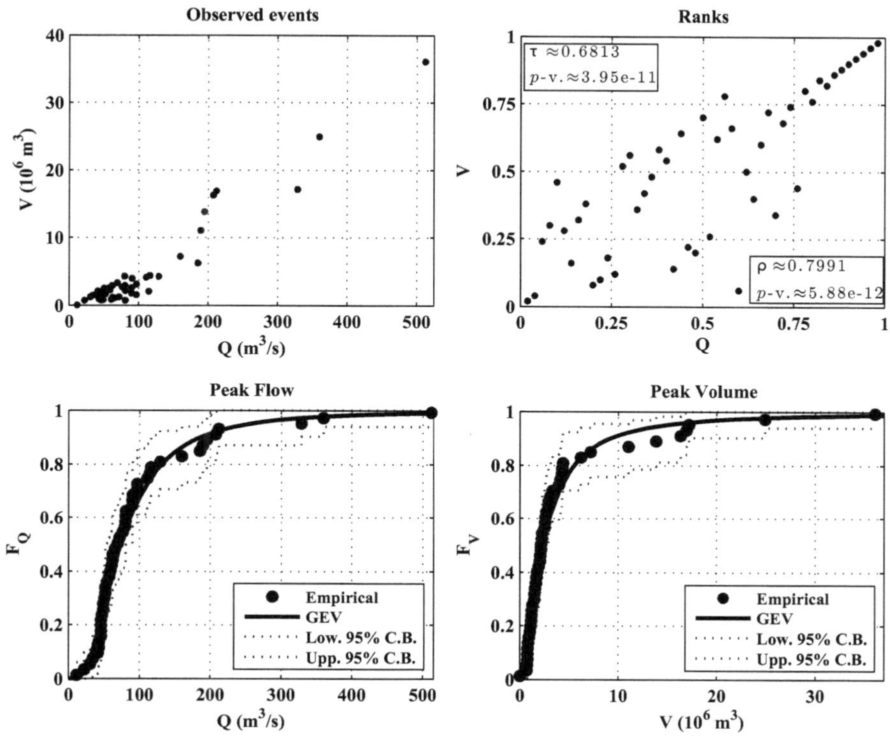

Fig. 5.1 Plot of the available (Q, V) observations, and fits of the marginal distributions—see text. Also shown is the rank-plot of (Q, V), as well as the estimates of the Kendall's τ and the Spearman's ρ, and the corresponding p-values (derived from non-parametric tests of independence based on rank statistics)

were the same: viz., they happened during the same flood event. Concerning the univariate analysis of the marginals, both F_Q and F_V are fitted by means of GEV distributions. In Fig. 5.1 we plot the available observations (Q, V), and the marginal fits (including 95 % confidence bands): the results are valuable, since standard Goodness-of-Fit tests (namely, Kolmogorov-Smirnov and Anderson-Darling) are passed at all usual levels (viz., 1 %, 5 %, and 10 %).

The very first step of the copula analysis concerns the non-parametric study of the joint behavior of the variables (Q, V): traditionally, the study is carried out (Genest and Favre 2007) by showing the ranks of the data normalized into the interval (0, 1). The bivariate rank-plot shown in Fig. 5.1 provides some rough indications about the global dependence structure (i.e., the copula) linking the two variables (Q, V). Note that this graph is not equivalent to the simple bivariate plot of the original pairs of interest shown in Fig. 5.1, since it is not affected by the behavior of the marginals F_Q and F_V.

Fitting copulas to empirical observations is still an open problem in Statistics, and several *Goodness-of-Fit* procedures have recently been proposed. Apparently, at present the most feasible, practical, and realistic solution is represented by multivariate Goodness-of-Fit tests based on the empirical copula process (Genest et al. 2009; Berg 2009; Kojadinovic et al. 2011). However, the research is in progress, and further methods are expected in the near future.

Another notion of interest is represented by the *Kendall's measure function* $K_{\mathbf{C}}$ Genest and Rivest (1993) and Genest and Rivest (2001) given by

$$K_{\mathbf{C}}(t) = \mathbf{P}\{W \leq t\} = \mathbf{P}\{\mathbf{C}(U_1, \ldots, U_d) \leq t\}, \tag{5.21}$$

where $t \in \mathbf{I}$ is a probability level, $W = \mathbf{C}(U_1, \ldots, U_d)$ is a univariate r.v. taking value on $\mathbf{I}$, and the U_i's are Uniform r.v.s on $\mathbf{I}$ with copula $\mathbf{C}$. In the bivariate Extreme Value case, $K_{\mathbf{C}}$ is given by Ghoudi et al. (1998)

$$K_{\mathbf{C}}(t) = t - (1 - \tau_{\mathbf{C}})\, t \, \ln t, \tag{5.22}$$

where $\tau_{\mathbf{C}}$ is the value of the Kendall's τ associated with the copula $\mathbf{C}$ (see Sect. 5.2.2). Clearly, bivariate Extreme Value copulas with the same value of τ share the same function $K_{\mathbf{C}}$.

Unfortunately, at present no useful expressions similar to Eq. (5.22) are known for the general multivariate case $d > 2$, and it is necessary to make recourse to simulations. For the sake of convenience, we report below the algorithm explained in Salvadori et al. (2010) for the calculation of $K_{\mathbf{C}}$ (see also Genest and Rivest 1993), which yields a consistent Maximum-Likelihood estimator of $K_{\mathbf{C}}$. Here we assume that the copula model is well specified, i.e. it is available in a parametric form.

• ALGORITHM 1.

1. Generate a sample $\mathbf{u}_1, \ldots, \mathbf{u}_m$ from copula $\mathbf{C}$.
2. For $i = 1, \ldots, m$ calculate $v_i = \mathbf{C}(\mathbf{u}_i)$.
3. For $t \in \mathbf{I}$ estimate $\widehat{K}_{\mathbf{C}}(t) = \frac{1}{m} \sum_{i=1}^{m} \mathbf{1}(v_i \leq t)$.

The Kendall's measure $K_{\mathbf{C}}$ is a fundamental tool for introducing a mathematically consistent (copula-based) definition of the return period for multivariate events (see also the discussion in Salvadori (2004), Salvadori et al. (2004), Salvadori et al. (2007), Durante and Salvadori (2010) and Salvadori et al. (2010)). In fact, Eq. (5.21) represents a multivariate quantile relationship, since it corresponds to a multidimensional Probability Integral Transform (Genest et al. 2006). Since empirical estimators of $K_{\mathbf{C}}$ are available (Genest et al. 2009) (simply use Algorithm 1, where the sample $\mathbf{u}_1, \ldots, \mathbf{u}_m$ in Step 1 is the one given by the available observations), we shall see later how to use the Kendall's measure function to perform return period analyses of practical utility. As an illustration, in Fig. 5.4a we plot the function $K_{\mathbf{C}}$ associated with the copula $\mathbf{C}_{QV}$, as well as its empirical estimate—see the discussion in Sect. 5.4.1.

5.2.2 Multivariate Association Measures

The problem of measuring the amount of association between the variables involved is a central issue when modeling multivariate phenomena. For instance, in literature the pair-wise dependence is generally measured via the canonical Pearson's correlation coefficient. However, it may not be the best measure of dependence when dealing with extremes (Joe 1997), since it does not exist for heavy-tailed variables with infinite variance, and only involves a linear kind of dependence. Indeed, other quantities can be considered (Nelsen 2006) to measure the association between pairs of r.v.'s: among others, Kendall's τ and Spearman's ρ rank correlation coefficients, or the Blomqvist's β medial correlation coefficient. These measures always exist (being based on the ranks), and model several types of association (for a practical discussion see, e.g., the case studies illustrated in Salvadori et al. (2007)).

A numerical *measure of association* is a statistical summary of the degree of relationship between variables. For the ease of comparison, coefficients of association are usually constructed to vary between -1 and $+1$. Their absolute values increase as the strength of the relationship increases, with a $+1$ (or -1) value when there is perfect positive (or negative) association. Each coefficient of association measures a special type of relationship: for instance, Pearson's product-moment correlation coefficient measures the amount of linear relationship. Most importantly, some of these measures are *scale-invariant*, i.e. they remain unchanged under strictly increasing transformations of the variables of interest: thus, they are expressible in terms of the copula linking these variables (Schweizer and Wolff 1981).

The most widely known (and used), scale-invariant, measures of association are the *Kendall's* τ and the *Spearman's* ρ, both of which measure a form of dependence known as *concordance*. These two measures also play an important role in applications, since the practical fit of a copula to the available data is often carried out via the estimate of τ or ρ (Genest and Favre 2007) (as in the dam case study used in this work).

Mathematically speaking (Scarsini 1984), τ and ρ measure the *concordance* between two r.v.'s: roughly, X and Y are *concordant* if small values of one are likely to be associated with small values of the other, and large values of one are likely to be associated with large values of the other. More precisely, let (x_i, y_i) and (x_j, y_j) be two observations from a vector (X, Y) of continuous r.v.'s. Then, the pairs (x_i, y_i) and (x_j, y_j) are concordant if

$$(x_i - x_j)(y_i - y_j) > 0, \tag{5.23}$$

and discordant if

$$(x_i - x_j)(y_i - y_j) < 0. \tag{5.24}$$

The population version of Kendall's τ (Kendall 1937; Kruskal 1958; Rohatgi 1976; Nelsen 2006) is defined as the difference between the probability of concordance and the probability of discordance, which gives it a practical meaning.

Definition 4 (Kendall's τ). Let (X_1, Y_1) and (X_2, Y_2) be i.i.d. vectors of continuous r.v.'s. The difference

$$\tau = \mathbf{P}\{(X_1 - X_2)(Y_1 - Y_2) > 0\} - \mathbf{P}\{(X_1 - X_2)(Y_1 - Y_2) < 0\} \tag{5.25}$$

defines the population version of Kendall's τ.

As shown in Nelsen (2006), if $\mathbf{C}$ is the copula of X and Y, then

$$\tau = \int_{\mathbf{I}^2} \mathbf{C}(u, v)\, d\mathbf{C}(u, v) - 1 = 4\,\mathbf{E}\big(\mathbf{C}(U, V)\big) - 1. \tag{5.26}$$

The sample version t of Kendall's τ is easy to calculate:

$$t = \frac{c - d}{c + d}, \tag{5.27}$$

where c (d) represent the number of concordant (discordant) pairs in a sample of size n from a vector of continuous r.v.'s (X, Y). Unfortunately, in applications it often happens that continuous variables are "discretely" sampled, due to a finite instrumental resolution. For instance, the rainfall depth could be returned as an integer multiple of 0.1 mm, or the storm duration could be expressed in hours and rounded to an integer value. Clearly, this procedure introduces repetitions in the observed values (called *ties* in statistics), which may adversely affect the estimation of τ. However, corrections to Eq. (5.27) are specified for solving the problem (see, e.g., the formulas in Press et al. (1992)).

As with Kendall's τ, also the population version of Spearman's ρ (Kruskal 1958; Rohatgi 1976; Nelsen 2006) is based on concordance and discordance.

Definition 5 (Spearman's ρ). Let (X_1, Y_1), (X_2, Y_2) and (X_3, Y_3) be i.i.d. vectors of continuous r.v.'s. The difference

$$\rho = 3\left(\mathbf{P}\{(X_1 - X_2)(Y_1 - Y_3) > 0\} - \mathbf{P}\{(X_1 - X_2)(Y_1 - Y_3) < 0\}\right) \tag{5.28}$$

defines the population version of Spearman's ρ.

Also Spearman's ρ can be written in terms of a suitable expectation (Nelsen 2006):

$$\rho = 12 \int_{\mathbf{I}^2} \left[\mathbf{C}(u, v) - uv\right] du dv. \tag{5.29}$$

Thus, ρ is proportional to the signed volume between the graphs of $\mathbf{C}$ and the independence copula Π_2. Roughly, ρ measures the "average distance" between the joint distribution of X and Y (as represented by $\mathbf{C}$), and independence (given by Π_2).

The sample version r of Spearman's ρ is easy to calculate:

$$r = 1 - \frac{6 \sum_{i=1}^{n} (R_i - S_i)^2}{n^3 - n}, \tag{5.30}$$

where $R_i = \text{Rank}(X_i)$, $S_i = \text{Rank}(Y_i)$, and n is the sample size. As already mentioned before, instrumental limitations may adversely affect the estimation of ρ in practice due to the presence of ties. However, corrections to Eq. (5.30) are specified for solving the problem (see, e.g., the formulas in Press et al. (1992)).

A further measure of association based on concordance is given by Blomqvist's medial correlation coefficient β (Blomqvist 1950).

Definition 6 (Blomqvist's $\boldsymbol{\beta}$). Let (X, Y) be a r.v., and let $\tilde{x}, \tilde{y}$ be, respectively, the medians of X and Y. The difference

$$\beta = \mathbf{P}\{(X - \tilde{x})(Y - \tilde{y}) > 0\} - \mathbf{P}\{(X - \tilde{x})(Y - \tilde{y}) < 0\} \tag{5.31}$$

defines the population version of Blomqvist's medial correlation coefficient β.

The point is that, if $\mathbf{C}$ is the copula of (X, Y), then β can be written as Nelsen (2006)

$$\beta = 4\,\mathbf{C}\,(1/2, 1/2) - 1, \tag{5.32}$$

which may be easy to estimate and calculate.

Example 3. As an illustration, consider the case study presented in Example 2. The ranks plotted in Fig. 5.1 show that the variables (Q, V) are significantly positively associated (i.e., concordant), and thus Q and V are not independent: the estimates of both the Kendall's τ and the Spearman's ρ are large, and the corresponding p-values are negligible (see the values reported in Fig. 5.1).

The more general notion of cluster-type dependence, when the size of the cluster is larger than two (i.e., beyond the simple pair-wise case), has only been partially explored. Generalizations of Kendall's τ (Nelsen 1996), Spearman's ρ (Schmid and Schmidt 2007a,c), and Blomqvist's medial correlation coefficient β (see Durante et al. (2007) and Schmid and Schmidt (2007b), and references therein) to the d-variate case ($d > 2$) were only recently introduced. These extensions may be of practical importance: on the one hand, they provide useful tools to quantify the dependence within clusters; on the other hand, they can be used to estimate the parameters of the multivariate model at play. However, at present the application of these measures in actual case studies is still quite limited: for an illustration, see Salvadori and De Michele (2011). As a general remark, it must be stressed that it may be difficult (if not impossible or meaningless) to summarize all the dependencies of the variables in cluster by a single number. A short summary of the generalized association measures involving a generic d-copula $\mathbf{C}$ is shown below.

$$\tau_d = \frac{1}{2^{d-1}-1}\left(2^d \int_{\mathbf{I}^d} \mathbf{C}(\mathbf{u})\, d\mathbf{C}(\mathbf{u}) - 1\right), \tag{5.33}$$

$$\rho_{d,1} = h(d)\left(2^d \int_{\mathbf{I}^d} \mathbf{C}(\mathbf{u})\, d\mathbf{u} - 1\right), \tag{5.34}$$

$$\rho_{d,2} = h(d)\left(2^d \int_{\mathbf{I}^d} \Pi_d(\mathbf{u})\, d\mathbf{C}(\mathbf{u}) - 1\right), \tag{5.35}$$

$$\rho_{d,3} = h(2)\left(2^2 \sum_{i<j} \binom{d}{2}^{-1} \int_{\mathbf{I}^2} \mathbf{C}_{ij}(u,v)\, du\, dv - 1\right), \tag{5.36}$$

where $h(d) = (d+1)/(2^d - (d+1))$ is a normalizing function, and $\mathbf{C}_{ij}$ is the bivariate (i,j)-margin of $\mathbf{C}$. Note that $\rho_{d,3}$ is essentially the average Spearman's ρ for all the pairs in a set of d variables. Concerning the Blomqvist's medial correlation coefficient, a possible extension is as follows:

$$\beta_d = \frac{2^{d-1}\left(\mathbf{C}(\mathbf{1/2}) + \overline{\mathbf{C}}(\mathbf{1/2})\right) - 1}{2^{d-1}-1}, \tag{5.37}$$

where $\overline{\mathbf{C}}$ is the survival function associated with $\mathbf{C}$, given by $\overline{\mathbf{C}}(\mathbf{u}) = \mathbf{P}\{\mathbf{C} > \mathbf{u}\}$, and $\mathbf{1/2} = (1/2, \ldots, 1/2)$. Clearly, also β_d is invariant with respect to the distributions of the margins. As pointed out in Schmid and Schmidt (2007b), β_d has some advantages over competing measures such as τ_d or $\rho_{d,i}$'s. In fact, it can explicitly be derived whenever the copula is of explicit form, which is often not possible for other measures, and its estimation requires a low computational complexity. Thus, β_d may represent a fast alternative for estimating the copula parameters.

5.2.3 Asymptotic Dependence

Besides the problem of measuring the amount of association between the variables at play, a further central theme is represented by the way of quantifying the dependence between them. In fact, two situations are possible (Coles et al. 1999): one where, in a limiting sense, the extremes are dependent; the other where, in the same sense, the extremes are independent.

In the context of extrapolation in multivariate frequency analysis, it may be of great importance to be able to model the possible dependence of the extremes, i.e. the *tail dependence*. This quantity is a fundamental ingredient in order to estimate the risk adequately. As will be shown shortly, tail dependence is essentially a characteristic of the copula underlying a random vector. In turn, tail dependence, which relates to dependencies of extreme events, can be considered as a "scale-invariant" dependence measure, since the copula separates the dependence structure of multivariate distributions from its marginals. Clearly, the notion of tail dependence may provide useful indications for choosing a suitable family of copulas for modeling a given phenomenon (Poulin et al. 2007). Below we shall only consider the bivariate case: generalizations to the multivariate case can be found in Charpentier (2006), Frahm (2006) and Schmid and Schmidt (2007c).

The notion of tail dependence (Joe 1993) for bivariate distributions relates to the amount of dependence in the upper-right-quadrant tail or lower-left-quadrant tail. Usually it is measured via the *tail dependence coefficients*, introduced by Sibuya (1960). They reflect the limiting proportion of exceedance of one marginal over a quantile of a certain level, given that the other marginal has exceeded the same quantile.

We now present one of the possible definitions of tail dependence (Joe 1997). A thorough exposition can be found in Schmidt (2003); for a survey on various estimators of tail dependence coefficients within a parametric, semiparametric, and non-parametric framework see Frahm et al. (2005). A survey on other measures of tail dependence can be found in Coles et al. (1999) and Charpentier (2006)—see also Frahm (2006) and Schmid and Schmidt (2007c).

Definition 7. Let $Z = (X, Y)$. The random vector Z is upper tail dependent if

$$\lambda_U = \lim_{t \to 1^-} \mathbf{P}\left\{X > F_X^{[-1]}(t) \mid Y > F_Y^{[-1]}(t)\right\} > 0, \tag{5.38}$$

provided that the limit exists. If $\lambda_U = 0$ then Z is upper tail independent. λ_U is called the upper tail dependence coefficient.

Similarly, the lower tail dependence coefficient λ_L is defined as

$$\lambda_L = \lim_{t \to 0^+} \mathbf{P}\left\{X \le F_X^{[-1]}(t) \mid Y \le F_Y^{[-1]}(t)\right\}, \tag{5.39}$$

provided that the limit exists. If $\lambda_L = 0$ then Z is lower tail independent, and is lower tail dependent if $\lambda_L > 0$.

As an example, consider the Gumbel-Hougaard copula introduced later in Example 4: then, $\lambda_L = 0$ and $\lambda_U = 2 - 2^{1/\theta}$, where $\theta \ge 1$ is the parameter of the family. The following result shows that tail dependence is a copula property.

Proposition 3. *Let* $Z = (X, Y)$ *have copula* $\mathbf{C}$. *Then*

$$\lambda_U = \lim_{t \to 1^-} \frac{1 - 2t + \mathbf{C}(t, t)}{1 - t}, \tag{5.40}$$

provided that the limit exists. Similarly,

$$\lambda_L = \lim_{t \to 0^+} \frac{\mathbf{C}(t, t)}{t}, \tag{5.41}$$

provided that the limit exists.

Since the tail dependence coefficients can be expressed via copulas, many properties of copulas apply to these coefficients: for instance, they are invariant under strictly increasing transformations of the marginals.

As explained thoroughly in Nelsen (1997), *Archimedean* copulas play an important role in the study of tail dependence. Briefly, a 2-copula $\mathbf{C}$ is *Archimedean* (Nelsen 2006; Salvadori et al. 2007) if

$$\mathbf{C}\,(u,v) = \gamma^{[-1]}\,(\gamma(u) + \gamma(v))\,, \tag{5.42}$$

where $\gamma\colon \mathbf{I} \to [0,\infty]$ is called *generator*, and is continuous, convex, and strictly decreasing, with $\gamma(1) = 0$. For the sake of simplicity we only consider strict Archimedean copulas (ie., $\gamma(0) = \infty$). Note that, if $\mathbf{C}$ is a strict copula generated by γ, then $\mathbf{C}\,(t,t) = \gamma^{-1}(2\gamma(t))$.

Proposition 4. *Let* $\mathbf{C}$ *be a strict Archimedean copula generated by* γ. *Then*

$$\lambda_U = 2 - \lim_{t\to 1^-} \frac{1-\gamma^{-1}(2\gamma(t))}{1-t} = 2 - \lim_{t\to 0+} \frac{1-\gamma^{-1}(2t)}{1-\gamma^{-1}(t)}, \tag{5.43}$$

provided that the limit exists. Similarly,

$$\lambda_L = \lim_{t\to 0+} \frac{\gamma^{-1}(2\gamma(t))}{t} = \lim_{t\to\infty} \frac{\gamma^{-1}(2t)}{\gamma^{-1}(t)}, \tag{5.44}$$

provided that the limit exists.

Within the framework of tail dependence for Archimedean copulas, the following result is important (Schmidt 2003). Note that the one-sided derivatives of the generator at the domain boundaries exist, since γ is convex.

Theorem 2. *Let* $\mathbf{C}$ *be an Archimedean copula generated by* γ. *Then*

1. *Upper tail dependence implies* $\gamma'(1) = 0$, *and* $\lambda_U = 2 - (\gamma^{-1} \circ 2\gamma)'(1)$;
2. $\gamma'(1) < 0$ *implies upper tail independence;*
3. $\gamma'(0) > -\infty$ *or a non-strict generator implies lower tail independence;*
4. *Lower tail dependence implies* $\gamma'(0) = -\infty$, *a strict generator, and* $\lambda_L = (\gamma^{-1} \circ 2\gamma)'(0)$.

Given a generator γ, a simple way to construct new generators (and, consequently, families of Archimedean copulas) is to consider *interior* and *exterior power families*, as shown here.

Proposition 5. *Let* γ *be a generator. Then*

1. ***(Interior power)*** $\gamma_\alpha(t) = \gamma(t^\alpha)$ *is a generator for all* $\alpha \in (0,1]$;
2. ***(Exterior power)*** $\gamma_\beta(t) = [\gamma(t)]^\beta$ *is a generator for all* $\beta \ge 1$.

The families constructed via Proposition 5 are usually called the α- or β-family associated with γ. Most importantly, using the *interior* and *exterior power families*, it is shown in Nelsen (1997) how to generate families of Archimedean copulas with arbitrary (positive) values of λ_L and λ_U.

Proposition 6. *Let* $\mathbf{C}$ *be an Archimedean copula generated by* γ*, with lower and upper tail dependence parameters* λ_L *and* λ_U*. Then:*

1. *The lower and upper tail dependence parameters for the copula* $\mathbf{C}_\alpha$ *generated by* γ_α*, with* $\alpha \in (0, 1]$*, are, respectively,* $\lambda_L^{1/\alpha}$ *and* λ_U*;*
2. *The lower and upper tail dependence parameters for the copula* $\mathbf{C}_\beta$ *generated by* γ_β*, with* $\beta \geq 1$*, are, respectively,* $\lambda_L^{1/\beta}$ *and* $2 - (2 - \lambda_U)^{1/\beta}$*.*

As already mentioned (see Eq. (5.18)), a convex linear combination of copulas is itself a copula. As a consequence, due to the linearity of the limit operator, the tail dependence coefficients of the resulting copula are simply the convex linear combinations of those of the mixing copulas.

As for any parameter associated with the asymptotic behavior of a phenomenon, the estimation of the tail dependence coefficients is an involved task. Both parametric and non-parametric are available. In the former case, these can be estimated either by assuming a specific bivariate distribution (Embrechts et al. 2002), or a class of distributions (Serfling 2002), or a specific copula or a class of copulas (Juri and Wüthrich 2002). For the non-parametric case see Poulin et al. (2007) (and references therein), where a case study is also presented: this latter approach is based on the empirical copula (see Definition 3), and is quite general, for no assumptions are made about the copula and the marginals. For a thorough review see Frahm et al. (2005).

5.2.4 Simulation

Copulas have primary and direct applications in the simulation of dependent variables. We now present general procedures to simulate bivariate vectors, as well as multivariate ones. The mathematical kernel for simulating copulas is represented by *conditional distributions*, whose calculation is greatly facilitated by using copulas, as shown below (Nelsen 2006):

$$\mathbf{P}\{U \leq u \mid V = v\} = \frac{\partial}{\partial v}\mathbf{C}(u, v), \tag{5.45}$$

$$\mathbf{P}\{U \leq u \mid V \leq v\} = \frac{\mathbf{C}(u, v)}{v}, \tag{5.46}$$

$$\mathbf{P}\{U \leq u \mid V > v\} = \frac{u - \mathbf{C}(u, v)}{1 - v}, \tag{5.47}$$

and similar expressions hold for the conditional distributions of V given U.

A general algorithm for generating observations (x, y) from a pair of r.v.'s (X, Y) with marginals F_X, F_Y, joint distribution F_{XY}, and 2-copula $\mathbf{C}$ is as follows. By virtue of Sklar's Theorem, we need only to generate a pair (u, v) of observations of r.v.'s (U, V), Uniform on $\mathbf{I}$ and having the 2-copula $\mathbf{C}$. Then, using the *Probability Integral Transform*, we can transform (u, v) into (x, y). In order to generate the pair (u, v) we use the conditional distribution of U given the event $\{V = v\}$:

$$c_v(u) = \mathbf{P}\{U \le u \mid V = v\} = \frac{\partial}{\partial v} \mathbf{C}(u, v) . \qquad (5.48)$$

A possible algorithm is as follows.

• ALGORITHM 2.

1. Generate independent variates v, t Uniform on $\mathbf{I}$.
2. Set $u = c_v^{(-1)}(t)$.
3. The desired pair is then $\left(x = F_X^{(-1)}(u), y = F_Y^{(-1)}(v)\right)$.

The general d-dimensional case ($d > 2$) can be dealt with as follows. Let $\mathbf{F}$ be a multivariate distribution with continuous marginals $F_1, \ldots, F_d$, and suppose that $\mathbf{F}$ can be expressed in a unique way via a d-copula $\mathbf{C}$ by virtue of Sklar's Theorem. In order to simulate a vector $(X_1, \ldots, X_d) \sim \mathbf{F}$, it is sufficient to simulate a vector $(U_1, \ldots, U_d) \sim \mathbf{C}$, where the r.v.'s U_i's are Uniform on $\mathbf{I}$, and then apply the Probability Integral Transform. We now show how to simulate a sample extracted from $\mathbf{C}$. For the sake of simplicity, we assume that $\mathbf{C}$ is absolutely continuous. Here the notation $\partial_{u_1,\ldots,u_k}$ indicates the partial derivative with respect to $\{u_1, \ldots, u_k\}$.

• ALGORITHM 3.

1. To simulate the first variable U_1, it suffices to sample from a r.v. U_1' Uniform on $\mathbf{I}$. Let us call u_1 the simulated sample.
2. To obtain a sample u_2 from U_2, consistent with the previously sampled u_1, we need to know the distribution of U_2 conditional on the event $\{U_1 = u_1\}$. Let us denote this law by $G_2(\cdot \mid u_1)$, given by:

$$\begin{aligned} G_2(u_2 \mid u_1) &= \mathbf{P}\{U_2 \le u_2 \mid U_1 = u_1\} \\ &= \frac{\partial_{u_1} \mathbf{C}(u_1, u_2, 1, \ldots, 1)}{\partial_{u_1} \mathbf{C}(u_1, 1, \ldots, 1)} \\ &= \partial_{u_1} \mathbf{C}(u_1, u_2, 1, \ldots, 1) , \end{aligned} \qquad (5.49)$$

   being $\mathbf{C}(u_1, 1, \ldots, 1) = u_1$, since a copula has Uniform marginals. Then we take $u_2 = G_2^{-1}(u_2' \mid u_1)$, where u_2' is the realization of a r.v. U_2' Uniform on $\mathbf{I}$, that is independent of U_1'.
3. In general, to simulate a sample u_k from U_k, consistent with the previously sampled $u_1, \ldots, u_{k-1}$, we need to know the distribution of U_k conditional on the events $\{U_1 = u_1, \ldots, U_{k-1} = u_{k-1}\}$. Let us denote this law by $G_k(\cdot \mid u_1, \ldots, u_{k-1})$, given by:

$$G_k(u_k \mid u_1, \dots, u_{k-1}) = \mathbf{P}\{U_k \le u_k \mid U_1 = u_1, \dots, U_{k-1} = u_{k-1}\}$$
$$= \frac{\partial_{u_1,\dots,u_{k-1}} \mathbf{C}(u_1, \dots, u_k, 1, \dots, 1)}{\partial_{u_1,\dots,u_{k-1}} \mathbf{C}(u_1, \dots, u_{k-1}, 1, \dots, 1)}. \tag{5.50}$$

Then we take $u_k = G_k^{-1}(u'_k \mid u_1, \dots, u_{k-1})$, where u'_k is the realization of a r.v. U'_k Uniform on $\mathbf{I}$, that is independent of $U'_1, \dots, U'_{k-1}$.

Other simulation algorithms can be found in Devroye (1986) and Johnson (1987), in many of the exercises proposed by Nelsen (2006), in Salvadori et al. (2007, Appendix C), and in Sect. 5.3.2 below.

5.3 Multivariate Extreme Value Models

As outlined in the previous Section, copulas represent a fundamental tool to describe the structure of multivariate distributions (essentially via Sklar's Theorem). In turn, the results of classical Multivariate Extreme Value theory can be rewritten in terms of copulas: as we shall see, this approach has quite a few advantages. The main references to this subject are Marshall and Olkin (1983), Galambos (1987), Marshall and Olkin (1988), Joe (1997), Nelsen (2006) and Salvadori et al. (2007), where further bibliography is indicated.

In multivariate Extreme Value analysis it is a standard practice to investigate vectors of *componentwise extremes*, as defined below. However, alternative approaches are available (among others, multivariate Peaks-Over-Threshold methods, or Point-Process approaches), though they will not be considered here: for a thorough introduction see Coles (2001), and references therein.

Definition 8. Let $\{(X_{i,1}, \dots, X_{i,d})\}$, $i = 1, \dots, n$, be a sample of size n of i.i.d. d-variate r.v.'s with joint distribution $\mathbf{F}$. The corresponding vector of componentwise maxima $\mathbf{M}_n$ is defined as

$$\mathbf{M}_n = (M_{n,1}, \dots, M_{n,d}) = \left(\max_{1 \le i \le n} \{X_{i,1}\}, \dots, \max_{1 \le i \le n} \{X_{i,d}\} \right). \tag{5.51}$$

Note 2. Henceforth we concentrate on the analysis of maxima only, since there is no need to study separately the behavior of the minima. In fact, let $Y_i = -X_i$; then, if $X_{(1)} = \min\{X_1, \dots, X_n\}$ and $Y_{(n)} = \max\{Y_1, \dots, Y_n\}$, it is clear that

$$X_{(1)} = -Y_{(n)}. \tag{5.52}$$

Thus, the analysis of the maximum of the Y_i's suffices.

As in the univariate case, the extremal behavior of multivariate maxima is based on the limiting behavior of "block" maxima (see, e.g., Salvadori et al. 2007, and references therein). However, it is important to realize that the maximum of each of the d different marginal sequences $X_{i,1}, \dots, X_{i,d}$ may occur for different indices, say $i_1^*, \dots, i_d^*$. As a consequence, $\mathbf{M}_n$ does not necessarily correspond to an observed sample value in the original series.

A standard way to operate is to look for the existence of sequences of real constants, $\{a_{n,i}\}$ and $\{b_{n,i} > 0\}$, $1 \le i \le d$, such that, for all $\mathbf{x} \in \mathbf{R}^d$, the function

$$\mathbf{G}(x_1, \dots, x_d) = \lim_{n\to\infty} \mathbf{P}\left\{ \frac{M_{n,1} - a_{n,1}}{b_{n,1}} \le x_1, \dots, \frac{M_{n,d} - a_{n,d}}{b_{n,d}} \le x_d \right\}, \qquad (5.53)$$

which is equivalent to the following limit of the joint distribution function $\mathbf{F}$

$$\lim_{n\to\infty} \mathbf{F}^n(a_{n,1} + b_{n,1}x_1, \dots, a_{n,d} + b_{n,d}x_d),$$

is a proper distribution with non-degenerate marginals. Then $\mathbf{G}$ is a Multivariate Extreme Value (hereinafter, MEV) law. Via the *Pickands' representation* (Marshall and Olkin 1983), it can be shown that MEV distributions are *continuous* (but not always absolutely continuous). Furthermore, setting all the x_i's but one to $+\infty$ in Eq. (5.53) yields

$$\lim_{n\to\infty} F_i^n(a_{n,i} + b_{n,i}x_i) = G_i(x_i), \quad i = 1, \dots, d, \qquad (5.54)$$

where F_i and G_i are, respectively, the i-th marginals of $\mathbf{F}$ and $\mathbf{G}$. In turn, G_i is a member of the *Generalized Extreme Value* (hereinafter, GEV) family. For a representation in terms of the so-called *spectral measure* see, e.g., De Haan and Resnick (1977), and also Gudendorf and Segers (2010) and references therein.

5.3.1 Extreme Value Copulas

Following Eq. (5.54), if $\mathbf{F}$ is a MEV distribution, then its marginals belong to the continuous GEV family. Here we focus our attention on the copula $\mathbf{C}$ associated with $\mathbf{F}$. Note that, since the marginals F_i's are continuous, the copula representation is unique. The following notion is fundamental.

Definition 9 (Extreme Value Copula). A d-copula $\mathbf{C}$ is max-stable if it satisfies the relationship

$$\mathbf{C}\left(u_1^t, \dots, u_d^t\right) = \mathbf{C}^t\left(u_1, \dots, u_d\right) \qquad (5.55)$$

for all $t > 0$ and $\mathbf{u} \in \mathbf{I}^d$. Max-stable copulas are called Extreme Value (hereinafter, EVC or MEV).

A characterization of MEV distributions can be given as follows.

Theorem 3 (MEV Distribution). *Let* $\mathbf{F}$ *be a* d*-variate distribution with copula* $\mathbf{C}$*. Then* $\mathbf{F}$ *is a MEV distribution if, and only if,*

1. *The marginals* F_i*'s have a GEV distribution, and*
2. $\mathbf{C}$ *is max-stable.*

As a consequence of Theorem 3, if $\mathbf{C}$ is an EVC, then the multivariate distribution $\mathbf{F}(\mathbf{x}) = \mathbf{C}(F_1(x_1), \ldots, F_d(x_d))$, where each marginal F_i's belongs to the GEV class, is an EV distribution. Clearly, it is necessary that the marginals are GEV laws.

Example 4. As an illustration, consider the case study presented in Example 2. The bivariate Gumbel-Hougaard copula (see Eq. (5.61)) was used to model the joint behavior of the two variables (Q, V). Clearly, $\mathbf{C}_{QV}$ is an Extreme Value copula, being max-stable. As a consequence, since both F_Q and F_V are GEV distributions, then $F_{QV} = \mathbf{C}_{QV}(F_Q, F_V)$ is a bivariate EV law. In addition, $\mathbf{C}_{QV}$ is *Archimedean* (Nelsen 2006; Salvadori et al. 2007) (see Eq. (5.42)), with generator

$$\gamma(t) = (-\ln t)^{\theta}, \tag{5.56}$$

where $\theta \geq 1$ is the parameter identifying the copula. Note that θ can be estimated via the relation

$$\widehat{\theta} = \frac{1}{1-\widehat{\tau}}, \tag{5.57}$$

where $\widehat{\tau}$ is an estimate of the Kendall's τ calculated using the (Q, V) sample (see Fig. 5.1). In the present case, $\widehat{\tau} \approx 0.6813$, and hence $\widehat{\theta} \approx 3.1378$. The multivariate Goodness-of-Fit tests outlined in Genest et al. (2009), Berg (2009) and Kojadinovic et al. (2011) can then be used to check whether or not $\mathbf{C}_{QV}$ is an acceptable copula model for the data of interest (actually, it is).

The following result emphasizes an important feature of EVC's.

Theorem 4. *If* $\mathbf{C}^*$ *is an EVC, then*

$$\mathbf{C}^*(\mathbf{u}) \geq \Pi_d(\mathbf{u}) \tag{5.58}$$

for all $\mathbf{u} \in \mathbf{I}^d$.

Example 5. There are several families of copulas suitable for representing EVC's.

1. **(Independence):** for the product d-copula one sees that

$$\Pi_d(\mathbf{u}^t) = \prod_{i=1}^{d} u_i^t = \Pi_d^t(\mathbf{u}). \tag{5.59}$$

2. **(Co-monotone):** for the co-monotone d-copula one sees that

$$\mathbf{M}_d(\mathbf{u}^t) = \min\left\{u_1^t, \dots, u_d^t\right\} = \mathbf{M}_d^t(\mathbf{u}). \tag{5.60}$$

3. **(Gumbel-Hougaard):** for the Gumbel-Hougaard d-copula $\mathbf{C}_\theta$, $\theta \geq 1$, it follows that

$$\mathbf{C}_\theta\left(\mathbf{u}^t\right) = e^{-\left[(-t \ln u_1)^\theta + \cdots + (-t \ln u_d)^\theta\right]^{1/\theta}} = \mathbf{C}_\theta^t\left(\mathbf{u}\right). \tag{5.61}$$

It is worth noting that the Gumbel-Hougaard family is the *only* Archimedean class of EV copulas (Genest and Rivest 1989; Nelsen 2006; Salvadori et al. 2007).

As already mentioned in Proposition 1, 2-copulas (and, more generally, d-copulas) are *invariant* under strictly increasing transformations of the marginals. More specifically: if $g_i : \mathbf{R} \to \mathbf{R}$, $i = 1, \dots, d$, are all strictly increasing functions, then the vector $\mathbf{Y} = (g_1(X_1), \dots, g_d(X_d))$ has the same copula as the vector $\mathbf{X} = (X_1, \dots, X_d)$. The interesting point is that such an invariance property holds also for the limiting EVC of $\mathbf{Y}$.

Proposition 7 (Invariance). *Let* $\mathbf{X} = (X_1, \dots, X_d)$*, and set* $\mathbf{Y} = (g_1(X_1), \dots, g_d(X_d))$*, where the functions* $g_i : \mathbf{R} \to \mathbf{R}$*,* $i = 1, \dots, d$*, are all strictly increasing. If* $\mathbf{C}_\mathbf{X}^*$ *and* $\mathbf{C}_\mathbf{Y}^*$ *denote, respectively, the EVC of* $\mathbf{X}$ *and* $\mathbf{Y}$*, then*

$$\mathbf{C}_\mathbf{Y}^* = \mathbf{C}_\mathbf{X}^*. \tag{5.62}$$

The invariance property mentioned above shows that the limiting EVC only depends upon the copula $\mathbf{C}$ of the multivariate distribution $\mathbf{F}$, and is independent of the marginals of $\mathbf{F}$.

As already mentioned in Sect. 5.2.1, any convex linear combination of copulas is itself a copula. It is then of interest to calculate the limiting copula of such a combination. Let us consider a d-dimensional copula $\mathbf{C}$ given by a convex linear combination as in Eq. (5.18). Then, the EVC $\mathbf{C}^*$ of $\mathbf{C}$ is given by

$$\mathbf{C}^*\left(\mathbf{u}\right) = \prod_{i=1}^{k} \left(\mathbf{C}_i^*\left(\mathbf{u}\right)\right)^{\lambda_i}, \tag{5.63}$$

where $\mathbf{C}_i^*$ denotes the EVC associated with the copula $\mathbf{C}_i$. It must be pointed out that, in general, even if all the mixing copulas $\mathbf{C}_i$'s are EVC, this does not imply that their convex linear combination $\mathbf{C}$ is also an EVC, as shown below.

Example 6. Let $\mathbf{C}_1 = \Pi_d$ and $\mathbf{C}_2 = \mathbf{M}_d$: both these multivariate copulas are EVC. Set $\mathbf{C} = (1-\lambda)\mathbf{C}_1 + \lambda\mathbf{C}_2$, where $\lambda \in (0, 1)$. Then, using Eq. (5.63), the EVC $\mathbf{C}^*$ of $\mathbf{C}$ is given by

$$\mathbf{C}^*\left(\mathbf{u}\right) = \left(\Pi_d\left(\mathbf{u}\right)\right)^{1-\lambda}\left(\mathbf{M}_d\left(\mathbf{u}\right)\right)^{\lambda} = \left(u_1 \cdots u_d\right)^{1-\lambda} \min\left\{u_1^\lambda, \dots, u_d^\lambda\right\}. \tag{5.64}$$

Thus, $\mathbf{C}^*$ is a member of the Cuadras-Augé family, a subfamily of the Marshall-Olkin family of copulas (Nelsen 2006; Salvadori et al. 2007). Evidently, $\mathbf{C}^* \neq \mathbf{C}$, and $\mathbf{C}$ is not an EVC. This EVC was used in Durante and Salvadori (2010), in order to model the joint behavior of a set of three certified gauge stations recording annual maximum flood data in northwestern Apennines and Thyrrhenian Liguria basins (see also Example 8).

A notion of interest when dealing with EV 2-copulas is represented by Pickands' dependence function A (De Haan and Resnick 1977; Pickands 1981; Mai and Scherer 2010). A bivariate copula $\mathbf{C}$ is MEV if, and only if, there exists a convex function $A : \mathbf{I} \to [1/2, 1]$, satisfying the constraint $\max\{t, 1-t\} \leq A(t) \leq 1$ for all $t \in \mathbf{I}$, such that

$$\mathbf{C}(u, v) = \exp\left[\ln(uv)\, A\left(\frac{\ln v}{\ln(uv)}\right)\right] \tag{5.65}$$

for all $(u, v) \in \mathbf{I}^2$. In particular, if $A(t) \equiv 1$ then $\mathbf{C} = \Pi_2$, and if $A(t) = \max\{t, 1-t\}$ then $\mathbf{C} = \mathbf{M}_2$. Conversely, given a bivariate MEV copula $\mathbf{C}$, the corresponding Pickands' dependence function A is given by

$$A(t) = -\ln \mathbf{C}\left(e^{-(1-t)}, e^{-t}\right), \tag{5.66}$$

where $t \in \mathbf{I}$. It is worth noting that the value of the Kendall's τ associated with $\mathbf{C}$, as well as that of the Spearman's ρ, can be expressed in terms of A via (Nelsen 2006; Salvadori et al. 2007)

$$\tau = \int_0^1 \frac{t(1-t)}{A(t)}\, dA'(t), \tag{5.67}$$

where $A'(t)$ is the derivative of $A(t)$, and

$$\rho = 12 \int_0^1 \frac{1}{(1 + A(t))^2}\, dt - 3. \tag{5.68}$$

A generalization of Pickands' dependence function to the multivariate case is shown in Falk and Reiss (2005). Since A can be estimated via empirical data (a thorough review can be found in Genest and Segers (2009)), then it may be used to check the adequacy of different bivariate models (see, e.g., the hydrological case study investigated in Salvadori and De Michele (2011)). However, it must be stressed that the empirical estimates of the true (but unknown) dependence functions do not generally respect the convexity constraint (the estimator is not intrinsic, viz. not convex by design), and the construction of a valuable estimator is still an open problem in Statistics.

Example 7. As an illustration, we consider the Gumbel-Hougaard 2-copula $\mathbf{C}_\theta$ introduced in Eq. (5.61). As is well known (Nelsen 2006), if $\theta = 1$, then $\mathbf{C}_1 = \Pi_2$, while $\mathbf{C}_\theta \to \mathbf{M}_2$ as $\theta \to \infty$. Such a behavior can be easily spotted by studying the corresponding Pickands' dependence function given by, for $t \in \mathbf{I}$,

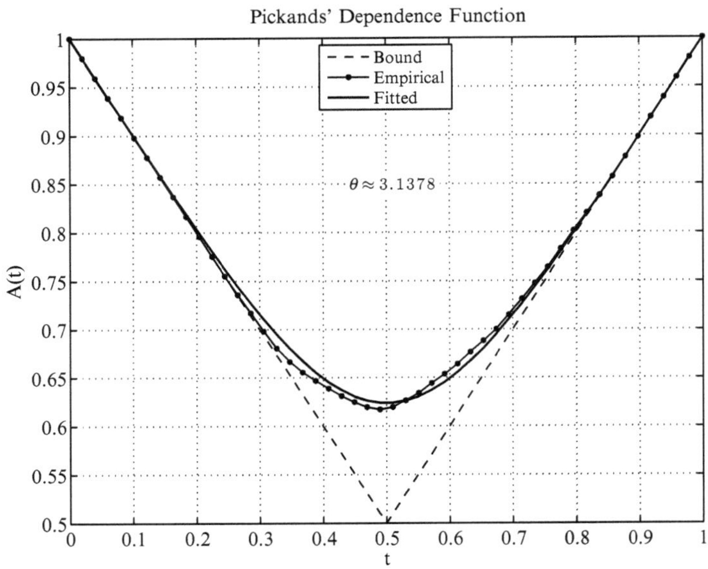

Fig. 5.2 Plot of the Pickands' dependence function of the Gumbel-Hougaard 2-copula for the dam data shown in Fig. 5.1, with $\theta \approx 3.1378$, as well as an empirical estimate—see text. Also shown is the *lower bound* given by the co-monotone 2-copula $\mathbf{M}_2$

$$A(t) = \left(t^{\theta} + (1-t)^{\theta}\right)^{1/\theta}. \tag{5.69}$$

In Fig. 5.2 we plot the function A for the dam case study investigated in this work (see Example 2), as well as an empirical estimate: overall, the agreement is valuable.

5.3.2 *Special Construction of MEV Distributions*

The MEV copula $\mathbf{C}^*$ shown in Eq. (5.64) provides a typical example of a situation found in applications. In fact, d-copulas may sometimes have limited applications, due to the fact that only one (or few) *global* parameters rule the whole dependence structure (in this case, λ). In order to improve the modeling features, further dependence parameters should be introduced: a simple example of an "extra-parametrization" technique is provided by Proposition 5 for the case of Archimedean 2-copulas.

Indeed, in practical applications, a suitable number of parameters may be necessary to capture the inter-dependencies between all the variables at play. As we shall see, the new parameters introduced via the techniques outlined below may describe and rule how each variable affects the others, and/or how each variable is affected by the others. Note that the approach presented here may be applied to any copula (for more details, see Durante and Salvadori 2010).

To handle the extra-parametrization problem, we use the following construction method for copulas, which can be derived from the results outlined in Liebscher (2008).

Theorem 5. *Let $d \geq 2$, and let* **A** *and* **B** *be two d-copulas. For $i = 1, \dots, d$, let $F_i : \mathbf{I} \to \mathbf{I}$ be a continuous and strictly increasing function, with $F_i(0) = 0$ and $F_i(1) = 1$. Moreover, suppose that, for $i = 1, \dots, d$, $G_i(t) = t/F_i(t)$ is strictly increasing, with $G_i(0^+) = 0$. Then the mapping $\mathbf{C}: \mathbf{I}^d \to \mathbf{I}$ given by*

$$\mathbf{C}(u_1, \dots, u_d) = \mathbf{A}(F_1(u_1), \dots, F_d(u_d)) \cdot \mathbf{B}(G_1(u_1), \dots, G_d(u_d)) \tag{5.70}$$

is a d-copula.

The following corollary is of interest in practical applications (Durante and Salvadori 2010).

Corollary 2. *Let* **A** *and* **B** *be d-copulas, and let $\mathbf{a} = (a_1, \dots, a_d) \in \mathbf{I}^d$. Then*

$$\mathbf{C}_{a_1,\dots,a_d}(\mathbf{u}) = \mathbf{A}\left(\mathbf{u}^{\mathbf{a}}\right) \cdot \mathbf{B}\left(\mathbf{u}^{\mathbf{1}-\mathbf{a}}\right) = \mathbf{A}\left(u_1^{a_1}, \dots, u_d^{a_d}\right) \cdot \mathbf{B}\left(u_1^{1-a_1}, \dots, u_d^{1-a_d}\right) \tag{5.71}$$

defines a family of d-copulas. In particular, if **A** *and* **B** *are MEV, then so is $\mathbf{C}_{a_1,\dots,a_d}$.*

The copulas defined by Eq. (5.71) belong to the *Khoudraji family*, since this method of construction was first proposed in Khoudraji's PhD thesis (Khoudraji 1995; Genest et al. 1998). A bivariate generalization has been discussed in Durante (2009). The MEV model given by Eq. (5.71) is quite general and rich: an infinite number of variants is possible via a suitable selection of **A** and **B**, and new families of MEV copulas can be easily introduced.

Interestingly enough, should further parameters be needed (say, another set $\mathbf{b} = (b_1, \dots, b_d) \in \mathbf{I}^d$), the above procedure can be easily iterated as follows. First, simply consider the copula $\mathbf{C}_{a_1,\dots,a_d}$ given by Eq. (5.71), and a new copula **D**. Then, construct

$$\mathbf{C}_{a_1,\dots,a_d;b_1,\dots,b_d}(\mathbf{u}) = \mathbf{C}_{a_1,\dots,a_d}\left(\mathbf{u}^{\mathbf{b}}\right) \cdot \mathbf{D}\left(\mathbf{u}^{\mathbf{1}-\mathbf{b}}\right). \tag{5.72}$$

Again, if the copulas $\mathbf{C}_{a_1,\dots,a_d}$ and **D** are MEV, then so is $\mathbf{C}_{\mathbf{a};\mathbf{b}}$. Clearly, the actual number of new parameters introduced with the extra-parametrization technique may depend upon the specific analytical structure of the copulas of interest. The above technique for constructing copulas yields the following simulation algorithm.

• ALGORITHM 4.

1. Generate the variates $s_1, \dots, s_d$ from copula **A**.
2. Generate the variates $t_1, \dots, t_d$ from copula **B**.
3. For $i = 1, \dots, d$ set $v_i = F_i^{-1}(s_i)$ and $z_i = G_i^{-1}(t_i)$.
4. For $i = 1, \dots, d$ return $u_i = \max\{v_i, z_i\}$.

Example 8. As an illustration, we consider the hydrological case study reported in Durante and Salvadori (2010): a set of three certified gauge stations recording annual maximum flood data in northwestern Apennines and Thyrrhenian Liguria basins (Italy) was investigated. A trivariate MEV copula was used for modeling the dependence between these stations. Note that it was not possible to assume a priori the independence of the stations, since the corresponding estimates of the Kendall's τ's were significantly different from zero. In order to provide a flexible multi-parametric dependence structure, Corollary 2 was applied to the pair of copulas $\mathbf{A} = \mathbf{C}^*$ (as calculated in Example 6, Eq. (5.64)) and $\mathbf{B} = \Pi_d$, and the new set of d parameters $\mathbf{a} = (a_1, \ldots, a_d) \in \mathbf{I}^d$ was introduced (here $d = 3$):

$$\begin{aligned}\widetilde{\mathbf{C}}^*(\mathbf{u}) &= \mathbf{C}^*(\mathbf{u}^{\mathbf{a}}) \cdot \Pi_d\left(\mathbf{u}^{\mathbf{1}-\mathbf{a}}\right) \\ &= \Pi_d\left(\mathbf{u}^{\mathbf{1}-\mathbf{a}\lambda}\right) \mathbf{M}_d\left(\mathbf{u}^{\mathbf{a}\lambda}\right) \\ &= u_1^{1-\lambda a_1} \cdot \ldots \cdot u_d^{1-\lambda a_d} \cdot \min\{u_1^{\lambda a_1}, \ldots, u_d^{\lambda a_d}\}. \end{aligned} \tag{5.73}$$

Note that every new family $\widetilde{\mathbf{C}}^*$ includes the "basic" MEV copula $\mathbf{C}^*$ as a special case (it is enough to take $\mathbf{a} = \mathbf{1}$), which provides a sort of a "latent" global dependence structure. Evidently, $\widetilde{\mathbf{C}}^*$ is a MEV copula with d free parameters $\mathbf{b} = \lambda \cdot \mathbf{a}$, and bivariate margins $\widetilde{\mathbf{C}}^*_{ij}$'s given by

$$\widetilde{\mathbf{C}}^*_{ij}(u_i, u_j) = \begin{cases} u_i\, u_j^{1-a_j\lambda}, & \text{if } u_i^{a_i} \le u_j^{a_j} \\ u_i^{1-a_i\lambda}\, u_j, & \text{if } u_i^{a_i} \ge u_j^{a_j} \end{cases} = \begin{cases} u_i\, u_j^{1-b_j}, & \text{if } u_i^{b_i} \le u_j^{b_j} \\ u_i^{1-b_i}\, u_j, & \text{if } u_i^{b_i} \ge u_j^{b_j} \end{cases}, \tag{5.74}$$

which belong to the *Marshall-Olkin* family of 2-copulas (Marshall and Olkin 1967; Nelsen 2006; Salvadori et al. 2007). In particular, these copulas are not absolutely continuous for $\lambda > 0$, and are asymmetric for $a_i \neq a_j$ (viz., they may model non-exchangeable variables, a feature of great importance in applications (Grimaldi and Serinaldi 2006).

The Spearman's association measure ρ_{ij} for the generic pair $\{X_i, X_j\}$ is given by

$$\rho_{ij} = \frac{3\, a_i a_j \lambda}{2a_i + 2a_j - a_i a_j \lambda} = \frac{3\, b_i b_j}{2b_i + 2b_j - b_i b_j} \ge 0. \tag{5.75}$$

In the limiting case, $\rho_{ij} = 0$ whenever $a_i = 0$, or $a_j = 0$, or $\lambda = 0$; also, $\rho_{ij} = 1$ if $a_i = a_j = \lambda = 1$. Actually, $\rho_{ij} \in [0, 1]$, i.e. it can take on all the admissible values for a MEV copula.

Analogously, Kendall's association measure τ_{ij} for the generic pair $\{X_i, X_j\}$ is given by Fredericks and Nelsen (2007) and Salvadori et al. (2007)

$$\tau_{ij} = \frac{a_i a_j \lambda}{a_i + a_j - a_i a_j \lambda} = \frac{b_i b_j}{b_i + b_j - b_i b_j} \ge 0. \tag{5.76}$$

In the limiting case, $\tau_{ij} = 0$ whenever $a_i = 0$, or $a_j = 0$, or $\lambda = 0$; also, $\tau_{ij} = 1$ if $a_i = a_j = \lambda = 1$. Actually, also $\tau_{ij} \in [0, 1]$, i.e. it can take on all the admissible values for a MEV copula.

5.4 Multivariate Return Periods and Design

In order to provide a consistent theory of multivariate return periods and design, it is first necessary to precisely define the abstract framework where to embed the question. Preliminary studies can be found in Salvadori (2004), Salvadori et al. (2004), Durante and Salvadori (2010), Salvadori et al. (2010) and Salvadori et al. (2011), and some applications are presented in De Michele et al. (2007), Salvadori et al. (2010) and Vandenberghe et al. (2010).

Hereinafter, we shall consider as the object of our investigation a sequence $\mathscr{X} = \{\mathbf{X}_1, \mathbf{X}_2, \ldots\}$ of independent and identically distributed d-dimensional random vectors, with $d \geq 1$: thus, each $\mathbf{X}_i$ has the same multivariate distribution $\mathbf{F}$ as of the random vector $\mathbf{X} \sim \mathbf{F} = \mathbf{C}(F_1, \ldots, F_d)$ describing the phenomenon under investigation, with suitable marginals F_i's and d-copula $\mathbf{C}$. For example, we may think of a set of flood observations given by the pairs of non-independent r.v.'s Flood Peak—Flood Volume, joined by the copula $\mathbf{C}$ (see, e.g., Example 4).

It must be stressed that most of the results outlined below are valid in a stationary context, but are much trickier in case of non-stationarity, and will not be discussed here. As general references, see Cox (1962) and Feller (1971).

5.4.1 *Multivariate Return Periods*

In applications, usually, the event of interest is of the type $\{\mathbf{X} \in \mathscr{D}\}$, where $\mathscr{D}$ is a non-empty Borel set in $\mathbf{R}^d$ collecting all the values judged to be "dangerous" according to some suitable criterion. Note that the Borel family includes all the sets of interest in practice (like, e.g., the intervals $(-\infty, x_1), (x_1, x_2), (x_2, \infty)$, as well as the corresponding multivariate versions). Let $\mu > 0$ be the expectation of the random inter-arrival time of the realizations in $\mathscr{X}$ (viz., μ is the average time elapsing between $\mathbf{X}_i$ and $\mathbf{X}_{i+1}$). Following, e.g., Embrechts et al. (2003), and given the fact that the sequence $\mathscr{X}$ is i.i.d. (and, thus, stationary), the univariate r.v.'s $\{B_i = \mathscr{I}_{\mathscr{D}}(\mathbf{X}_i)\}$ form a Bernoulli process (where $\mathscr{I}_{\mathscr{D}}$ is the indicator function of the set $\mathscr{D}$), with positive probability of "success" $p_{\mathscr{D}}$ given by

$$p_{\mathscr{D}} = \mathbf{P}\{\mathbf{X} \in \mathscr{D}\}, \tag{5.77}$$

where we assume that $0 < p_{\mathscr{D}} < 1$. Then, it makes sense to calculate the first random time $A_{\mathscr{D}}$ that the sequence $\mathscr{B} = \{B_1, B_2, \ldots\}$, generated by $\mathscr{X}$, takes on the value 1 (viz., the first random time that $\mathscr{X}$ enters $\mathscr{D}$):

$$A_{\mathscr{D}} = \mu \cdot \min\{i : \mathbf{X}_i \in \mathscr{D}\}. \tag{5.78}$$

Clearly, the r.v. $A_{\mathscr{D}}/\mu$ follows a Geometric distribution with parameter $p_{\mathscr{D}}$, and therefore the expected value of $A_{\mathscr{D}}$ is

$$\mathbf{E}(A_{\mathscr{D}}) = \mu_{\mathscr{D}} = \mu/p_{\mathscr{D}}. \tag{5.79}$$

Assuming again a stationary context, both the well known "memoryless property" of the Geometric distribution, as well as the features of the Bernoulli process, hold (see, e.g., Cox 1962; Feller 1971). Therefore, $\mu_{\mathscr{D}}$ also corresponds to the average inter-arrival time between two successive realizations of the event $\{\mathbf{X} \in \mathscr{D}\}$. Evidently, $\mu_{\mathscr{D}}$ ranges in $[\mu, +\infty)$: for example, if annual maxima are investigated, then $\mu = 1$ year, and hence $\mu_{\mathscr{D}} = 1/p_{\mathscr{D}} \geq \mu$. We are now ready to introduce a consistent notion of return period (hereinafter, *RP*) in a stationary framework.

Definition 10 (Return Period). The RP associated with the event $\{\mathbf{X} \in \mathscr{D}\}$ (or the dangerous set $\mathscr{D} \subset \mathbf{R}^d$) is given by

$$\mu_{\mathscr{D}} = \mu/\mathbf{P}\{\mathbf{X} \in \mathscr{D}\}. \tag{5.80}$$

Note 3. Definition 10 is a very general one: the set $\mathscr{D}$ may be constructed in order to satisfy broad requirements, useful in different applications. Indeed, most of the approaches already present in literature are particular cases of the one outlined above.

Example 9. As a univariate illustration, let X be a r.v. with distribution F_X, and let x^* denote a prescribed critical design value. Then, e.g., in hydrology, if droughts are of concern, x^* may represent a small value of river flow, and the critical realizations of interest are those for which $X < x^*$ (viz., $\mathscr{D} = [0, x^*)$). Instead, if floods are of concern, x^* may indicate a large value of river flow, and the critical realizations of interest are those for which $X > x^*$ (viz., $\mathscr{D} = (x^*, \infty)$). According to Definition 10 and Eq. (5.79), the corresponding RP's are $\mu/F_X(x^*)$ in the former case, and $\mu/(1 - F_X(x^*))$ in the latter one.

It is important to stress that the RP is a quantity associated with a proper event. However, with a slight abuse of language, we may also speak of "the RP of a realization" (viz., x^* in the example given above), meaning in fact "the RP of the event $\{X$ belongs to the dangerous region $\mathscr{D}_{x^*}$ identified by the given realization $x^*\}$". Indeed, in a univariate framework, usually the assignment of x^* uniquely specifies the corresponding region $\mathscr{D}_{x^*}$. Actually, also in a multivariate framework it is possible to associate a given multi-dimensional realization $\mathbf{x}^* \in \mathbf{R}^d$ with a dangerous region $\mathscr{D}_{\mathbf{x}^*} \subset \mathbf{R}^d$.

Example 10. As an illustration, consider the two different bivariate dangerous regions constructed in Salvadori (2004) and Salvadori et al. (2004). In these papers the joint behavior of the vector $(X, Y) \sim \mathbf{F} = \mathbf{C}(F_X, F_Y)$ was analysed: for instance, in terms of variables of hydrological interest, think of the pairs Flood

Peak–Volume, or Storm Intensity–Duration. In particular, great attention was paid to the following two sets:

1. ("OR" case) the dangerous region is

$$\mathscr{D}^{\vee}_{\mathbf{z}^*} = \{(x, y) \in \mathbf{R}^2 : x > x^* \vee y > y^*\}, \tag{5.81}$$

   where at least one of the components exceeds a prescribed threshold (roughly, it is enough that one of the variables is too large);
2. ("AND" case) the dangerous region is

$$\mathscr{D}^{\wedge}_{\mathbf{z}^*} = \{(x, y) \in \mathbf{R}^2 : x > x^* \wedge y > y^*\}, \tag{5.82}$$

   where both the components exceed a prescribed threshold (roughly, it is necessary that both variables are too large).

Here $\mathbf{z}^* = (x^*, y^*)$ is a prescribed vector of thresholds, and $\vee, \wedge$ are the "(inclusive) OR" and "AND" operators. We shall see later in Example 13 how to deal with these two dangerous regions.

In this Chapter we adopt an original approach (Salvadori et al. 2011), which generalizes the univariate one. First of all we need to introduce the following notion.

Definition 11 (Critical Layer). Let $\mathbf{F} = \mathbf{C}(F_1, \ldots, F_d)$ be a d-dimensional distribution. Given $t \in (0, 1)$, the *critical layer* $\mathscr{L}_t^{\mathbf{F}}$ of level t is defined as

$$\mathscr{L}_t^{\mathbf{F}} = \{\mathbf{x} \in \mathbf{R}^d : \mathbf{F}(\mathbf{x}) = t\}. \tag{5.83}$$

Clearly, $\mathscr{L}_t^{\mathbf{F}}$ is the iso-hyper-surface (having dimension $d - 1$) where $\mathbf{F}$ equals the constant value t: thus, $\mathscr{L}_t^{\mathbf{F}}$ is a (iso)line for bivariate distributions, a (iso)surface for trivariate ones, and so on. In Fig. 5.5 we show a selected critical iso-line on the 2-copula $\mathbf{C}_{QV}$ of interest here (see below for an explanation).

Evidently, for any given $\mathbf{x} \in \mathbf{R}^d$, there exists a *unique* critical layer $\mathscr{L}_t^{\mathbf{F}}$ supporting $\mathbf{x}$ (say, using a quick-and-dirty notation, $\mathscr{L}_{\mathbf{x}}^{\mathbf{F}}$): namely, the one identified by the level $t = \mathbf{F}(\mathbf{x})$. Note that, thanks to the Probability Integral Transform, there exists a one-to-one correspondence between the two iso-hyper-surfaces $\mathscr{L}_t^{\mathbf{C}}$ (pertaining to $\mathbf{C}$ in $\mathbf{I}^d$) and $\mathscr{L}_t^{\mathbf{F}}$ (pertaining to $\mathbf{F}$ in $\mathbf{R}^d$).

The critical layer $\mathscr{L}_t^{\mathbf{F}}$ partitions $\mathbf{R}^d$ into three non-overlapping and exhaustive regions:

1. $\mathscr{R}_t^{<} = \{\mathbf{x} \in \mathbf{R}^d : \mathbf{F}(\mathbf{x}) < t\}$;
2. $\mathscr{L}_t^{\mathbf{F}}$, the critical layer itself;
3. $\mathscr{R}_t^{>} = \{\mathbf{x} \in \mathbf{R}^d : \mathbf{F}(\mathbf{x}) > t\}$.

Practically, at any occurrence of the phenomenon, only three mutually exclusive things may happen: either a realization of $\mathbf{X}$ lies in $\mathscr{R}_t^{<}$, or over $\mathscr{L}_t^{\mathbf{F}}$, or it lies in $\mathscr{R}_t^{>}$. Note that all these three regions are Borel sets.

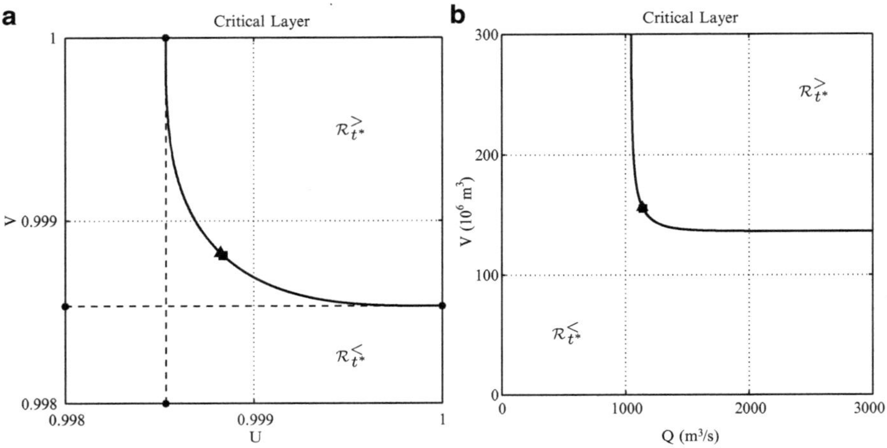

Fig. 5.3 Plot of the iso-line (**a**) $\mathbf{C}_{QV}(u,v) = t^*$ and (**b**) $\mathbf{F}_{QV}(x,y) = t^*$ for the critical level $t^* \approx 0.998533$ (*circles*) corresponding to a millenary RP—see text. Also emphasized are the regions $\mathscr{R}^{<}_{t^*}$ and $\mathscr{R}^{>}_{t^*}$. The *triangle* marker indicates the Component-Wise Excess design realization, and the *square* marker indicates the Most-Likely design realization—see Sect. 5.4.3 and Table 5.1

Example 11. As an illustration, in Fig. 5.3 we plot the critical layers (actually, isolines) $\mathscr{L}^{\mathbf{C}_{QV}}_{t^*}$ and $\mathscr{L}^{\mathbf{F}_{QV}}_{t^*}$ for the critical level $t^* \approx 0.998533$ and the pair (Q, V) of interest here (see below for an explanation).

Thanks to the above discussion, it is now clear that the following (multivariate) notion of RP is meaningful, and coincide with the one used in the univariate framework.

Definition 12. Let $\mathbf{X}$ be a multivariate r.v. with distribution $\mathbf{F} = \mathbf{C}(F_1, \ldots, F_d)$. Also, let $\mathscr{L}^{\mathbf{F}}_t$ be the critical layer supporting a realization $\mathbf{x}$ of $\mathbf{X}$ (i.e., $t = \mathbf{F}(\mathbf{x})$). Then, the RP $T_{\mathbf{x}}$ associated with $\mathbf{x}$ is defined as

1. For the region $\mathscr{R}^{>}_t$,
$$T^{>}_{\mathbf{x}} = \mu / \mathbf{P}\{\mathbf{X} \in \mathscr{R}^{>}_t\}, \tag{5.84}$$
2. For the region $\mathscr{R}^{<}_t$,
$$T^{<}_{\mathbf{x}} = \mu / \mathbf{P}\{\mathbf{X} \in \mathscr{R}^{<}_t\}. \tag{5.85}$$

In the sequel we shall concentrate only upon $\mathscr{R}^{>}_t$: the corresponding formulas for $\mathscr{R}^{<}_t$ could easily be derived. Note that $\mathscr{R}^{>}_t$ may be of interest, e.g., when floods are investigated, while $\mathscr{R}^{<}_t$ may be appropriate if droughts are of concern.

Now, in view of the results outlined in Nelsen et al. (2001) and Nelsen et al. (2003), it is immediate to show that

$$T^{>}_{\mathbf{x}} = \frac{\mu}{1 - K_{\mathbf{C}}(t)}, \tag{5.86}$$

where $K_{\mathbf{C}}$ is the Kendall's distribution function associated with $\mathbf{C}$ (see Eq. (5.21) and the ensuing discussion). Clearly, $T^{>}_{\mathbf{x}}$ is a function of the level t identified by the

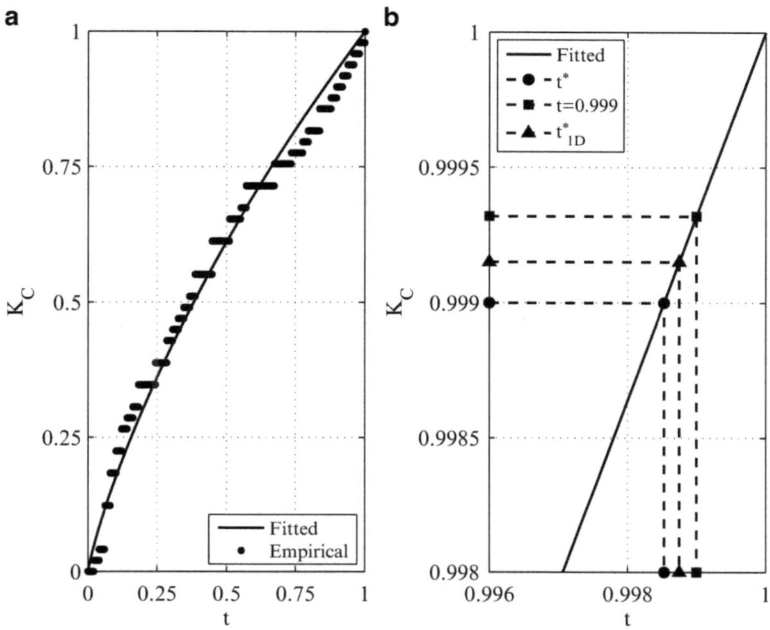

Fig. 5.4 (**a**) Plot of the function K_{C} (*continuous line*) associated with the copula $\mathbf{C}_{QV}$; also shown is its empirical estimate (*markers*) calculated by using the available observations—see text. (**b**) Plot of the (millenary KRP) multivariate quantile $t^* \approx 0.998533$ (*circle*) associated with the critical probability level $p = 0.999$; plot of the multivariate quantile $t^*_{1D} \approx 0.998753$ (*triangle*) corresponding to a KRP of about 1,170 years; plot of the value $K_{\mathrm{C}}(0.999) \approx 0.999319$ (*square*) corresponding to a KRP of about 1,470 years—see text

relation $t = \mathbf{F}(\mathbf{x})$. It is then convenient to denote the above RP via a special notation as follows.

Definition 13 (Kendall's RP). The quantity $\kappa_{\mathbf{x}} = T^{>}_{\mathbf{x}}$ is called the *Kendall's RP* of the realization $\mathbf{x}$ belonging to $\mathscr{L}^{\mathbf{F}}_t$ (hereinafter, *KRP*).

An advantage of the approach outlined in this work is that realizations lying over the same critical layer do always generate the same dangerous region. Evidently, this is not the case considering the "OR–AND" approach discussed previously. Instead, in the approach outlined here, all the realizations $\mathbf{y}$ having a KRP $\kappa_{\mathbf{y}} < \kappa_{\mathbf{x}}$ must lie in $\mathscr{R}^{<}_t$, whereas all those $\mathbf{y}$ having a KRP $\kappa_{\mathbf{y}} > \kappa_{\mathbf{x}}$ must lie in $\mathscr{R}^{>}_t$, and all the realizations lying over $\mathscr{L}^{\mathbf{F}}_t$ share the same KRP $\kappa_{\mathbf{x}}$.

Example 12. As an illustration, in Fig. 5.4a we plot an estimate of the function K_{C} associated with the copula $\mathbf{C}_{QV}$: here Eq. (5.22) is used, and τ_{C} is as reported in Fig. 5.1. Also shown is the empirical estimate of K_{C} calculated by using the available observations (see Algorithm 1 and the ensuing discussion): the horizontal ties are simply due to the small sample size.

In simple words, if $\mathscr{R}_t^>$ is used to identify the potentially destructive occurrences (viz., those involving the events "exceeding" the prescribed critical layer, as in the approach outlined above), then, the events in $\mathscr{R}_t^<$ are safe realizations, whereas the ones over $\mathscr{L}_t^{\mathbf{F}}$ are alert events. As a consequence, a notion of *criticality pre-order* "$\preceq_c$" can be introduced in $\mathbf{R}^d$ via the following formula:

$$\mathbf{x} \preceq_c \mathbf{y} \quad \text{if} \quad \kappa_{\mathbf{x}} \leq \kappa_{\mathbf{y}}. \tag{5.87}$$

In simple words, $\mathbf{x}$ is "less critical" than $\mathbf{y}$ if its KRP is smaller than the one of $\mathbf{y}$: this exactly corresponds to the intuitive meaning of the concept of return period. Conversely, $\mathbf{y}$ is "more critical" than $\mathbf{x}$ if it is more improbable to exceed the critical layer associated with $\mathbf{y}$ than the one associated with $\mathbf{x}$.

5.4.2 Multivariate Quantiles

Traditionally, in the univariate framework, once a RP (say, T) is fixed (e.g., by design or regulation constraints), the corresponding critical probability level p is calculated as $1-p = \mathbf{P}\{X > x_p\} = \mu/T$, and by inverting F_X it is then immediate to obtain the quantile $x_p = F_X^{(-1)}(p)$, which is usually unique. A similar rationale holds if, instead, the dangerous region is $\{X < x_p\}$. Then, x_p is used in practice for design purposes and rational decision making. As shown below, the same approach can also be adopted in a multivariate environment (to be compared with Belzunce et al. 2007).

Definition 14 (Multivariate Quantile). Given a d-dimensional distribution $\mathbf{F} = \mathbf{C}(F_1, \ldots, F_d)$ with d-copula $\mathbf{C}$, and a probability level $p \in \mathbf{I}$, the *Kendall's quantile* $q_p \in \mathbf{I}$ of order p is defined as

$$q_p = \inf\{t \in \mathbf{I}\colon K_{\mathbf{C}}(t) = p\} = K_{\mathbf{C}}^{(-1)}(p), \tag{5.88}$$

where $K_{\mathbf{C}}^{(-1)}$ (generalized) is the inverse of $K_{\mathbf{C}}$.

Note 4. Definition 14 provides a close analogy with the definition of univariate quantile: indeed, recall that $K_{\mathbf{C}}$ is a univariate distribution function (see Eq. (5.21)), and hence q_p is simply the quantile of order p of $K_{\mathbf{C}}$. Thanks to the Probability Integral Transform, it is clear that the critical layer $\mathscr{L}_{q_p}^{\mathbf{F}}$ is the iso-hyper-surface in $\mathbf{R}^d$ where $\mathbf{F}$ takes on the value q_p, while $\mathscr{L}_{q_p}^{\mathbf{C}}$ is the corresponding one in $\mathbf{I}^d$ where the related copula $\mathbf{C}$ equals q_p.

Now, let $\mathscr{L}_{q_p}^{\mathbf{F}}$ be fixed. Then, according to Eq. (5.21),

$$p = K_{\mathbf{C}}(q_p) = \mathbf{P}\{\mathbf{C}(F_1(X_1), \ldots, F_d(X_d)) \leq q_p\}.$$

Therefore, p is the probability measure induced by $\mathbf{C}$ on the region $\mathscr{R}^{<}_{q_p}$, while $(1-p)$ is the one of $\mathscr{R}^{>}_{q_p}$. From a practical point of view this means that, in a simulation of n independent d-dimensional vectors extracted from $\mathbf{F}$, np realizations are expected to lie in $\mathscr{R}^{<}_{q_p}$, and the others in $\mathscr{R}^{>}_{q_p}$.

Note 5. It is worth stressing that a common error is to confuse the value of the copula $\mathbf{C}$ with the probability induced by $\mathbf{C}$ on $\mathbf{I}^d$ (and, hence, on $\mathbf{R}^d$ via the formula $\mathbf{F} = \mathbf{C}(F_1, \ldots, F_d)$): on the critical layer $\mathscr{L}^{\mathbf{C}}_{q_p}$ it is $\mathbf{C} = q_p$, but the corresponding region $\mathscr{R}^{<}_{q_p}$ has probability $p = K_{\mathbf{C}}(q_p) \neq q_p$, since $K_{\mathbf{C}}$ is usually non-linear (the same rationale holds for the region $\mathscr{R}^{>}_{q_p}$). In other words, while in the univariate case the values of F_X correspond to the probabilities induced on the Real line, this is not so in the multivariate case.

Since $K_{\mathbf{C}}$ is a probability distribution, and q_p is the corresponding quantile of order p, we could use a standard bootstrap technique (see, e.g., Davison and Hinkley 1997) to estimate q_p if it cannot be calculated analytically. The idea is simple, and stems directly from the very definition of q_p: viz., to look for the value q_p of $\mathbf{C}$ such that, in a simulation of size n, np realizations show a copula value less than q_p. Then, by performing a large number of independent simulations of size n, the sample average of the estimated q_p's is expected to converge to the true value of q_p by virtue of the Law of Large Numbers (see also Genest and Rivest 1993; Barbe et al. 1996). A possible algorithm is given below, most suitable for vectorial software. Here we assume that the copula model is well specified, i.e. it is available in a parametric form.

• **ALGORITHM 5.** First of all, choose a sample size n, a critical probability level p, the total number of simulations N, and fix the critical index $k = \lfloor np \rfloor$.

```
for i = 1 : N
   S = sim(C; n); % simulate n d-vectors from copula C
   C = C(S); % calculate C for simulated vectors
   C = sort(C); % sort-ascending simulated C values
   E(i) = C(k); % store new estimate of q_p into vector E
end
q_p = Mean(E); % calculate the estimate of q_p
```

Once the loop is completed, q_p provides a consistent estimate of the critical multivariate quantile of order p. Practically, Algorithm 5 does the "inverse" task of Algorithm 1. The bootstrap method may also yield an approximate confidence interval for q_p (see DiCiccio and Efron (1996) for more refined solutions): for instance, at a 10 % level, the random interval $(q_{0.05}, q_{0.95})$ can be used, where $q_{0.05}$ and $q_{0.95}$ are, respectively, the quantiles of order 5 % and 95 % extracted from the vector E.

Example 13. As an illustration, in Fig. 5.4b we show an estimate of the function $K_{\mathbf{C}}$, associated with the copula $\mathbf{C}_{QV}$, at the critical quantile $t^* \approx 0.998533$ (corresponding to a millenary KRP): as expected, the value is almost exactly equal to 99.9 %.

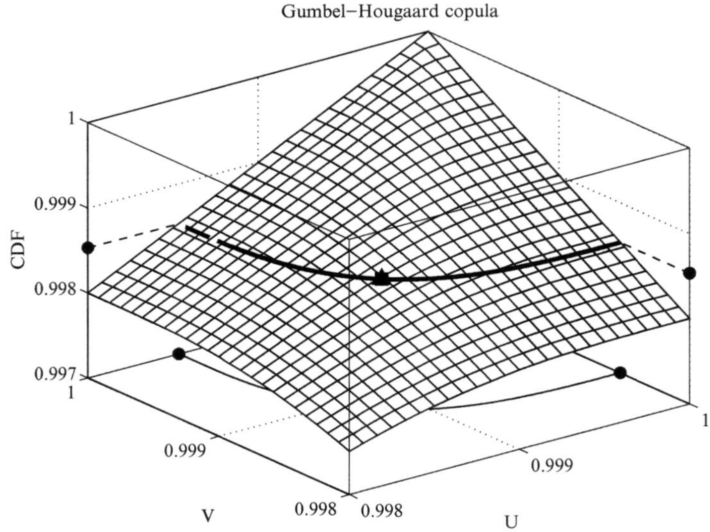

Fig. 5.5 Critical iso-line $\mathscr{L}_{t^*}^{\mathbf{C}}$ of the copula $\mathbf{C}_{QV}$ corresponding to the (millenary KRP) critical level $t^* \approx 0.998533$, indicated by the *circle* markers on the axes—see also Fig. 5.3. The *triangle* marker indicates the Component-Wise Excess design realization, and the *square* marker indicates the Most-Likely design realization—see Sect. 5.4.3 and Table 5.1

Furthermore, in Fig. 5.5 we plot the Gumbel-Hougaard copula $\mathbf{C}_{QV}$ in the upper-right corner of the unit square, and show the critical layer (iso-line) $\mathscr{L}_{t^*}^{\mathbf{C}}$ for the critical level $t^* \approx 0.998533$, corresponding to a regulation return period of 1,000 years (viz., all the realizations on $\mathscr{L}_{t^*}^{\mathbf{C}}$ have a KRP equal to 1,000 years)—see also Fig. 5.3. Then, $\mathbf{C}_{QV} = t^*$ for all points belonging to $\mathscr{L}_{t^*}^{\mathbf{C}}$. Instead, $\mathbf{C}_{QV} < t^*$ (and $\kappa_{\mathbf{X}} < 1000$ years) in the region $\mathscr{R}_{t^*}^{<}$ "below" $\mathscr{L}_{t^*}^{\mathbf{C}}$, the one containing the origin $\mathbf{0} = (0,0)$, whereas $\mathbf{C}_{QV} > t^*$ (and $\kappa_{\mathbf{X}} > 1000$ years) in the region $\mathscr{R}_{t^*}^{>}$ "above" $\mathscr{L}_{t^*}^{\mathbf{C}}$, the one containing the upper corner $\mathbf{1} = (1,1)$. On average, only 0.1 % of the realizations extracted from a simulation of $\mathbf{C}_{QV}$ are expected to lie in $\mathscr{R}_{t^*}^{>}$. However, the level of the critical layer is $t^* = q_{0.999} \approx 0.998533 < p = 0.999$, as indicated by the markers in the plot.

Moreover, let $q^* = 0.999$ be a fixed univariate (millenary) quantile, and set $\mathbf{z}^* = (q^*, q^*)$—see Example 10: essentially we assume here that Q and V are equally "important". Then, the "OR" dangerous region introduced in Eq. (5.81) has probability $1 - \mathbf{C}_{QV}(q^*, q^*) \approx 0.001247$, and a RP of about 800 years. Instead, the "AND" dangerous region introduced in Eq. (5.82) has probability $1 - 2 \cdot q^* + \mathbf{C}_{QV}(q^*, q^*) \approx 0.000753$, and a RP of about 1300 years. In order to have exact millenary regions, then the following quantiles should be used: $q_{\vee}^* \approx 0.999198 > q^*$ when considering $\mathscr{D}_{\mathbf{z}^*}^{\vee}$, and $q_{\wedge}^* \approx 0.998672 < q^*$ when considering $\mathscr{D}_{\mathbf{z}^*}^{\wedge}$.

Note 6. The region $\mathscr{R}_{0.999}^{>}$ identified by the critical layer $\mathscr{L}_{0.999}^{\mathbf{F}}$ (where the multivariate distribution $\mathbf{F}$—or, equivalently, the copula $\mathbf{C}$—takes on the value 0.999) has an estimated probability ≈ 0.000682, and a corresponding KRP of about 1,470 years: practically, only one realization of $\mathbf{C}_{QV}$ out of 1,470 simulations is expected to lie in such a region (instead of 1 out of 1,000). Evidently, if $\mathbf{F}$ (or $\mathbf{C}$) were substituted for $K_{\mathbf{C}}$ in Eq. (5.86) during the design phase, then the structure to be constructed might be over-sized.

5.4.3 Multivariate Design Events

The situation outlined in the previous section is generally similar to the one found in the study of univariate phenomena, where a single r.v. X with distribution F_X is used to model the stochastic dynamics. However, as already mentioned, the multivariate case generally fails to provide a natural solution to the problem of identifying a unique design realization. In fact, even if also the layer $\mathscr{L}_t^{\mathbf{F}}$ acts as a (multi-dimensional) critical threshold, there is no natural criterion to select which realization lying on $\mathscr{L}_t^{\mathbf{F}}$ (among the ∞^{d-1} possibilities) should be used for design purposes. In other words, in a multivariate environment, the sole tool provided by the RP may not be sufficient to identify a design realization, and additional considerations may be required in order to pick out a "characteristic" realization over the critical layer of interest. In the following, we outline possible ways to carry out such a selection. Clearly, several approaches can be proposed, each one possibly yielding a different solution: below, we only show two possible elementary strategies to deal with the problem.

The basic idea is simply to introduce a suitable function (say, $\mathbf{w}$) that "weighs" the realizations lying on the critical layer of interest. Following this approach, the practitioner can then freely choose the criterion (i.e., the function $\mathbf{w}$) that best fits the practical needs. Clearly, without loss of generality, $\mathbf{w}$ can be assumed to be non-negative. In turn, a "design realization" can be defined as follows.

Definition 15 (Design Realization). Let $\mathbf{w}\colon \mathscr{L}_t^{\mathbf{F}} \to [0,\infty)$ be a weight function. The design realization $\delta_{\mathbf{w}} \in \mathscr{L}_t^{\mathbf{F}}$ is defined as

$$\delta_{\mathbf{w}}(t) = \operatorname{argmax}_{\mathbf{x}\in\mathscr{L}_t^{\mathbf{F}}} \mathbf{w}(\mathbf{x}), \tag{5.89}$$

provided that the argmax exists and is finite.

Definition 15 deserves some comments.

- In general, the unicity of the maximum may not be guaranteed. When this happens, a recourse to physical/phenomenological considerations, or to additional procedures (like, e.g., Maximum Information/Entropy schemes (Jaynes 2003)), may help solving the problem.
- Different copulas may share the same Kendall's measure $K_{\mathbf{C}}$, and hence the same KRP (e.g., all the bivariate Extreme Value copulas with the same Kendall's τ (Ghoudi et al. 1998)). However, in general, the critical layers of such copulas will have different geometries, and, in turn, will provide different design realizations.

- The search of the point of maximum in Eq. (5.89) can be subjected to additional constraints, in order to take into account the possible sensitivity of the structure under design to the behavior of specific marginals (see also the discussion in Note 7 below): for instance, a Bayesian approach might be advisable.
- Sometimes it could be more appropriate to select a set of possible design realizations (i.e., an ensemble, rather than a single one) that should be used, together with experts' opinions, in order to better evaluate the features of the phenomenon affecting the structure under design. This procedure can be carried out by using a suitable step function in Eq. (5.89).

Note 7. A delicate problem may arise when adopting the approach outlined above: to make the point clear, consider the following example. Suppose that we use the duration of a storm and the storm intensity as the two variables of interest. In a fast responding system (e.g., a sewer structure), a storm having short duration but high intensity may cause a failure, whereas the same storm may not cause any problem at a catchment level. In the catchment, however, a storm with long duration and intermediate to low intensity may cause a flood event, whereas the same storm does not cause any problem to the sewer system. Now, as a matter of principle, the design realization $\delta_{\mathbf{w}}$ for the given return period (i.e., the "typical" storm calculated according to the strategy illustrated here) may not cause any problem in both systems, and therefore these would be wrongly designed. Practically, the sewer system should be designed using storms of short durations and high intensities, whereas a structure in the main river of the watershed should be designed using storms of long durations and intermediate to low intensities. However, the problem is more apparent than real. In fact, there are neither theoretical nor practical limitations to restrict the search for the maxima in Eq. (5.89) over a suitable sub-region of $\mathscr{L}_{t^*}^{\mathbf{F}}$: remember that all the realizations on the critical layer share the same prescribed KRP. Thus, when a sewer system is of concern, only storms having short durations and high intensities could be considered, whereas a design storm for a structure in the main river could be spotted by restricting the attention to storms of long durations and intermediate to low intensities. Roughly speaking, in the approach outlined here, the calculation of the design realization can be made dependent on both the environment in which a structure should be designed, as well as on the stochastic dynamics of the phenomenon under investigation.

Overall, the procedure to identify the design realization could be described as follows. Let $\mathbf{X}$ be a random vector with distribution $\mathbf{F} = \mathbf{C}(F_1, \ldots, F_d)$.

1. Fix a RP T.
2. Calculate the corresponding probability level $p = 1 - \mu/T$.
3. Compute the Kendall's quantile q_p as in Eq. (5.88), either analytically or by using Algorithm 5.
4. Fix a suitable weight function $\mathbf{w}$.
5. Calculate the point(s) of maximum $\delta_{\mathbf{w}}$ of $\mathbf{w}$ on the critical layer $\mathscr{L}_{q_p}^{\mathbf{F}}$ (or on a sub-region of it).

The resulting $\delta_{\mathbf{w}}$ represents a "typical" realization in $\mathbf{R}^d$ with a given KRP. Roughly speaking, it denotes the design realization obtained by considering the very stochastic dynamics of the phenomenon. Note that, in general, $\delta_{\mathbf{w}}$ (or, better, the corresponding critical layer) should be considered together with other information (e.g., the physical features of the structure) in order to be correctly used in practice.

As a further consideration, the framework presented in this work is also most suitable for the use of a different strategy, namely a simulation-based approach. In fact, a large number of multivariate events could be generated, and the associated "cost of failure" of each event could be computed. Then, the structure could be designed, for instance, in order to minimize the expected cost, e.g. via a Bayesian approach to decision making (Berger 1985; Smith 1988; Robert 2007).

For the sake of illustration, below we introduce two elementary weight functions.

5.4.3.1 Component-Wise Excess Design Realization

A realization lying on the critical layer $\mathscr{L}_t^{\mathbf{F}}$ may be marked as critical when all of its marginal components are exceeded with the largest probability. In simple words, we suggest to look for the point(s) $\mathbf{x} = (x_1, \ldots, x_d) \in \mathscr{L}_t^{\mathbf{F}}$ such that it is maximum the probability that a realization $\mathbf{y} = (y_1, \ldots, y_d)$ satisfies all the following componentwise inequalities:

$$y_1 \geq x_1, \ldots, y_d \geq x_d, \tag{5.90}$$

or $\mathbf{y} > \mathbf{x}$ using a simplified notation. The next definition is immediate.

Definition 16. The Component-Wise Excess weight function $\mathbf{w}_{CE}$ is defined as

$$\mathbf{w}_{CE}(\mathbf{x}) = \mathbf{P}(\mathbf{X} \in [\mathbf{x}, \infty)), \tag{5.91}$$

where $\mathbf{X}$ has distribution $\mathbf{F} = \mathbf{C}(F_1, \ldots, F_d)$, and $[\mathbf{x}, \infty)$ is the hyper-rectangle in $\mathbf{R}^d$ whose points satisfy all the inequalities stated in Eq. (5.90).

Then, by restricting our attention to the critical layer $\mathscr{L}_t^{\mathbf{F}}$, the following definition is immediate.

Definition 17. The Component-wise Excess design realization δ_{CE} of level t is defined as

$$\delta_{CE}(t) = \operatorname{argmax}_{\mathbf{x} \in \mathscr{L}_t^{\mathbf{F}}} \mathbf{w}_{CE}(\mathbf{x}), \tag{5.92}$$

where $t \in (0, 1)$.

Note 8. Via the Probability Integral Transform and Sklar's Theorem, it is easy to show that

$$\mathbf{w}_{CE}(\mathbf{x}) = \mathbf{P}(\mathbf{U} \in [\mathbf{u}(\mathbf{x}), \mathbf{1}]), \tag{5.93}$$

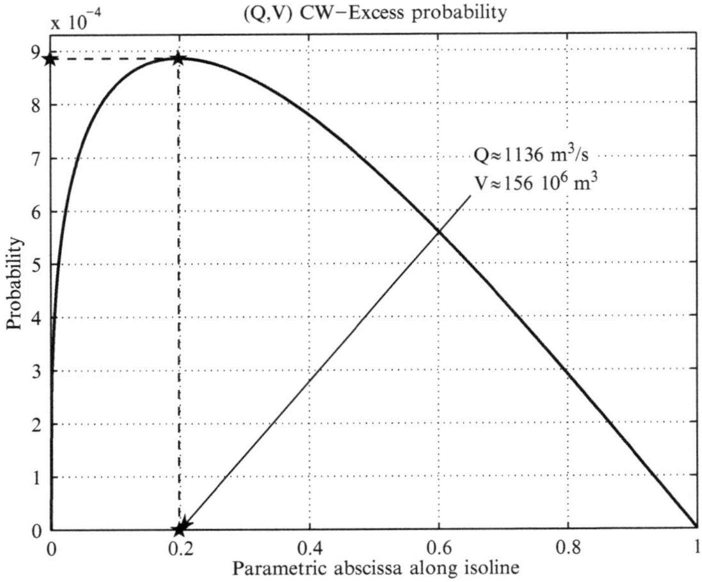

Fig. 5.6 Plot of the Component-Wise Excess weight function $\mathbf{w}_{CE}$ over the critical iso-line $\mathscr{L}_{t^*}^{\mathbf{C}_{QV}}$, corresponding to the (millenary KRP) critical level $t^* \approx 0.998533$. The *star* marker indicates where the maximum is attained—see text and Table 5.1, and also Fig. 5.3

where $\mathbf{U}$ has the same copula $\mathbf{C}$ as of $\mathbf{X}$ and Uniform marginals, $[\mathbf{u}, \mathbf{1}]$ is the hyper-rectangle in $\mathbf{I}^d$ with "lower" corner $\mathbf{u}$ and "upper" corner $\mathbf{1}$, and $\mathbf{u} = \mathbf{u}(\mathbf{x}) = (F_1(x_1), \ldots, F_d(x_d))$. Thus, the probabilities of interest can be directly computed in the unit hyper-cube (see, e.g., Joe 1997) by working directly on the critical layer $\mathscr{L}_t^{\mathbf{C}}$ (instead of $\mathscr{L}_t^{\mathbf{F}}$), a solution numerically more convenient. Note that, for large d-dimensional problems, the CPU time involved may become prohibitive, though clever solutions have been proposed for $d \gg 1$ (see, e.g., Cherubini and Romagnoli 2009). In some cases, δ_{CE} can be calculated analytically; otherwise, it can be empirically estimated (e.g., by performing a suitable sampling over $\mathscr{L}_t^{\mathbf{C}}$ or $\mathscr{L}_t^{\mathbf{F}}$).

Example 14. In Fig. 5.6 we show the behavior of $\mathbf{w}_{CE}$ over $\mathscr{L}_{t^*}^{\mathbf{C}_{QV}}$, as well as the Component-wise Excess design realization $\delta_{CE}(t^*)$ calculated for the case study investigated here. This point has the largest probability to be componentwise exceeded by a dangerous realization, and therefore it should be regarded as a sort of (statistical) "safety lower-bound": viz., the structure under design should, at least, withstand realizations having (multivariate) size $\delta_{CE}(t^*)$, as reported in Table 5.1.

As a test, using the design realization $\delta_{CE}(t^*)$, we operated the reservoir routing of the corresponding flood hydrograph. We use here an initial water level $W = 780.91$ m a.s.l., corresponding to the empirical mode of the distribution of the W's (i.e., the most frequent initial water level). Then, we checked whether or

Table 5.1 Estimates of the design realizations, for a millenary KRP, according to different strategies, and assuming an initial water level W=780.91 m a.s.l.–see text. The right-most column shows the maximum water level of the dam associated with the flood event (Q, V) reported on the corresponding row

Strategy	**Q** (m^3/s)	**V** ($10^6 m^3$)	M.W.L. (m a.s.l.)
C.-E.	1,136	156	784.58
M.-L.	1,141	155	784.59
$F_{\bullet}^{(-1)}(0.999)$	1,209	173	784.82

not the reservoir level exceeds the crest level of the dam (at 784 m a.s.l.). The column"M.W.L." in Table 5.1 reports the value 784.58 m a.s.l.: thus, the overtopping may occur, i.e. the dam is apparently at risk against Component-Wise Excess millenary bivariate realizations.

5.4.3.2 Most-Likely Design Realization

A further approach to the definition of a characteristic design event consists in taking into account the density of the multivariate distribution describing the overall statistics of the phenomenon investigated: in fact, assuming that the density **f** of **F** is well defined over $\mathscr{L}_t^{\mathbf{F}}$, we may think of using it as a weight function.

Clearly, the restriction $\mathbf{f}_t$ of **f** over $\mathscr{L}_t^{\mathbf{F}}$ is not a proper density, since it does not integrate to one. However, it may provide useful information, since it induces a (weak) form of likelihood over $\mathscr{L}_t^{\mathbf{F}}$: in fact, it can be used to weigh the realizations lying on $\mathscr{L}_t^{\mathbf{F}}$, and spot those that are (relatively) "more likely" than others. Indeed, $\mathbf{f}_t$ inherits all the features of interest here directly from the true global density **f**. The next definition is immediate.

Definition 18. The Most-Likely weight function $\mathbf{w}_{ML}$ is defined as

$$\mathbf{w}_{ML}(\mathbf{x}) = \mathbf{f}(\mathbf{x}), \tag{5.94}$$

where **f** is the density of $\mathbf{F} = \mathbf{C}(F_1, \dots, F_d)$.

Then, by restricting our attention to the critical layer $\mathscr{L}_t^{\mathbf{F}}$, the following definition is immediate.

Definition 19. The Most-Likely design realization δ_{ML} of level t is defined as

$$\delta_{ML}(t) = \operatorname{argmax}_{\mathbf{x} \in \mathscr{L}_t^{\mathbf{F}}} \mathbf{w}_{ML}(\mathbf{x}) = \operatorname{argmax}_{\mathbf{x} \in \mathscr{L}_t^{\mathbf{F}}} \mathbf{f}(\mathbf{x}), \tag{5.95}$$

where $t \in (0, 1)$.

Note 9. As a rough interpretation, δ_{ML} plays the role as of a "characteristic realization", i.e. the one that has to be expected if a dangerous event with given KRP happens. In some cases, δ_{ML} can be calculated analytically; otherwise, it can be empirically estimated (e.g., by performing a suitable sampling of **f** over $\mathscr{L}_t^{\mathbf{F}}$).

In general, provided that weak regularity conditions are satisfied, **f** can be calculated by using the density $\mathbf{c}(\mathbf{u}) = \frac{\partial^d}{\partial u_1 \cdots \partial u_d} \mathbf{C}(u_1, \ldots, u_d)$ of the copula **C** and the marginal densities f_i's of **X**:

$$\begin{aligned} \mathbf{f}(\mathbf{x}) &= \frac{\partial^d}{\partial x_1 \cdots \partial x_d} \mathbf{C}(F_1(x_1), \ldots, F_d(x_d)) \\ &= \mathbf{c}(F_1(x_1), \ldots, F_d(x_d)) \cdot \prod_{i=1}^{d} f_i(x_i). \end{aligned} \tag{5.96}$$

Since our target is to compare the "weight" of different realizations, from a computational point of view it may be better to minimize $-\ln(\mathbf{f})$ over $\mathscr{L}_t^{\mathbf{F}}$ (since the maxima are preserved).

Example 15. In the present (absolutely continuous) case, the expression of the bivariate density $\mathbf{f}_{QV}$ is given by

$$\mathbf{f}_{QV}(x, y) = \mathbf{c}_{QV}(F_Q(x), F_V(y)) \cdot f_Q(x) \cdot f_V(y), \tag{5.97}$$

where $(x, y) \in \mathbf{R}^2$, and $\mathbf{c}_{QV}$ is the density of the Gumbel-Hougaard copula modeling the pair (Q, V). In Fig. 5.7 we show the behavior of (the logarithm of) $\mathbf{w}_{ML}$ (i.e., $\mathbf{f}_{QV}$) over $\mathscr{L}_{t^*}^{\mathbf{F}}$, as well as the Most-Likely design realization $\delta_{ML}(t^*)$ calculated for the case study investigated here. The actual values of the function $\mathbf{w}_{ML}$ are irrelevant (since they do not represent a true density): in fact, we are only interested in spotting where $\mathbf{f}_{QV}$ is maximal. Therefore, the Most-Likely design realization could be regarded as the "typical" realization: viz., the structure under design should be expected to withstand events having (multivariate) size $\delta_{ML}(t^*)$, as reported in Table 5.1.

Again, as a test, using the design realization $\delta_{ML}(t^*)$, we operated the reservoir routing of the corresponding flood hydrograph, and checked whether or not the reservoir level exceeds the crest level of the dam. The column "M.W.L." in Table 5.1 reports the value 784.59 m a.s.l.: thus, again, the over-topping may occur, i.e. the dam is apparently at risk against Most-Likely millenary bivariate realizations.

5.4.3.3 Further Notes About Design

Concerning design realizations, another interesting test can be carried out. In fact, as a further possible strategy, suppose that a design realization $\delta_{1D} = (x_{0.999}, y_{0.999})$ is defined in terms of the millenary univariate quantiles of the two variables (Q, V)

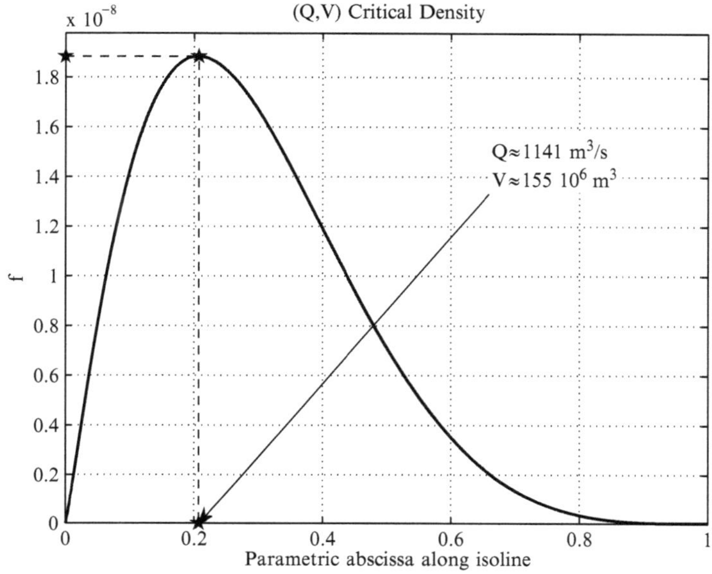

Fig. 5.7 Plot of the (log) Most-Likely weight function $\mathbf{w}_{ML}$ over the critical iso-line $\mathscr{L}^{\mathbf{F}}_{t^*}$, corresponding to the (millenary KRP) critical level $t^* \approx 0.998533$. The *star* marker indicates where the maximum is attained—see text and Table 5.1, and also Fig. 5.3

of interest here (see the last row of Table 5.1). In turn, $t^*_{1D} = \mathbf{C}_{QV}(0.999, 0.999) \approx 0.998753$ and $K_{\mathbf{C}}(t^*_{1D}) \approx 0.999150$, corresponding to a KRP of about 1,170 years (see Fig. 5.4b). It is then immediate to realize that, in order to provide a true millenary multivariate design realization, it may not be enough (or necessary) to rely upon millenary univariate quantiles. Also, operating the reservoir routing using δ_{1D}, yields a reservoir level of about 784.82 m a.s.l. (see Table 5.1), which may cause an over-topping.

The behavior of the Ceppo Morelli dam was used throughout this Chapter in order to illustrate the practical application of the techniques outlined. It is interesting to note that the results reported above about the safety of the dam are apparently in contrast to those presented in De Michele et al. (2005) (bivariate analysis) and Salvadori et al. (2011) (trivariate analysis).

In De Michele et al. (2005), a simulation of 1,000 pairs of the variables (Q, V) was used to check the adequacy of the dam spillway, where the initial water level of the reservoir was chosen at random (instead of fixing it at the modal value $W = 780.91$ m a.s.l.), by using the empirical distribution of the water levels. As a conclusion, no over-topping occurred. Clearly, the analysis of only a 1,000-year sample (viz., a too small sample size), might have hidden possible dangerous realizations: thus, the different conclusions presented in De Michele et al. (2005) may simply be due to an inadequate simulation approach. Alternatively, one could operate the reservoir routing for, say, 10^6 simulations, and provide an estimate of the over-topping probability.

Table 5.2 Estimates of the critical design realizations, for a millenary KRP, according to different strategies as calculated in Salvadori et al. (2011)—to be compared with Table 5.1

Strategy	**Q** (m^3/s)	**V** ($10^6 m^3$)	W (m a.s.l.)	M.W.L. (m a.s.l.)
C.-E.	352.76	25.21	781.25	782.08
M.-L.	316.23	19.64	781.29	781.98

In Salvadori et al. (2011), a trivariate model was used by considering the triplet (Q, V, W), i.e. by including the water level W as a random variable. None of the design realizations δ_{CE} and δ_{ML} calculated in Salvadori et al. (2011) caused an over-topping. Actually, the corresponding marginal design values of Q and V (see Table 5.2) were much smaller than the ones reported in Table 5.1: at first sight, this fact may be puzzling. However, this apparent inconsistency may be due to a general phenomenon related to a *dimensionality paradox*: an explanation can be found in the following illustration.

Example 16 (Dimensionality paradox). Let the Return Period T be given by Eq. (5.80) in Definition 10. Without loss of generality, we assume here that $\mu = 1$. Let X_1 be a r.v. with Uniform law over $(0, 1)$, and suppose that the event of interest is $E_1 = \{X_1 > x_1\}$, where $x_1 = 0.9$ is the quantile of X_1 of order 0.9. Then, the Return Period T_1 of E_1 is 10, since $\mathbf{P}\{E_1\} = 0.1$.

Now, let X_2 be an i.i.d. copy of X_1, i.e. assume that $\mathbf{C}_{X_1 X_2} = \Pi_2$. Suppose that the event of interest is

$$E_2 = \{X_1 > x_2 \cap X_2 > x_2\},$$

where the quantile x_2 has to be chosen in order to yield a Return Period T_2 for E_2 equal to 10. Given the fact the X_i's are i.i.d. and Uniform over $(0, 1)$, it is immediate to realize that E_2 is a small square in the right-upper corner of the unit square, and that its area must be 0.1. This yields a quantile $x_2 = 1 - \sqrt{0.1} \approx 0.683772 < x_1 = 0.9$.

In general, let $d > 1$ be the dimension of the problem under investigation, and let $X_1, \ldots, X_d$ be i.i.d. r.v.'s Uniform over $(0, 1)$, i.e. assume that $\mathbf{C}_{X_1 \ldots X_d} = \Pi_d$. Suppose that the event of interest is

$$E_d = \bigcap_{i=1}^{d} \{X_i > x_d\},$$

where the quantile x_d has to be chosen in order to yield a Return Period T_d for E_d equal to 10. As above, $x_d = 1 - (0.1)^{1/d} < x_1 = 0.9$.

As a conclusion, increasing the dimensionality of the problem may yield a reduction of the univariate marginal quantiles, when embedded in a multivariate framework. Note that x_d is a decreasing function of d. As a consequence, the discrepancies mentioned above between the values of the design realizations in two and three dimensions should not be considered as anomalies, but rather have to be expected. Which of the approaches (bivariate rather than trivariate) is the "right" one is at present an open question.

5.5 Discussion and Perspectives

In this Chapter we provide a short overview of some of the main aspects of MEV theory and applications. Although some problems may find a consistent solution within the MEV framework, the wealth of challenges that are still ahead when dealing with multivariate extremes makes this area an open and interesting field for future researches. Overall, it should be kept in mind that MEV methods only have asymptotic justifications. As thoroughly explained in Coles (2001), to which we make reference, the following issues need to be considered.

- **Methods of estimation.** Estimating the parameters of multivariate models is always a difficult task: for instance, as soon as the dimension increases, it is more and more improbable that the numerical algorithms adopted converge towards the exact solution. Using copulas may help, since the estimate of the parameters of the univariate marginals is carried out separately from the calculation of the ones of the copula (see, e.g., Genest and Favre 2007). However, this may raise statistical questions about the robustness of the estimates, and no ultimate solutions have been proposed yet.
- **Quantification of uncertainty.** Given the available historical data, the statistical estimates (e.g., those of the parameters, or the return periods, or the design realizations) should only be intended as "best guesses" at the truth. In other words, whether the results of the statistical analysis should be trusted or not may be an "act of faith". For instance, it may be enough to consider different samples of the same phenomenon to find different estimates of the same parameters. Thus, measures of uncertainty should be used in order to deal with the sample variability. This issue is more and more important in a MEV context, since extrapolations to very large (or small) levels may yield unreliable results.
- **Model diagnostic.** Once a model is adopted, it should always be validated via suitable Goodness-of-Fit tests. Unfortunately, the construction of powerful tests is still an open research area in multivariate analysis (see, e.g., Genest et al. (2009), Berg (2009) and Kojadinovic et al. (2011), and references therein).
- **Non-stationarity.** The stationarity assumption is almost always at the basis of the statistical techniques used for analysing historical data. However, it is a hypothesis that rarely can be verified. Some techniques are available for dealing with non-stationary phenomena (see, e.g., Wigley 1988; Cooley 2009; Katz 2010), but the theory gets quite tricky. In addition, the number of parameters involved may remarkably increase, and it often happens that the available sample

size may not be enough to calculate reliable estimates. For this reasons, the research on suitable methods for dealing with non-stationarity in a multivariate context is still open. We report here the final remark pointed out in Katz (2010), with which we agree: "Finally, the question remains of the terms in which to most effectively convey the risk of extreme events under a changing climate. The familiar, if often misunderstood, concepts of return period and return level (e.g., the proverbial 100-year flood), strictly speaking, no longer apply in a non-stationary climate. The most appropriate way to extend these concepts is an open question."

Acknowledgements The Authors thank C. Sempi (Università del Salento, Lecce, Italy) and F. Durante (Free University of Bozen–Bolzano, Bolzano, Italy) for invaluable helpful discussions and suggestions, and M. Pini (Università di Pavia, Pavia, Italy) for software recommendations. The excellent and constructive review of an anonymous Referee is also gratefully acknowledged. [G.S.] The research was partially supported by the Italian M.I.U.R. via the project "Metodi stocastici in finanza matematica". The support of "Centro Mediterraneo Cambiamenti Climatici" (CMCC – Lecce, Italy) is acknowledged.

References

Alexander GN (1954) Some aspects of time series in hydrology. J Inst Eng Aust 26:196

Bacchi B, Becciu G, Kottegoda N (1994) Bivariate exponential model applied to intensities and durations of extreme rainfall. J Hydrol 155:225–236

Barbe P, Genest C, Ghoudi K, Rémillard B (1996) On Kendall's process. J Multivar Anal 58(2):197–229

Belzunce F, Castaño A, Olvera-Cervantes A, Suárez-Llorens A (2007) Quantile curves and dependence structure for bivariate distributions. Comput Stat Data Anal 51(10):5112–5129

Berg D (2009) Copula goodness-of-fit testing: an overview and power comparison. Eur J Financ 15(7–8):675–701

Berger JO (1985) Statistical decision theory and bayesian analysis, 2nd edn. Springer, New York

Blomqvist N (1950) On a measure of dependence between two random variables. Ann Math Stat 21:593–600

Charpentier A (2006) Dependence structures and limiting results, with applications in finance and insurance. Ph.D. thesis, Katholieke Universiteit Leuven (Belgium)

Chebana F, Ouarda TBMJ (2009a) Index flood-based multivariate regional frequency analysis. Water Resour Res 45:W10,435. doi: 10.1029/2008WR007490

Chebana F, Ouarda TBMJ (2009b) Multivariate quantiles in hydrological frequency analysis. Environmetrics. doi: 10.1002/env.1027

Chebana F, Ouarda T (2011) Multivariate extreme value identification using depth functions. Environmetrics 22(3):441–455. doi: 10.1002/env.1089

Cherubini U, Romagnoli S (2009) Computing the volume of n-dimension copulas. Appl Math Financ 16(4):307–314

Coles S (2001) An introduction to statistical modeling of extreme values. Springer, London/New York

Coles S, Tawn J (1991) Modeling extreme multivariate events. J R Stat Soc B 53(2):377–392

Coles S, Tawn J (1994) Statistical methods for multivariate extremes: an application to structural design. Appl Stat 43:1–48

Coles S, Heffernan J, Tawn J (1999) Dependence measures for extreme value analyses. Extremes 2(4):339–365

Cooley D (2009) Extreme value analysis and the study of climate change: a commentary on Wigley 1988. Clim Chang 97:77–83
Cox DR (1962) Renewal theory. Wiley, New York
Davison AC, Hinkley DV (1997) Bootstrap methods and their application. Cambridge series in statistical and probabilistic mathematics, vol 1. Cambridge University Press, Cambridge
De Haan L, Resnick S (1977) Limit theory for multivariate sample extremes. Zeitschrift für Wahrscheinlichkeitstheorie und verwandte Gebiete 40(4):317–337
Deheuvels P (1979) La fonction de dépendence empirique et ses propriétés. un test non paramétrique d'indépendence. Acad R Belg Bull Cl Sci 65(5):274–292
De Michele C, Salvadori G, Canossi M, Petaccia A, Rosso R (2005) Bivariate statistical approach to check adequacy of dam spillway. ASCE J Hydrol Eng 10(1):50–57
De Michele C, Salvadori G, Passoni G, Vezzoli R (2007) A multivariate model of sea storms using copulas. Coast Eng 54:734–751
Devroye L (1986) Non-uniform random variate generation. Springer, New York
DiCiccio TJ, Efron B (1996) Bootstrap confidence intervals. Stat Sci 11(3):189–228
Durante F (2009) Construction of non-exchangeable bivariate distribution functions. Stat Papers. 50(2):383–391
Durante F, Salvadori G (2010) On the construction of multivariate extreme value models via copulas. Environmetrics 21:143–161
Durante F, Quesada-Molina J, Úbeda-Flores M (2007) On a family of multivariate copulas for aggregation processes. Inf Sci 177(24):5715–5724
Embrechts P, McNeil A, Straumann D (2002) Correlation and dependence in risk management: properties and pitfalls. In: Risk management: value at risk and beyond. Cambridge University Press, Cambridge
Embrechts P, Klüppelberg C, Mikosch T (2003) Modelling extremal events for insurance and finance, 4th edn. Springer, Berlin
Falk M, Reiss RD (2005) On Pickands coordinates in arbitrary dimensions. J Multivar Anal 92:426–453
Favre AC, El Adlouni S, Perreault L, Thiémonge N, Bobée B (2004) Multivariate hydrological frequency analysis using copulas. Water Resour Res 40(1):W01101. doi: 10.1029/2003WR002456
Feller W (1971) An introduction to probability and its applications, vol 1, 3rd edn. Wiley, New York
Frahm G (2006) On the extremal dependence coefficient of multivariate distributions. Stat Probab Lett 76:1470–1481
Frahm G, Junker M, Schmidt R (2005) Estimating the tail-dependence coefficient: properties and pitfalls. Insur Math Econ 37:80–100
Fredericks G, Nelsen R (2007) On the relationship between Spearman's rho and Kendall's tau for pairs of continuous variables. J Stat Plan Inference 137:2143–2150
Galambos J (1987) The asymptotic theory of extreme order statistics, 2nd edn. Kreiger, Malabar
Genest C, Favre A (2007) Everything you always wanted to know about copula modeling but were afraid to ask. J Hydrol Eng 12(4):347–368
Genest C, Rivest L (1989) A characterization of Gumbel's family of extreme value distributions. Stat Probab Lett 8:207–211
Genest C, Rivest LP (1993) Statistical inference procedures for bivariate Archimedean copulas. J Am Stat Assoc 88(423):1034–1043
Genest C, Rivest LP (2001) On the multivariate probability integral transformation. Stat Probab Lett 53(4):391–399
Genest C, Segers J (2009) Rank-based inference for bivariate Extreme Value copulas. Ann Stat 37(5B):2990–3022
Genest C, Ghoudi K, Rivest LP (1998) Understanding relationships using copulas, by Edward Frees and Emiliano Valdez, January 1998. North Am Act J 2(3):143–149
Genest C, Quessy JF, Rémillard B (2006) Goodness-of-fit procedures for copula models based on the probability integral transformation. Scand J Stat 33:337–366

Genest C, Rémillard B, Beaudoin D (2009) Goodness-of-fit tests for copulas: a review and a power study. Insur Math Econ 44:199–213

Ghoudi K, Khoudraji A, Rivest L (1998) Propriétés statistiques des copules de valeurs extrêmes bidimensionnelles. Can J Stat 26:187–197

Grenney W, Heyse E (1985) Suspended sediment – river flow analysis. J Environ Eng 111(6): 790–803

Grimaldi S, Serinaldi F (2006) Asymmetric copula in multivariate flood frequency analysis. Adv Water Resour 29(8):1155–1167

Gudendorf G, Segers J (2010) Extreme-value copulas. In: Jaworski P, Durante F, Härdle W, Rychlik T (eds) Copula theory and its applications. Lecture notes in statistics – proceedings, vol 198. Springer, Berlin/Heidelberg, pp 127–145

Herr H, Krzysztofowicz R (2005) Generic probability distribution of rainfall in space: the bivariate model. J Hydrol 306(1–2):234–263

Hosking J, Wallis J (1988) The effect of intersite dependence on regional flood frequency analysis. Water Resour Res 24(2):588–600

Hutchinson T, Lai C (1990) Continuous bivariate distributions, emphasising applications. Rumsby Scientific Publishing, Adelaide

Jaworski P, Durante F, Haerdle W, Rychlik T (eds) (2010) Copula theory and its applications. Lecture notes in statistics – proceedings, vol 198. Springer, Berlin/Heidelberg

Jaynes ET (2003) Probability theory: the logic of science. Cambridge University Press, Cambridge

Joe H (1993) Parametric families of multivariate distributions with given margins. J Multivar Anal 46:262–282

Joe H (1994) Multivariate extreme-value distributions with applications to environmental data. Can J Stat 22:47–64

Joe H (1997) Multivariate models and dependence concepts. Chapman and Hall, London

Johnson M (1987) Multivariate statistical simulation. Wiley, New York

Journel A, Alabert F (1988) Non-gaussian data expansion in the earth sciences. TERRA Nova 1:123–134

Juri A, Wüthrich M (2002) Copula convergence theorems for tail events. Insur Math Econ 30(3):405–420

Katz R (2010) Statistics of extremes in climate change. Clim Chang 100:71–76

Keef C, Svensson C, Tawn J (2009) Spatial dependence in extreme river flows and precipitation for Great Britain. J Hydrol 378(3–4):240–252

Kendall M (1937) A new measure of rank correlation. Biometrika 6:83–93

Khoudraji A (1995) Contributions à l'étude des copules et à la modélisation des valeurs extrêmes bivariées. Ph.D. thesis, Université de Laval, Québec (Canada)

Kim TW, Valdes J, Yoo C (2003) Nonparametric approach for estimating return periods of droughts in arid regions. ASCE J Hydrol Eng 8(5):237–246

Kojadinovic I, Yan J, Holmes M (2011) Fast large-sample goodness-of-fit tests for copulas. Stat Sin 21(2):841–871

Kottegoda N, Natale L (1994) Two-component log-normal distribution of irrigation-affected low-flows. J Hydrol 158:187–199

Kruskal W (1958) Ordinal measures of association. J Am Stat Assoc 53:814–861

Liebscher E (2008) Construction of asymmetric multivariate copulas. J Multivar Anal 99: 2234–2250

Mai J, Scherer M (2010) Bivariate extreme-value copulas with discrete pickands dependence measure. Extremes 14:311–324

Marshall A, Olkin I (1967) A multivariate exponential distribution. J Am Stat Assoc 62:30–44

Marshall A, Olkin I (1983) Domains of attraction of multivariate Extreme Value distributions. Ann Probab 11(1):168–177

Marshall A, Olkin I (1988) Families of multivariate distributions. J Am Stat Assoc 83(403): 834–841

Nelsen R (1996) Nonparametric measures of multivariate association. In: Distributions with fixed marginals and related topics (Seattle, WA, 1993). IMS lecture notes monograph series, vol 28. Institute of mathematical statistics. Hayward, pp 223–232
Nelsen R (1997) Dependence and order in families of Archimedean copulas. J Multivar Anal 60:111–122
Nelsen R (2006) An introduction to copulas, 2nd edn. Springer, New York
Nelsen RB, Quesada-Molina JJ, Rodríguez-Lallena JA, Úbeda-Flores M (2001) Distribution functions of copulas: a class of bivariate probability integral transforms. Stat Probab Lett 54(3):277–282
Nelsen RB, Quesada-Molina JJ, Rodríguez-Lallena JA, Úbeda-Flores M (2003) Kendall distribution functions. Stat Probab Lett 65(3):263–268
Pickands J (1981) Multivariate Extreme Value distributions. Bull Int Stat Inst 49:859–878
Pons F (1992) Regional flood frequency analysis based on multivariate lognormal models. Ph.D. thesis, Colorado State University, Fort Collins
Poulin A, Huard D, Favre A, Pugin S (2007) On the importance of the tail dependence in bivariate frequency analysis. ASCE J Hydrol Eng 12(4):394–403. doi: 10.1061/(ASCE)1084-0699(2007)12:4(394). (Special Issue on Copulas in Hydrology)
Press W, Flannery B, Teukolsky S, Vetterling W (1992) Numerical recipes in C, 2nd edn. Cambridge University Press, Cambridge
Raynal-Villasenor J, Salas J (1987) Multivariate extreme value distributions in hydrological analyses. In: IAHS Press Institute of Hydrology (ed) Water for the future: hydrology in perspective, vol 164, pp 111–119. Wallingford, Oxfordshire, UK
Robert C (2007) The bayesian choice, 2nd edn. Springer, New York
Rohatgi V (1976) An introduction to probability theory and mathematical statistics. Wiley, New York
Rootzén H, Tajvidi N (2006) Multivariate generalized Pareto distributions. Bernoulli 12(5): 917–930
Russo D (2009) On probability distribution of hydraulic conductivity in variably saturated bimodal heterogeneous formations. J Vadose Zone 8:611–622
Salvadori G (2004) Bivariate return periods via 2-copulas. Stat Methodol 1:129–144
Salvadori G, De Michele C (2004) Frequency analysis via Copulas: theoretical aspects and applications to hydrological events. Water Resour Res 40:W12511
Salvadori G, De Michele C (2010) Multivariate multiparameter extreme value models and return periods: a copula approach. Water Resour Res 46:W10501
Salvadori G, De Michele C (2011) Estimating strategies for multiparameter multivariate extreme value copulas. Hydrol Earth Syst Sci 15:141–150
Salvadori G, De Michele C, Kottegoda N, Rosso R (2007) Extremes in nature. An approach using copulas. Water science and technology library, vol 56. Springer, Dordrecht
Salvadori G, De Michele C, Durante F (2011) On the return period and design in a multivariate framework. Hydrol Earth Syst Sci Discuss 15:3293–3305. doi: 10.5194/hess-15-3293-2011
Scarsini M (1984) On measures of concordance. Stochastica 8:201–218
Schmid F, Schmidt R (2007a) Multivariate extensions of Spearman's rho and related statistics. Stat Probab Lett 77(4):407–416
Schmid F, Schmidt R (2007b) Nonparametric inference on multivariate versions of Blomqvist's beta and related measures of tail dependence. Metrika 66:323–354
Schmid F, Schmidt R (2007c) Multivariate conditional versions of Spearman's rho and related measures of tail dependence. J Multivar Anal 98(6):1123–1140
Schmidt R (2003) Dependencies of extreme events in finance, modelling, statistics and data analysis. Ph.D. thesis, Universität Ulm, Fakultät für Mathematik und Wirtschaftswissenschaften
Schweizer B, Wolff E (1981) On nonparametric measures of dependence for random variables. Ann Stat 9:879–885
Serfling R (2002) Quantile functions for multivariate analysis: approaches and applications. Stat Neerl 56(2):214–232
Sibuya M (1960) Bivariate extreme statistics. Ann Inst Stat Math Tokyo 11:195–210

Singh V (ed) (1986) Hydrologic frequency modeling. Reidel, Dordrecht

Singh V, Jain S, Tyagi A (2007) Risk and reliability analysis. ASCE Press, Reston

Sklar A (1959) Fonctions de répartition à n dimensions et leurs marges. Public Inst Stat Univ Paris 8:229–231

Smith J (1988) Decision analysis: a bayesian approach. Chapman and Hall, London/New York

Stedinger J (1983) Estimating a regional flood frequency distribution. Water Resour Res 19(2): 503–510

Tawn J (1990) Modelling multivariate extreme value distributions. Biometrika 77(2):245–253

Titterington D, Smith A, Makov U (1985) Statistical analysis of finite mixture distributions. Wiley, New Jersey

Vandenberghe S, Verhoest NEC, Buyse E, Baets BD (2010) A stochastic design rainfall generator based on copulas and mass curves. Hydrol Earth Syst Sci 14:2429–2442

Wigley T (1988) The effect of changing climate on the frequency of absolute extreme events. Clim Monit 17:44–55. Reprinted in (2009) Clim Chang 97:67–76

Wilks D (1998) Multisite generalization of a daily stochastic precipitation generation model. J Hydrol 210:178–191

Yue S (2000) The Gumbel logistic model for representing a multivariate storm event. Adv Water Resour 24:179–185

Zhang L (2005) Multivariate hydrological frequency analysis and risk mapping. Ph.D. thesis, Louisiana State University, Baton Rouge

Chapter 6
Methods of Tail Dependence Estimation

Amir AghaKouchak, Scott Sellars, and Soroosh Sorooshian

Abstract Characterization and quantification of climate extremes and their dependencies are fundamental to the studying of natural hazards. This chapter reviews various parametric and nonparametric tail dependence coefficient estimators. The tail dependence coefficient describes the dependence (degree of association) between concurrent extremes at different locations. Accurate and reliable knowledge of the spatial characteristics of extremes can help improve the existing methods of modeling the occurrence probabilities of extreme events. This chapter will review these methods and use two case studies to demonstrate the application of tail dependence analysis.

6.1 Introduction

Weather and climate extremes are of particular importance due to their impacts on the economy, environment and human life. Understanding the spatial dependence structure of rare events is fundamental in risk assessment and decision making. Most measures of dependence (e.g., Pearson linear correlation, Spearman (1904) and Kendall (1962) correlation) are designed to describe the dependence of random variables over their distributions. Most commonly used measures are not able to correctly capture the dependence of the upper or lower parts (extremes) of the distribution (Kotz and Nadarajah 2000). For example, the Pearson correlation coefficient may not exist for random variables above a certain high (extreme) threshold (De Michele et al. 2003). The Pearson linear correlation describes how

---

A. AghaKouchak (✉) • S. Sellars • S. Sorooshian
Department of Civil and Environmental Engineering, University of California Irvine, 92697 Irvine, CA, USA
e-mail: amir.a@uci.edu; scott.sellars@uci.edu; soroosh@uci.edu

A. AghaKouchak et al. (eds.), *Extremes in a Changing Climate*, Water Science and Technology Library 65, DOI 10.1007/978-94-007-4479-0_6,

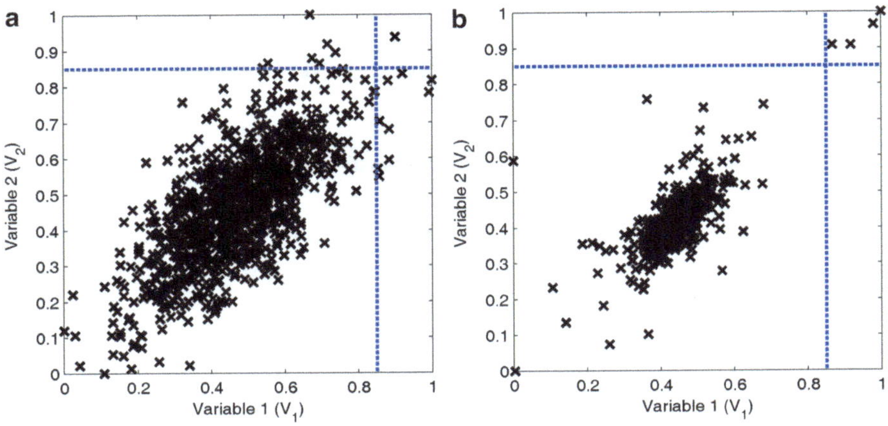

Fig. 6.1 (*left*): Upper tail values (upper right quadrant – above *dotted lines*) of V_1 and V_1 are locally independent; (*right*): Upper tail values of V_1 and V_2 seem to be locally correlated (Modified after AghaKouchak et al. (2010c))

well two random variables are linearly correlated with respect to their entire distribution. However, this information cannot be used to understand how the extremes of two random variables are dependent (Serinaldi 2008).

In general, most dependence measures associate the entire distribution of two or more random variables. However, the dependence between the upper part of the distribution may be different than the mid-range and/or lower part of the distribution (Embrechts et al. 2002). For example, two random variables with low dependence between mid-range values, but strong association among high (low) values.

In extreme value analysis, the tail dependence coefficient describes the association between the upper or lower part (tail) of two or more random variables (Schmidt 2005; Frahm et al. 2005; Ledford and Tawn 1997). The tail dependence coefficient is first introduced by Sibuya (1959) as the dependence in the upper-right and lower-left quadrants of a bivariate distribution function. Put differently, in a bivariate distribution, the tail dependence refers to the limiting proportion that one variable's marginal distribution exceeds a certain threshold given the other variable's margin has already exceeded the same threshold.

Figure 6.1 explains the concept of tail dependence using an example. The figure displays two generated random variables with the same linear correlation coefficient of approximately 0.7. Figure 6.1 (left) is simulated using the bivariate normal distribution, and Fig. 6.1 (right) is generated using the bivariate t-distribution. In both cases the simulated variables are transformed to uniform [0-1] distribution. The variables (V_1 and V_2) in both figures show positive Pearson linear correlation coefficient (≈ 0.7). However, the upper right quadrant (above the dotted lines) is different in the left and right panels of Fig. 6.1. As shown, in Fig. 6.1 (left) the values in the upper right quadrant (upper tails of V_1 and V_1) are locally independent, while in Fig. 6.1 (right) the upper tail values seem to be locally correlated (compare the upper right corners of both panels). This indicates that the probability of occurrence of V_2 above a given high threshold (e.g., dotted line in the figure), assuming V_1 exceeds the same threshold is higher in the right panel compared to the left panel in Fig. 6.1. For additional information and graphical examples, the interested reader is referred to Fisher and Switzer (2001) and Abberger (2005).

Parametric methods are frequently used for univariate extreme value analysis (e.g., Fisher and Tippett 1928; Gumbel 1958). On the other hand, in multivariate extreme value analysis, the joint probabilities of multiple random variables is considered. This includes the probability occurrence (risk) of each variable based on its univariate marginal distribution and the dependence of multiple probability occurrences. Depending on the marginal distribution of random variables and their dependence structure, a parametric model may or may not be sufficient to model the characteristics of the joint extremes. Thus far, many parametric and nonparametric methods have been developed for analysis of tail dependence of random variables (Schmidt and Stadtmüller 2006; Malevergne and Sornette 2004; Poon et al. 2004; Ledford and Tawn 2003; Malevergne and Sornette 2002; Ledford and Tawn 1996).

In the past three decades, most applications of tail dependence models have been in financial risk management and dependence analysis of between extreme assets (e.g., Schmidt (2005), Frahm et al. (2005) and Embrechts et al. (2002) and references therein). In particular, many non-parametric methods are introduced based on the concept of empirical copula. Copulas are multivariate distribution functions that can describe the dependence of two or more random variables independent of their marginal distrubtions. In recent years, multivariate copulas have been applied in numerous hydrologic applications (Nazemi and Elshorbagy 2011; AghaKouchak et al. 2010b; Bárdossy and Li 2008; Serinaldi 2009; Zhang et al. 2008; AghaKouchak et al. 2010a; Favre et al. 2004; De Michele and Salvadori 2003; Kelly and Krzysztofowicz 1997). Renard and Lang (2007) investigated the usefulness of the multivariate normal copula in extreme value analysis. With several case studies, Renard and Lang (2007) demonstrated that the multivariate normal copula can be reasonably used for extreme value analysis. However, the authors acknowledge that the low probabilities can be significantly underestimated if asymptotically dependent random variables are described by the normal copula, which is an asymptotically independent model. Serinaldi (2008) investigated the association of rainfall data using the non-parametric Kendall rank correlation. The study suggests a copula-based mixed model for modeling the dependence structure and marginal distributions of variables.

In general, the tail dependence between variables may strongly depend on the choice of model or estimation technique (Frahm et al. 2005). This chapter reviews several parametric and non-parametric tail dependence estimators. Various aspects of modeling tail dependence between variables are discussed in detail, including the choice of extreme value threshold, and advantages and disadvantages of tail dependence models. The chapter is organized into seven sections. After the introduction, the concept of tail dependence is reviewed. The third section is devoted to parametric tail dependence analysis and copulas. In section four, non-parametric methods are discussed. Section five provides additional insight into the choice of extreme value threshold. The last section highlights two case studies using the tail dependence estimators.

6.2 Tail Dependence: Basic Definitions

Let $X_1, \ldots, X_n$ be n random variables. The the upper tail (λ_{up}) for a multivariate distribution with n random variables $X(X_1, \ldots, X_n)$ is defined as (Joe 1997; Melchiori 2003):

$$\lambda_{up} = \lim_{u \to 1^-} Pr\,(F_1(X_1) > t | F_2(X_2) > t \ldots F_n(X_n) > t) \tag{6.1}$$

In Eq. 6.1, $F_1, \ldots, F_n$ are the cumulative distribution functions for the random variables $X_1, \ldots, X_n$, and t is the extreme value threshold. The equation expresses the probability (Pr) of occurrence of extremes (values above the threshold t) in X_1, conditioned on the occurrence of extremes (above the same threshold) in $X_2, \ldots, X_n$. Similarly, the lower tail dependence coefficient (λ_{lo}) is described as:

$$\lambda_{lo} = \lim_{u \to 1^-} Pr(F_1(X_1) \leq t | F_2(X_2) \leq t \ldots F_n(X_n) \leq t) \tag{6.2}$$

The multivariate distribution function is said to be upper tail dependent if $0 < \lambda_{up} \leq 1$ and upper (lower) tail independent if $\lambda_{up} = 0$ ($\lambda_{lo} = 0$). For example, in Fig. 6.1 (left), the upper tail coefficient is approximately zero ($\lambda_{up} \approx 0$), while for Fig. 6.1 (right) the upper tail coefficient is approximately 0.8 ($\lambda_{up} \approx 0.8$). For a more comprehensive discussion on the theoretical concept of tail independence, the interested reader is referred to Draisma et al. (2004) and Husler and Li (2009).

6.3 Copulas and Tail Dependence

The upper (lower) tail coefficient can also be defined using copulas. Copulas are joint cumulative distribution functions that describe dependencies among variables independent of their marginals (Joe 1997; Nelsen 2006):

$$C^n(u_1, \ldots, u_n) = F(F_1^{-1}(u_1), \ldots, F_n^{-1}(u_n)) \tag{6.3}$$

where C^n is an n-dimensional joint cumulative distribution function (CDF) of a multivariate random variable $(U(U_1, \ldots, U_n))$ and whoses marginals are $u[0, 1]$. Equation 6.1 can be alternatively presented as:

$$\lambda_{up} = \lim_{u \to 1^-} Pr\,(X_1 > F_{X_1}^{-1}(u) | X_2 > F_{X_2}^{-1}(u) \ldots X_n > F_{X_n}^{-1}(u)) \tag{6.4}$$

where $F_1^{-1}, \dots, F_n^{-1}$ are the inverse CDF of the random variables $X_1, \dots, X_n$. Notice that the conditional probability, given in Eq. 6.4, can be described as:

$$
\begin{aligned}
&Pr\,(X_1 > F_{X_1}^{-1}(u) | X_2 > F_{X_2}^{-1}(u) \dots X_n > F_{X_n}^{-1}(u)) \\
&= \frac{1 - Pr(X_1 \le F_{X_1}^{-1}(u)) - \dots - Pr(X_n \le F_{X_n}^{-1}(u)) + Pr(X_1 \le F_{X_1}^{-1}(u), \dots, X_n \le F_{X_n}^{-1}(u))}{1 - Pr(X_2 \le F_{X_2}^{-1}(u)) - \dots - Pr(X_n \le F_{X_n}^{-1}(u))}
\end{aligned}
\tag{6.5}
$$

Substituting Eq. 6.3 into Eq. 6.5 with some algebraic manipulation yields the following formulation for the upper tail (Joe 1997; Frahm et al. 2005):

$$
\lambda_{up} = \lim_{u \to 1} \frac{1 - nu + C^{(n)}(u, \dots, u)}{1 - (n-1)u} \tag{6.6}
$$

Similarly, the lower tail dependence coefficient (λ_{lo}) can be expressed as (Joe 1997):

$$
\lambda_{lo} = \lim_{u \to 1} \frac{C^{(n)}(u, \dots, u)}{(n-1)u} \tag{6.7}
$$

There are various copulas families, which have been developed for different purposes. One major difference between different copula families is the upper (lower) tail association they represent. For example, copula families may differ in the upper and lower tail of the distribution, where the dependence is strongest (weakest). In this study, two elliptical copulas, namely a normal copula and t-copula, as well as a non-Gaussian (v-transformed) copula, are used for simulations. In the following section, a number of copula families and their tail dependence behavior is discussed.

6.3.1 Gaussian Copula

One of the most commonly used copula families is the multivariate Gaussian (normal) copula, which is obtained from the multivariate normal distribution (Nelsen 2006):

$$
C_\rho(u_1, \dots, u_n) = F_\rho^n(F^{-1}(u_1), \dots, F^{-1}(u_n)) \tag{6.8}
$$

Equation 6.8 describes an n-dimensional multivariate Gaussian copula with correlation matrix $\rho_{n \times n}$ whose density function is:

$$
c(u_1, \dots, u_n) = \frac{1}{\sqrt{det\rho}} exp\left(-\frac{1}{2} y(u)' (\rho^{-1} - I) y(u)\right) \tag{6.9}
$$

where: F^n = Multivariate Gaussian CDF
$y(u_i) = F^{-1}(u_i)$

6.3.2 t-Copula

The t-copula (alternatively known as Student copula), is an elliptical copula derived from the Student distribution:

$$C_{\nu,\rho}(u_1,\dots,u_n) = t^n_{\nu,\rho}(t_\nu^{-1}(u_1),\dots,t_\nu^{-1}(u_n)) \tag{6.10}$$

where: t^n = Multivariate Student CDF
ρ = shape matrix and
ν = degrees of freedom

$$t^n_{\nu,\rho}(x) = \frac{1}{\sqrt{det\rho}}\frac{\Gamma\left(\frac{\nu+n}{2}\right)}{\Gamma\left(\frac{\nu}{2}\right)(\pi\nu)^{n/2}} \times \int_{-\infty}^{x_1}\cdots\int_{-\infty}^{x_n}\frac{dx}{\left(1+\frac{x'\rho^{-1}x}{\nu}\right)^{(\nu+n)/2}} \tag{6.11}$$

For $\nu > 2$, the shape matrix (ρ) is proportional to the correlation matrix (Malevergne and Sornette 2003). The density function of the t-copula is expressed as (Malevergne and Sornette 2003):

$$c(u_1,\dots,u_n) = \frac{1}{\sqrt{det\rho}}\frac{\Gamma\left(\frac{\nu+n}{2}\right)\left(\Gamma\left(\frac{\nu}{2}\right)\right)^{n-1}}{\left(\Gamma\left(\frac{\nu+1}{2}\right)\right)^n} \times \frac{\Pi_{k=1}^n\left(1+\frac{y_k^2}{\nu}\right)^{(\nu+1)/2}}{\left(1+\frac{y'\rho^{-1}y}{\nu}\right)^{(\nu+n)/2}} \tag{6.12}$$

where: $y_k = t_\nu^{-1}(u_k)$
t_ν = univariate Student distribution with ν degrees of freedom

Both Gaussian and t-copulas are elliptical; however, they represent different tail dependencies. The Gaussian copula is upper (lower) tail independent ($\lambda_{up} \approx 0$) regardless of the correlation coefficient among variables (Coles 2001; Renard and Lang 2007; Mikosch and Resnick 2006). This indicates that the extreme values from the different random variables occur independently, even if the random variables exhibit a high correlation. It is worth pointing out that for independent variables, one could expect $\lambda_{up} = 0$. Note that the converse is not necessarily true, meaning that $\lambda_{up} = 0$ does not indicate that the random variables are necessarily independent (Malevergne and Sornette 2003).

Contrary to the Gaussian copula, the t-copula can capture the upper (lower) tail dependence (if exists) among two or more random variables. The t-copula can capture the asymptotic dependence even when the variables are negatively (inversely) associated (Embrechts et al. 2001). In t-copula formulation, as ν increases, the tail dependence weakens, and thus, the probability of occurrence of extreme values reduces. Figure 6.2a displays the tail behavior of the bivariate t-copula with $\nu = 1 - 10$. The Figure presents occurrences of $x > 0.8$ (percentage) in both

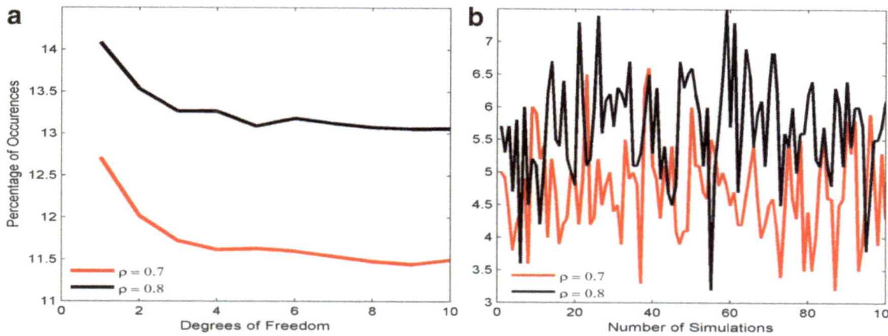

Fig. 6.2 Tail dependence behavior from bivariate random variable simulated using (**a**) t-copula and (**b**) Gaussian copula. The y-axes show the occurrences of joint extremes (percentage) in dependent random variables simulated using t-copula and Gaussian copula (modified after AghaKouchak et al. 2010b)

random vectors of the bivariate t-copula. One can see that an increase in ν results in less occurrences of extremes (values above the threshold t in Eq. 6.1). Figure 6.2b shows the tail dependence of Gaussian copula. The occurrence of joint extremes in Gaussian copulas is considerably less than t-copula (threshold: 0.8). This indicates that if strong dependence exist among multiple variables, using the Gaussian copula may not be suitable for modeling dependence of extremes. In fact, the multivariate Gaussian distribution is upper (lower) tail independent ($\lambda_{up} \approx 0$) meaning it cannot be used to describe dependencies of extremes (Coles 2001; Renard and Lang 2007). It is noted that the tail behavior of a multivariate model depends solely on the type of copula and not on the marginal distribution of individual variables. Therefore, in modeling the dependencies of extremes, the choice of copula family plays a significant role.

6.3.3 Gumbel-Hougaard Copula

In the following section, a heavy upper tailed Archimedean copula (Nelsen 2006), known as the Gumbel-Hougaard copula, is introduced. Unlike many copula families, Archimedean copulas are not derived from standard multivariate distributions. Generally, the multivariate Archimedean copulas can be expressed as:

$$C(x_1, \ldots, x_n) = \Psi^{-1}\left(\sum_{i=1}^{n} \Psi(F_i(x_i))\right) \tag{6.13}$$

where Ψ is the so-called generator function. For the Gumbel-Hougaard copula, the generator function can be expressed as:

$$\Psi(x) = (-\ln(x))^{\theta} \tag{6.14}$$

where: $\theta \geq 1 =$ copula parameter.

By substituting Eq. 6.14 into Eq. 6.13, the general formulation of the bivariate Gumbel-Hougaard copula can be described as (Venter 2002):

$$C(u,\nu) = \exp\left(-\left((-\ln u)^{\theta} + (-\ln \nu)^{\theta}\right)^{\frac{1}{\theta}}\right) \tag{6.15}$$

Equation 6.15 represents the bivariate Gumbel-Hougaard copula with variables $u[0,1]$ and $\nu[0,1]$. The Gumbel-Hougaard copula is parameterized through a single parameter θ. The copula parameter θ is to be estimated based on available data. By substituting Eq. 6.15 into Eq. 6.6, the upper tail dependence coefficient for the Gumbel-Hougaard copula can be derived as: $\lambda_{up} = 2 - 2^{\theta}$ (Salvadori et al. 2007; Frahm et al. 2005; Nelsen 2006). A discussion on copula parameter estimation techniques is beyond the scope of this chapter. The interested reader is pointed to Genest et al. (1995), Salvadori et al. (2007) and Nelsen (2006) for more detailed discussions on parameter estimation.

6.4 Nonparametric Tail Dependence Methods

There are different nonparametric tail dependence estimators that can be used to evaluate the significance of tail behavior. The first nonparametric approach introduced here is based on the concept of the empirical copula (C^m) – AghaKouchak et al. (2010c):

$$C^m(u,\nu) = F_{(m)}\left(F_{(m)1}^{-1}(u), F_{(m)2}^{-1}(\nu)\right) \tag{6.16}$$

where $F_{(m)}$ refers to the empirical distribution of random variables. The tail dependence estimator $\lambda_{up}^{(1)}$ is then expressed as (Schmidt and Stadtmüller 2006):

$$\lambda_{up}^{(1)} = \frac{T}{k} C^m\left(\left(1 - \frac{k}{T}, 1\right] \times \left(1 - \frac{k}{T}, 1\right]\right) \tag{6.17}$$
$$= \frac{1}{k}\Sigma_{j=1}^{T} I(R_1^j > T-k, R_2^j > T-k)$$

where:

where:	T	=	sample size
	k	=	threshold rank
	R_1^j	=	rank of u^j
	R_2^j	=	rank of ν^j
	I	=	indicator function

It is worth pointing out that Eq. 6.17 is the empirical copula with the interval $(1 - \frac{k}{T}, 1] \times (1 - \frac{k}{T}, 1]$. That is, $\lambda_{up}^{(1)}$ is derived using the empirical tail-copula introduced by Genest et al. (1995).

Haung (1992) suggested another tail dependence measure (here, $\lambda_{up}^{(2)}$) based on empirical copulas and extreme value theory:

$$\lambda_{up}^{(2)} = 2 - \frac{T}{k}\left(1 - C^m\left(1 - \frac{k}{T}, 1 - \frac{k}{T}\right)\right)$$
$$= 2 - \frac{1}{k}\Sigma_{j=1}^{T} I(R_1^j > T - k \text{ OR } R_2^j > T - k) \tag{6.18}$$

Coles et al. (1999) proposed a different nonparametric tail dependence measure ($\lambda_{up}^{(3)}$) as follows (Frahm et al. 2005):

$$\lambda_{up}^{(3)} = 2 - \frac{\log C^m\left(\frac{T-k}{T}, \frac{T-k}{T}\right)}{\log\left(\frac{T-k}{T}\right)} \tag{6.19}$$

where

$$C^m(u, v) = \frac{1}{T}\Sigma_{j=1}^{T} I\left(\frac{R_1^j}{T} \le u, \frac{R_2^j}{T} \le v\right) \tag{6.20}$$

Another nonparametric tail dependence estimator (here, $\lambda_{up}^{(4)}$) is proposed by Joe et al. (1992):

$$\lambda_{up}^{(4)} = 2 - \frac{1 - C^m\left(\frac{T-k}{T}, \frac{T-k}{T}\right)}{1 - \frac{T-k}{T}} \tag{6.21}$$

where the term C^m is the empirical copula as described in Eq. 6.20. It should be noted that nonparametric methods of estimating tail dependence is not limited to the ones mentioned above (see for example Capéraa et al. 1997).

6.5 Extreme Value Threshold

Estimation of the extreme value threshold requires assuming a threshold above (or below) values that are considered as extreme (Frahm et al. 2005). For tail dependence analysis, one can use a fixed (e.g., above 95 % of data) or variant threshold approach. The so-called optimal threshold approach (Frahm et al. 2005; Peng 1998) uses a kernel plateau-finding algorithm to estimate the optimal extreme value threshold. In this method, the optimal plateau is estimated in four steps: (1) a kernel box with a bandwidth of b (e.g., $b = int(0.05n)$) is selected; (2) the mean values of the coefficients that fall within each box results in $n - 2b\ \overline{\lambda}$ values; (3) for a moving plateau with a length of $l = \sqrt{n - 2b}$, the corresponding λ values are calculated ($\lambda_k, \ldots, \lambda_{k+l+1}$ where k=1,..., $n - 2b - m + 1$); (IV) the optimal plateau (extreme value threshold) is the first one that fulfills the following condition (for more detailed description, the reader is referred to Frahm et al. (2005) and Peng (1998)):

$$\Sigma_{i=k+1}^{k+l-1}|\overline{\lambda}_i - \lambda_k| \le 2\sigma \tag{6.22}$$

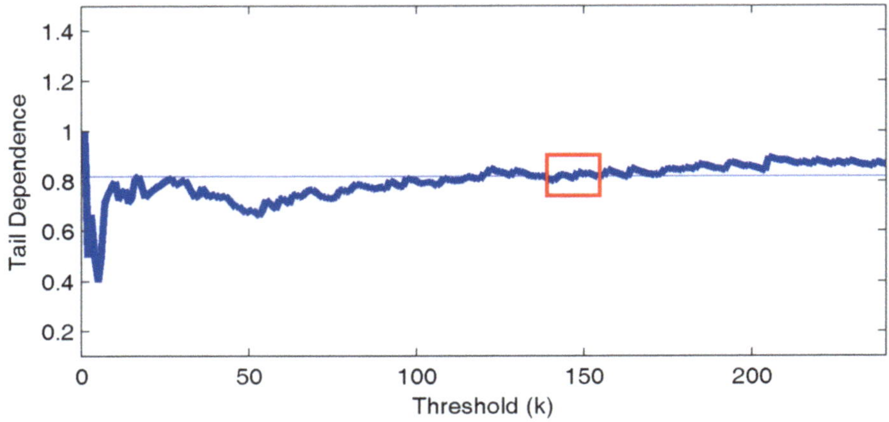

Fig. 6.3 Variability of the tail dependence coefficient with respect to threshold (After AghaKouchak et al. (2010c))

where σ is the standard deviation of the $\overline{\lambda}_i$ values (means of coefficients that fall within each box). The optimal tail dependence coefficient is then expressed as:

$$\lambda = \frac{1}{T}\Sigma_{i=1}^{T}\lambda_{k+i-1} \tag{6.23}$$

Figure 6.3 displays an example of tail dependence coefficient variability versus the choice of extreme value threshold. In this figure, the box refers to the plateau that satisfies the condition mentioned above (Eq. 6.22) and its corresponding TDC. Note that the box size in Fig. 6.3 is not scaled, and the box size is placed on the figure for illustration. For other methods of extreme value threshold estimation, the interested reader is pointed to Tancredi et al. (2006).

6.6 Case Studies

Case Study A: In the following example, the tail dependence coefficient is used for analysis of anisotropy of spatial dependencies of extremes. Figure 6.4 displays the rainfall accumulations above 95 % threshold normalized to [0-1]. The precipitation data used in this example is from the Precipitation Estimation from Remotely Sensed Information using Artificial Neural Networks (PERSIANN, Sorooshian et al. 2000; Hsu et al. 1997) data set, which is an infrared-based microwave-adjusted precipitation product. The tail dependence coefficient is estimated using Eq. 6.17 ($\lambda_{up}^{(1)}$) for the two perpendicular directions shown in Fig. 6.4. In this example, the data is smoothed with a moving-average window with bandwidth of 2 pixels and 80 % overlap. Figure 6.5 indicates that heavy precipitation rates are dependent over

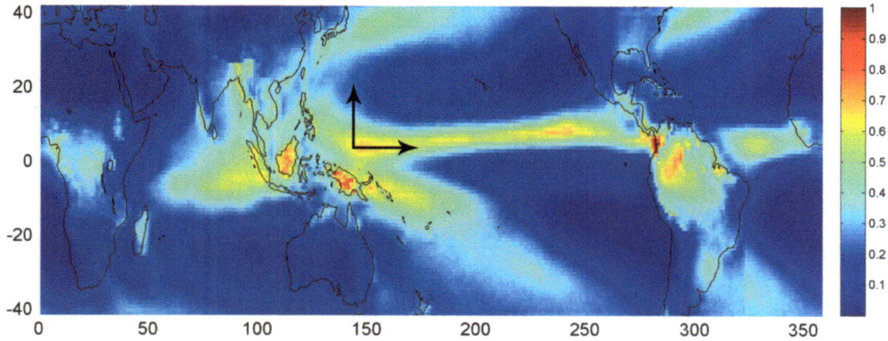

Fig. 6.4 Sum of rainfall exceedance above 95 % normalized to [0-1]

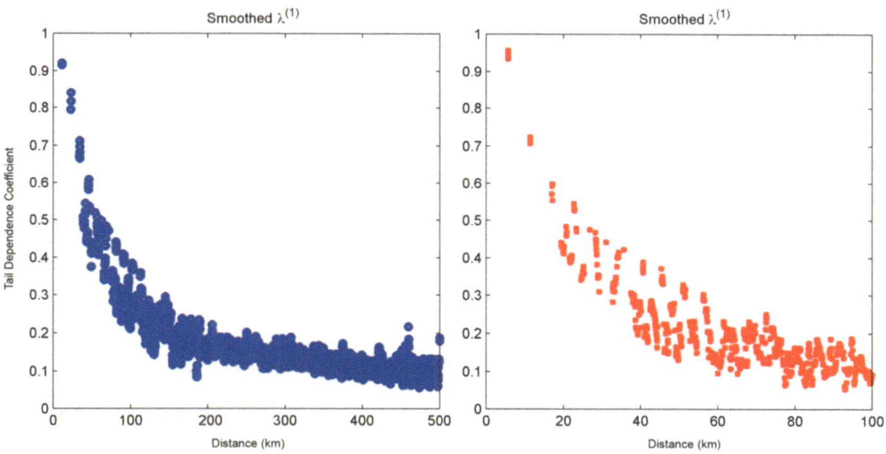

Fig. 6.5 Tail dependence behavior of rainfall data in (*left*) E-W, and (*right*) N-S directions versus distance (km)

longer distances in the horizontal direction (see Fig. 6.5 (left)) as opposed to the vertical direction (see Fig. 6.5 (right)). This indicates that the spatial dependence structure of heavy rainfall rates is asymmetrical, and in this example, heavy rainfall rates are spatially more dependent in the horizontal direction.

Case Study B: Many earth science variables provide excellent data for studying spatial dependencies of extreme events. This example demonstrates a nonparametric approach to evaluating the dependence structure of the extreme precipitation values over a region in the southern part of the United States. Understanding extreme precipitation spatial dependencies and behavior on the local, regional and global scale will provide enhanced insight in the spatial dependence structure of precipitation in different regions of the world. This information can then be used to assist in planning and decision making purposes.

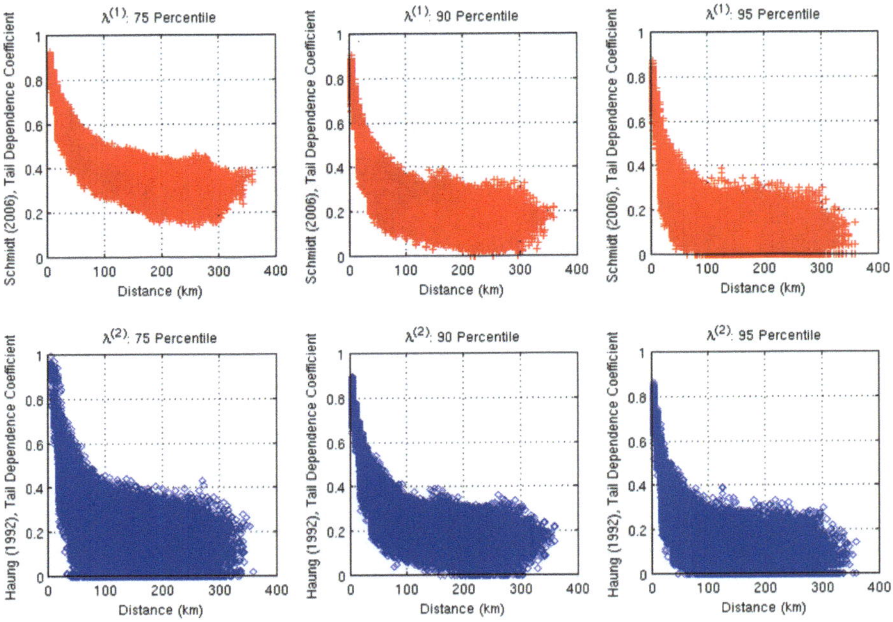

Fig. 6.6 Two nonparametric methods for calculating tail dependence coefficients. (*Top*) $\lambda_{up}^{(1)}$ for three percentile groups, 75th, 90th and 95th as described in Eq. 6.17; and (*Bottom*) $\lambda_{up}^{(2)}$ for three percentile groups, 75th, 90th and 95th as described by 6.18

Two nonparametric tail dependence methods based on empirical copulas are used to derive tail dependence estimators: (1) $\lambda_{up}^{(1)}$ introduced in Eq. 6.17; and (2) $\lambda_{up}^{(2)}$ introduced in Eq. 6.18. In both cases, the tail dependence estimator helps describe the dependent structure or degree of association between concurrent rainfall extremes at different locations. High spatial and temporal resolution precipitation data can be analyzed using these nonparametric tail dependence methods, which allows for solving for the tail dependence coefficient and thus describing the dependence structure of the extreme precipitation events.

The study region is over Mississipp, which is located in the southern part of the United States with a latitude of 38N to 35.5N and longitude of −110W to −107.5W from January 1st, 2005 to December 31st, 2008. The precipitation data used in this example is the National Center for Environmental Prediction (NCEP)'s Stage 4 mosaic multi-sensor national precipitation analysis, which has a 4 km spatial resolution and hourly temporal resolution (Lin and Mitchell 2005).

Solving for $\lambda_{up}^{(1)}$ (Eq. 6.17) and $\lambda_{up}^{(2)}$ introduced in Eq. 6.18, and smoothing the results for display purposes, one can see the dependence structure of precipitation for the data in this region. Similar to the previous example, the data is smoothed using a moving-average window with a bandwidth of 2 pixels and 80 % overlap. Figure 6.6 shows three different precentile groupings of the extreme precipitation events: 75th percentile, 90th precentile and 95th precentile for $\lambda_{up}^{(1)}$ and $\lambda_{up}^{(2)}$

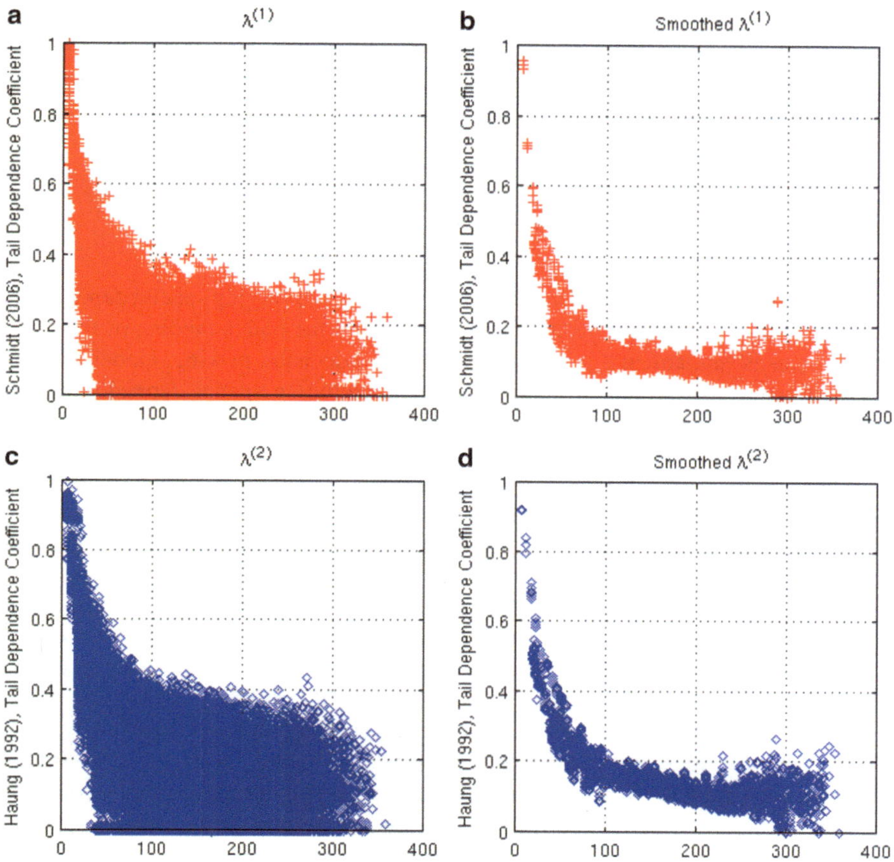

Fig. 6.7 Tail dependence analysis using $\lambda_{up}^{(1)}$ and $\lambda_{up}^{(2)}$ and based on the concept of optimal threshold introduced in Eqs. 6.22 and 6.23

tail dependence methods. Both nonparametric tail dependence methods from $\lambda_{up}^{(1)}$ (Fig. 6.6, top) and $\lambda_{up}^{(2)}$ (Fig. 6.6, bottom) are consistent with each other showing the expected decrease in dependence with distance across each of the different precentile groups. The figure indicates that the spatial dependence of extreme convective precipitation will decrease rapidly as distance increases. This is expected because with extreme precipitation events, spatial dependence is typically the highest near the region of convective activity, which produces the largest observed precipitation. This is not always the case, for example, extra-tropical cyclones can also produce extreme precipitation events and can have spatial dependence up to a few hundred kilometers.

Figure 6.7 displays $\lambda_{up}^{(1)}$ (Fig. 6.7, top) and $\lambda_{up}^{(2)}$ (Fig. 6.7, bottom) methods using the concept of optimal threshold (Eqs. 6.22 and 6.23) to determine the percentile for calculating the tail dependence coefficient as well as a smoothed version for illustration purposes. Contrary to commonly used method (Fig. 6.6), this approach of tail dependence analysis is independent of a fixed (constant) threshold. In other words, this method provides a tail dependence analysis that does not require additional decisions regarding the choice of extreme value threshold (Frahm et al. 2005).

6.7 Summary and Conclusions

Extreme events (e.g., floods, droughts, heat waves) have varying spacial dependence structures across different geographic locations and the understanding of these dependencies is fundamental to risk assessment and decision making. Understanding the different characteristics of extreme events, including spatial dependence, will provide regional planners and policy makers with information and knowledge of extreme events that impact their the local and regional communities. Spatial characteristics of extreme events can be investigated through estimation of tail dependence coefficient for different locations. This chapter reviewed several nonparametric and parametric tail dependence coefficient estimators. The tail dependence coefficient describes the degree of association between concurrent extremes. The presented nonparametric methods are based on the concept of bivariate empirical copula of random variables, whereas the parametric approach is based on the concept of Gumbel-Hougaard Copula. The chapter also reviewed different aspects of modeling tail dependence such as the choice of extreme value threshold.

In the first case study, the tail dependence coefficient is used for analysis of anisotropy of spatial dependencies of extremes. The results showed that the spatial dependence structure of heavy rainfall rates was asymmetrical. In the second example, the tail dependence coefficient is used to investigate spatial dependencies of precipitation extremes on a local scale revealing the spatial dependence structure of the extreme convective precipitation as described by the tail dependence coefficient. Extreme precipitation impacts many aspects of human society, such as loss of property and life due to flooding and area destruction from severe storms.

In the case studies, a kernel plateau-finding algorithm is used to obtain tail dependence coefficients, avoiding a fixed extreme value threshold. The results of previous studies (e.g., AghaKouchak et al. 2010c) reveal that using the kernel plateau-finding algorithm for tail dependence is superior to the fixed threshold approaches. This method, also known as the optimal threshold approach, can obtain a measure of tail dependence that does not require additional decisions regarding the choice of extreme value threshold.

The tail dependence coefficient has numerous applications including: validation and verification of weather and climate models in reproducing extreme events; analysis of simultaneous extremes; probabilistic assessment of occurrences of extremes, and understanding climate variability. For example, by deriving tail dependence coefficients for simulations of a numerical weather prediction model or a climate model, one can evaluate whether these models produces dependencies as seen in the observations. These approaches are not limited to precipitation, but also a wide variety of earth science variables. This study of extremes tail dependence on local, regional and global scales can assist in planning and policy making as well as validating numerical models, thus providing a valuable tool for understanding how extreme events impact society.

References

Abberger K (2005) A simple graphical method to explore tail-dependence in stock-return pairs. Appl Financ Econ Taylor Fr J 15(1):43–51

AghaKouchak A, Bárdossy A, Habib E (2010a) Conditional simulation of remotely sensed rainfall data using a non-gaussian v-transformed copula. Adv Water Resour 33(6):624–634

AghaKouchak A, Bárdossy A, Habib E (2010b) Copula-based uncertainty modeling: application to multi-sensor precipitation estimates. Hydrol Process 24(15):2111–2124

AghaKouchak A, Ciach G, Habib E (2010c) Estimation of tail dependence coefficient in rainfall accumulation fields. Adv Water Resour 33(9):1142–1149

Bárdossy A, Li J (2008) Geostatistical interpolation using copulas. Water Resour Res 44(7):W07412.1–W07412.15

Capéraa P, Fougéres A-L, Genest C (1997) A nonparametric estimation procedure for bivariate extreme value copulas. Biometrika 84(3):567–577

Coles S (2001) An introduction to statistical modeling of extreme values. Springer, London

Coles S, Heffernan J, Tawn J (1999) Dependence measures for extreme value analyses. Extremes 2:339–365

De Michele C, Salvadori J (2003) A generalized pareto intensity-duration model of storm rainfall exploiting 2-copulas. J Geophy Res Atmos 108(D2):1–11

De Michele C, Salvadori J, Canossi M, Petaccia A, Rosso R (2003) Bivariate statistical approach to check adequacy of dam spillway. J Hydrol Eng 10(1):50–57

Draisma G, Drees H, Ferreira A, de Haan L (2004) Bivariate tail estimation: dependence in asymptotic independence. Bernoulli 10:251–280

Embrechts P, Lindskog F, McNeil A (2001) Modelling dependence with copulas and applications to risk management. Department of Mathematics, ETHZ, Zurich, Switzerland

Embrechts P, Mcneil A, Straumann D (2002) Correlation and dependence in risk management: properties and pitfalls. In: Risk management: value at risk and beyond. Cambridge University Press, Cambridge, pp 176–223

Favre A-C, El Adlouni S, Perreault L, Thiémonge N, Bobée B (2004) Multivariate hydrological frequency analysis using copulas. Water Resour Res 40:1–12

Fisher N, Switzer P (2001) Graphical assessment of dependence: is a picture worth 100 tests? Am Stat 55:233–239

Fisher R, Tippett L (1928) Limiting forms of the frequency distribution of the largest or smallest member of a sample. Proc Camb Philos Soc 24:180–190

Frahm G, Junker M, Schmidt R (2005) Estimating the tail-dependence coefficient: properties and pitfalls. Insur Math Econ 37(1):80–100

Genest C, Ghoudi K, Rivest L-P (1995) A semiparametric estimation procedure of dependence parameters in multivariate families of distributions. Biometrika 82:543–552

Gumbel E (1958) Statistics of extremes. Columbia University Press, New York, p 23

Haung X (1992) Statistics of bivariate extreme values. Thesis Publishers, Amsterdam

Hsu K, Gao X, Sorooshian S, Gupta H (1997) Precipitation estimation from remotely sensed information using artificial neural networks. J Appl Meteorol 36:1176–1190

Husler J, Li D (2009) Testing asymptotic independence in bivariate extreme. J Stat Plan Inference 139(3):990–998

Joe H (1997) Multivariate models and dependence concepts. Chapman Hall, London
Joe H, Smith R, Weissman I (1992) Bivariate threshold models for extremes. J R Stat Soc Ser B 54:171–183
Kelly k, Krzysztofowicz R (1997) A bivariate meta-gaussian density for use in hydrology. Stoch Environ Res Risk Assess 11(1):17–31
Kendall M (1962) Rank correlation methods. Griffin, London
Kotz S, Nadarajah S (2000) Extreme value distributions: theory and applications. Imperial College Press, London. ISBN 1860942245
Ledford A, Tawn J (1996) Statistics for near independence in multivariate extreme values. Biometrika 83(1):169–187
Ledford A, Tawn J (1997). Modelling dependence within joint tail regions. J R Stat Soc Ser B-Methodol 59(2):475–499
Ledford A, Tawn J (2003) Diagnostics for dependence within time series extremes. J R Stat Soc Ser B-Stat Methodol 65(Part 2):521–543
Lin Y, Mitchell K (2005) The ncep stage ii/iv hourly precipitation analyses: development and applications. In: Preprints, 19th conference on hydrology, American Meteorological Society, San Diego, CA, 9–13 Jan 2005, Paper 1.2
Malevergne Y, Sornette D (2002) Minimising extremes. Risk 15(11):129–133
Malevergne Y, Sornette D (2003) Testing the gaussian copula hypothesis for financial assets dependences. Quant Financ Vol 3:231–250
Malevergne Y, Sornette D (2004) How to account for extreme co-movements between individual stocks and the market. Risk 6:71–116
Melchiori MR (2003) Which Archimedean Copula is the Right One? Yield Curve 37:1–20
Mikosch T, Resnick S (2006). Activity rates with very heavy tails. Stoch Process Their Appl 116(2):131–155
Nazemi A, Elshorbagy A (2011) Application of copula modelling to the performance assessment of reconstructed watersheds. Stoch Environ Res Risk Assess 1–17. doi: 10.1007/s00477-011-0467-7
Nelsen R (2006) An introduction to copulas. Springer series in statistics. Springer, New York
Peng L (1998) Second order condition and extreme value theory. Tinbergen Institute Research Series, vol 178. Thesis Publishers and Tinbergen Institute, Amsterdam
Poon S, Rockinger M, Tawn J (2004) Extreme value dependence in financial markets: diagnostics, models, and financial implications. Rev Financ Stud 17(2):581–610
Renard B, Lang M (2007) Use of a gaussian copula for multivariate extreme value analysis: some case studies in hydrology. Adv Water Resour 30:897–912
Salvadori G, DeMichele C, Kottegoda N, Rosso R (2007) Extremes in Nature: an approach using copulas. Springers, Berlin
Schmidt R (2005) Tail dependence. In: Cizek P, Häardle W, Weron R (eds) Statistical tools in finance and insurance. Springer, New York
Schmidt R, Stadtmüller U (2006) Non-parametric estimation of tail dependence. Scand J Stat 33(2):307–335
Serinaldi F (2008) Analysis of inter-gauge dependence by kendall's τ, upper tail dependence coefficient, and 2-copulas with application to rainfall fields. Stoch Environ Res Risk Assess 22:671–688
Serinaldi F (2009) Copula-based mixed models for bivariate rainfall data: an empirical study in regression perspective. Stoch Environ Res Risk Assess 23(5):677–693
Sibuya M (1959) Bivariate extreme statistics I. Ann Inst Stat Math 11(2):195–210
Sorooshian S, Hsu K, Gao X, Gupta H, Imam B, Braithwaite D (2000) Evolution of the persiann system satellite-based estimates of tropical rainfall. Bull Am Meteorol Soc 81(9):2035–2046
Spearman C (1904) The proof and measurement of association between two things. Am J Psychol 15:72–101

Tancredi A, Anderson C, O'Hagan A (2006) Accounting for threshold uncertainty in extreme value estimation. Extremes 9:87–106
Venter G (2002) Tails of copulas. In: Proceedings of the casualty actuarial society, LXXXIX 2002, Boston, MA, p 68–113
Zhang LSR, Singh V (2008) Bivariate rainfall frequency distributions using archimedean copulas. J Hydrol 32(1–2):93–109

Chapter 7
Stochastic Models of Climate Extremes: Theory and Observations

Philip Sura

Abstract One very important topic in climatology, meteorology, and related fields is the detailed understanding of extremes in a changing climate. There is broad consensus that the most hazardous effects of climate change are due to a potential increase (in frequency and/or intensity) of extreme weather and climate events. Extreme events are by definition rare, but they can have a significant impact on people and countries in the affected regions. Here an extreme event is defined in terms of the non-Gaussian tail (occasionally also called a weather or climate regime) of the data's probability density function (PDF), as opposed to the definition in extreme value theory, where the statistics of time series maxima (and minima) in a given time interval are studied. The non-Gaussian approach used here allows for a dynamical view of extreme events in weather and climate, going beyond the solely mathematical arguments of extreme value theory. Because weather and climate risk assessment depends on knowing the tails of PDFs, understanding the statistics and dynamics of extremes has become an important objective in climate research. Traditionally, stochastic models are extensively used to study climate variability because they link vastly different time and spatial scales (multi-scale interactions). However, in the past the focus of stochastic climate modeling hasn't been on extremes. Only in recent years new tools that make use of advanced stochastic theory have evolved to evaluate the statistics and dynamics of extreme events. One theory attributes extreme anomalies to stochastically forced dynamics, where, to model nonlinear interactions, the strength of the stochastic forcing depends on the flow itself (multiplicative noise). This closure assumption follows naturally from the general form of the equations of motion. Because stochastic

P. Sura (✉)
Department of Earth, Ocean and Atmospheric Science, The Florida State University, 1017 Academic Way, Tallahassee, FL 32306-4520, USA

Center for Ocean-Atmospheric Prediction Studies, The Florida State University, 2035 E. Paul Dirac Drive, Tallahassee, FL 32306-2840, USA
e-mail: psura@fsu.edu

A. AghaKouchak et al. (eds.), *Extremes in a Changing Climate*, Water Science and Technology Library 65, DOI 10.1007/978-94-007-4479-0_7,

theory makes clear and testable predictions about non-Gaussian variability, the multiplicative noise hypothesis can be verified by analyzing the detailed non-Gaussian statistics of atmospheric and oceanic variability. This chapter discusses the theoretical framework, observational evidence, and related developments in stochastic modeling of weather and climate extremes.

7.1 Introduction

In this introductory section we first review some fundamental ideas about extreme events in climate, before focusing on novel stochastic approaches. In particular, we qualitatively introduce the fundamental concept of stochastic climate models driven by state-dependent (multiplicative) noise, where the strength of the stochastic forcing depends on the flow itself, resulting in non-Gaussian statistics. This becomes our basic paradigm for extreme events in climate, to be discussed in detail in the remainder of this chapter.

7.1.1 Extreme Events: Definition, Relevance, and Sampling

Extreme events in nature and society are by definition scarce, but they can have a significant physical and socioeconomic impact on people and countries in the affected regions (Albeverio et al. 2006). Here we deal with extreme events in weather and climate (i.e., nature), keeping in mind that there exists a strong connection to society, because an extreme event in nature often triggers an extreme socioeconomic event. For example, a natural disaster is often followed by a financial crisis. Alongside our intuitive knowledge that a hurricane, tornado, or earthquake might qualify, how can we define extreme events more quantitatively? Let's recall two common definitions:

- One often-used (e.g., by the Intergovernmental Panel on Climate Change (IPCC)) definition of an extreme event is based on the tails of its climatological (i.e., reference) probability density function (PDF) at a particular geographical location. An extreme event would normally be as rare as or rarer than the 5th or 95th percentiles. Note that this definition does not depend on the particular shape (e.g., Gaussian or non-Gaussian) of the PDF. Yet the Gaussian distribution is very often used to estimate the odds of extreme events, neglecting the non-Gaussianity (higher probability of extreme events) of real world observations.
- Another widely accepted definition of an extreme event is that of a phenomenon that is beyond our *normal* expectations (e.g., Brooks and Carruthers 1953; White 1980; Nakamura and Wallace 1991; Peinke et al. 2004; Newman 2005; Sornette 2006; Taleb 2010). When we say that an event is normally distributed, we mean that the PDF follows the Gaussian bell curve. Thus, an extreme event

can also be defined as the non-normal (i.e., non-Gaussian) tail of the data's PDF. This definition is akin to more non-technical (or popular) views, where an extreme event is a high-impact, hard-to-predict phenomenon that is beyond our normal expectations. In popular culture an extreme event is sometimes also called a "Black Swan", because a Black Swan provides a metaphor for a highly improbable incident (e.g., Taleb 2010); before the discovery of Australia, people in the Old World were convinced that all swans where white.

The latter definition of an extreme as the non-Gaussian tail of the data's PDF is the one we are using here. However, note that the non-Gaussian definition does not exclude the percentile view, because percentiles do not depend on the particular shape of the PDF. We employ the non-Gaussian approach because, as we will see, real world observations are almost always non-Gaussian (if not averaged too heavily; then the data become close to Gaussian because of the central limit theorem). That is, we want to specify and understand the PDFs of climate variability beyond empirically determining percentiles. In addition, the information included in the non-Gaussian statistics allows us to learn more about the actual dynamics of extreme climate variability. Therefore, here an event is only considered *extreme* if its probability of occurrence is governed by *non-Gaussian* statistics. That implies, that a high amplitude event does not qualify as extreme if it is described by Gaussian statistics. It is important to recognize that in the following we, therefore, use the terms *extreme event* and *non-Gaussian* statistics (and related phrases) synonymously.

Understanding extremes has become an important objective in weather/climate research, because weather and climate risk assessment depends on knowing and understanding the tails of PDFs. At this point we need to define what we mean by weather and climate a bit more. Here we typically think of representative atmospheric variables (for example, pressure, wind, vorticity, temperature etc.) as varying on weather timescales of hours, days, to a few weeks, and the ocean (for example, sea surface temperatures, sea level heights, currents etc.) varying on longer climate timescales of weeks, months, years, and decades. There is, of course, certain overlap and we use the terms weather and climate in a loose way, specifying the timescales as needed for particular applications. It should also be noted that nonlinear multi-scale interactions make a strict separation of timescales impossible.

There is broad consensus that the most hazardous effects of climate change are related to a potential increase (in frequency and/or intensity) of extreme weather and climate events (e.g., Houghton 2009; Brönnimann et al. 2008; Alexander et al. 2006; Easterling et al. 2000). The overarching goal of studying extremes is, therefore, to understand and then manage the risks of extreme events and related disasters to advance strategies for efficient climate change adaptation. While numerous important studies have focused on changes in mean values under global warming, such as mean global temperature (one of the key variables in almost every discussion of climate change; see, for example, reports from the IPCC available at http://www.ipcc.ch), the interest in how extreme values are altered by a changing climate is a relatively recent topic in climate research. The reasons for that are primarily twofold. First, we need high-quality, high-resolution (in space and time) observational data sets for a comprehensive analysis of non-Gaussian climate variability. It is only recently that global high-quality daily observations became readily available to the international research community (Alexander et al. 2006). Second, we need extensive simulations of high-resolution climate models to (hopefully) simulate realistic non-Gaussian climate variability. Again, only recently long enough high-resolution numerical simulations of climate variability became feasible to study global non-Gaussian and higher-order statistics in some detail (e.g., Kharin and Zwiers 2005; Kharin et al. 2007).

The general problem of understanding extremes is, of course, their scarcity: it is very hard to obtain reliable (if any) statistics of those events from a finite observational record. Therefore, we have to somehow extrapolate from the well sampled center of a PDF to the scarcely or unsampled tails. The extrapolation into the more or less uncharted tails of a distribution can be roughly divided into three major, by no means mutually exclusive categories. In fact, the study of extreme events in weather and climate is most often done by combining the strategies of the following methods (Garrett and Müller 2008).

7.1.2 Common Methods to Study Extreme Events

The *statistical* approach (extreme value theory) is solely based on mathematical arguments (e.g., Coles 2001; Garrett and Müller 2008; Wilks 2006; Gumbel 1942, 1958). It provides methods to extrapolate from the well sampled center to the scarcely or unsampled tails of a PDF using mathematical tools. The key point of the statistical approach is that, in place of an empirical or physical basis, asymptotic arguments are used to justify the extreme value model. In particular, the generalized extreme value distribution (GEV) is a family of PDFs for the maximum (or minimum) of a large sample of independent random variables drawn from the same arbitrary distribution. While the statistical approach is based on sound mathematical arguments, it does not provide much insight into the physics of extreme events. Extreme value theory is, however, widely used to explore climate extremes (Katz and Naveau 2010). In fact, the foundation of extreme value theory is very closely related to the study of extreme values in meteorological data (Gumbel 1942, 1958). Nowadays this is very often done in conjunction with the numerical modeling approach discussed below. That is, model output is analyzed using extreme value theory to see if statistics are altered in a changing climate.

The *empirical-physical* approach uses physical reasoning based on empirical knowledge to provide a basis for an extreme value model. The key point here is that, in contrast to the purely statistical method that primarily uses asymptotic mathematical arguments, physical reasoning is employed to perform the extrapolation into the scarcely sampled tails of the PDF. The empirical-physical method can itself be further split into either empirical or physical strategies, focusing on the empirical or physical aspects of the problem respectively. The empirical-physical method lacks the mathematical rigor of the statistical method, but it provides valuable physical insight into relevant real world problems. An example for an empirical-physical application is the Gamma distribution which is often used to describe atmospheric variables that are markedly asymmetric and skewed to the right (Wilks 2006). Well known applications are precipitation and wind speed, which are physically constrained to be non-negative. It should be noted that the empirical-physical approach can be, in principle, put on a more rigorous foundation using the principle of maximum entropy (Jaynes 1957a,b, 2003). That is, given some physical information (i.e., constraints) of a process, the PDF that maximizes the information entropy under the imposed constraints is the one most likely found in nature.

The *numerical modeling* approach aims to estimate the statistics of extreme events (the tails of the PDF) by integrating a general circulation model (GCM) for a very long period (e.g., Easterling et al. 2000; Kharin and Zwiers 2005; Kharin et al. 2007). That is, this approach tries to effectively lengthen the limited observational record with proxy data from a GCM, filling the unsampled tails of the observed PDF with probabilities from model data. Numerical modeling allows for a detailed analysis of the physics (at least model-physics) of extreme events. In addition, the statistical and empirical-physical methods can also be applied to model data, validating (or invalidating) the quality of the model. It is obvious that the efforts by the IPCC to understand and forecast the statistics of extreme weather and climate events in a changing climate fall into this category.

What do these approaches have in common? Every approach effectively extrapolates from the known to the scarcely known (or unknown) using certain assumptions and, therefore, requires a leap of faith. For the statistical approach the assumptions are purely mathematical. For example, the assumption of classical extreme value theory, that the extreme events are independent and drawn from the same distribution, and that sufficient data is available for convergence to a limiting distribution (the generalized extreme value distribution) may not be met (e.g., Coles 2001; Wilks 2006). The potential drawback of the empirical-physical approach is its lack of mathematical rigor (with the exception of the principle of maximum entropy); it primarily depends on empirical knowledge and physical arguments. The weakness of numerical modeling lies in the largely unknown ability of a model to reproduce the correct statistics of extreme events. Currently, GCMs are calibrated to reproduce the observed first and second moments (mean and variance) of the general circulation of the ocean and atmosphere. Very little is known about the credibility of GCMs to reproduce non-Gaussian statistics, that is, extreme events.

We conclude that the common methods to study extreme events have some limitations, and that it is fair to say that until recently the study of extreme meteorological events has been largely empirical. In particular, there exists no physical theory on what kind of non-Gaussian PDFs we should expect for climate variability (we don't have a null hypothesis for the shape of non-Gaussian climate PDFs). Most investigators use observations or model output to estimate the probabilities of weather and climate extremes without actually addressing the detailed dynamical-physical reason for the shape of the PDFs beyond the mathematical arguments of extreme value theory. In addition, many investigators typically study non-Gaussian statistics in a phase-space spanned by the two or three leading empirical orthogonal functions (EOFs) (e.g., Mo and Ghil 1988, 1993; Molteni et al. 1990; Corti et al. 1999; Smyth et al. 1999; Monahan et al. 2001; Berner 2005; Berner and Branstator 2007; Majda et al. 2003, 2008; Franzke et al. 2005). Significant exceptions are White (1980), Trenberth and Mo (1985), Nakamura and Wallace (1991) and Holzer (1996) who present maps of observed skewness and (partly) kurtosis of Northern and Southern Hemisphere geopotential heights. More recently Petoukhov et al. (2008) calculated skewnesses and mixed third-order statistical moments for observed synoptic variation of horizontal winds, temperature, vertical velocity and the specific humidity. Again, none of those papers provide complete dynamical explanations for the observed non-Gaussian structures. One partial exception is Holzer (1996) who attributes negative midlatitude skewness bands to the rectification of velocity fluctuations by the advective nonlinearity. However, the dynamics of the general non-Gaussian structures and extreme events remain largely unexplained.

7.1.3 Novel Stochastic Approaches to Study Extreme Events

In recent years, new tools that make use of advanced stochastic theory have evolved to evaluate extreme events and the physics that govern these events (e.g., Berner 2005; Berner and Branstator 2007; Franzke et al. 2005; Majda et al. 2003, 2008; Monahan 2004, 2006a,b; Peinke et al. 2004; Sura 2003; Sura et al. 2005, 2006; Sura and Newman 2008; Sura and Sardeshmukh 2008; Sardeshmukh and Sura 2009). These tools take advantage of the non-Gaussian structure of the PDF by linking a stochastic model to the observed non-Gaussianity. The novel feature of those models is that the stochastic component is allowed to be state dependent (or multiplicative). The physical significance of multiplicative noise is that it has the potential to produce non-Gaussian statistics in linear systems. Because that phenomenon is the foundation of this chapter, the basic physical principle is explained first. In Sect. 7.2 we will explore the related mathematics in more detail. Figure 7.1 illustrates the following discussion, using a non-Gaussian bimodal PDF as an accessible example (note that in reality we rarely observe bimodal PDFs).

Suppose climate dynamics are split into a slow (i.e., slowly decorrelating) contribution and a fast (i.e., rapidly decorrelating) contribution; this is an assumption well known in turbulence theory. The fast part is then approximated as noise. That is, we consider the dynamics of an n-dimensional system whose state vector $\mathbf{x}$ is governed by the stochastic differential equation (SDE)

$$\frac{d\mathbf{x}}{dt} = \mathbf{A}(\mathbf{x}) + \mathbf{B}(\mathbf{x})\eta(t), \tag{7.1}$$

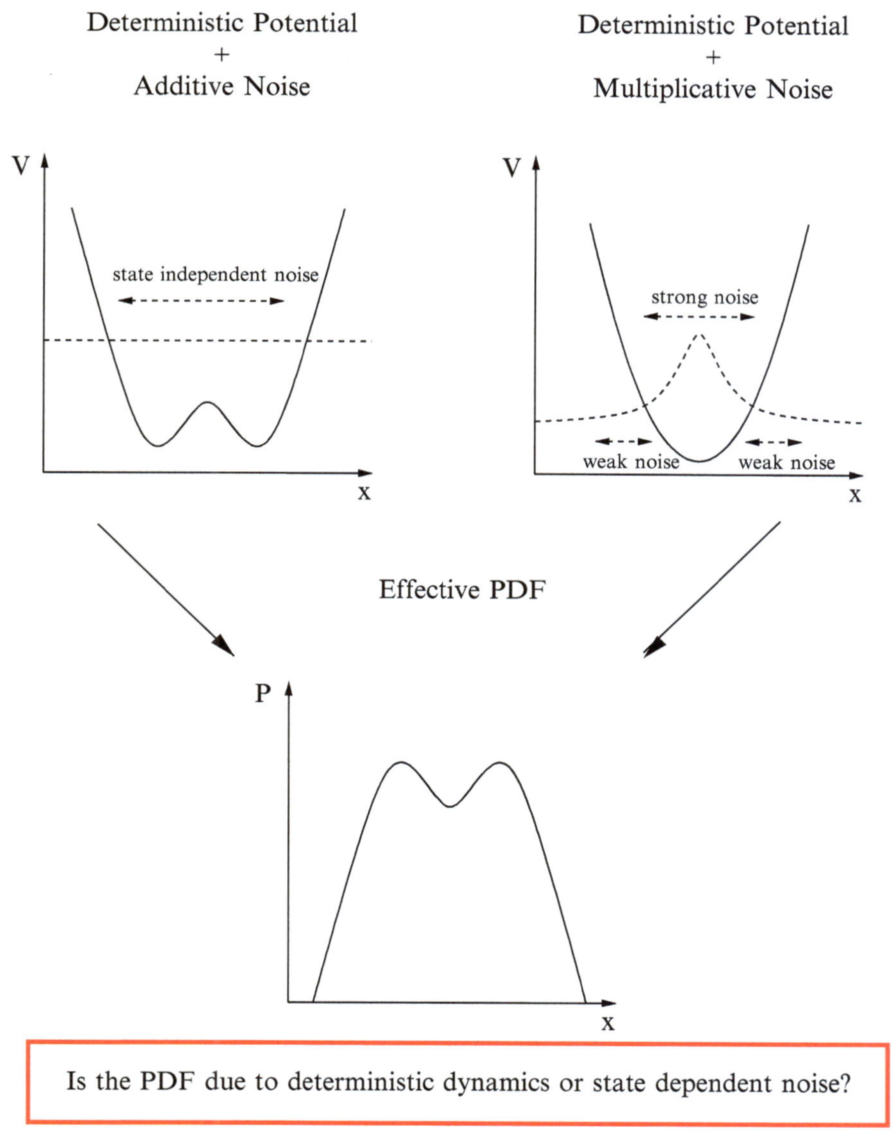

Fig. 7.1 A schematic drawing to illustrate the fundamental dynamical difference between deterministically and stochastically induced PDFs. The effective PDF (denoted P; *solid line* in *lower panel*) of a trajectory in a deterministic double-well potential driven by state-independent (additive) noise (*dashed line* in *left upper panel*) will be bimodal (the potential is denoted V; *solid lines* in *upper panels*). The same effective PDF can be produced by a trajectory in a unimodal potential forced by state-dependent (multiplicative) noise (*dashed line* in *right upper panel*). Because of the larger noise amplitudes near the center of the monomodal potential, as compared to the strength of the noise to the *right* and *left*, the system's trajectory is more often found on either side of the central noise maximum. Thus, the PDF becomes bimodal (non-Gaussian) (Adapted from Sura et al. (2005))

where the vector $\mathbf{A}(\mathbf{x})$ represents all slow, deterministic processes. The product of the matrix $\mathbf{B}(\mathbf{x})$ and the noise vector $\boldsymbol{\eta}(t)$, $\mathbf{B}(\mathbf{x})\boldsymbol{\eta}$, represents the stochastic approximation to the fast phenomena. The stochastic components η_i are assumed to be independent Gaussian white-noise processes: $\langle \eta_i(t) \rangle = 0$ and $\langle \eta_i(t)\eta_i(t') \rangle = \delta(t-t')$, where $\langle \ldots \rangle$ denotes the time average (or, assuming ergodicity, an ensemble average) and δ Dirac's delta function. In general, $\mathbf{x}$ in Eq. (7.1) will have non-Gaussian statistics and is, therefore, well suited to study extreme events. This is well known to be the case if $\mathbf{A}(\mathbf{x})$ is nonlinear, even if $\mathbf{B}(\mathbf{x})$ is constant (i.e., the noise is state-independent or additive; see left branch of Fig. 7.1). It is also true, yet less well known, if the deterministic dynamics are linear, represented for example as $\mathbf{A}_0\mathbf{x}$ with the matrix $\mathbf{A}_0$, as long as $\mathbf{B}(\mathbf{x})$ is not constant (i.e., the noise is state-dependent or multiplicative; see right branch of Fig. 7.1).

Let us look into the physics of each branch in more detail, starting with the left one. There it is obvious that transitions from one potential well to the other driven by additive noise will result in a bimodal PDF, as long as the additive noise is not too strong (in that case we get a monomodal PDF because the potential barrier does not inhibit the motion of the trajectory). This is not, however, the only dynamical system which can produce such a PDF. Consider instead a linear system (the right branch), represented by a unimodal deterministic potential, in which the trajectories are perturbed by multiplicative noise. Because of the larger noise amplitudes near the center of the monomodal potential, as compared to the strength of the noise to the right and left, the system's trajectory is more often found on either side of the central noise maximum, and this system will have a bimodal PDF as well. Thus the same non-Gaussian PDF can result from either a slow (deterministic) nonlinear dynamical system or a fast (stochastic) nonlinear dynamical system.

There are different ways to study and use Eq. (7.1) to understand non-Gaussian statistics. One is to derive an SDE of type (7.1), including a nonlinear deterministic and a multiplicative noise term, directly from the equations of motion. This method has been pioneered over the last decade by applied mathematicians (e.g., Majda et al. 2003, 2008; Franzke et al. 2005). It is sometime called the MTV method, after the names (Majda, Timofeyev, and Vanden-Eijnden) of its proponents (Majda et al. 1999, 2001, 2003). The drawback of this method is, while mathematically rigorous and scientifically useful, that it requires detailed knowledge of the underlying equations and processes. It is also unclear if it can be applied to very complex, state-of-the-art climate models. Most important, the MTV method cannot be used to analyze observational or model data directly.

That drawback can be overcome by trying to estimate the terms of Eq. (7.1) directly from data. Unfortunately, it is nontrivial to estimate both the nonlinear deterministic and multiplicative noise terms for higher (larger than 2) dimensional systems. However, for 1-d and 2-d systems both terms can be estimated empirically (e.g., Berner 2005; Berner and Branstator 2007; Crommelin and Vanden-Eijnden 2006; Lind et al. 2005; Siegert et al. 1998; Friedrich et al. 2000; Sura and Barsugli 2002; Sura 2003; Sura and Gille 2003; Sura et al. 2005, 2006; Sura and Newman 2008; Sura and Sardeshmukh 2009). For higher dimensions the multiplicative noise term poses the largest problem, whereas the deterministic contribution can still be empirically estimated using nonlinear regression techniques. Therefore, one approach is to simply use additive noise, which is relatively easy to estimate, to close the system. For example, an empirical multivariate nonlinear (with quadratic nonlinearities) stochastic model with additive (state-independent) noise captures non-Gaussian tropical SST and extratropical atmospheric variability remarkably well (Kravtsov et al., 2005, 2010; Kondrashov et al., 2006).

There is a obviously a gap between purely mathematical approaches on the one hand, and predominantly empirical methods on the other hand. An approach, introduced by Sura and Sardeshmukh (2008) and Sardeshmukh and Sura (2009), to study non-Gaussian statistics in climate combines mathematical and empirical methods in an effort to fill the aforementioned gap. It attributes extreme anomalies to stochastically forced linear dynamics, where the strength of the stochastic forcing depends linearly on the flow itself (i.e., linear multiplicative noise). Most important, because the theory makes clear and testable predictions about non-Gaussian variability, it can be verified by analyzing the detailed non-Gaussian statistics of oceanic and atmospheric variability. In fact, Sura and Sardeshmukh (2008) and Sardeshmukh and Sura (2009) confirm the multiplicative noise hypothesis of extreme events for sea surface temperature and atmospheric geopotential height and vorticity anomalies. The theoretical framework, the observational evidence, and the implications on how to explore and interpret extreme events in weather and climate are discussed in the remainder of this chapter.

In Sect. 7.2 the theoretical underpinnings of stochastically describing Gaussian and non-Gaussian climate variability are presented. The observational evidence for multiplicative noise dynamics in the ocean and atmosphere (and very briefly, in plasma turbulence) is discussed in Sect. 7.3. Finally, Sect. 7.4 provides a summary and discussion of the status quo, focusing on outstanding issues, challenges, and perspectives of future research on extreme events in weather and climate.

7.2 Theory

7.2.1 Stochastic Dynamics in a Nutshell

This subsection reviews a few basic ideas of stochastic dynamics used in this paper. More comprehensive treatments may be found in many textbooks (e.g., Gardiner 2004; Øksendal 2007; Paul and Baschnagel 1999; van Kampen 2007; Horsthemke and Léfèver 1984). As already mentioned in the introduction, we split climate dynamics into slowly deccorelating deterministic, and a rapidly deccorelating stochastic component: $d\mathbf{x}/dt = \mathbf{A}(\mathbf{x}) + \mathbf{B}(\mathbf{x})\eta(t)$. One important step to understand stochastic dynamics is to first clarify how to interpret this SDE, before introducing the related equation for the PDF $p(\mathbf{x}, t)$ of the stochastic variable $\mathbf{x}(t)$.

7.2.1.1 Interpretation of SDEs

Let us consider, without loss of generality, the stochastic dynamics of a scalar x governed by

$$\frac{dx}{dt} = A(x) + B(x)\eta(t), \tag{7.2}$$

with the delta-correlated white noise η: $\langle \eta(t) \rangle = 0$ and $\langle \eta(t)\eta(t') \rangle = \delta(t - t')$ (the n-dimensional case is a simple component-wise extension of the scalar equation; see Eq. (7.1)). What do we make of this equation? The underlying problem is that, while the white noise $\eta(t)$ is defined for every time t, it is not continuous and, therefore, not differentiable. This means that, mathematically speaking, the SDE (7.2) is not well defined. An alternative interpretation is called for. The solution is to interpret (7.2) in its integral form, and hence expect the noise term to be integrable. That is, we integrate (7.2) to obtain the integral equation

$$x(t) - x(t_0) = \int_{t_0}^{t} A[x(t')]dt' + \int_{t_0}^{t} B[x(t')]\eta(t')dt', \tag{7.3}$$

which can be interpreted consistently. In addition, the discontinuous white noise $\eta(t)$ is usually replaced with its continuous integral, the Wiener process $W(t)$, given by

$$W(t) = \int_{0}^{t} \eta(t')dt', \tag{7.4}$$

or

$$\frac{dW(t)}{dt} = \eta(t). \tag{7.5}$$

Informally we may write $dW(t) = \eta(t)dt$, keeping in mind that this expression is rigorously defined only within an integral. From the definition of the Wiener process its autocovariance can also be easily calculated: $\langle W(t)W(s) \rangle = min(t, s)$. Using the definition of the Wiener process, the integral equation (7.3) becomes

$$x(t) - x(t_0) = \int_{t_0}^{t} A[x(t')]dt' + \int_{t_0}^{t} B[x(t')]dW(t'). \tag{7.6}$$

To summarize, the SDE (7.2) has to be interpreted in its mathematically consistent integral form. As a reminder of the integral definition, the SDE (7.2) is often written as

$$dx = A(x)dt + B(x)dW(t). \tag{7.7}$$

The more rigorous (in the Itô sense) integral expression of the increment of the Wiener process is

$$
\begin{aligned}
dW(t) &= W(t+dt) - W(t) \\
&= \int_0^{t+dt} \eta(t')dt' - \int_0^t \eta(t')dt' \\
&= \int_t^{t+dt} \eta(t')dt' \,.
\end{aligned} \tag{7.8}
$$

Using this expression we can calculate the autocovariance $\langle dW(t)dW(s)\rangle$ as

$$
\begin{aligned}
\langle dW(t)dW(s)\rangle &= \langle \int_t^{t+dt} \int_s^{s+dt} \eta(t')\eta(s')\,dt'ds'\rangle \\
&= \int_t^{t+dt} \int_s^{s+dt} \langle \eta(t')\eta(s')\rangle\,dt'ds' \\
&= \int_t^{t+dt} \int_s^{s+dt} \delta(t'-s')\,dt'ds' \\
&= dt\,\delta_{t,s},
\end{aligned} \tag{7.9}
$$

where $\delta_{t,s}$ denotes the Kronecker delta. For $t = s$ we obtain $\langle dW(t)^2\rangle = dt$, stating that the variance of white noise (the so called Wiener process W) scales with the time increment dt; in fact, a more detailed discussion would show that $dW(t)^2 = dt$. That is, the increment of the Wiener process effectively scales with $\sqrt{dt}$ in contrast to the dt scaling of the deterministic terms. It is this $\sqrt{dt}$ scaling of the stochastic terms that is responsible for the so called *stochastic calculus*, in contrast to the dt scaling of the ordinary *deterministic calculus*.

7.2.1.2 SDE Versus Fokker-Planck Equation

Another important point to bear in mind for the following discussion is that an SDE describes a single realization of the stochastic trajectory $\mathbf{x}(t)$ governed by the, in general, n-dimensional SDE $d\mathbf{x}/dt = \mathbf{A}(\mathbf{x}) + \mathbf{B}(\mathbf{x})\boldsymbol{\eta}$. For sufficiently smooth and bounded $\mathbf{A}(\mathbf{x})$ and $\mathbf{B}(\mathbf{x})$ the PDF $p(\mathbf{x},t)$ of the stochastic variable $\mathbf{x}(t)$ is governed by the corresponding Fokker-Planck equation

$$
\begin{aligned}
\frac{\partial p(\mathbf{x},t)}{\partial t} &= -\sum_i \frac{\partial}{\partial x_i}\left[A_i + \hat{\alpha}\sum_{j,k}\left(\frac{\partial}{\partial x_j}B_{ik}\right)B_{jk}\right]p(\mathbf{x},t) \\
&+ \frac{1}{2}\sum_{i,j}\frac{\partial^2}{\partial x_i \partial x_j}(\mathbf{B}\mathbf{B}^T)_{ij}\,p(\mathbf{x},t)\,,
\end{aligned} \tag{7.10}
$$

which describes the conservation of $p(\mathbf{x},t)$ under both deterministic and diffusive influences. Two different values of $\hat{\alpha}$ yield two physically important stochastic calculi: the Itô ($\hat{\alpha} = 0$) and the Stratonovich calculus ($\hat{\alpha} = 1/2$). On the right hand side, the first term within square brackets describes the dynamics of the deterministic system and is called the deterministic drift. The second term within square brackets, which does not occur in Itô systems ($\hat{\alpha} = 0$), is called the noise-induced drift. The remaining term (i.e., the sum including the derivatives of $(\mathbf{BB}^T)_{ij}\, p(\mathbf{x},t)$) is associated with the diffusion of the probability density by noise. For a detailed discussion of stochastic integration and the differences between Itô and Stratonovich SDEs see for example Gardiner (2004), Øksendal (2007), or van Kampen (2007). The key point here is that the Stratonovich calculus is relevant for continuous physical systems, such as the atmosphere, in which rapidly fluctuating quantities with small but finite correlation times are approximated as white noise. Thus, simplified stochastic models constructed from atmospheric dynamical equations may assume Stratonovich calculus. However, if instead a stochastic model is indirectly estimated from observed discrete data, then the inferred drift will be the *sum* of the deterministic and the noise-induced drifts. In this case using the Itô framework may be preferable, where now $\mathbf{A}(\mathbf{x})$ represents not just the deterministic drift but rather this sum, or the "effective drift". For example (and for future use) in the 1-d case (7.10) simplifies to

$$\frac{\partial p(x,t)}{\partial t} = -\frac{\partial}{\partial x} A_{\mathit{eff}}\, p(x,t) + \frac{1}{2}\frac{\partial^2}{\partial x^2} B^2 p(x,t)\ , \tag{7.11}$$

where A_{eff} is now the effective drift.

7.2.2 Stochastic Dynamics of Climate Variability

Next we outline the basic underlying principle of the stochastic approach to study climate variability in general, and non-Gaussian statistics and extreme events in particular. The main point is that we assume that climate dynamics can be split into a slowly decorrelating contribution and a rapidly decorrelating contribution, and that the fast part can be approximated as white noise. While this is, at first sight, an oversimplification, the white noise approximation is appropriate because it is actually sufficient to require that the fast motions decorrelate more rapidly than the slow motions. This is the general requirement discussed and used by Hasselmann (1976) (see Sect. 2.3 below). Note that the decomposition into fast and slow modes is also a common approximation made in turbulence theory (e.g., Monin and Yaglom, 1971, 1975; Pope, 2000).

Thus the starting point is again Eq. (7.1). We now define $\mathcal{B}(\mathbf{x}) \equiv \mathbf{B}(\mathbf{x})\eta$ (to use the vector/matrix form of the coming expansion), let $\mathbf{x} = \langle\mathbf{x}\rangle + \mathbf{x}'$, and linearize around the basic state $\langle\mathbf{x}\rangle$ through a Taylor expansion to get

$$\frac{d\mathbf{x}'}{dt} = \mathbf{J}_A(\langle\mathbf{x}\rangle)\mathbf{x}' + \mathcal{B}(\langle\mathbf{x}\rangle) + \mathbf{J}_{\mathcal{B}}(\langle\mathbf{x}\rangle)\mathbf{x}' - \langle\mathbf{J}_{\mathcal{B}}(\langle\mathbf{x}\rangle)\mathbf{x}'\rangle + \mathbf{r}\ , \tag{7.12}$$

with the Jacobian $\mathbf{J}_f(\langle\mathbf{x}\rangle)$ (for the function $\mathbf{f(x)}$) evaluated at $\langle\mathbf{x}\rangle$ and assuming that the tendency of the mean is forced by the deterministic drift of the mean, $d\langle\mathbf{x}\rangle/dt = \mathbf{A}(\langle\mathbf{x}\rangle)$. Because the time mean of the multiplicative noise term $\langle\mathbf{J}_{\mathscr{B}}(\langle\mathbf{x}\rangle)\mathbf{x}'\rangle$ is, in general, not zero we have to introduce an additional mean forcing $-\langle\mathbf{J}_{\mathscr{B}}(\langle\mathbf{x}\rangle)\mathbf{x}'\rangle$ to ensure that $\langle\mathbf{x}'\rangle = 0$. We also included a supposedly small residual $\mathbf{r}$, representing the higher-order terms of the Taylor expansion and all other neglected phenomena. Thus, although the governing equations for weather and climate are nonlinear, for many applications the anomalies are well approximated by a linear SDE of the general component form

$$\frac{dx_i}{dt} = A_{ij}x_j + (G_{ip} + E_{ijp}x_j)\eta_p - D_i + r_i\ , \tag{7.13}$$

where we dropped the primes, lumped together all the derivatives and coefficients into new matrices and tensors, and used Einstein's summation convention (sum over repeated indices) to highlight the functional form of, specifically, the noise terms. Therefore, besides the linear deterministic term $A_{ij}x_j$, $G_{ip}\eta_p$ represents the additive (state-independent), and $E_{ijp}x_j\eta_p$ the multiplicative (state-dependent) noise contribution. Note that the additive and multiplicative noise terms are correlated because each term is multiplied by the same noise η_p. In addition, D_i represents the correlations of the noise forcing with the state variable itself (the "noise-induced drift" in Stratonovich systems), which can be shown to be related to the noise parameters as $D_i = -(1/2)E_{ijp}G_{jp}$ (e.g., Sardeshmukh and Sura, 2009).

Equation (7.13) *without* multiplicative noise (i.e., $E_{ijp} = 0$) already plays a central role in studying climate variability. For example, the matrices A_{ij} and $Q_{ij} = G_{im}G_{jm}$ (the product of G_{im} with its transpose) can be estimated directly from data. This technique is called linear inverse modeling and has been successfully applied to study many climate phenomena (e.g., Penland 1989; Penland and Ghil 1993; Penland and Matrosova 1994; Penland and Sardeshmukh 1995; Winkler et al. 2001). In addition, dynamical studies of quasigeostrophic turbulence often rely on multivariate linear models with additive noise (e.g. Farrell and Ioannou 1995, 1996; Whitaker and Sardeshmukh 1998; DelSole 2004). While the multivariate linear models with additive noise are very useful for many applications, they have one inherent shortcoming: they can only represent Gaussian statistics (Gardiner, 2004) and are, therefore, more often than not inconsistent with observations (we will present and discuss several examples in Sect. 7.3).

This brings us to the main theme of this chapter. In the general case *with* multiplicative noise (i.e., $E_{ijp} \neq 0$) Eq. (7.13) will exhibit non-Gaussian statistics. In the remainder of this chapter we will discuss the non-Gaussian statistics of (7.13) and compare those with observations. As we will see, recent detailed studies demonstrated that certain types of observed non-Gaussian statistics are consistent with linear stochastically forced dynamics with correlated additive and multiplicative (CAM) noise forcing. The consistency follows from the fact that the PDF and closed expressions for the moments, such as skewness and kurtosis, can be analytically derived from the stationary Fokker-Planck equation for a system such as (7.13). In particular, for a variable x (with zero mean and standard deviation σ) drawn from such a PDF $p(x)$ we obtain

- The striking property that the (excess) kurtosis $K \equiv \langle x^4 \rangle / \sigma^4 - 3$ is always greater than 1.5 times the square of the skewness $S \equiv \langle x^3 \rangle / \sigma^3$ (minus an adjustment constant r, discussed later), $K \geq (3/2)S^2 - r$;
- Power-law tails $p(x) \propto x^{-\alpha}$ (with the power-law exponent α) for large $\pm x$.

Remember that skewness represents the asymmetry of a PDF. It is positive if the right tail contains more data than the left tail, and negative if the opposite is true. Symmetric distributions, such as the classic Gaussian, have zero skewness. Kurtosis measures the probability (fatness) in the tails of a distribution. The definition of K here is sometimes referred to as "excess kurtosis", because we have subtracted 3, corresponding to the kurtosis for a Gaussian distribution. Here we exclusively use "excess kurtosis" in most calculations and plots, and simply call it "kurtosis" most (but not all) of the times. Thus, in this paper "excess kurtosis" and "kurtosis" are equivalent. (Only in one later example (Fig. 7.15) "excess kurtosis +3" is used and denoted by $\tilde{K}$.)

Remarkably, the theoretically derived K-S inequality and the power-law behavior is found to be consistent with observed non-Gaussian variability of many variables in the ocean and the atmosphere. However, before we move on to present the theoretical foundation and the related observations in some detail, we will briefly rehash the theory of the Gaussian model (Hasselmann's paradigm), as it facilitates the understanding of the non-Gaussian extension.

7.2.3 *Stochastic Models of Gaussian Variability: Hasselmann's Paradigm and the Red Climate Spectrum*

Hasselmann (1976) uses the univariate version of (7.13) with additive noise as the simplest quantitative model of climate variability:

$$\frac{dx}{dt} = -\lambda\, x + \sigma \eta \,, \tag{7.14}$$

where x represents the "slow" climate driven by "fast" weather noise $\sigma\eta$. Here λ is a positive damping constant, and σ denotes the constant amplitude of the Gaussian white-noise forcing η, satisfying $\langle \eta(t) \rangle = 0$ and $\langle \eta(t)\eta(t') \rangle = \delta(t - t')$. From (7.14) it is straightforward (e.g., Gardiner 2004; Paul and Baschnagel 1999, or see the discussion below) to calculate the autocorrelation function $\rho(t')$ for lag t',

$$\rho(t') = \frac{\langle x(t)x(t + t') \rangle}{\langle x^2 \rangle} = \exp(-t'\lambda) \,, \tag{7.15}$$

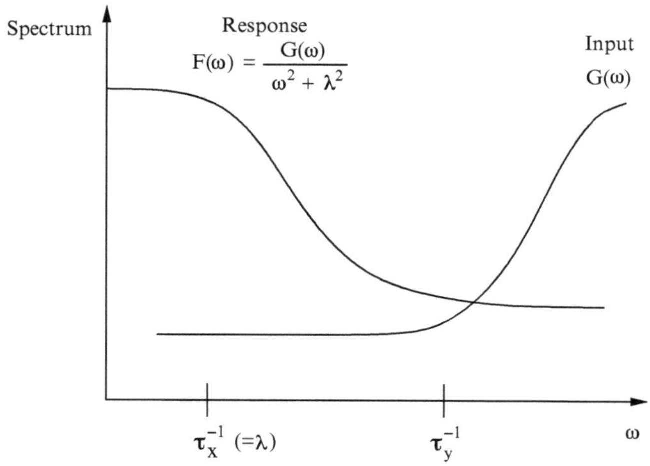

Fig. 7.2 Spectral input $G(\omega)$ and response $F(\omega)$ of a stochastically forced univariate (single component) climate model with additive noise, where ω denotes the angular frequency. The forcing $G(\omega)$ has to be white only below a cut-off frequency $\omega_y = 1/\tau_y$. The high-frequency part of the forcing can have any spectral form

and $\langle x^2 \rangle = \sigma^2/2\lambda$. This is a substantial result because we now know the most important and commonly used characteristic of climate variability, namely the frequency spectrum $F(\omega)$ (where ω denotes the angular frequency). Recall that the spectrum is defined as the Fourier transform of the autocovariance function $\gamma(t) = \langle x^2 \rangle \rho(t)$. The Fourier transform of common functions is tabulated. For example, if $\gamma(t) = \delta(t)$ (autocovariance of white noise), then its Fourier transform $F(\omega) = const$. That is, white noise has a flat spectrum: every frequency has the same weight. If $\gamma(t) = \langle x^2 \rangle \exp(-t\,\lambda)$ (autocovariance of (7.14), see above), then $F(\omega) \propto 1/(\omega^2 + \lambda^2)$ and the response has a very different character from that of the white-noise forcing. We see that $\omega_c = \lambda$ defines a critical frequency. At frequencies $\omega > \omega_c$ the response to white-noise forcing decreases rapidly with frequency. On the other hand, at frequencies $\omega < \omega_c$ the response levels out and becomes independent of ω. It is called a red-noise spectrum because there is more power in lower frequencies (as in the electromagnetic spectrum of red light). This simple model is the generally accepted paradigm of how inertia introduced by sluggish elements of the climate system damps high frequencies to yield a slowly decorrelating response. This is sometimes called a “reddening” of the spectrum of climate variability. The universal spectral forcing $G(\omega)$ and response $F(\omega)$ representation of the stochastic climate model is summarized in Fig. 7.2. Note that, in general, the forcing $G(\omega)$ has to be white only below a cut-off frequency $\omega_y = 1/\tau_y$. The high-frequency part of the forcing can have any spectral form. The relevance of this simple model is, of course, that it explains the ubiquitous red spectrum of many climate time series.

Besides the spectrum we can also easily derive the steady-state PDF $p(x)$ of x governed by (7.14). The related time-independent Fokker-Planck equation is

$$\frac{d}{dx}\lambda\, x\, p(x) + \frac{1}{2}\frac{d^2}{dx^2}\sigma^2\, p(x) = 0. \tag{7.16}$$

It is straightforward to verify that the solution is a centered Gaussian distribution with variance $\langle x^2\rangle = \sigma^2/2\lambda$,

$$p(x) = \frac{1}{\sqrt{2\pi\,(\sigma^2/2\lambda)}}\exp\left[-\frac{x^2}{2\,(\sigma^2/2\lambda)}\right]. \tag{7.17}$$

Therefore, while Hasselmann's paradigm is successful in explaining the ubiquitous red spectrum of climate variability, it fails to account for observed non-Gaussian statistics. At this point the thoughtful reader may object that perhaps the stochastic forcing is non-Gaussian itself, inducing a non-Gaussian response in x. While this is, of course, the obvious question to ask, a simple scaling argument shows that non-Gaussian additive noise alone has no significant effect on the distribution of x. For reasonable climate applications we can assume that λ is a small parameter (representing slowly decorrelating, i.e., low-frequency climate variability), overshadowed by a strong stochastic forcing of amplitude σ. That means, in a first-order approximation the evolution of x is, on short timescales, given by $dx/dt \approx \sigma\eta$ so $x \approx \sum \sigma\eta\Delta t$, and the central limit theorem applies. In other words, for sufficiently small λ, non-Gaussian additive noise will result in a close to Gaussian distribution of x. It is easy (and instructive) to verify this simple scaling argument with numerical integrations of (7.14) with a strongly non-Gaussian stochastic forcing. The bottom line is that, while the linear model with additive noise (7.14) is very useful as a paradigm for the general shape of observed climate spectra, it fails to explain non-Gaussian statistics. As we will see now, this shortcoming can be addressed by considering multiplicative noise models.

7.2.4 *Stochastic Models of Non-Gaussian Variability: A Null Hypothesis for the Statistics of Extreme Events*

Let us restate the question. How can we extend Hasselmann's paradigm (or null hypothesis) for climate variability to include non-Gaussian statistics? One obvious approach is to retain the, so far neglected, multiplicative noise contribution in Eq. (7.13). Thus we take the univariate version of (7.13) as a physically plausible null hypothesis for non-Gaussian climate variability; as we will see in Sect. 7.3, observations support this approach. Therefore, the starting point is the 1-d Stratonovich SDE

$$\frac{dx}{dt} = -\tilde{\lambda}\, x + \phi\, x\, F' + F' + R' - \phi\,\langle xF'\rangle, \tag{7.18}$$

with the effective drift $-\tilde{\lambda}$, a constant ϕ, and the Gaussian white-noise forcing terms F' and R' with amplitudes $\sigma_{F'}$ and $\sigma_{R'}$: $\langle F'(t)F'(t')\rangle = (\sigma_{F'})^2\delta(t-t')$ and $\langle R'(t)R'(t')\rangle = (\sigma_{R'})^2\delta(t-t')$. The effective drift equals the sum of the deterministic drift and the noise-induced drift: $-\tilde{\lambda} = -\lambda + (1/2)(\phi\sigma_{F'})^2$. Note that we changed the nomenclature of the noise to F' and R' (from η used beforehand) to clearly distinguish between the correlated additive and multiplicative (CAM) noise forcing $(\phi\, x + 1)F'$ and the purely additive noise residual R'.

Equation (7.18) is a natural extension of Hasselmann's paradigm (7.14), adding a correlated linear state-dependent noise term to the traditional state-independent one. It is this CAM noise term that is responsible for the specific non-Gaussian statistics of our stochastic model (7.18). Before we move on to discuss the non-Gaussian statistics, let us answer another important question. Will our CAM noise model (7.18) still generate a red spectrum? Remember that the main success of the Hasselmann model (7.14) is that it explains the ubiquitous red spectrum of many climate time series. The answer is *yes* because (7.18) still has an exponentially decaying autocorrelation function $\rho(t) = \exp(-t\tilde{\lambda})$. The only difference to the autocorrelation function (7.15) of Hasselmann's additive noise model is that the decay time is now given by the inverse of the effective drift, and thus is now $1/\tilde{\lambda}$. This can be easily seen by calculating the ensemble average of x from (7.18) for a given initial condition $x(0)$:

$$\langle x(t)\rangle = x(0)\exp(-t\tilde{\lambda}). \qquad (7.19)$$

If we now multiply by $x(0)$ and average over initial conditions we get

$$\langle x(t)x(0)\rangle = \langle x(0)^2\rangle \exp(-t\tilde{\lambda}), \qquad (7.20)$$

and the autocorrelation function becomes

$$\rho(t) = \exp(-t\tilde{\lambda}). \qquad (7.21)$$

Therefore, the spectrum of x governed by (7.18) has the characteristic form $F(\omega) \propto 1/(\omega^2+\tilde{\lambda}^2)$, and the introduction of CAM noise does not alter the spectral behavior of Hasselmann's paradigm. As we will see next, however, CAM noise significantly alters the PDF of the stochastic null hypothesis.

7.2.4.1 Skewness-Kurtosis Link

For x governed by (7.18), the Fokker-Planck equation for the stationary PDF $p(x)$ may be written

$$\frac{d}{dx}\tilde{\lambda}\, x\, p(x) + \frac{1}{2}\frac{d^2}{dx^2}\left[(\sigma_{F'}^2+\sigma_{R'}^2+\phi^2\,\sigma_{F'}^2\, x^2+2\,\phi\,\sigma_{F'}^2\, x)p(x)\right] = 0. \qquad (7.22)$$

Equations for the moments can be obtained by multiplying (7.22) by powers of x and integrating by parts. This yields $\langle x \rangle = 0$ for the first moment, and for the second, third, and fourth moments we get

$$\begin{aligned}
\langle x^2 \rangle &= \frac{(\sigma_{F'}^2 + \sigma_{R'}^2)}{[2\tilde{\lambda} - (\phi\sigma_{F'})^2]}, \\
\langle x^3 \rangle &= \frac{2\phi\sigma_{F'}^2 \langle x^2 \rangle}{[\tilde{\lambda} - (\phi\sigma_{F'})^2]}, \\
\langle x^4 \rangle &= \frac{[3\phi\sigma_{F'}^2 \langle x^3 \rangle + (3/2)(\sigma_{F'}^2 + \sigma_{R'}^2)\langle x^2 \rangle]}{[\tilde{\lambda} - (3/2)(\phi\sigma_{F'})^2]}.
\end{aligned} \tag{7.23}$$

Note that for $\phi > 0$ the additive and multiplicative noises are positively correlated (have the same sign), resulting in positive skewness of x. For $\phi < 0$ the reverse is true, and for no multiplicative noise at all ($\phi = 0$) x obeys, of course, Gaussian statistics. Using the definitions of skewness $S \equiv \langle x^3 \rangle / \sigma^3$ and kurtosis $K \equiv \langle x^4 \rangle / \sigma^4 - 3$ (for an x with zero mean and standard deviation $\sigma = \sqrt{\langle x^2 \rangle}$) we can derive the relation $K = A\,S^2 + B$, with

$$A = \frac{3}{2} \frac{\left[-\tilde{\lambda} + (\phi\sigma_{F'})^2\right]}{\left[-\tilde{\lambda} + (3/2)(\phi\sigma_{F'})^2\right]}, \tag{7.24}$$

and

$$B = 3\frac{\left[-\tilde{\lambda} + (1/2)(\phi\sigma_{F'})^2\right]}{\left[-\tilde{\lambda} + (3/2)(\phi\sigma_{F'})^2\right]} - 3. \tag{7.25}$$

For stable conditions this results in the general relation between skewness and kurtosis

$$K \geq \frac{3}{2} S^2, \tag{7.26}$$

where we used the fact, that, for the fourth moment to exist, there is an upper limit for the strength of the multiplicative noise: $(\phi\sigma_{F'})^2 < (2/3)\tilde{\lambda}$. Note that for weak multiplicative noise we obtain the equality $K \approx (3/2)S^2$, and that the weak-multiplicative-noise approximation is not equivalent to having no multiplicative noise at all. For vanishing multiplicative noise $S = 0$, and $K = A\,S^2 + B$ collapses to $K = 0$.

Above we already mentioned that the CAM noise parabola might include a small vertical offset: $K \geq (3/2)S^2 - r$, with the positive constant r. We now briefly discuss the dynamical reason for this offset. Sardeshmukh and Sura (2009) showed that the above-mentioned one-dimensional model becomes progressively better at representing higher-order statistics of multivariate systems through a principle of increasing "diagonal dominance" in the higher-order moment equations. Diagonal dominance refers to the progressively greater importance of the self-correlation terms in the higher-order moment equations of multivariate systems. In physical terms this means that local correlations are dominating the moment equations, while spatial correlations over long distances are less important. That makes sense for rapidly varying high-amplitude day-to-day atmospheric variability, as long as we are neglecting low-frequency (and amplitude) teleconnection patterns in our first-order approximation. The increasing importance of self-correlation terms in the higher-order moment equations is the basic reason for the relevance of the one-dimensional model in the dynamics of the higher-order moments even in multivariate systems. The diagonal dominance of the higher-order moment equations now helps explain the vertical offset relative to the prediction of the one-dimensional theory. The key to understanding this, qualitatively, is to recognize that diagonal dominance is stronger for the higher moments. That is, the error we make using a one-dimensional system to approximate a multivariate system is largest for the second moment (the mean is zero by construction of our anomaly equations), and successively smaller for the higher moments. In addition, since the multivariate system has larger variance, we know the sign of the error we make in the second moments by using a univariate system to approximate a multivariate system. This relies on the fact that a linear operator A is, in almost all geophysical contexts, a "non-normal" (here "non-normal" refers to the non-orthogonality of the eigenfunctions of A and should not be confused with non-Gaussianity) that does not commute with its transpose. This non-normality of the linear dynamical system leads to a greater variance of x than for a "normal" A with the same eigenvalues (Ioannou 1995); a one-dimensional system is, of course, always "normal". In most cases, this non-normality is associated with the ability of anomalies to draw energy from a background state, of which there is pervasive evidence and which is indeed one of the cornerstones of dynamical meteorology and oceanography. Now it can be seen from the definition of kurtosis, $K = \langle x^4 \rangle / \sigma^4 - 3$, that an increased variance in combination with a negligible error in the fourth moment will lead to a decrease of K: the parabola is effectively shifted downward. Therefore, the multivariate system, results in an inequality $K \geq (3/2)S^2 - r$, with a small positive constant r.

Before we move on to discuss the PDF of CAM noise dynamics, let us extract the physical significance of the skewness-kurtosis link. For our purpose the key point to recognize is that the constraint $K \geq (3/2)S^2 - r$ is stronger than the lower bound for data drawn from any PDF, where $K \geq S^2 - 2$ (e.g., Wilkins 1944). Another general constraint states that for a unimodal PDF (i.e., a distribution with a single peak), $K \geq S^2 - 186/125 \approx S^2 - 1.5$ (Klaassen et al. 2000). Thus the factor $3/2$ is clearly due to the presence of the CAM noise forcing, and will be used to identify CAM noise dynamics in the climate system. While the K-S link can be used to identify CAM noise dynamics, for the study of extreme events the knowledge of the tail of a PDF is more crucial. As we will see next, CAM noise systems will produce power-law tails. This is important because power-law tails are very closely related to our discussion of *non-Gaussian* statistics (i.e., extreme events) as power-law tails decay much slower than the tails of Gaussian PDFs. In fact, power-law PDFs are often considered an archetype for the statistics of extreme events (e.g., Newman 2005; Taleb 2010).

7.2.4.2 PDF and Power-Law Tails

Besides being used to obtain the moments, the Fokker-Planck equation (7.22) can also be analytically solved for the full PDF $p(x)$ (e.g., van Kampen 2007; Müller 1987; Sardeshmukh and Sura 2009). After one integration we obtain

$$\tilde{\lambda}\, x\, p(x) + \frac{1}{2}\frac{d}{dx}\left[(\sigma_{F'}^2 + \sigma_{R'}^2 + \phi^2\, \sigma_{F'}^2\, x^2 + 2\phi\, \sigma_{F'}^2\, x)\, p(x)\right] = \text{const.} \equiv 0, \tag{7.27}$$

where the constant of integration has to be zero to fulfill the boundary conditions of vanishing probability at $x = \pm\infty$. It is now straightforward to verify that the solution of (7.27) is

$$p(x) = \mathcal{N}\left[(\phi\sigma_{F'}\, x + \sigma_{F'})^2 + \sigma_{R'}^2\right]^{-(\tilde{\lambda}/\phi^2\sigma_{F'}^2 + 1)} \exp\left[\frac{2\tilde{\lambda}}{\phi^2\sigma_{F'}\sigma_{R'}} \arctan\left(\frac{\phi\sigma_{F'}\, x + \sigma_{F'}}{\sigma_{R'}}\right)\right], \tag{7.28}$$

with the normalization constant $\mathcal{N}$. The PDF (7.28) is clearly skewed if CAM noise is present, whereas for vanishing multiplicative noise it becomes Gaussian. In addition to skew, (7.28) has power-law tails with exponent α,

$$\alpha = 2\left(\frac{\tilde{\lambda}}{\phi^2\sigma_{F'}^2} + 1\right), \tag{7.29}$$

because for large magnitudes of x we get $p(x) \propto x^{-\alpha}$. That is, besides the skewness-kurtosis link, another important result is that Eq. (7.18) will produce power-law tails. The exponent (7.29) tells us to expect heavier tails (smaller α) for a weaker damping (or stronger multiplicative noise forcing $\phi^2\sigma_{F'}^2$) and vice versa.

That the CAM noise system (7.18) will produce power-law tails can also be directly seen from the Fokker-Planck equation (7.27) without having the general solution (7.28). For large x and small noise amplitudes (i.e., keeping only the quadratic term) the Fokker-Planck equation (7.27) simplifies to

$$\tilde{\lambda}\, x\, p + \frac{1}{2}\frac{d}{dx}(\phi^2\, \sigma_{F'}^2\, x^2\, p) = 0. \tag{7.30}$$

Using the power-law ansatz $p(x) \propto x^{-\alpha}$ one easily finds that $p(x)$ is a solution of (7.30) if $\alpha = 2(\tilde{\lambda}/\phi^2\sigma_{F'}^2 + 1)$. It should be noted, though, that the limit of very small $\phi^2\sigma_{F'}^2$ (resulting in $\alpha \to \infty$) is not allowed within the approximation we made to simplify the Fokker-Planck equation: $\phi^2\sigma_{F'}^2 \to 0$ counteracts the approximation of large x. The limit $\phi^2\sigma_{F'}^2 \to 0$ simply means that the multiplicative noise is negligible, and the PDF $p(x)$ becomes Gaussian.

7.2.4.3 Synthesis

To summarize our hitherto discussed results, a CAM noise SDE of type (7.18) is potentially a good candidate to explain non-Gaussian statistics of weather and climate phenomena because (7.18) is derived as a simple Taylor expansion from a very general dynamical system (7.1). Therefore, we have reason to believe that the statistics of observed climate variability might be, approximately, explained by multiplicative noise dynamics. In fact, as we will see in Sect. 7.3, many relevant weather and climate phenomena in the atmosphere and ocean follow the statistics (S-K constraint and power-law tails) predicted by (7.18), strongly suggesting the relevance of CAM noise dynamics in statistically explaining extreme events in weather and climate. At this point we are not aware of another dynamical mechanism capable of producing the constraint skewness-kurtosis link $K \geq (3/2)S^2 - r$ and power-law statistics $p(x) \propto x^{-\alpha}$ for large $\pm x$.

7.3 Observations and Applications

In this section we present and discuss recent observational examples and applications of our non-Gaussian stochastic framework. We will see that several relevant weather and climate phenomena in the atmosphere (Sardeshmukh and Sura 2009; Sura 2010; Sura and Perron 2010) and ocean (Sura and Sardeshmukh 2008; Sura and Gille 2010) conform to the non-Gaussian skewness-kurtosis and power-law statistics predicted by (7.18), allowing us to attribute the statistics of extreme events to a correlated additive and multiplicative noise forcing. We will also briefly discuss the successful application of non-Gaussian stochastic dynamics to other fields such as plasma turbulence (Krommes 2008; Sandberg et al. 2009), pointing to the general character of our theory.

7.3.1 Oceanic Examples

7.3.1.1 Sea Surface Temperature

We start with the discussion of sea surface temperature (SST) because Hasselmann's Gaussian stochastic climate model (7.14) was tested (and also inspired) using SST variability (Frankignoul and Hasselmann 1977). The high-resolution (daily, on a 0.25-degree latitude/longitude grid) observational dataset used by Sura and Sardeshmukh (2008) consists of a blended analysis of daily SST fields based on infrared satellite data from the Advanced Very High Resolution Radiometer

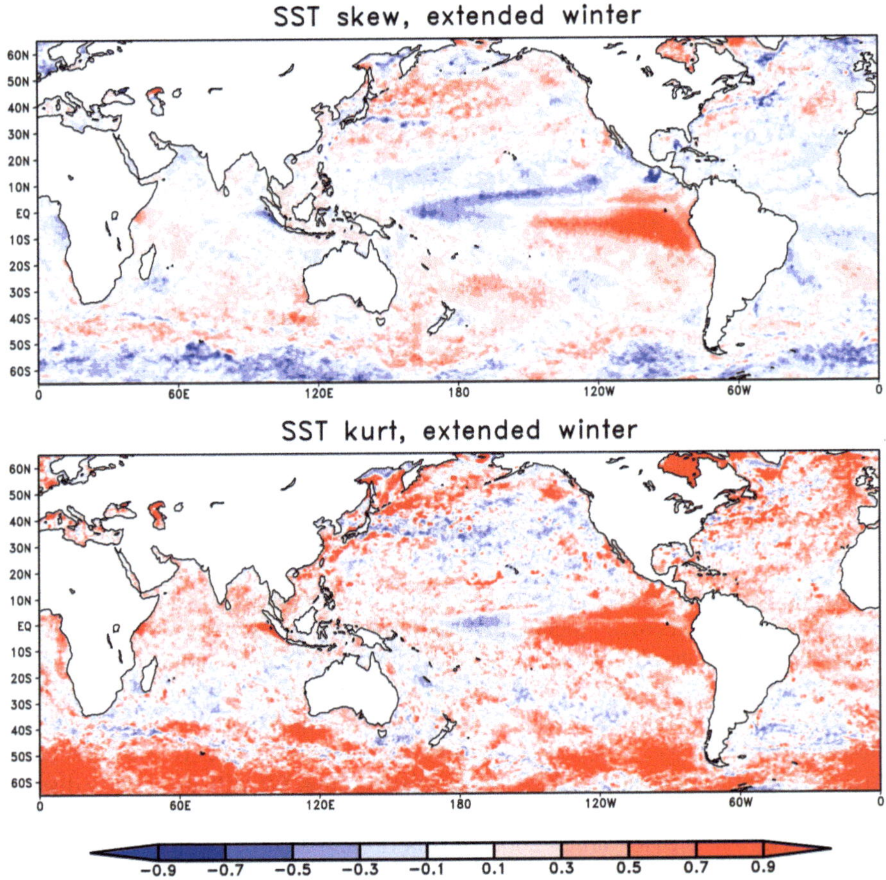

Fig. 7.3 Skewness and kurtosis of daily SST anomalies (1985–2005) for extended winter (Adapted from Sura and Sardeshmukh (2008))

(AVHRR) and in situ data from ships and buoys, and is compiled and provided by Reynolds et al. (2007). A more detailed description of the dataset and analysis procedure can be found in Reynolds et al. (2007). In Sura and Sardeshmukh (2008) 21 years of daily SST data (January 1985 to December 2005) were used, and anomalies were calculated by subtracting the daily climatology and linear trend from the full daily values. Then the extended summer (May-October) and extended winter (November-April) seasons were analyzed.

The skewness (upper panel) and kurtosis (lower panel) of extended winter SST anomalies are shown in Fig. 7.3. Note again that we are using "excess kurtosis" in most calculations and plots, and simply call it "kurtosis" most of the times. Results for the extended summer season look very similar and are, therefore, not shown here (but can be found in Sura and Sardeshmukh 2008). In fact, here we are not so

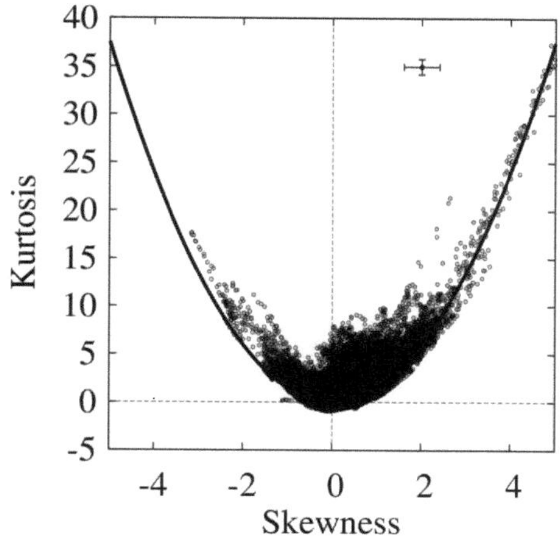

Fig. 7.4 Scatter plot of excess kurtosis versus skewness for full-year SST anomalies equatorward of 65° North and South. The *solid line* denotes the function $K = (3/2)S^2$. The estimated local 95 % confidence intervals on the values are indicated in the *upper right* corner of the figure (Adapted from Sura and Sardeshmukh (2008))

much interested in the rich spatial structure but the global skewness-kurtosis link. Figure 7.4 shows a scatterplot of kurtosis as a function of skewness for all data points equatorward of 65° North and South for extended summer and winter. The solid line in Fig. 7.4 clearly shows a parabolic bound on kurtosis in the SST data: $K \geq (3/2)S^2$ (while many points fall outside the bound, they do so within the error bars). Thus, the distribution of SST anomalies is obviously (and statistically significant) non-Gaussian and also follows our analytically derived constraint (7.26).

What about the predicted power-law tails? In Fig. 7.5 we present two representative examples from locations with strong skewness and kurtosis (taken from Sura 2010). The sample time series are full year SST anomalies from a location off Cape Hatteras (35°N, 75°W), presented in the upper row, and a location in the North Atlantic Current (40°N, 50°W), presented in the bottom row. The PDFs are presented on a log-log scale to highlight the scaling properties of the data. For example, a highly non-Gaussian power-law tail (that is, a PDF $P(x) \propto x^{-\alpha}$ with the exponent α) will appear as a straight line on a log-log plot. Because of the log scale on the x-axis absolute SST anomalies $|T'|$ are presented. In addition, because the emphasis is on extreme events, the center of the PDFs ($\pm$ one standard deviation) is not shown. Thus, the tails of the PDFs of negative/positive SST anomalies at Cape Hatteras are shown from left to right in the upper row, and the PDFs of negative/positive SST anomalies in the North Atlantic Current are shown from left to right in the bottom row of Fig. 7.5. In all plots the solid line denotes a Gaussian distribution. Note the distinct heavy-tail power-law behavior for negative anomalies giving rise to negative skewness. However, even the positive anomalies follow a power-law (weaker than Gaussian, though). The straight dashed lines are maximum likelihood estimates of the power-law behavior (given by the power-law exponent α also included in each plot) above a systematically estimated lower bound $|T'|_{min}$. The lower bounds $|T'|_{min}$ are shown by the dotted-dashed vertical lines. The procedure to estimate the power-law parameters and the related goodness-of-fit test (a Monte-Carlo approach using Kolmogorov-Smirnov statistics) are described

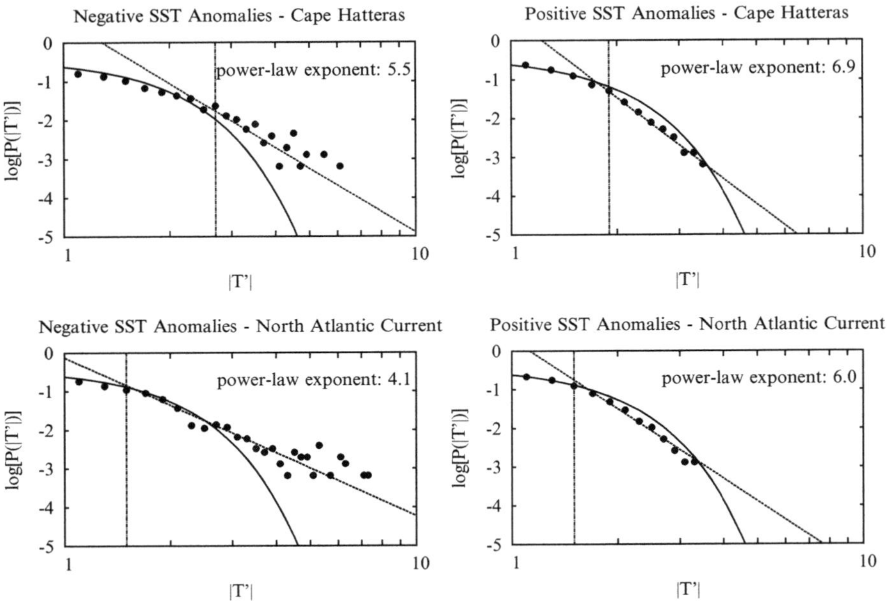

Fig. 7.5 Log-log scale PDFs (*circles*) of absolute values $|T'|$ of full year (*left*) negative and (*right*) positive SST anomalies (*upper row*) off Cape Hatteras (35° N, 75° W) and (*lower row*) in the North Atlantic Current (40° N, 50° W). The *straight dashed lines* are Maximum Likelihood Estimates of the power-law behavior and the *solid lines* denote Gaussian distributions. The lower bounds of the power-law scaling are shown by the *dotted-dashed vertical lines* (Adapted from Sura (2010))

in Clauset et al. (2009) and Sura (2010). Here it is sufficient to know that in all cases the power-law hypotheses are statistically significant (cannot be rejected) and that, therefore, we conclude that PDFs of anomalous SST variability in the Gulf Stream system indeed obey a power-law. At other locations around the globe we also observe power-law statistics (Sura 2010).

To summarize, we have ample evidence to attribute non-Gaussian SST statistics to a correlated additive and multiplicative noise forcing. While, in Sect. 7.2, we predominantly used mathematical arguments to justify the CAM noise equation (7.18), in the case of SST variability the physical aspect of the state-dependent noise forcing is easy to understand. If we (realistically) assume that SST variability is mainly driven by the vertical heat flux F_Q through the air-sea boundary, the typical bulk heat flux parameterization takes the form $F_Q = \beta(T_o - T_a)|\mathbf{U}|$ (positive flux upward), where T_o and T_a are the sea surface and air temperatures, respectively. $|\mathbf{U}|$ is the wind speed, and β is a positive parameter including the densities of seawater, air, specific heats, the Bowen ratio, mixed-layer depth, and other physical processes. That is, the local rate of change of T_o is given by

$$\frac{\partial T_o}{\partial t} = F_Q, \tag{7.31}$$

where F_Q is a function of location and time. For small sea surface temperature anomalies T_o' a Taylor expansion of the heat flux F_Q with respect to $T_o = \overline{T}_o + T_o'$ yields

$$\frac{\partial T_o'}{\partial t} = \frac{\partial \overline{F}_Q}{\partial T_o} T_o' + \frac{\partial F_Q'}{\partial T_o} T_o' + F_Q' + R', \tag{7.32}$$

where it is assumed that the evolution of the mean temperature $\overline{T}_o$ is balanced by the mean heat flux $\overline{F}_Q$, using $F_Q = \overline{F}_Q + F_Q'$, and that the derivative is evaluated at $\overline{T}_o$. We also added a residual term R' to represent higher-order terms of the Taylor expansion and other processes not included in our simple picture. In addition we use $\overline{F}_Q = \beta(\overline{T_a} - \overline{T_o})\overline{|\mathbf{U}|}$ and $F_Q' = \beta(\overline{T_a} - \overline{T_o})|\mathbf{U}|'$, allowing for rapidly varying, effectively stochastic, wind speed fluctuations $|\mathbf{U}|'$. In other words, it is assumed that $|\mathbf{U}|'$ can be approximated by Gaussian white noise. This assumption is reasonable since daily wind speed anomalies are almost uncorrelated and have a distribution that is nearly Gaussian (Sura et al. 2006). For example, at Ocean Weather Station P (in the Gulf of Alaska; 50°N, 145°W; 1949–1981) wind speed anomalies are almost uncorrelated after 2–3 days (Fig. 7.6a) and deviations from Gaussianity are relatively small (Fig. 7.6b). Note that the mean SST $\overline{T_o}$ is still allowed to vary very slowly. We then obtain $\partial \overline{F}_Q/\partial T_o = -\beta\overline{|\mathbf{U}|}$ and $\partial F_Q'/\partial T_o = -\beta|\mathbf{U}|'$, or, equivalently, $\partial F_Q/\partial T_o = -\beta|\mathbf{U}| = -\beta(\overline{|\mathbf{U}|} + |\mathbf{U}|')$. It is the constant term that justifies the introduction of the constant feedback parameter $-\lambda$ in the general formulation (7.14). However, the rapidly varying term $\partial F_Q'/\partial T_o$ cannot be neglected as done in many studies (e.g., in Frankignoul and Hasselmann 1977). Therefore, if we allow for rapidly varying, effectively stochastic, winds we immediately see that the rapid wind component multiplies the slowly varying SST, giving rise to an effectively multiplicative noise forcing. In fact, we recover the by now familiar CAM noise equation (7.18) (Sura and Newman 2008; Sura and Sardeshmukh 2009; Sura 2010). Therefore, in the SST case the CAM noise predominantly (but not exclusively) originates from rapidly varying winds forcing the turbulent heat flux through the air-sea interface. In strong currents like the Gulf Stream random advection by surface currents plays a (mathematically) similar role.

7.3.1.2 Sea Surface Height

Next we consider sea surface height (SSH) anomalies along the TOPEX/Poseidon satellite altimeter ground tracks. We use data distributed by Archiving, Validation and Interpretation of Satellite Oceanographic (AVISO) that are archived at regular 10 km intervals along satellite ground tracks, with standard tidal and atmospheric corrections applied. Here we analyze data from 25 September 1992 to 11 August 2002, corresponding to the duration of the TOPEX/Poseidon mission along its original ground tracks. We subtract the time mean from each observation, so that at each measurement point SSH anomalies have a zero mean.

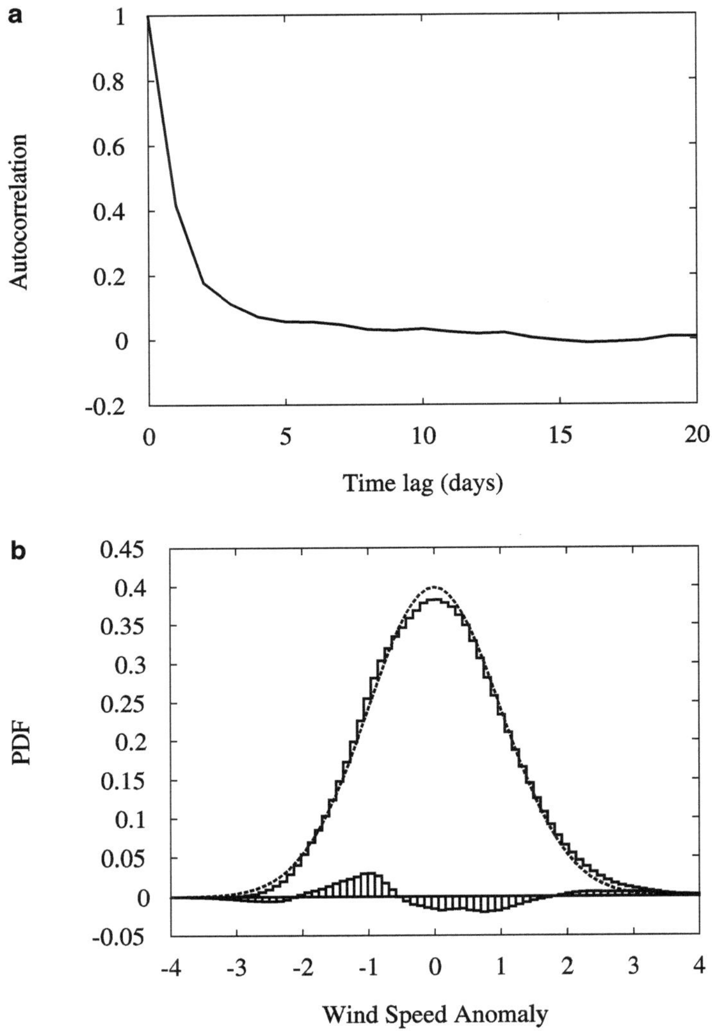

Fig. 7.6 (**a**) Autocorrelation function and (**b**) PDF of daily wind speed anomalies (m/s) at Ocean Weather Station P (Gulf of Alaska; 50°N, 145°W; 1949–1981). In (**b**) the PDF is denoted by steps. The *dashed line* is the related Gaussian distribution, whereas the boxes denote deviations from Gaussianity (Adapted from Sura et al. (2006))

At first sight, it seems very likely that SSH and SST dynamics should be governed by similar stochastic dynamics, because surface quasigeostrophic (SQG) theory predicts that both quantities should resemble each other on oceanic mesoscales (approximately $<$ 400 km) (e.g., Lapeyre and Klein 2006; Isern-Fontanet et al. 2008). For example, we might expect skewness maps of SSH and SST to be almost identical on scales of meandering boundary current systems, such as the Gulf

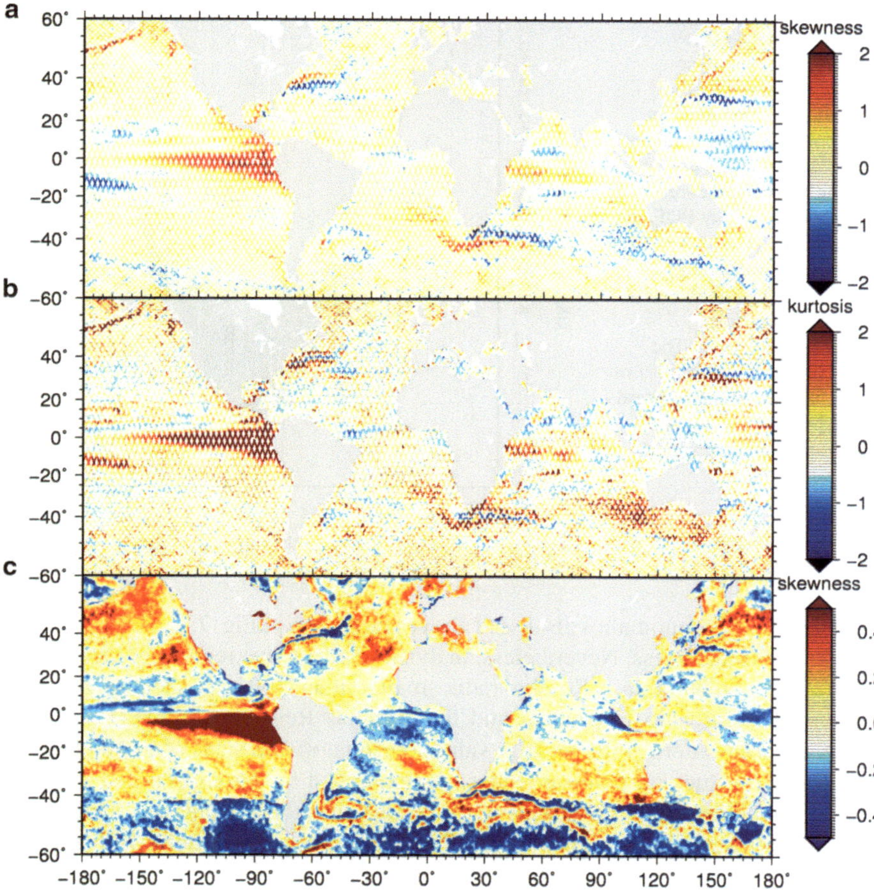

Fig. 7.7 (**a**) Skewness from TOPEX/Poseidon mission, computed for each data point along ground tracks. (**b**) Excess kurtosis, similarly computed along TOPEX/Poseidon ground tracks. (**c**) Skewness from full-year sea surface temperature anomalies, computed from daily Reynolds data (Adapted from Sura and Gille (2010))

Stream, with first baroclinic Rossby radii smaller than 400 km. (Note that the typical Rossby radius in the Gulf Stream system is about 50–100 km, and the typical width of the Gulf Stream is of the same order.) However, as we will see, a comparison of SSH and SST skewness maps in those regions shows that their spatial distributions are not identical on those scales. This implies that the non-Gaussian extremes of SSH and SST variability are driven by different physical processes, and we have no a priori reason to expect the non-Gaussian statistics of SSH and SST to be governed by similar dynamics.

Figure 7.7a shows skewness, and Fig. 7.7b shows kurtosis of SSH. In some regions, the SSH maps show structures that resemble skewness and kurtosis maps

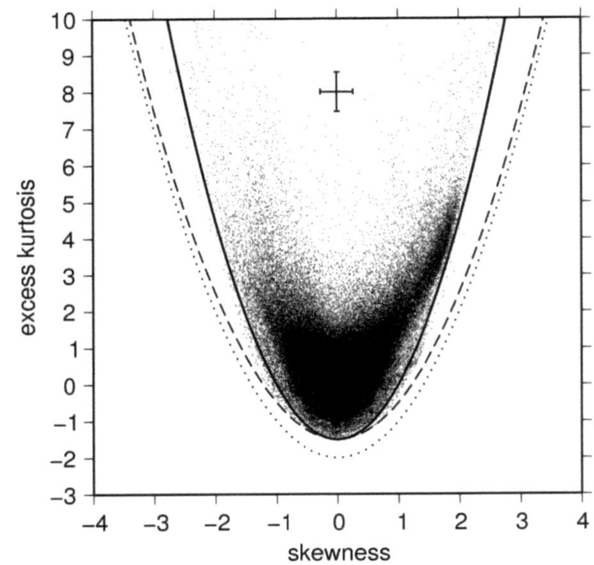

Fig. 7.8 Scatter plot of skewness versus excess kurtosis for sea surface heights measured by the TOPEX/Poseidon mission. Dots come from all latitudes and longitudes. The *dotted line* indicates the theoretical lower bound for any PDF, the *dashed line* indicates the theoretical lower bound for a unimodal PDF, and the *solid line* corresponds to $K = (3/2)S^2 - 1.5$. The error bars in the *upper part* of the figure denote the standard errors in K and S (Adapted from Sura and Gille (2010))

produced for SST anomalies discussed above and shown in Fig. 7.3; see Fig. 7.7c for full-year SST skewness. Nevertheless, in dynamically active regions of meandering jets (with first baroclinic Rossby radii smaller than 400 km) such as the Gulf Stream, the Kuroshio Extension, and the Agulhas Retroflection, SSH skewness forms a dipole pattern, with negative values on equatorward sides of frontal features and positive values on poleward sides. As explained by Thompson and Demirov (2006), this structure is consistent with the fact that these fronts are meandering jets. In contrast, SST anomalies show a very different tripole pattern, with negative skewness in the mesoscale (or Rossby radius scale) core of the current, sandwiched between regions of positive skewness (for details, see Sura 2010; Sura and Gille 2010). The related SSH skewness-kurtosis scatterplot is presented in Fig. 7.8. The solid line denotes the limit of $K \geq (3/2)S^2 - 1.5$. Note that the theoretical lower bound for any PDF, $K \geq S^2 - 2$, is included as the dotted line, and that the dashed line indicates the theoretical lower bound for a unimodal PDF, $K \geq S^2 - 1.5$. Presumably, the larger vertical offset of the skewness-kurtosis bound for SSH anomalies, as compared to SST, is due to the fact that SSH variability is impacted more strongly by non-local effects.

As before for SST (Fig. 7.5), we also present PDFs for two representative regions within the ocean interior and two areas within the Agulhas region (Fig. 7.9). In the ocean interior we contrast an area with weak positive skewness (Interior I: 25.0–30.0°S, 350.0–360.0°E; Fig. 7.9a) against an area with strong negative skewness (Interior II: 35.0–40.0°S, 320.0–330.0°E; Fig. 7.9b). In the Agulhas region we compare an area with strong positive skewness (Agulhas I: 40.0–45.0°S, 10.0–20.0°E; Fig. 7.9c) to an area with strong negative skewness (Agulhas II: 34.8–3.5°S,

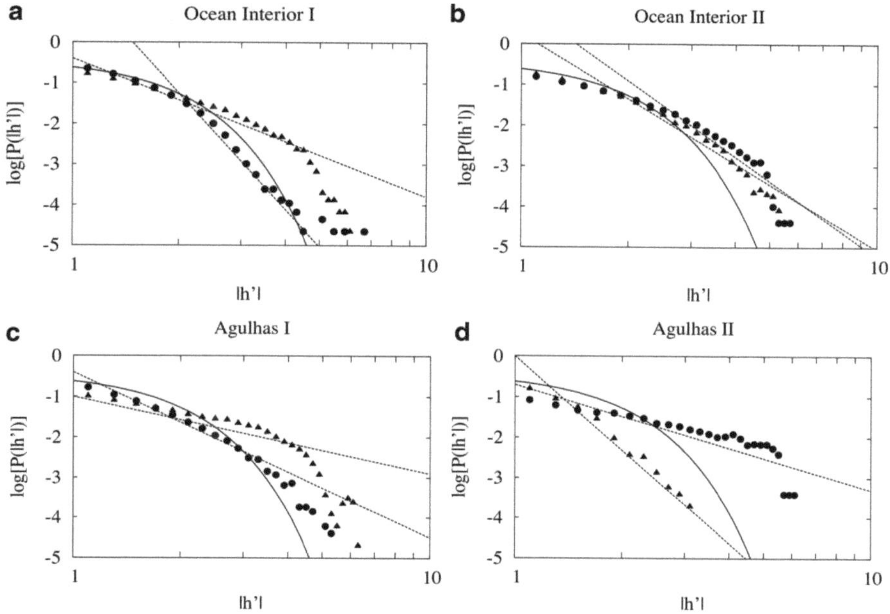

Fig. 7.9 Log-log scale PDFs of the absolute values $|h'|$ of positive (*triangles*) and negative (*circles*) normalized SSH anomalies in four different regions. (**a**) Interior I: 25.0–30.0 °S, 350.0–360.0 °E. (**b**) Intererior II: 35.0–40.0 °S, 320.0–330.0 °E. (**c**) Agulhas I: 40.0–45.0 °S, 10.0–20.0 °E. (**d**) Agulhas II: 34.8–36.5 °S, 23.0–26.1 °E. The straight dashed lines are Maximum Likelihood Estimates of the power-law behavior, and the solid lines denote Gaussian distributions (Adapted from Sura and Gille (2010))

23.0–26.1°E; Fig. 7.9d). Height anomalies within each box are normalized by their standard deviation, and PDFs are plotted on log-log scales. In Fig. 7.9 positive and negative normalized height anomalies h' are indicated by triangles and circles, respectively. Because the emphasis is on extreme events, the center of the PDFs ($\pm$ one standard deviation) is not shown. In fact, the center of the PDFs is close to being Gaussian, consistent with previous studies of SSH and geostrophic velocities. In all plots the solid line denotes a Gaussian distribution. The straight dashed lines are Maximum Likelihood Estimates of the power-law behavior (given by the exponent α) above a systematically estimated lower bound $|h'|_{min}$. The tails decay linearly in large regions of the PDF domain, providing compelling evidence for power-law behavior of SSH variability. However, the power-law scaling is not constant throughout the PDF domain, but instead shows an approximately piecewise linear behavior (in log-log coordinates). That means that the PDFs show a different scaling above a certain value of h' (a scale break) in some PDFs. Nevertheless, in all cases the power-law behavior above a scale break is statistically significant as tested by Kolmogorov-Smirnov statistics. The different scalings may point to mechanisms not included in our simple model.

The bottom line is that the tails of SSH variability approximately follow a power-law distribution. This, together with the observed skewness-kurtosis link, allows us to also attribute non-Gaussian SSH statistics to a correlated additive and multiplicative noise forcing. Because we have different scalings of the power-law, the attribution is, however, only a first-order approximation. Yet it is astonishing that the basic non-Gaussian features of SSH variability can be captured by our simple CAM noise model. The fundamental reason is, of course, that the CAM noise model is a linearization of a more general stochastic-dynamical system.

7.3.2 Atmospheric Examples

We will now turn our attention to the atmosphere, presenting non-Gaussian attributes from daily National Centers for Environmental Prediction (NCEP)-National Center for Atmospheric Research (NCAR) reanalysis data. Here we focus on the dynamically relevant variables: relative vorticity, potential vorticity, and geopotential height. Anomalies were calculated by subtracting the daily climatology and linear trend from the daily averages.

Figure 7.10 shows (a) skewness and (b) kurtosis of the observed daily-averaged 300 hPa relative vorticity anomalies ζ' in the northern winters (DJF) of 1970–1999. Both quantities are large in the hemispheric jet stream waveguide and have a coherent, almost zonally symmetric structure. Another prominent feature is the observed change of sign in ζ' skewness and kurtosis across the storm tracks, with positive skewness (and kurtosis) to the south and negative skewness (and kurtosis) to the north. This feature is known in terms of the skewness of geopotential height anomalies Φ' (White 1980; Nakamura and Wallace 1991; Trenberth and Mo 1985). Remember that geopotential height and (quasigeostrophic) vorticity anomalies Φ' and ζ' are linked through the horizontal Laplacian equation $\zeta' = (1/f)\nabla^2\Phi'$, where f is the Coriolis parameter. That is, in the Northern Hemisphere ($f > 0$) ζ' and Φ' are expected to have opposite signs and, therefore, opposite signs of skewness. It is often speculated that the sign of the skewness is related to cut-off lows at low, and from blocking anticyclones at high latitudes (e.g., White 1980; Nakamura and Wallace 1991; Trenberth and Mo 1985). In addition, Holzer (1996) attributes the band of positive ζ' skewness (approximately equivalent to negative skewness of Φ') to the rectification of near-symmetric velocity fluctuations by nonlinear advective effects. Figure 7.10c presents the relationship between skewness and kurtosis in form of a scatterplot; the solid curve denotes the parabola $K = 1.5\,S^2 - 0.6$. It is again obvious that there exists a lower parabolic bound on kurtosis in the vorticity data: $K \geq 1.5\,S^2 - 0.6$. The vertical offset of the skewness-kurtosis bound for vorticity anomalies is presumably due to the non-local character of atmospheric variability.

Figure 7.11 verifies yet again the prediction of power-law tails. As a representative example, Figs. 7.11a, b present observed (NCEP-NCAR) wintertime (DJF) 300 hPa vorticity anomalies ζ' at 20°N, 180°W, and 500 hPa geopotential height

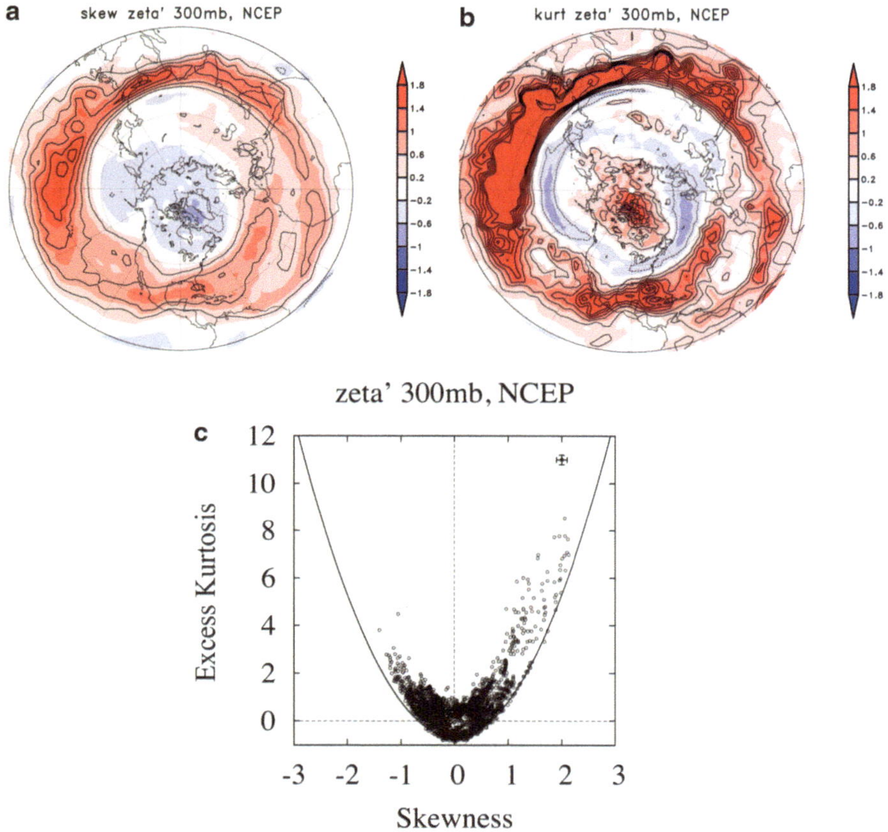

Fig. 7.10 (**a**) Skewness S and (**b**) excess kurtosis K of daily 300 hPa vorticity anomalies ζ' during northern winters (DJF) of 1970–1999, calculated from NCEP/NCAR reanalysis data. (**c**) Scatterplot of S and K values. The *solid curve* denotes the parabola $K = 1.5\,S^2 - 0.6$. The error bars indicate the 95% confidence intervals (approximately ± 2 standard deviations) (Adapted from Sardeshmukh and Sura (2009))

anomalies Φ' at 15°N, 180°W. Negative anomalies are denoted by circles, and positive anomalies by triangles. Thin solid curves denote a Gaussian distribution. Note the non-Gaussian power-law tails of positive ζ' and negative Φ', as highlighted by the straight lines (for ζ' we see that the power-law scaling is not constant throughout the PDF domain, but instead shows an approximately piecewise linear behavior (in log-log coordinates)). To confirm the observed distributions, model results were also analyzed (the model used is the Portable University Model of the Atmosphere, available from the Department of Meteorology, University of Hamburg, at http://www.mi.uni-hamburg.de/TheoMet.6.0.html). In Fig. 7.11c, d the modeled PDFs of the same variables at identical locations as above are shown, underpinning the power-law statistics of non-Gaussian atmospheric variability. Note

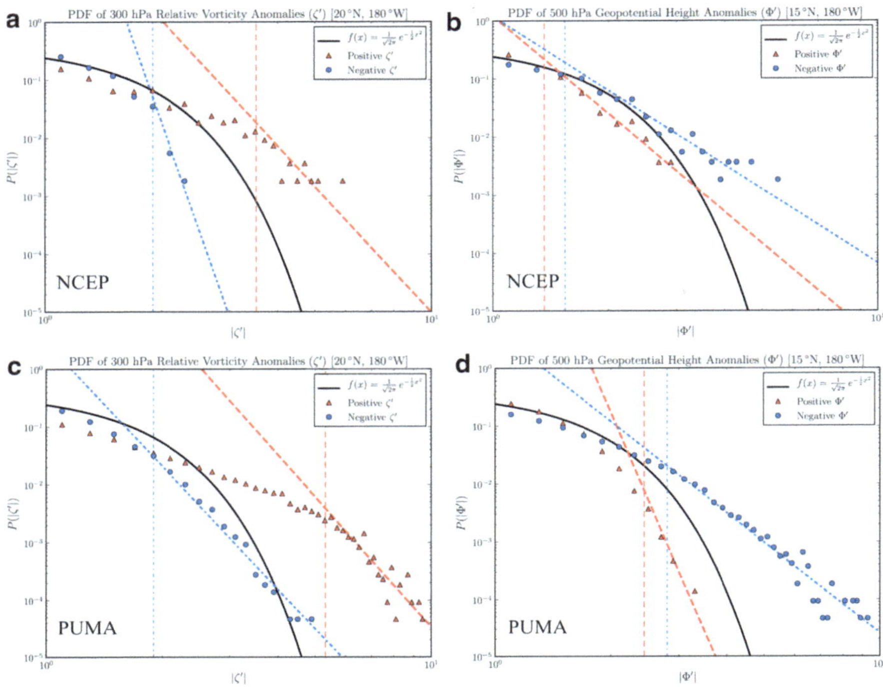

Fig. 7.11 PDFs of the absolute values of (**a–b**) observed (NCEP/NCAR; 1970–1999) and (**c–d**) modeled (PUMA) 300 hPa vorticity anomalies ζ' at 20°N, 180°W, and 500 hPa geopotential height anomalies z' at 15°N, 180°W. Negative anomalies are denoted by *blue circles*, and positive anomalies by *red triangles*. Thin *solid curves* denote a Gaussian distribution. Note the non-Gaussian power-law tails of positive ζ' and negative Φ' (highlighted by the *blue/red straight lines*). All plots for DJF (winter). The lower bounds of the power-law scaling are shown by the *vertical lines*

that the lower bound of the power-law scaling is estimated objectively from data (using Kolmogorov-Smirnonov statistics) and, therefore, might change from dataset to dataset.

The vertical structure of non-Gaussian atmospheric variability is shown in terms of zonally averaged skewness and kurtosis of daily quasigeostrophic potential vorticity anomalies q' for wintertime (DJF) data from 1948–2007 (Fig. 7.12). (In the following we equate the terminology *quasigeostrophic potential vorticity* with *potential vorticity*, keeping in mind that they are physically not strictly identical; see Hoskins et al. 1985.) Relative vorticity anomalies ζ' and potential vorticity anomalies q' are, of course, closely related: q' is the sum of the eddy quasigeostrophic relative vorticity and the eddy stretching vorticity (e.g., Holton 1992). That is, in the quasi-horizontal atmosphere, where the anomalous stretching term is locally less important than the anomalous quasigeostrophic relative vorticity contribution, we expect q' and ζ' (and their statistics) to be somewhat similar. In

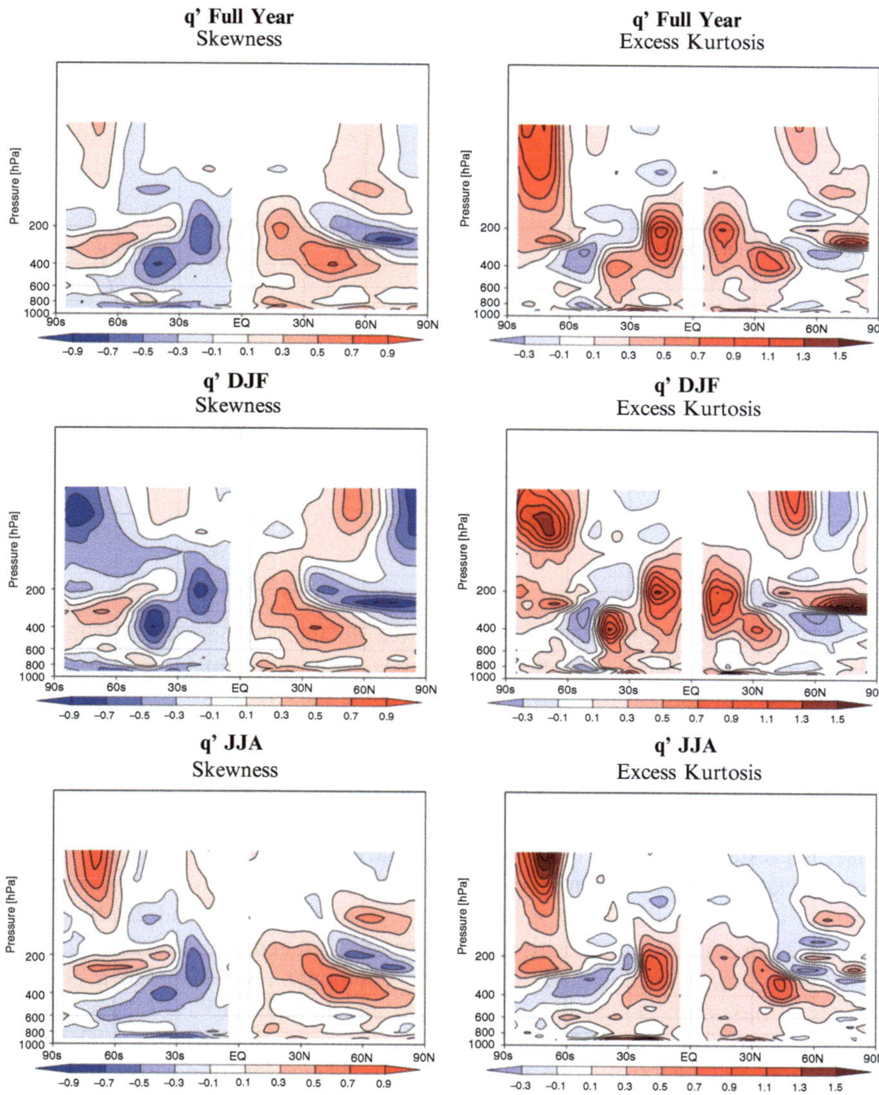

Fig. 7.12 Zonally averaged skewness (*left column*) and excess kurtosis (*right column*) of daily potential vorticity anomalies q' for (from *top* to *bottom*) full year, winter (DJF), and summer (JJA) subsets for data from 1948–2007 (Adapted from Sura and Perron (2010))

fact, comparing the zonal structures of Figs. 7.10 and 7.12, the overall skewness and kurtosis patterns of q' resemble those of ζ' (Sura and Perron 2010). In a nutshell, Fig. 7.12 reveals that in the troposphere equatorward of the Northern Hemisphere storm track potential/relative vorticity skewness is positive. Poleward of the same

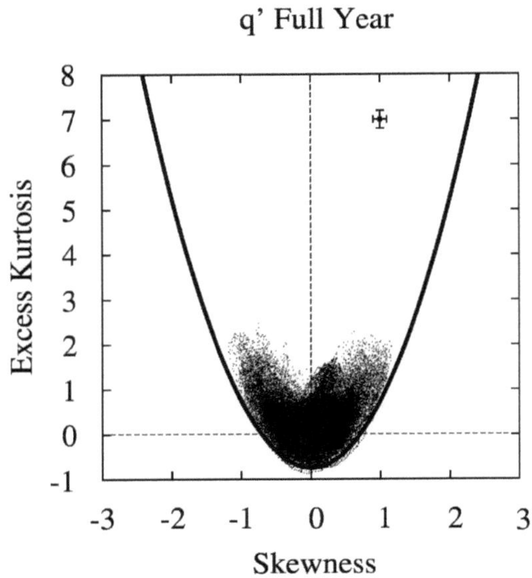

Fig. 7.13 Scatterplot of excess kurtosis vs. skewness from full year potential vorticity anomalies q'. The *solid curve* denotes the parabola $K = 1.5\,S^2 - 0.8$. The error bars indicate the 95% confidence intervals (approximately ± 2 standard deviations) (Adapted from Sura and Perron (2010))

storm track the vorticity skewness is negative. In the Southern Hemisphere the relation is reversed. We also see that in the troposphere the overall large scale pattern of skewness and kurtosis remains the same over the seasons.

Figure 7.13 visualizes the global (horizontally and vertically) relationship of q' skewness and kurtosis in a scatterplot; the solid curve denotes the parabola $K = 1.5\,S^2 - 0.8$. It is obvious that there also exists a lower parabolic bound on kurtosis in the potential vorticity anomaly data. As mentioned before, the vertical offset is presumably due to the non-local character of atmospheric variability. We also checked if non-Gaussian q' variability follows a power-law. A representative example is given in Fig. 7.14 for full year q' at 160°W, 15°N, and 250 hPa. This is a location with strong positive skewness. As, in general, atmospheric non-Gaussianity has a large zonally symmetric component, we also see a positive skewness maximum at that height (250 hPa) and latitude (15°N) in the zonally averaged field (see Fig. 7.12). In both plots the solid line denotes a Gaussian distribution. Note the distinct heavy-tail power-law behavior for positive anomalies giving rise to positive skewness. However, even the negative anomalies follow a power-law (weaker than Gaussian, though). Thus, we again observe (statistically significant) power-law tails, strongly suggesting that potential vorticity variability is also governed by stochastic dynamics with CAM noise.

To summarize the atmospheric results, we have strong evidence that the non-Gaussian statistics of atmospheric variability can be well described by CAM noise dynamics.

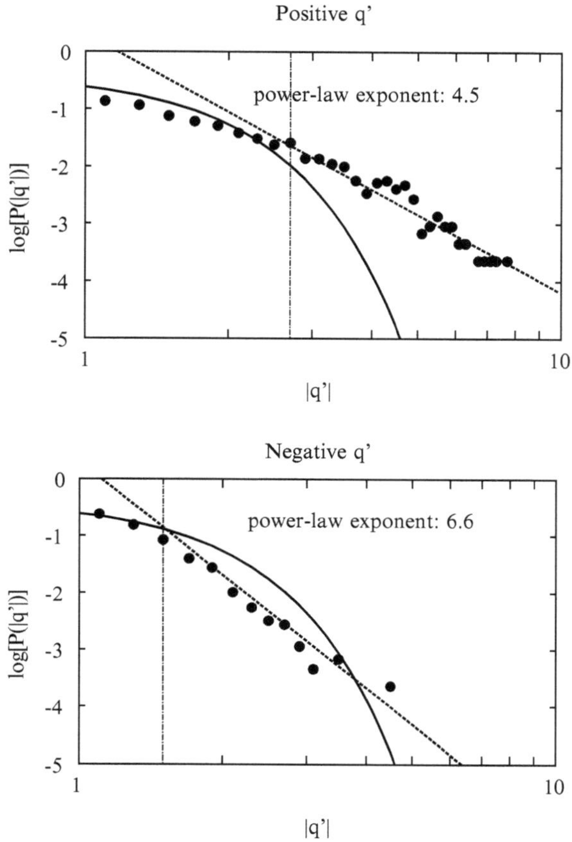

Fig. 7.14 Log-log scale PDFs (*circles*) of the absolute values $|q'|$ of full year (*top*) positive and (*bottom*) negative potential vorticity anomalies at 160° W, 15° N and 250 hPa. The straight *dashed lines* are Maximum Likelihood Estimates of the power-law behavior and the *solid lines* denote Gaussian distributions. The lower bounds of the power-law scaling are shown by the *dotted-dashed vertical lines* (Adapted from Sura and Perron (2010))

7.3.3 Other Applications

At this point an alert reader might ask if we observe the parabolic skewness-kurtosis constraint $K \geq (3/2)S^2 - r$ and power-law tails in other turbulent flows as well. The argument behind this line of thought is that the derivation of our simple CAM noise equation (7.18) only requires quadratic nonlinearities, and those are, of course, present as advective terms in the general Navier-Stokes equations. Therefore, the governing equations of any fluid dynamical system can be always expressed in the general form (e.g., Salmon 1998)

$$\frac{dX_i}{dt} = L_{ij}X_j + N_{ijk}X_jX_k + F_i, \qquad (7.33)$$

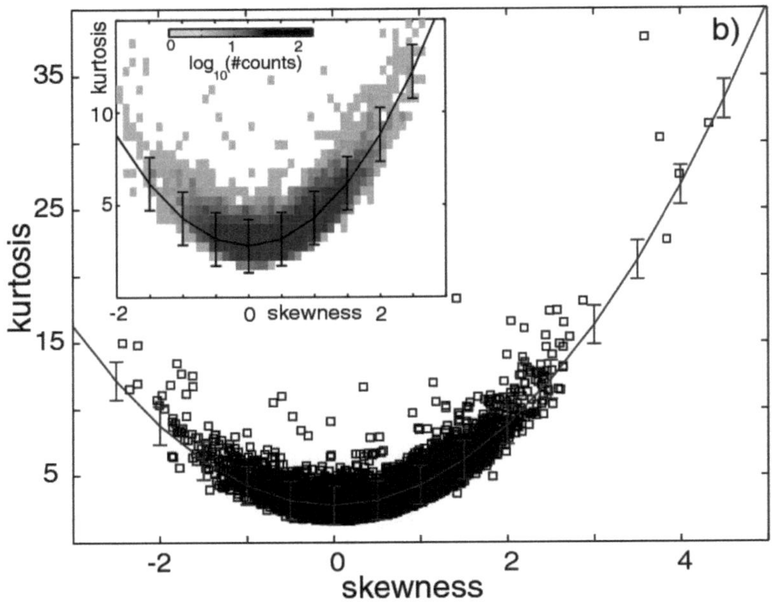

Fig. 7.15 Scatterplot of skewness S vs. kurtosis $\tilde{K}$ of density fluctuations in a turbulent plasma. The *solid line* is the parabola $\tilde{K} = 1.5S^2 + 2.78$. Note that in this plot the kurtosis $\tilde{K}$ on the ordinate is defined as our "excess kurtosis +3", or $\tilde{K} = K + 3$. Therefore, in our terminology we see the parabola $K = 1.5S^2 - 0.22$, and the approximate constraint $K \geq 1.5S^2 - 1.25$. The error bars denote the standard errors in $\tilde{K}$ (Adapted from Labit et al. (2007))

where the state of the fluid is given by the vector $\mathbf{X}$ with components X_i. The first and second terms on the right are the linear and quadratically nonlinear tendencies (including linear and quadratically nonlinear damping terms), and F_i denotes a forcing term. If X_i in (7.33) is expanded into a mean and anomaly, and the anomaly is then split into a slowly and a rapidly decorrelating contribution, the multivariate CAM noise equation (7.13) follows (Sardeshmukh and Sura 2009). Therefore, we should see our parabolic constraint (and power-law tails) in many turbulent flow situations. Indeed, the constraint $K \geq (3/2)S^2 - r$ has been also observed and discussed in plasma turbulence (Labit et al. 2007; Krommes 2008; Sandberg et al. 2009). In particular, Labit et al. (2007) found that the non-Gaussian density fluctuations in plasmas follow the approximate constraint $K \geq (3/2)S^2 - 1.25$ (see Fig. 7.15; note that in that plot the kurtosis $\tilde{K}$ on the ordinate is defined as our "excess kurtosis +3", or $\tilde{K} = K + 3$). Therefore, there exists some evidence that the CAM noise approach might be able to explain a general characteristic of turbulence. More research is needed to explore the skewness-kurtosis link in a variety of turbulent fluids.

7.4 Conclusions

7.4.1 Where Do We Stand?

Knowing the tails of PDFs is an important goal in the atmospheric and ocean sciences because weather and climate risk assessment depends on understanding extremes. In recent years new tools that make use of advanced stochastic theory have evolved to evaluate extreme events and the physics that govern these events. Stochastic methods are ideal to study extreme events because they link vastly different time and spatial scales. We have seen that non-Gaussian statistics of extreme anomalies can be attributed to stochastically forced dynamics, where, to model nonlinear interactions, the strength of the stochastic forcing depends on the flow itself (multiplicative noise). In particular, we have provided theoretical and observational evidence suggesting that a simple linear stochastic differential equation with correlated additive and multiplicative (CAM) noise is an excellent candidate to explain the non-Gaussian statistics and extreme events of numerous weather and climate phenomena. The evidence is based on the fact that the CAM noise theory makes clear and testable predictions about non-Gaussian variability that can be verified (or falsified) by analyzing the detailed statistics of atmospheric and oceanic variability. That is, we have seen that many dynamically relevant weather and climate phenomena in the atmosphere and ocean follow the predicted non-Gaussian CAM noise statistics (skewness-kurtosis constraint, $K \geq (3/2)S^2 - r$, and power-law tails, $p(x) \propto x^{-\alpha}$), strongly suggesting the relevance of CAM noise dynamics in statistically explaining extreme events in weather and climate. We are not aware of another dynamical mechanism capable of accomplishing that.

The somewhat astonishing general character (or applicability) of CAM noise dynamics in capturing non-Gaussian statistics in a wide range of applications can be understood by recognizing that the theory is essentially a linearization (Taylor expansion) of a general stochastic dynamical system. In addition, it can also be derived from the equations of motion of any fluid dynamical system, as long as we can justify a fast-slow timescale separation. Then the multiplicative noise appears naturally from the nonlinear advection terms. Thus, non-Gaussian behavior is an inherent property of geophysical flows, and stochastic theory provides an excellent tool to model the statistics of extreme events in weather and climate. In fact, because of its general character, our approach provides a natural null hypothesis for non-Gaussian climate variability, as it is capturing the ubiquitous red climatic spectrum *and* the statistics of extreme events. Therefore, the outlined univariate CAM noise approach is a natural *non-Gaussian* extension of Hasselmann's Gaussian red-noise null hypothesis.

Going back to the classification and terminology we used in the introduction, we now have a general null hypothesis of extreme events that enables us to extrapolate into the tails of a distribution based on mathematical-dynamical arguments.

Thus, we have a dynamically based idea of the general shape we expect in the tail of a PDF, allowing us to understand and potentially predict the probability of extreme events in weather in climate. This is a crucial step beyond the predominantly empirical (and numerical) studies widely used in this field of research. The research presented here is, of course, only the first leg of a journey into largely uncharted waters. There are numerous issues and challenges to be discussed and, hopefully, resolved. In the following we go through a, not necessarily complete, list of open issues and speculate how they might be addressed in the future.

7.4.2 *Outstanding Issues and Challenges*

One point that comes to mind is that of potential applications. It is obvious that the knowledge of the power-law exponent could be used to predict the probability of extreme events in the unsampled part of the PDF. So we could indeed extrapolate into the scarcely sampled tail of a distribution. To establish such an application we also need to look deeper into the validity of power-law statistics of dynamically relevant variables. That is, we have to identify, in detail, variables and regions where our null hypothesis works and where it fails. In particular, we have to ascertain the physical reasons why our null hypothesis is potentially breaking down. For example, so far we do not understand why we observe scale breaks in some power-law distributions.

Closely related is the technical difficulty of applying multiplicative noise models. In general, it is non-trivial to estimate coefficients of SDEs from limited data (e.g., Kloeden and Platen 1992; Sura and Barsugli 2002). Even in our particular case we have not come up with a stable and reliable method to estimate all parameters of the CAM noise SDE (7.18) from relatively short records in order to directly compare modeled and observed PDFs and power-law exponents. One method we are exploring is to use the observational estimates of $\tilde{\lambda}$ and the moments $\langle x^2 \rangle$, $\langle x^3 \rangle$, $\langle x^4 \rangle$ to solve the nonlinear set of equations (7.23) for the remaining noise parameters $\sigma_{F'}^2$, $\sigma_{R'}^2$, and ϕ. Preliminary results show a satisfactory agreement of the power-law exponent α calculated from (7.29) with the direct estimate from data. However, the method is fraught with the potential for error and uncertainty due to the nonlinearity of the equations involved.

Another important point is that the discussed stochastic theory makes several simplifying assumptions, such as that of stationary statistics. This is a common approximation in statistical climate research. However, its validity is not yet studied within the non-Gaussian stochastic framework. The main contributor to non-stationary statistics is, of course, the seasonal cycle. Thus, there is a need to study the impact of the seasonal cycle on extreme events within the multiplicative noise framework.

In the future we might be able to develop a conceptual, yet physically reasonable model to study how climate change might affect extreme events. At this point we do not have a paradigm at hand that allows us to physically understand how a PDF might change in a warming climate. Many climate projections just look into the change of the mean and the variance, that is, assuming Gaussian statistics. However, the non-Gaussian statistics (the shape of the distribution) will most likely also be altered in a changing climate.

We hope that this chapter (and book) will inspire students and researchers to address the plethora of outstanding question and challenges.

Acknowledgements The author thanks the anonymous reviewer whose comments greatly improved the chapter. This project was in part funded by the National Science Foundation through awards ATM-840035 "The Impact of Rapidly-Varying Heat Fluxes on Air-Sea Interaction and Climate Variability" and ATM-0903579 "Assessing Atmospheric Extreme Events in a Stochastic Framework".

References

Albeverio S, Jentsch V, Kantz H (eds) (2006) Extreme events in nature and society. Springer, Berlin, 352 pp

Alexander LV et al (2006) Global observed changes in daily extremes of temperature and precipitation. J Geophys Res 111. doi:10.1029/2005JD006 290

Berner J (2005) Linking nonlinearity and non-gaussianity of planetary wave behavior by the Fokker-Planck equation. J Atmos Sci 62:2098–2117

Berner J, Branstator G (2007) Linear and nonlinear signatures in the planetary wave dynamics of an AGCM probability density function. J Atmos Sci 64:117–136

Brönnimann S, Luterbacher J, Ewen T, Diaz HF, Stolarski RS, Neu U (eds) (2008) Climate variability and extremes during the past 100 years. Springer, Dordrecht 364 pp

Brooks CEP, Carruthers N (1953) Handbook of statistical methods in meteorology. Her Majesty's Stationery Office, London, 412 pp

Clauset A, Shalizi CR, Newman MEJ (2009) Power-law distributions in empirical data. SIAM Rev 51:661–703

Coles S (2001) An introduction to statistical modeling of extreme values. Springer, London, 208 pp

Corti S, Molteni F, Palmer TN (1999) Signature of recent climate change in frequencies of natural atmospheric circulation regimes. Nature 29:799–802

Crommelin DT, Vanden-Eijnden E (2006) Reconstruction of diffusions using spectral data from timeseries. Comm Math Sci 4:651–668

DelSole T (2004) Stochastic models of quasigeostrophic turbulence. Surv Geophys 25:107–149

Easterling DR, Meehl GA, Parmesan C, Changnon SA, Karl TR, Mearns LO (2000) Climate extremes: observations, modeling, and impacts. Science 289:2068–2074

Farrell BF, Ioannou PJ (1995) Stochastic dynamics of the midlatitude atmospheric jet. J Atmos Sci 52:1642–1656

Farrell BF, Ioannou PJ (1996) Generalized stability theory. Part I: autonomous operators. J Atmos Sci 53:2025–2040

Frankignoul C, Hasselmann K (1977) Stochastic climate models. Part II. application to sea-surface temperature anomalies and thermocline variability. Tellus 29:289–305

Franzke C, Majda AJ, Vanden-Eijnden E (2005) Low-order stochastic mode reduction for a realistic barotropic model climate. J Atmos Sci 62:1722–1745

Friedrich R et al (2000) Extracting model equations from experimental data. Phys Lett A 271:217–222

Gardiner CW (2004) Handbook of stochastic methods for physics, chemistry and the natural science, 3rd edn. Springer, Berlin/New York, 415 pp

Garrett C, Müller P (2008) Extreme events. Bull Am Meteor Soc 89:ES45–ES56
Gumbel EJ (1942) On the frequency distribution of extreme values in meteorological data. Bull Am Meteor Soc 23:95–105
Gumbel EJ (1958) Statistics of extremes. Columbia University Press, New York, 375 pp
Hasselmann K (1976) Stochastic climate models. Part I. Theory. Tellus 28:473–484
Holton JR (1992) An introduction to dynamic meteorology, 3rd edn. Academic, San Diego, 507 pp
Holzer M (1996) Asymmetric geopotential height fluctuations from symmetric winds. J Atmos Sci 53:1361–1379
Horsthemke W, Léfèver R (1984) Noise-induced transitions: theory and applications in physics, chemistry, and biology. Springer, Berlin, 318 pp
Hoskins BJ, McIntyre ME, Robertson AW (1985) On the use and significance of isentropic potential vorticity maps. Quart J Roy Meteor Soc 111:877–946
Houghton J (2009) Global warming – the complete briefing, 4th edn. Cambridge University Press, Cambridge, 438 pp
Ioannou PJ (1995) Nonnormality increases variance. J Atmos Sci 52:1155–1158
Isern-Fontanet J, Lapeyre G, Klein P, Chapron B, Hecht MW (2008) Three-dimensional reconstruction of oceanic mesoscale currents from surface information. J Geophys Res 113:C09005
Jaynes ET (1957a) Information theory and statistical mechanics. Phys Rev 106:620–630
Jaynes ET (1957b) Information theory and statistical mechanics. ii. Phys Rev 108:171–190
Jaynes ET (2003) Probability theory: the logic of science. Cambridge University Press, Cambridge, 758 pp
Katz RW, Naveau P (2010) Editorial: special issue on statistics of extremes in weather and climate. Extremes 13. doi:10.1007/s10 687–010–0111–9
Kharin VV, Zwiers FW (2005) Estimating extremes in transient cimate change simulations. J Clim 18:1156–1173
Kharin VV, Zwiers FW, Zhang X, Hegerl GC (2007) Changes in temperature and precipitation extremes in the IPCC ensemble of global coupled model simulations. J Clim 20:1419–1444
Klaassen CAJ, Mokveld PJ, van Es B (2000) Squared skewness minus kurtosis bounded by 186/125 for unimodal distributions. Stat Prob Lett 50:131–135
Kloeden P, Platen E (1992) Numerical solution of stochastic differential equations. Springer, Berlin, 632 pp
Kondrashov D, Kravtsov S, Ghil M (2006) Empirical mode reduction in a model of extratropical low-frequency variability. J Atmos Sci 63:1859–1877
Kravtsov S, Kondrashov D, Ghil M (2005) Multi-level regression modeling of nonlinear processes: derivation and applications to climate variability. J Clim 18:4404–4424
Kravtsov S, Kondrashov D, Ghil M (2010) Empirical model reduction and the modelling hierarchy in climate dynamics and the geosciences. In: Palmer T, Williams P (eds) Stochastic physics and climate modelling. Cambridge University Press, Cambridge, pp 35–72
Krommes JA (2008) The remarkable similarity between the scaling of kurtosis with squared skewness for TORPEX density fluctuations and sea-surface temperature fluctuations. Phys Plasma 15:030703
Labit B, Furno I, Fasoli A, Diallo A, Müller SH, Plyushchev G, Podestà M, Foli FM (2007) Universal statistical properties of drift-interchange turbulence in TORPEX plasmas. Phys Rev Lett 98:255002
Lapeyre G, Klein P (2006) Dynamics of the upper oceanic layers in terms of surface quasi-geostrophic theory. J Phys Oceanogr 36:165–176
Lind PG, Mora A, Gallas JAC, Haase M (2005) Reducing stochasticity in the north atlantic oscillation index with coupled langevin equations. Phys Rev E 72:056706
Majda AJ, Timofeyev I, Vanden-Eijnden E (1999) Models for stochastic climate prediction. Proc Natl Acad Sci 96:14687–14691
Majda AJ, Timofeyev I, Vanden-Eijnden E (2001) A mathematical framework for stochastic climate models. Commun Pure Appl Math 54:891–974
Majda AJ, Timofeyev I, Vanden-Eijnden E (2003) Systematic strategies for stochastic mode reduction in climate. J Atmos Sci 60:1705–1722

Majda AJ, Franzke C, Khouider B (2008) An applied mathematics perspective on stochastic modelling for climate. Phil Trans R Soc. 366:2429–2455
Mo K, Ghil M (1988) Cluster analysis of multiple planetary flow regimes. J Geophys Res 93:10927–10952
Mo K, Ghil M (1993) Multiple flow regimes in the Northern Hemisphere winter: Part I: methodology and hemispheric regimes. J Atmos Sci 59:2625–2643
Molteni F, Tibaldi S, Palmer TN (1990) Regimes in the wintertime circulation over northern extratropics. I. Observational evidence. Quart J R Meteor Soc 116:31–67
Monahan AH (2004) A simple model for the skewness of global sea-surface winds. J Atmos Sci 61:2037–2049
Monahan AH (2006a) The probability distributions of sea surface wind speeds Part I: theory and SSM/I observations. J Clim 19:497–520
Monahan AH (2006b) The probability distributions of sea surface wind speeds Part II: dataset intercomparison and seasonal variability. J Clim 19:521–534
Monahan AH, Pandolfo L, Fyfe JC (2001) The prefered structure of variability of the Northern Hemisphere atmospheric circulation. Geophys Res Lett 27:1139–1142
Monin AS, Yaglom AM (1971) Statistical fluid mechanics. mechanics of turbulence, vol I. MIT Press, Cambridge, 784 pp
Monin AS, Yaglom AM (1975) Statistical fluid mechanics. mechanics of turbulence, vol II. MIT Press, Cambridge, 896 pp
Müller D (1987) Bispectra of sea-surface temperature anomalies. J Phys Oceanogr 17:26–36
Nakamura H, Wallace JM (1991) Skewness of low-frequency fluctuations in the tropospheric circulation during the Northern Hemisphere winter. J Atmos Sci 48:1441–1448
Newman MEJ (2005) Power laws, Pareto distributions and Zipf's law. Contempor Phys 46:323–351
Øksendal B (2007) Stochastic differential equations, 6th edn. Springer, Berlin/New York, 369 pp
Paul W, Baschnagel J (1999) Stochastic processes: from physics to finance. Springer, Berlin/New York, 231 pp
Peinke J, Böttcher F, Barth S (2004) Anomalous statistics in turbulence, financial markets and other complex systems. Ann Phys 13:450–460
Penland C (1989) Random forcing and forecasting using principal oscillation pattern analysis. Mon Weather Rev 117:2165–2185
Penland C, Ghil M (1993) Forecasting Northern Hemisphere 700-mb geopotential height anomalies using empirical normal modes. Mon Weather Rev 121:2355–2372
Penland C, Matrosova L (1994) A balance condition for stochastic numerical models with application to El Niño – the southern oscillation. J Clim 7:1352–1372
Penland C, Sardeshmukh PD (1995) The optimal growth of tropical sea surface temperature anomalies. J Clim 8:1999–2024
Petoukhov V, Eliseev A, Klein R, Oesterle H (2008) On statistics of the free-troposphere synoptic component: an evaluation of skewnesses and mixed third-order moments contribution to the synoptic-scale dynamics and fluxes of heat and humidity. Tellus A 60:11–31
Pope SB (2000) Turbulent flows. Cambridge University Press, Cambridge, 771 pp
Reynolds RW, Smith TM, Liu C, Chelton DB, Casey KS, Schlax MG (2007) Daily high-resolution blended analyses for sea surface temperature. J Clim 20:5473–5496
Salmon R (1998) Lectures on geophysical fluid dynamics. Oxford University Press, New York, 378 pp
Sandberg I, Benkadda S, Garbet X, Ropokis G, Hizanidis K, del Castillo-Negrete D (2009) Universal probability distribution function for bursty transport in plasma turbulence. Phys Rev Lett 103:165001
Sardeshmukh PD, Sura P (2009) Reconciling non-Gaussian climate statistics with linear dynamics. J Clim 22:1193–1207
Siegert S, Friedrich R, Peinke J (1998) Analysis of data sets of stochastic systems. Phys Lett A 243:275–280

Smyth P, Ide K, Ghil M (1999) Multiple regimes in norther hemisphere height fields via mixture model clustering. J Atmos Sci 56:3704–3732
Sornette D (2006) Critical phenomena in natural sciences. Springer, Berlin/New York, 528 pp
Sura P (2003) Stochastic analysis of Southern and Pacific Ocean sea surface winds. J Atmos Sci 60:654–666
Sura P (2010) On non-Gaussian SST variability in the gulf stream and other strong currents. Ocean Dyn 60:155–170
Sura P, Barsugli JJ (2002) A note on estimating drift and diffusion parameters from timeseries. Phys Lett A 305:304–311
Sura P, Gille ST (2003) Interpreting wind-driven Southern Ocean variability in a stochastic framework. J Mar Res 61:313–334
Sura P, Gille ST (2010) Stochastic dynamics of sea surface height variability. J Phys Oceanogr 40:1582–1596
Sura P, Newman M (2008) The impact of rapid wind variability upon air-sea thermal coupling. J Clim 21:621–637
Sura P, Perron M (2010) Extreme events and the general circulation: observations and stochastic model dynamics. J Atmos Sci 67:2785–2804
Sura P, Sardeshmukh PD (2008) A global view of non-Gaussian SST variability. J Phys Oceanogr 38:639–647
Sura P, Sardeshmukh PD (2009) A global view of air-sea thermal coupling and related non-Gaussian SST variability. Atmos Res 94:140–149
Sura P, Newman M, Penland C, Sardeshmukh PD (2005) Multiplicative noise and non-Gaussianity: a paradigm for atmospheric regimes? J Atmos Sci 62:1391–1409
Sura P, Newman M, Alexander MA (2006) Daily to decadal sea surface temperature variability driven by state-dependent stochastic heat fluxes. J Phys Oceanogr 36:1940–1958
Taleb NN (2010) The black swan: the impact of the highly improbable, 2nd edn. Random House, New York, 480 pp
Thompson KR, Demirov E (2006) Skewness of sea level variability of the world's oceans. J Geophys Res 111:c05005. doi:10.1029/2004JC00283.
Trenberth KE, Mo KC (1985) Blocking in the southern hemisphere. Mon Weather Rev 113:3–21
van Kampen NG (2007) Stochastic processes in physics and chemistry, 3rd edn. Elsevier, North-Holland, 463 pp
Whitaker JS, Sardeshmukh PD (1998) A linear theory of extratropical synoptic eddy statistics. J Atmos Sci 55:237–258
White GH (1980) Skewness and kurtosis and extreme values of Northern Hemisphere geopotential heights. Mon Weather Rev 108:1446–1455
Wilkins JE (1944) A note on skewness and kurtosis. Ann Math Stat 15:333–335
Wilks DS (2006) Statistical methods in the atmospheric sciences, 2nd edn. Academic Press, Burlington, 627 pp
Winkler CR, Newman M, Sardeshmukh PD (2001) A linear model of wintertime low-frequency variability. Part I: formulation and forecast skill. J Clim 14:4474–4493

Chapter 8
Methods of Projecting Future Changes in Extremes

Michael Wehner

Abstract This chapter examines some selected methods of projecting changes in extreme weather and climate statistics. Indices of extreme temperature and precipitation provide measures of moderately rare weather events that are straightforward to calculate. Drought indices provide measures of both agricultural and hydrological drought that are especially suitable for constructing multi-model ensemble projections of future change. Extreme value statistical theories are surveyed and provide methodologies for projecting the changes in frequency and severity of very rare temperature and precipitation events.

Future changes in the average climate virtually guarantee that changes in extreme weather events will follow. Such rare events are best described statistically as it is difficult, but perhaps not impossible, to directly link individual disasters to human-induced climate change. Examples of extreme weather events with severe consequences to society that are amenable to projection include heat waves, cold spells, floods, droughts and tropical cyclones. Confidence in projections of future changes in the severity and frequency of such events is increased if the mechanisms of change can be identified and understood. Equally important, however, is the rigorous quantification of the uncertainties in these projections. These uncertainties include the inherent natural variability of the climate system as well as limitations in both the climate models' fidelity and the statistical methods used to analyze their output.

The discussions about future changes in extreme events in recent climate change assessment reports (including the IPCC 4th Assessment Report and the US national assessments) did not generally focus on sophisticated statistical analyses. Rather, extremes were presented in these documents by a series of "extreme indices".

M. Wehner (✉)
Lawrence Berkeley National Laboratory, 1 Cyclotron Road, MS50F, Berkeley, CA 94720, USA
e-mail: mfwehner@lbl.gov

A. AghaKouchak et al. (eds.), *Extremes in a Changing Climate*, Water Science and Technology Library 65, DOI 10.1007/978-94-007-4479-0_8,

Introduced first by Frich et al. (2002), they are often referred to as the Frich indices. While many of these represent significant departures from the mean climate, they are by no means descriptive of rare events or the far tails of the temperature or precipitation distributions. The fundamental difference between these index based treatments and formal Extreme Value Theory descriptions of rare events illustrates the difficulties in nomenclature when discussing climate and weather extremes. What constitutes "extreme" varies greatly in the literature and depends highly on the application of the final results. This chapter will survey some of these methods of projecting changes in climate and weather.

8.1 Extreme Indices

A set of extreme indices was part of the data output specifications for the Coupled Model Intercomparison Project (CMIP3, see www-pcmdi.llnl.gov). Table 8.1 lists these ten pre-calculated statistics that were specified to be calculated for each year of the simulations. Code was provided to the climate modeling groups to calculate these fields although they could also be replicated from the archived daily averaged surface air temperature and precipitation rates. Most of these indices are clearly motivated by their relevance to climate change impacts, e.g. the number of frost days, the growing season length and the number of consecutive dry days. However, for a more general interpretation of the effect of global climate change on

Table 8.1 The Frich indices saved as annualized quantities for the CMIP3 coordinated numerical experiment

Index name	Units	Description
fd	Day	Total number of frost days (days with absolute minimum temperature $< 0°$ C)
etr	Kelvin	Intra-annual extreme temperature range: difference between the highest temperature of any given calendar year (T_h) and the lowest temperature of the same calendar year (T_l)
gsl	Day	Growing season length: period between when $T_{day} > 5°$ C for > 5 days and $T_{day} < 5°$ C for > 5 days
hwdi	Day	Heat wave duration index: maximum period > 5 consecutive days with $T_{max} > 5°$ C above the 1961–1990 daily T_{max} normal
tn90	%	Fraction (expressed as a percentage) of time $T_{min} >$ 90th percentile of daily minimum temperature, where percentiles are for the 1961–1990 base period
r10	Day	No. of days with precipitation greater than or equal to 10 mm day^{-1}
cdd	Day	Maximum number of consecutive dry days ($R_{day} < 1$ mm)
r5d	$kg\ m^{-2}$	Maximum 5 days precipitation total
sdii	$kg\ m^{-2}\ s^{-1}$	Simple daily intensity index: annual total/number of R_{day} greater than or equal to 1 mm day^{-1}
r95t	%	Fraction (expressed as a percentage) of annual total precipitation due to events exceeding the 1961–1990 95th percentile

extreme events, these three indices and the others based on fixed threshold values are somewhat less useful (Alexander et al. 2006; Tebaldi et al. 2006). For instance, every day is a frost day in the very high latitudes but none are in the deep tropics. Similarly, 10 mm of precipitation in a single event is fairly common in tropical regions but impossible in many desert regions. Of more utility in this context are the percentile-based indices such as tn90 (hot nights) and r95t (very wet days). These two indices define base states (the 1961–1990 period) from which departures can be calculated anywhere on the planet. The bottom panel of Fig. 8.1 shows a CMIP3 multi-model projection of the change in r95t over land regions at the end of the twenty-first century under the SRES A1B forcing scenario. This index is defined as the percentage of annual total precipitation due to events exceeding the 1961–1990 95th percentile. During the base state period, this field would be uniformly 5%. End of twenty-first century values over land in this figure range from a low of 9% to a high of 54%. This increase in the index may be interpreted in the following sense: what might be currently considered very wet days (i.e. the top 5%) will occur from two to ten times more frequently in this future scenario. This also suggests that the shape of the distribution of daily precipitation must change in this scenario because the mean precipitation is not projected to change in a similar manner. Note that care should be exercised in interpreting such exceedance rate changes as sampling errors may play a role (Zhang et al. 2005).

In general, mean precipitation changes are a mix of increases and decreases and are smaller in magnitude as in the top panel of Fig. 8.1. In this multi-model example, an average projection is formed by equally weighting each climate model. Constructing weighted average projections based on model skill in replicating observed climate means or trends is a difficult task (Santer et al. 2009; Knutti et al. 2010a) and is presumably yet more difficult for extremes due to their less well characterized behavior. In Fig. 8.1, models with multiple realizations, if any, are ensemble averaged prior to inclusion into the multi-model result. Furthermore, the index is calculated on the models' native grids, then regridded to a common grid and masked prior to the multi-model averaging. These latter two points are the general practice in many climate change projection studies but have important implications for certain extremes, especially those related to precipitation, when evaluating model performance (Wehner et al. 2010).

Figure 8.2 shows a different way of representing the change in an extreme index. In this figure, a CMIP3 multi-model projection of the change in tn90 averaged over North American land regions under a variety of forcing scenarios is shown from the beginning of the twentieth century to the end of the twenty-first century. This index is defined as the percentage of time that daily minimum temperature exceeds the 90th percentile, of the 1961–1990 base period. This method of illustrating a projection, while lacking the spatial detail of the previous figure, allows the explicit depiction of projection uncertainty. The four major sources of projection uncertainty are the natural variability of the climate system, limited sample size (i.e. small ensembles and/or short time intervals), imperfect climate models (largely manifested by differences in climate model sensitivity to changes in greenhouse gas concentration but also realized in less well characterized ways for extremes), and

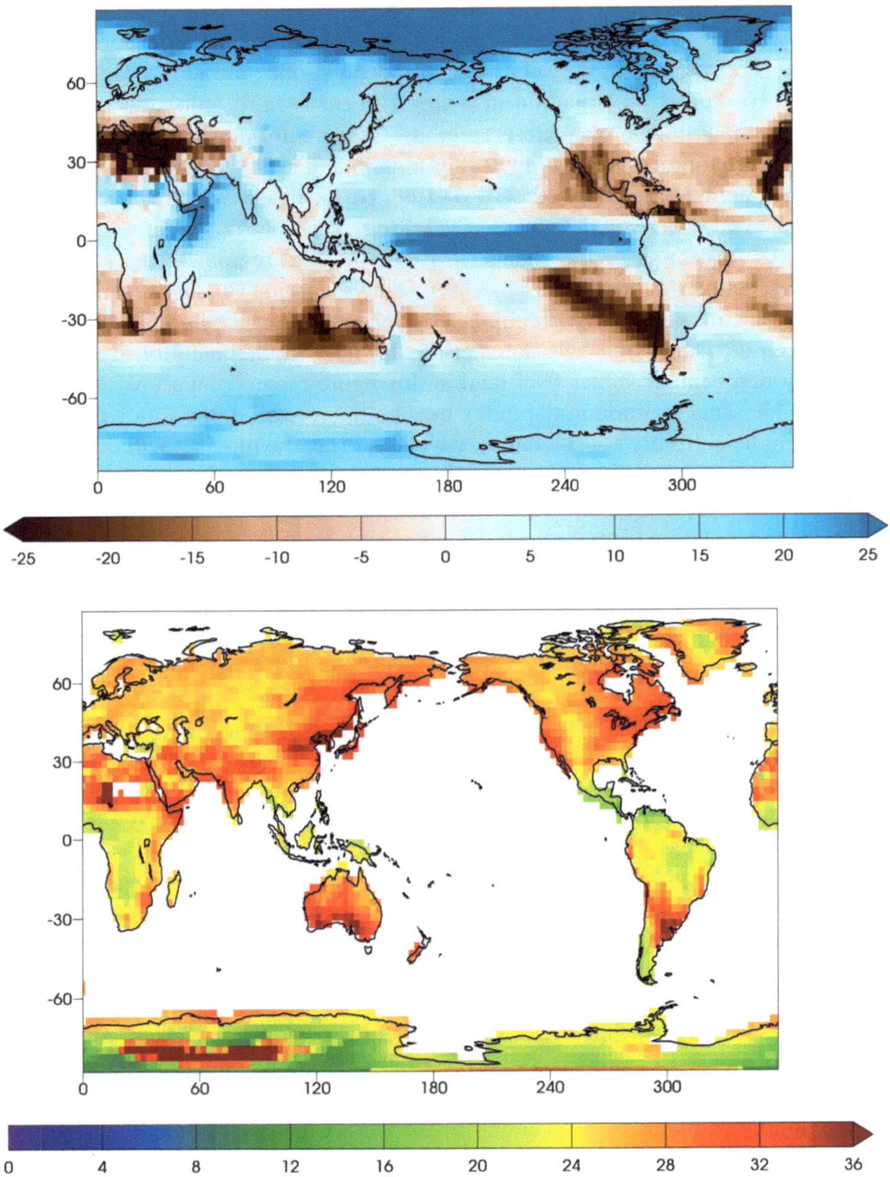

Fig. 8.1 A CMIP3 multi-model projection of changes in precipitation statistics at the end of the twenty-first century under the SRES A1B forcing scenario. (*Top panel*) Percent change of annual mean precipitation. (*Bottom panel*) Percentage of annual total precipitation due to events exceeding the 1961–1990 95th percentile (r95t). Ten different climate models were averaged with equal weighting in these projections (units: percent)

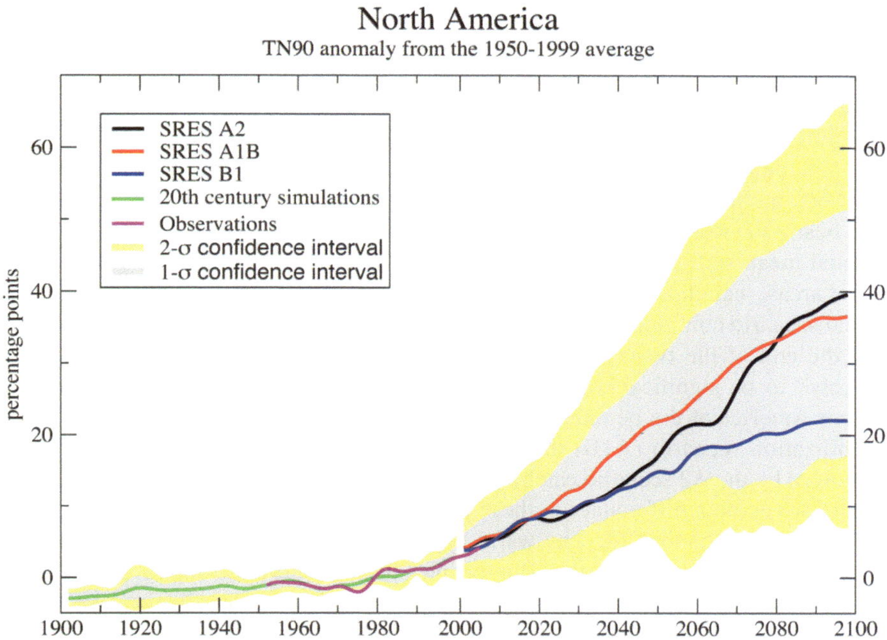

Fig. 8.2 A CMIP3 multi-model projection of the percentage of time the daily minimum temperature exceeds the 90th percentile of daily minimum temperature, calculated from 1961 to 1990 base period (tn90) at the end of the twenty-first century under the SRES A1B forcing scenario. Ten different climate models were averaged with equal weighting in this projection. A 13 point temporal filter is applied to all projections as in IPCC AR4 (units: percent)

the unpredictability of human behavior (i.e. the different scenarios). One method for quantifying the uncertainty from the imperfections of climate model is to calculate the variance in the projection across the ten climate models that provided this index to the CMIP3 database. In Fig. 8.2, one standard deviation across models is depicted by the gray shading and two standard deviations by the yellow shading. The envelopes plotted here are determined by the maximum spread across all three scenarios. One might also want to consider each scenario separately to base decisions on how significant the differences between the scenarios are. In this case, taken from the USGRCP report (Karl et al. 2009), this representation permitted usage of the "likelihood language" (Morgan et al. 2009). The gray shaded area represent the "likely" range of change (i.e. a 2 out of 3 chance of being a correct statement) while the yellow shaded bounds represent the "very likely" range of change (i.e. a 9 out of 10 chance of being a correct statement). However, given the limited set of available global models and that many of them are related, this measure of uncertainty does not completely sample the space of projection and underestimates to true uncertainty due to model deficiencies (Tebaldi and Knutti 2007; Knutti et al. 2010b).

As with projections of changes in mean quantities, the scenario uncertainty in the beginning of the twenty-first century is less than at the end (Hawkins and Sutton 2009; Yip et al. 2011). Comparison of Fig. 8.2 to similar figures for changes in mean temperature (Karl et al. 2009) reveals subtle differences in the timing of the separation of the low emissions scenario (B1, stabilizes at 550 ppm CO_2) from the high emissions scenarios (A2, business as usual). Furthermore, the relationship between the higher stabilization scenario (A1B, stabilizes at 720 ppm CO_2) from the business as usual scenario is quite different. In multi-model projections of the annual mean surface air temperature, the two scenarios are indistinguishable over most areas, including North America, until mid-century after which the business as usual scenario continues to increase and the stabilization scenario starts to stabilize. By the end of the twenty-first century, the differences between the scenarios are "likely" to be significant. In Fig. 8.2, warm nights increase in temperature over North America at the beginning of the twenty-first century at a greater rate in the stabilization scenario (A1B) than in the business as usual scenario (A2). This is followed by the A2 scenario catching up towards the end of the twenty-first century. There are enough differences in these forcing scenarios that one could hypothesize a plausible mechanism for why warm night temperatures might behave differently from annual mean temperatures. But the inter-model uncertainty in Fig. 8.2 is clearly large enough to prevent a conclusion that these differences are "likely" significant. Even a weaker statement about the significance of these differences is prevented by limitations in the sample size behind this index projection in comparison to that behind projection of mean temperature changes. For at the beginning of the twenty-first century, only about 10% of the daily temperature values are used in calculating the index as opposed to all of the values when calculating the annual mean. Although this fraction rises to about 50% towards the end of the century due to warming, the tn90 index remains a noisy quantity compared to annual mean temperatures. In order to ascertain, whether these tantalizing differences in the scenario behavior between the warm night index and the mean temperature are genuine, more realizations of each individual model are required. This will prove to be a recurring theme in ascertaining the significance of extreme changes. The exact details depend greatly on the variability of the quantity of interest and the magnitude of the differences (Wehner 2000).

Other extreme indices than that developed by Frich et al. (2002) can be useful tools in analyzing future climate change projections. In particular, there are a number of drought indices in wide use by the agricultural and other water intensive industries. Table 8.2 shows five drought indices that are provided to the public at regular intervals by the US National Climatic Data Center (NCDC) on their website, http://www.drought.noaa.gov/. A recent paper (Wehner et al. 2011) examined the performance and projections for the Palmer Drought Severity Index in the CMIP3 models finding wide variations between the models. In that study, the models simulated the observed PDSI much better after a bias correction procedure. Bias corrections can take many forms and can be useful in enhancing confidence in projections. Bias correction assumes that errors in the mean state may not influence trends or changes to the same degree. In many instances, this assumption can be

Table 8.2 The NCDC drought indices (see http://www.drought.noaa.gov/)

Drought index name	Units	Description
PDSI	Palmer Drought Severity Index	Duration and intensity of long-term drought
PHDI	Palmer Hydrological Drought Index	Similar to PDSI except measures longer term hydrological effects relevant to reservoir levels, groundwater levels, etc.
Z-index	Palmer Z Index	Short-term drought on a monthly scale
CMI	Crop Moisture Index	Short-term drought on a weekly scale
SPI	Standardized Precipitation Index	A normalized precipitation only index that is reported on time scales ranging from weeks to years.

tested by applying the correction over one part of an observational record and testing against another part. In the PDSI study, the input (monthly averaged temperature and precipitation) to the drought index calculation was corrected by applying a monthly varying climatological factor that altered the models' long term temperature and precipitation means to the observations but kept each models' particular variability intact. The PDSI is constructed to measure excursions from a neutral base state. Since the models' variability was not corrected, performance in replicating observed PDSI statistics ranged greatly. The simple land surface model contained in the PDSI algorithm is particularly sensitive to temperature leading to large projected changes in the severity and spatial extent of future drought in North America. However, this large temperature sensitivity caused large inter-model differences in these projections at the end of the century because of the large differences in climate model sensitivities to changes in atmospheric greenhouse gases.

This source of projection uncertainty can be reduced in a certain sense by rephrasing how the climate change question is asked. Most climate change projection questions ask something like: "What will happen at the end of the century?" Instead consider if a question such as the following is asked: "What will happen if the global mean temperature rises by 2.5 K?" In the former case, the time period is fixed but the different models exhibit vastly different warmings. In the latter case, the question of timing is foregone but at least the model states bear some resemblance to each other. In fact, under the SRES A1B scenario, the date at which the running decadal average global mean surface air temperature reaches 2.5 K over its preindustrial value ranges from 2038 in the most sensitive model to 2110 in the least sensitive model. The average date over all models to reach this amount of warming is 2070. Figure 8.3 shows maps of future North American PDSI under SRES A1B forcing and the associated inter-model uncertainty relevant to these two ways of posing future climate change questions. The upper two panels (a and c) show decadal averaged PDSI values and represent what the climatological values of PDSI would be relative to the current climatology. For interpretation of PDSI, drought is classified into the following categories: incipient ($-0.5 \geq \text{PDSI} > -1.0$), mild ($-1.0 \geq \text{PDSI} > -2.0$), moderate

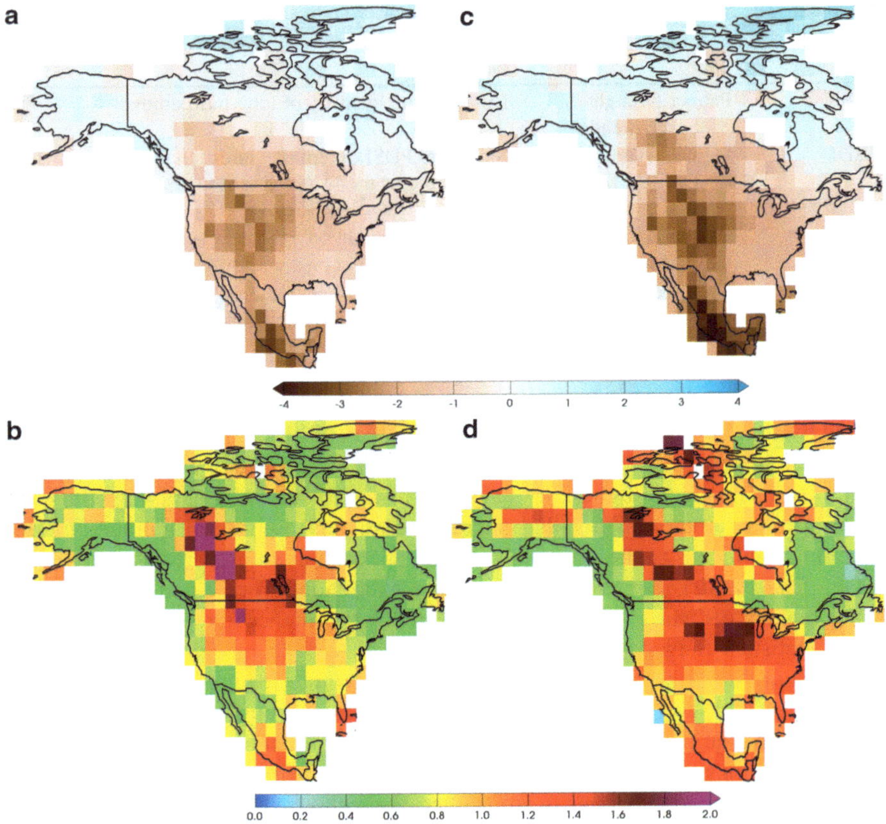

Fig. 8.3 (**a**) Multi-model average value of PDSI when the global average surface air temperature has increased 2.5 K over its 1900–1909 mean value (**b**) inter-model standard deviation of the values shown in panel (**a**, **c**) multi-model average value of PDSI for the decade centered at 2070 (**d**) inter-model standard deviation of the values shown in panel (**c**)

($-2.0 \geq$ PDSI > -3.0), severe ($-3.0 \geq$ PDSI > -4.0), and extreme ($-4.0 \geq$ PDSI). The upper right panel shows the PDSI averaged over all models for the decade centered around 2070 (with an average model global warming of 2.5 K). In this projection, conditions currently considered severe drought would become normal in the western US. In parts of Mexico, conditions currently considered extreme drought would become normal. However, uncertainty in this projection, shown as the inter-model standard deviation in the lower right panel (d), is large in these regions. By changing the climate change question to ask what the value of PDSI would be under a 2.5 K global warming (which would occur on average at 2070), this inter-model uncertainty is reduced in most areas as shown in the lower left panel (b). The actual projection of drought severity is also reduced as seen in the upper left panel (a), reflecting a nonlinear dependence of PDSI on temperature.

8.2 Extreme Value Theory Methods

Numerous studies in the literature as well as much of the contents of this book utilize sophisticated extreme value statistics to explore questions of climate change. These techniques differ from the index-based methods described in the previous section principally in their ability to quantify the statistical behavior of much rarer events. Rather than review the details of how and when extreme value theory methods may be applied to climate and weather datasets, this section discusses aspects of interpretation of results from these statistical formalisms in a context of climate change.

The parameters describing the generalized extreme value distribution and the generalized Pareto distribution can often offer interesting insight. However, these fields are not closely tied to observable quantities and are generally of limited utility to the users of climate change projections. Design engineers and other parties interested in climate change impacts are more concerned with how the limitations of their particular systems might be exceeded. Return value and/or return time often can provide the critical information necessary to make informed decisions about the impacts of rare weather and climate events. Whether the analysis takes a block maxima or threshold approach, these application relevant fields are readily calculated if the distribution parameters can be satisfactorily fit to the extreme data.

Extreme value theory (EVT) is often used to describe how extreme weather behaves in a changing climate by analyzing high frequency (i.e. daily) modeled or observed datasets. Return values from the fitted EVT distributions are defined over a fixed specified period, for instance, T with units in years. In a stationary climate, the return value can be interpreted as the value of the data that would be realized on average once every T years over a very long period of time. By introducing time as a covariate, EVT can be generalized to non-stationary datasets (Brown et al. 2008; Smith, Private communication, 2010). In a changing climate, this explanation loses meaning for a time dependent return value. Instead, a more appropriate alternative interpretation is that the return value at a given time represents the value that has a 1/T chance of occurring that year in the dataset.

Return time offers a slightly different way to express the same concepts. In a stationary climate, the return time is the average time between instances that the data take to reach or exceed a specified value over the course of a very long time. In a non-stationary climate, the return time for a fixed specified value would be a time dependent quantity. The inverse of the return time would be the chance that the specified value would be achieved in that year.

Uncertainty in return value and return time estimates depends on the magnitude of the time scale of interest in relation to the length of the datasets Wehner (2010). When this time scale is much less than the dataset length, the return values have likely been realized in the datasets and uncertainty is lower. When the time scales are much larger than the dataset lengths, the EVT estimates are extrapolations outside the datasets and uncertainty is higher. However, if the asymptotic assumptions of the EVT are valid, estimated return values and return times can be reasonable in extrapolated cases. The generalization of EVT to treat time dependent datasets can help reduce uncertainty by allowing the consideration of longer datasets. However, care must be exercised as these generalizations assume specific time dependences of the EVT distribution parameters. These can be linear, quadratic or even higher order in the Smith (Private communication, 2010) formalism but it is not always clear how to generate the best fits. In fact, the observed climate change has not been particularly linear and future changes may not even be monotonic if drastic remediation procedures are taken.

The actual climate system is of course limited to the single world that actually exists. Climate model simulations have no such limitations as they are routinely integrated in statistically independent realizations to be combined into large ensembles by varying initial conditions. If one assumes quasi-stationarity over short periods of time, these independent realizations can be combined into much longer datasets and stationary EVT used to provide accurate estimates of the distribution parameters. The length of such a period depends greatly on the variable of interest as well as the rate of climate change.

Stationarity would be guaranteed if linear detrending is applied over these short periods. In the literature (for instance Kharin et al. 2007), it is not uncommon to assume a decade or two. Although Santer et al. (2011) showed that any individual decade in the last century might exhibit observed positive or negative temperature trends, they also showed that over a large sample of decades, a statistically significant positive trend can be found. This suggests that detrending is prudent when combining intervals over individual realizations to construct a larger stationary dataset for EVT analysis. Ensemble sizes in the CMIP3 database ranged between three and eight, if multiple realizations were performed at all. In the CMIP5 specifications, a minimum of ten realizations is called for in the "Tier 1" experiments (Taylor et al. 2009). This then affords the opportunity to concatenate detrended decadal segments to build quasi-stationary datasets of about 100 years in length representing any time period during the integration.

Hence, there are two EVT methods that can be used to make projections of future changes in extreme weather event statistics. The first method is to fit non-stationary datasets with time dependent EVT distributions. The advantage in this approach is that the single realization of the observed climate system can be treated without any ad hoc assumptions of stationarity. The length of the record should be chosen carefully such that the trend is well fit by the specified time dependence. For multiple realizations of a single climate model, each realization should be treated separately in this method and ensemble mean return values and/or return times calculated. A continuous picture of change including trends is provided by this method. Additionally, a measure of the models' internal variability can be obtained by calculating the inter-realization variance to provide insight into this source of projection uncertainty. The second method is applicable to climate models with multiple realizations. In this approach, short intervals from each realization are concatenated to form a larger dataset. Detrending of the segments prior to concatenation is often desirable. Fitting a stationary EVT distribution to concatenated datasets formed at different times permits changes in return values and/or return times to be directly calculated. In both methods, the uncertainty from estimation of the fitted distribution parameters can be estimated by the scheme outlined by Hosking and Wallis (1997). This technique involves first estimating the distribution parameters for the actual dataset then generating random datasets distributed by the EVT distribution defined by those parameters. To estimate uncertainty, each of the random distributions is fit to an EVT with appropriate parameters and return fields estimated along with their variances. This forms a measure of uncertainty associated with the finiteness of the data. For the large ensembles promised with CMIP5, this uncertainty should be reduced.

Presentation of the results from an EVT analysis of climate change poses serious challenges as the concept of return value and return time may not be intuitive to the non-specialist. Changes in return value (for a fixed return period) can be expressed in similar forms to widely published changes in mean values. For instance, Fig. 8.4 shows a multi-model CMIP3 projection of the end of century changes under a business as usual scenario (SRES A1B) of the 20 year return value of the annual maximum daily average surface air temperature (upper panel) and the annual mean surface air temperature (lower panel). Exhibiting projected changes in pairs of figures such as these allows discussion of the differences between them. In this case, large changes in the 20-year return value are confined to land masses and are generally larger than changes in the annual mean. Not shown in these figures are the seasonal behaviors of projected changes that can reveal mechanistic insights. This seasonal aspect is particularly important in the interpretation of changes in precipitation extremes. Also not shown are changes in extremes of minima temperatures that exhibit different behaviors than the changes in maxima temperatures, again providing opportunities for understanding physical mechanisms of change.

Changes in severity of rare weather events is only a part of an EVT analysis. Changes in frequency may be yet more important. Figure 8.5 attempts to illustrate this point in the upper panel by showing the return time in the future for daily surface air temperatures exceeding the present day 20-year return value. In this case, the future return time is projected to become less than 20 years over most of the globe. Alternatively, the lower panel of Fig. 8.5 shows the number of times in a future 20 year period that this same temperature threshold can be expected to be exceeded on average. If the climate does not change, this number would be one. However, for annual maximum daily temperatures, the value is much greater than that over most of the globe. The EVT data used in both Figs. 8.4, 8.5 come from the same analysis. These figures reveal that warm weather events currently considered rare (once every 20 years) are projected to become relatively commonplace and that warm events of a fixed rarity are projected to become more severe.

8.3 Multi-Variate Climate and Weather Extremes

The literature of multi-variate extreme value statistics is well developed (see Chap. 7). However, it has not seen significant application to climate change projections or historical analyses despite an urgent need. For instance, consider

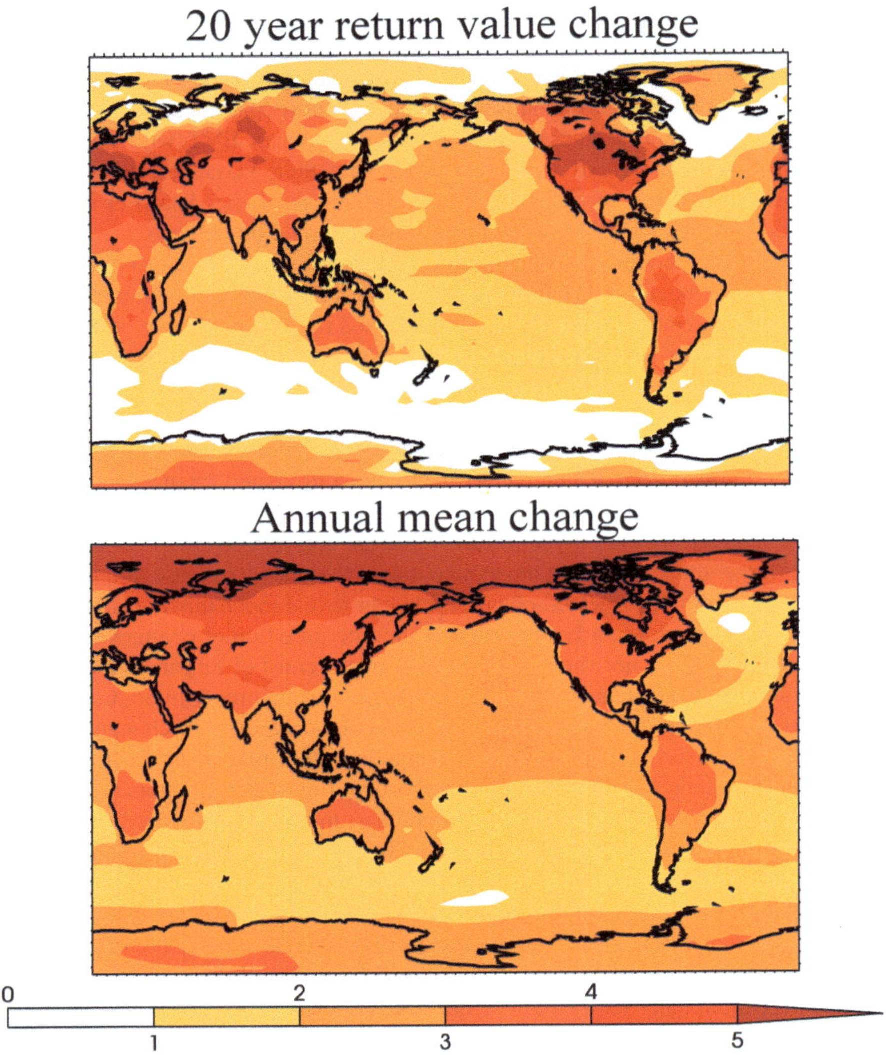

Fig. 8.4 Changes in the end of twenty-first century surface air temperature properties relative to the end of the twentieth century under SRES A1B forcing from the CMIP3 models. *Upper panel*: Change in 20-year return value of the annual maximum daily averaged temperature. *Lower panel*: Change in annual mean temperature (units: kelvin)

hot, dry and windy events versus hot, moist and stagnant events. The impacts of such events are very different. The former may lead to increased risk of fires while the latter may lead to increased human mortality through heatstroke or air quality issues. In both cases, at least one of the salient variables is not extreme in itself. In fact, it is often the combination of multiple events, each common in isolation, that is considered rare and/or dangerous.

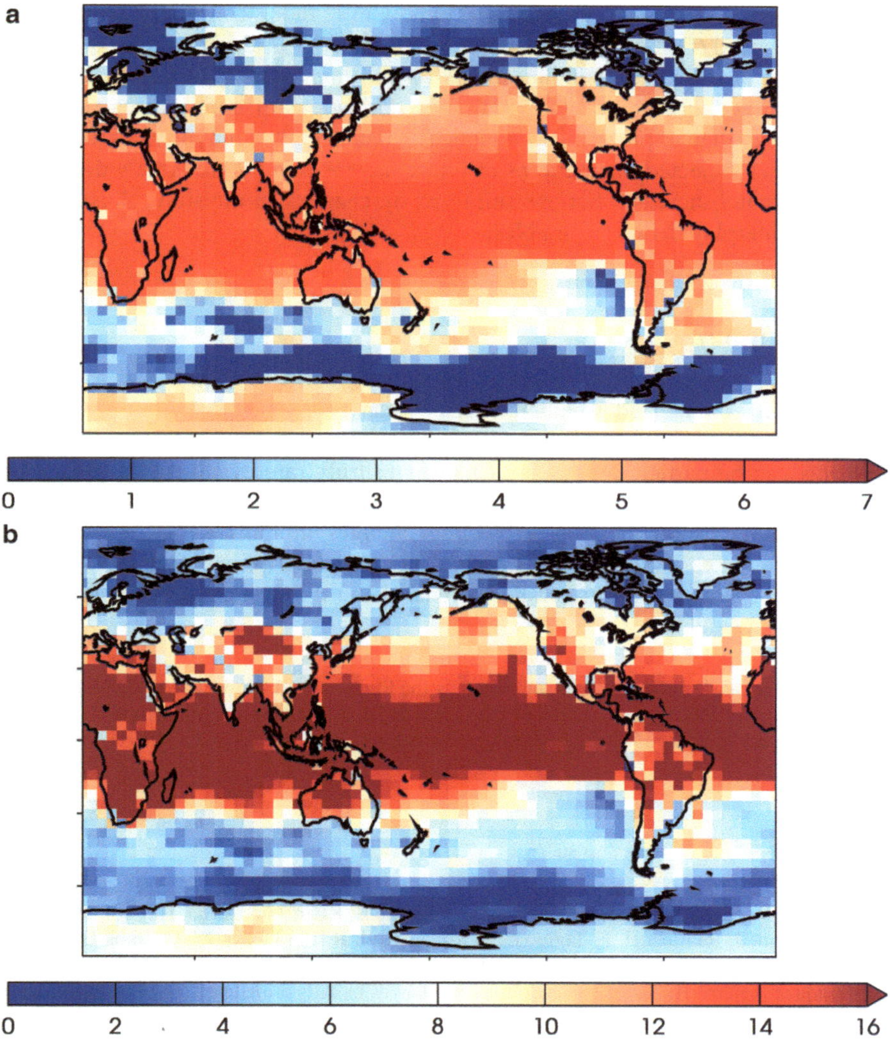

Fig. 8.5 Upper panel (**a**) The projected return time at the end of the twenty-first century under SRES A1B forcing associated with daily temperature threshold defined by the end of the twentieth century 20-year return value of the annual maximum daily averaged surface air temperature (units: years). Lower panel (**b**) The number of occurrences per 20 year period at the end of the twenty-first century when the daily averaged surface air temperature exceeds that same threshold. If the climate had not changed, this number would be one (units: dimensionless)

Two multivariate indices are in common usage in weather forecasting. Similar to the drought indices discussed above, they can be used to define the frequency and severity of extreme events in climate change projections. The first of these is the "Heat Index" (HI) that combines air temperature and relative humidity (Steadman

1979a, b). The second of these is the "Wind Chill" index combining temperature and wind speed (Osczevski and Bluestein 2005). Both of these indices, expressed in degrees, are used to estimate effects on the human body and are often said to describe how hot or cold "it feels". The derivations of both indices are rather involved and often implemented via tabular lookups or fitted polynomials. Delworth et al. (1999) projected that patterns of future increases in HI are largely dependent on temperature increases but are amplified by changes in moisture, illustrating a complex interplay between variables.

8.4 Summary

Changes in climate and weather extremes can be projected by a wide variety of methods. Indices and thresholds defined by their relevance to climate change impacts can be particularly useful. Changes in truly rare events, often associated with dire consequences, are well described by return value or return time changes using extreme value theories. Projecting changes in multi-variate climate and weather extremes is still a developing skill. The description of changes in rare compound events via multi-variate extreme value theory would be an important advance in the field.

Acknowledgement This work was supported by the Regional and Global Climate Modeling and the Earth System Modeling Programs of the Office of Biological and Environmental Research in the Department of Energy Office of Science under contract number DE-AC02-05CH11231.

References

Alexander LV, Zhang X, Peterson TC, Caesar J, Gleason B, Klein Tank AMG, Haylock M, Collins D, Trewin B, Rahimzdeh F, Tagipour A, Kumar Kolli R, Revadekar JV, Griffiths G, Vincent L, Stephenson DB, Burn J, Aguilar E, Brunet M, Taylor M, New M, Zhai P, Rusticucci M, Vazquez Aguirre JL (2006) Global observed changes in daily climate extremes of temperature and precipitation. J Geophys Res Atm 111:D05109. doi:10.1029/2005JD006290

Brown SJ, Caesar J, Ferro CAT (2008) Global changes in extreme daily temperature since 1950. J Geophys Res 113:D05115. doi:10.1029/2006JD008091, 2008

Delworth TL, Mahlman JD, Knutson TR (1999) Changes in heat index associated with CO_2-induced global warming. Clim Chang 43:369–386

Frich P, Alexander LV, Della-Manta P, Gleason B, Haylock M, Klein Tank AMG, Peterson T (2002) Observed coherent changes in climatic extremes during the second half of the twentieth century. Clim Res 19:193–212

Hawkins E, Sutton R (2009) The potential to narrow uncertainty in regional climate predictions. Bull Am Meteorol Soc 90(8):1095–1107

Hosking JRM, Wallis JR (1997) Regional frequency analysis, an approach based on L-moments. Cambridge University Press, Cambridge/New York

Karl TR, Mellilo JM, Peterson TC (eds) (2009) Global climate change impacts in the United States: a state of knowledge report Cambridge [England]. Cambridge University Press, Cambridge/New York. Available at www.globalchange.gov

Kharin VV, Zwiers FW, Zhang X, Hegerl GC (2007) Changes in temperature and precipitation extremes in the IPCC ensemble of global coupled model simulations. J Clim 20:1419–1444

Knutti R, Abramowitz G, Collins M, Eyring V, Gleckler PJ, Hewitson B, Mearns L (2010a) Good practice guidance paper on assessing and combining multi model climate projections. In: Stocker TF, Qin D, Plattner G-K, Tignor M, Midgley PM (eds) Meeting Report of the Intergovernmental Panel on Climate Change Expert Meeting on Assessing and Combining Multi Model Climate Projections. IPCC Working Group I Technical Support Unit, University of Bern, Bern, Switzerland

Knutti R, Furrer R, Tebaldi C, Cermak J, Meehl G (2010b) Challenges in combining projections from multiple climate models. J Clim 23:2739–2758

Morgan MG, Dowlatabadi H, Henrion M, Keith D, Lempert R, McBride S, Small M, Wilbanks T (eds) (2009) Best practice approaches for characterizing, communicating, and incorporating scientific uncertainty in decisionmaking. A report by the Climate Change Science Program and the Subcommittee on Global Change Research. National Oceanic and Atmospheric Administration, Washington, DC, 96pp

Osczevski R, Bluestein M (2005) The new wind chill equivalent temperature chart. Bull Am Meteorol Soc 86:1453–1458

Santer BD, Taylor KE, Gleckler PJ, Bonfils C, Barnett TP, Pierce DW, Wigley TML, Mears C, Wentz FJ, Brueggemann W, Gillett NP, Klein SA, Solomon S, Stott PA, Wehner MF (2009) Incorporating model quality information in climate change detection and attribution studies. Proc Natl Acad Sci. doi:10.1073/pnas.0901736106

Santer BD, Mears C, Doutriaux C, Gleckler P, Wigley T, Gillett N, Ivanova D, Karl T, Lanzante J, Meehl G, Stott P, Taylor K, Thorne P, Wehner M, Wentz F (2011) Separating signal and noise in atmospheric temperature changes: the importance of timescale. J Geophys Res Atm 116:D22105. doi:10.1029/2011JD016263

Steadman RG (1979a) The assessment of sultriness. Part I: A temperature-humidity index based on human physiology and clothing science. J Appl Meteorol 18:861–873

Steadman RG (1979b) The assessment of sultriness. Part II: Effects of wind, extra radiation and barometric pressure on apparent temperature. J Appl Meteorol 18:874–885

Taylor KE, Stouffer RJ, Meehl GA (2009) A summary of the CMIP5 experiment design. http://www-pcmdi.llnl.gov/

Tebaldi C, Knutti R (2007) The use of the multi-model ensemble in probabilistic climate projections. Phil Trans R Soc MathPhys Eng Sci 365:2053–2075. doi:2010.1098/rsta.2007.2076

Tebaldi C, Hayhoe K, Arblaster JM, Meehl GA (2006) Going to the extremes: an intercomparison of model-simulated historical and future changes in extreme events. Clim Chang 79:185–211. doi:10.1007/s10584-006-9051-4

Wehner MF (2000) Determination of the sampling size of AGCM ensemble simulations. Clim Dyn 16:321–331

Wehner MF (2010) Sources of uncertainty in the extreme value statistics of climate data. Extremes 13:205–217. doi:10.1007/s10687-010-0105-7

Wehner MF, Smith R, Duffy P, Bala G (2010) The effect of horizontal resolution on simulation of very extreme US precipitation events in a global atmosphere model. Clim Dyn 32:241–247. doi:10.1007/s00382-009-0656-y

Wehner MF, Easterling DR, Lawrimore JH, Heim RR Jr, Vose RS, Santer B (2011) Projections of future drought in the continental United States and Mexico. J Hydrometeorol 12:1359–1377. doi:10.1175/2011JHM1351.1

Yip S, Ferro CAT, Stephenson DB, Hawkins E (2011) A simple, coherent framework for partitioning uncertainty in climate predictions, to appear in. J Clim. doi:10.1175/2011JCLI4085.1

Zhang X, Hegerl G, Zwiers FW, Kenyon J (2005) Avoiding inhomogeneity in percentile-based indices of temperature extremes. J Clim 18:1641–1651

Chapter 9
Climate Variability and Weather Extremes: Model-Simulated and Historical Data

Siegfried D. Schubert and Young-Kwon Lim

Abstract The climate community is being challenged to provide increasingly more comprehensive societally-relevant information about the impacts of climate change that go well beyond broad statements about how much the global mean temperature will change. This in turn requires increasingly more comprehensive assessments of the quality of climate models to reproduce past regional climate impacts as well as the full spectrum of observed climate variability including those aspects (such as weather extremes) that are likely to have the greatest impact on society. This chapter examines the simulated and observed short-term climate variability and weather extremes that have occurred over the last three decades with a focus on the winter hemispheres.

9.1 Introduction

Extremes in weather and climate encompass a wide array of phenomena including tropical storms, mesoscale convective systems, snowstorms, floods, heat waves, and drought. Understanding how such extremes might change in the future requires an understanding of their past behavior including their connections to large-scale climate variability and trends.

S.D. Schubert (✉)
Global Modeling and Assimilation Office, NASA Goddard Space Flight Center (GSFC), Greenbelt, MD 20771, USA
e-mail: siegfried.d.schubert@nasa.gov

Y.-K. Lim
Global Modeling and Assimilation Office, NASA Goddard Space Flight Center (GSFC), Greenbelt, MD 20771, USA

Goddard Earth Sciences Technology and Research (GESTAR), 8800 Greenbelt Rd., Greenbelt, MD 20771, USA
e-mail: Young-Kwon.Lim@nasa.gov

A. AghaKouchak et al. (eds.), *Extremes in a Changing Climate*, Water Science and Technology Library 65, DOI 10.1007/978-94-007-4479-0_9,

Previous studies suggest that the most robust findings concerning changes in short-term extremes are those that can be most directly (though not completely) tied to the increase in the global mean temperatures. These include the findings that (IPCC 2007): "There has been a widespread reduction in the number of frost days in mid-latitude regions in recent decades, an increase in the number of warm extremes, particularly warm nights, and a reduction in the number of cold extremes, particularly cold nights." For North America in particular (CCSP SAP 3.3 2008): "There are fewer unusually cold days during the last few decades. The last 10 years have seen a lower number of severe cold waves than for any other 10-year period in the historical record that dates back to 1895. There has been a decrease in the number of frost days and a lengthening of the frost-free season, particularly in the western part of North America."

Other aspects of extremes such as the changes in storminess have a less clear signature of long term change, with considerable interannual, and decadal variability that can obscure any climate change signal. Nevertheless, regarding extratropical storms (CCSP SAP 3.3 2008): "The balance of evidence suggests that there has been a northward shift in the tracks of strong low pressure systems (storms) in both the North Atlantic and North Pacific basins." For North America: "Regional analyses suggest that there has been a decrease in snowstorms in the South and lower Midwest of the United States, and an increase in snowstorms in the upper Midwest and Northeast."

Despite the progress already made, our understanding of the basic mechanisms by which extremes vary is incomplete. As noted in IPCC (2007), "Incomplete global data sets and remaining model uncertainties still restrict understanding of changes in extremes and attribution of changes to causes, although understanding of changes in the intensity, frequency and risk of extremes has improved." Separating decadal and other shorter-term variability from climate change impacts on extremes requires a better understanding of the processes responsible for the changes. In particular, the physical processes linking sea surface temperature changes to regional climate changes, and a basic understanding of the inherent variability in weather extremes and how that is impacted by atmospheric circulation changes at subseasonal to decadal and longer time scales, are still inadequately understood.

Given the fundamental limitations in the time span and quality of global observations, substantial progress on these issues will rely increasingly on improvements in models, with observations continuing to play a critical role, though less as a detection tool, and more as a tool for addressing physical processes, and to insure the quality of the climate models and the verisimilitude of the simulations (CCSP SAP 1.3 2008).

In this chapter we examine the ability of the NASA Goddard Earth Observing System – Version 5 (GEOS-5) atmospheric-land general circulation model (AGCM), described in Appendix A, to reproduce cold-season weather extremes, the leading modes of climate variability and associated regional impacts, and the longer-term changes (including changes in extremes) that have occurred during the last three decades. Our focus on the boreal (January–March) and austral (July–September) winter hemispheres, avoids the more challenging warm and transition seasons for which much higher resolution than is typical of climate model simulations is required to adequately represent such extreme weather phenomena as hurricanes, and other mesoscale convective systems that are typical of many continental warm season extremes. Our focus on the last three decades is for the most part dictated by the availability of modern global atmospheric reanalyses (in particular NASA's high-resolution Modern-Era Reanalysis for Research and Applications – MERRA, Rienecker et al. 2011) and satellite observations, as well as the feasibility of producing an ensemble of multi-decadal simulations at relatively high (50 km) resolution with the GEOS-5 model. The simulations are done in a so-called AMIP-mode in which the GEOS-5 atmospheric general circulation model (AGCM) is forced with observed sea surface temperatures (SSTs). The name (AMIP) refers to the Atmospheric Model Inter-comparison Project (Gates 1992) in which such runs were, for the first time, done in a coordinated fashion. In the current experiments the model is also forced with the observed GHGs and ozone as described in Appendix A.

Section 9.2 examines the simulated and observed leading modes of variability on monthly to interannual time scales as well as the longer-term changes that have occurred over the last three decades, and how these changes impact regional weather and extremes. In Sect. 9.3, we examine more idealized simulations (e.g., doubled CO_2; spatially uniform increase in SST) to get a first order sense of how a warmer world might change the nature of climate variability and weather extremes compared to those in our current climate, and what they can tell us (if anything) about the longer-term changes that have occurred in the last three decades. The specific questions that we will address are:

- Does the GEOS-5 model reproduce the observed winter climatological fields?
- Does it have the correct low-frequency (monthly mean) modes of variability?
- Does it have the correct weather variability and extremes?
- Are the linkages between climate variability and regional weather simulated correctly?
- What are the longer terms changes that have occurred over the last three decades and are these reproduced in the model?
- What can idealized AGCM experiments tell us about the recent, and possible future changes in climate variability and weather extremes?

We begin by examining the GEOS-5 model's ability to reproduce the climate means and variations of the last three decades (1980–2009).

9.2 Observed and Simulated Climate Variability and Weather Extremes

9.2.1 *Boreal Winter (JFM)*

Numerous diagnostic studies of extratropical atmospheric variability on monthly and longer time scales (e.g., Quadrelli and Wallace 2004; Trenberth et al. 2005) have found that the atmospheric circulation is dominated by a relatively few patterns of variability that include the El Nino/Southern Oscillation (ENSO), the Pacific/North American Pattern (PNA), the North Atlantic Oscillation (NAO), the Northern and Southern Annular Modes (NAM and SAM), the Pacific Decadal Oscillation (PDO) and the Atlantic Multi-decadal Oscillation (AMO). As discussed below, these patterns of variability have impacts on weather throughout the world. Furthermore, as noted by Kenyon and Hegerl (2008): ". . . . for reliable attribution of changes in extremes as well as prediction of future changes, changes in modes of variability need to be accounted for." In particular, that study found considerable influences of ENSO, the NAO, and Pacific interdecadal variability on daily temperature extremes throughout the world.

We focus on January through March (JFM), since these are the winter months during which the Northern Hemisphere (NH) atmospheric response to ENSO SST and the variance of other large-scale middle latitude teleconnections appear to be most robust, including the NAM (e.g., Thompson and Wallace 2000). We also focus on the leading patterns of monthly-mean variability that occurs on interannual time scales and (in the next section) how this variability might change in a warmer world. While we cannot address decadal variability directly,[1] we do examine the impact of the longer time-scale modes of variability indirectly in Sect. 9.2.1.3 as they are reflected in the differences between the first and second half of the last three decades.

ENSO is known to impact weather during JFM in a number of regions throughout the world, primarily over the Pacific Rim and North America, but also parts of Europe and the high latitude Southern Hemisphere (e.g., Fraedrich and Muller 1992; Gershunov and Barnett 1998; Kenyon and Hegerl 2008; Schubert et al. 2008; Chang et al. 2012; Turner 2004). There is some evidence for an apparent change from eastern Pacific to central Pacific ENSOs (since about 1990 there have been more central Pacific ENSOs, Ashok et al. 2007; Kao and Yu 2009) and this appears to be consistent with global warming (a flattening of the thermocline – Yeh et al. 2009), although it cannot be excluded that these are natural variations of ENSO (Newman et al. 2011). The potential impacts of such changes in the characteristics of ENSO will be discussed in later sections.

The NAO impacts extremes over much of Eurasia and parts of North America (e.g., Hurrell et al. 2003a, b; Scaife et al. 2008; Kenyon and Hegerl 2008; Lim and Schubert 2011). The NAO is, in particular, linked to storm track changes over Europe (e.g., Hurrell 1995; Jones et al. 2003; Hurrell and Deser 2009; Mariotti and Dell'Aquila 2011). IPCC (2007) notes that "There is mounting evidence that the recent observed inter-decadal NAO variability comes from tropical and extratropical ocean influences (Hurrell et al. 2003a, b, 2004), land surface forcing (Gong et al. 2003; Bojariu and Gimeno 2003) and from other external factors (Gillett et al. 2003)." The trend in the NAO since 1950 is to more positive values (Hurrell et al. 2004; Raible et al. 2005) though, as we shall see below, that trend is less

[1] While the PDO and AMO are important components of climate variability on decadal and longer time scales and have impacts on extremes (e.g., Chang and Fu 2002), the limited time span (30 years) of our data does not allow us to consider these in any detail.

evident during the last three decades. Hurrell et al. (2004) indicate that tropical SST forcing (especially in the Indian Ocean) is key to understanding the trend since the 1950s. In fact, they assert that boreal winter North Atlantic climate change since 1950 is well described by the trend in the NAO.

The NAM impacts weather throughout the NH middle and high latitudes (Thompson and Wallace 2001). It is statistically linked with the NAO (Quadrelli and Wallace 2004). The NAM has exhibited a trend towards its positive phase between the 1960s and the 1990s, corresponding to a decrease in surface pressure over the Arctic and an increase over the subtropical North Atlantic (Hurrell 1996; Thompson et al. 2000; Gillett et al. 2003), with somewhat of a decrease after the mid-1990s. The positive phase of the NAM has been associated with a decrease (increase) in winter precipitation over southern (northern) Europe, due a northward shift of the storm track (Thompson et al. 2000).

The trend in the PNA during the period 1950–2000 is characterized by a general shift toward more negative heights over the Aleutians and the southeastern United States (e.g., as inferred from Raible et al. 2005). The more recent values of the PNA index,[2] however, suggest that the trend appears to end, if not reverse, in the 1980s (see also next section).

In the following, we assess the ability of the GEOS-5 AGCM to capture the above major modes of climate variability including their spatial structure, time variations, and their links to regional weather. We also assess the extent to which the model reproduces the basic observed weather variability and extremes. The model results are compared with the latest reanalysis data as well as various gridded station observations.

9.2.1.1 Climatology and Variability

In the following, we examine the results of three AMIP-style simulations with the GEOS-5 AGCM forced with observed SST, GHGs and ozone covering the period 1980–2009 (see Appendix A for details of the runs). Comparisons are made with MERRA, the NOAA Climate Forecast System Reanalysis (CFSR, Saha et al. 2010), and other gridded observations. The focus in this section is on assessing the quality of the model simulations of the mean climate, the large-scale monthly variability, and the daily weather variability for the 29 boreal winters (JFM) starting in 1981 (since the runs were started in early January of 1980, we discard JFM of 1980 to avoid any spin-up and incomplete months).

Figure 9.1 (top two rows) shows that GEOS-5 reproduces the observed (MERRA) boreal winter upper tropospheric stationary waves and their monthly variability quite well. Here we include a comparison with another reanalysis (CFSR) to get some sense of the uncertainties inherent in the reanalysis data. That comparison shows that the model stationary waves are, in fact, almost

[2]http://jisao.washington.edu/data/pna/

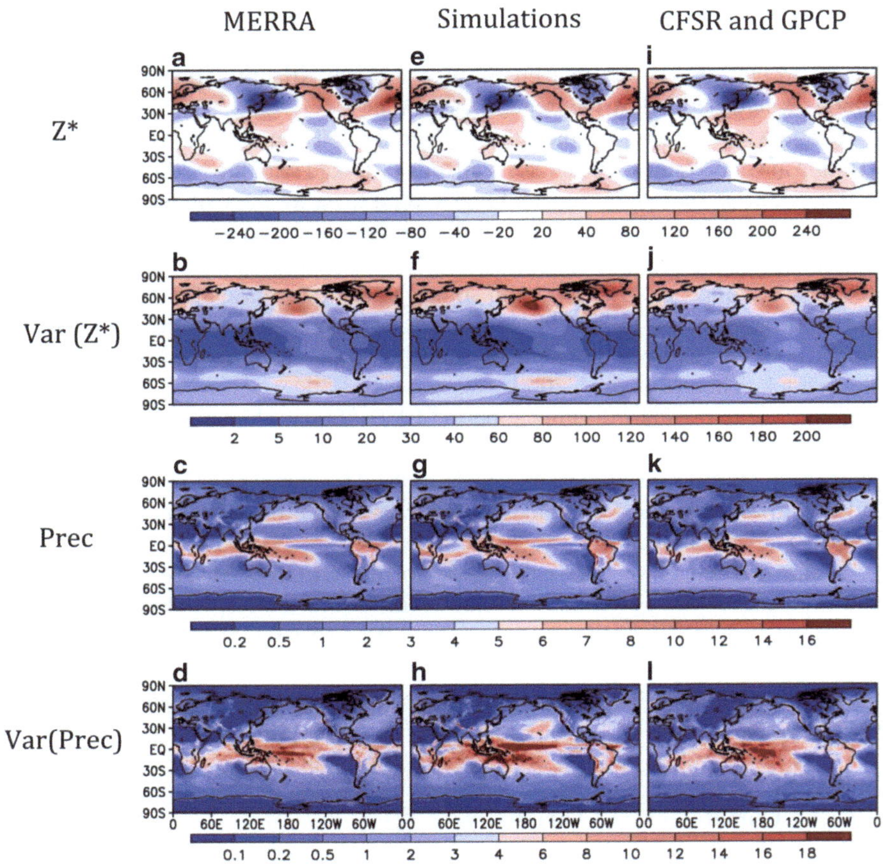

Fig. 9.1 The (JFM, 1981–2009) 250 hPa eddy height (Z^*) climatology (*top row*: **a**, **e**, **i**) and monthly variance (*second row*: **b**, **f**, **j**) fields from MERRA (*left panels*), GEOS-5 simulations (*middle panels*), and CFSR (*right panels*). Units are meters and meters squared times 0.01, respectively. The *third* and *fourth rows* are the same as the *first* and *second rows* except for precipitation (Prec) and the *right panels* (**k** and **l**) are from GPCP observations. Units are mm/day and $(\text{mm/day})^2$, respectively

indistinguishable from the reanalysis estimates. The model also reproduces the climatological precipitation pattern and its monthly variability (third and fourth rows of Fig. 9.1), although the simulated variance is somewhat larger than the observed (as estimated from GPCP or MERRA) in the tropics. In addition, the model reproduces the climatological middle latitude storm tracks, although with somewhat weaker amplitude than observed (top panels of Fig. 9.2).

The ability of the model to produce realistic extremes in daily precipitation is evaluated in the second half of Fig. 9.2, which shows the 10-year return values of the daily precipitation maximum (c and f). Also shown are the 10-year return values

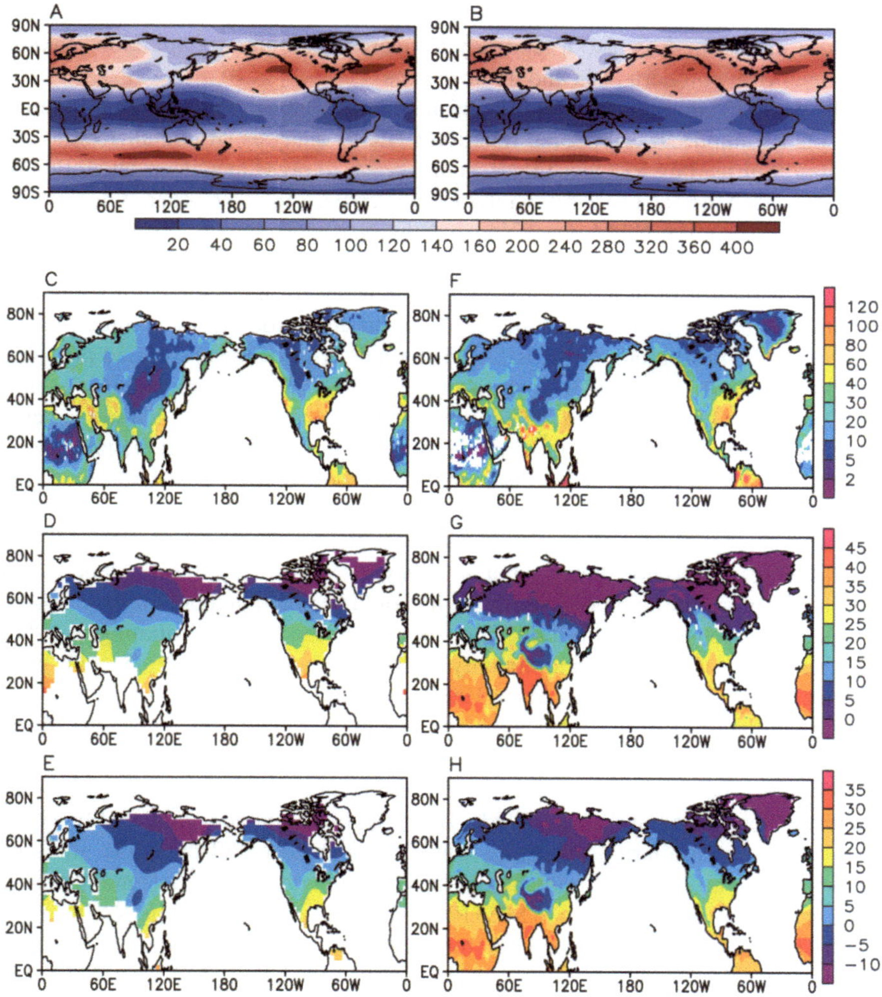

Fig. 9.2 *Upper panels*: The 250 hPa variance of the daily v-wind for JFM (1981–2009) from MERRA (**a**) and the simulations (**b**). *Lower panels*: Maps of 10-year return values for daily maximum precipitation (**c** and **f**, units are mm/day), warmest day (**d** and **g**, units are °C) and warmest night (**e** and **h**, units are °C). *Left* is for observations (GPCP or HadGHCN), and *right* is for the AMIP runs. *White regions* indicate missing or insufficient data to produce reliable fits to the GEVD

of the warmest days (d and g) and warmest nights (e and h). The return values are estimated by fitting the extremes to the Generalized Extreme Value Distribution (GEV, Coles 2001). Examples of the fits to the GEV distribution are given in Appendix B. Overall, the simulated extremes look reasonable. The greatest model deficiency appears to be the unrealistically cold day-time temperatures simulated over Canada and northern Eurasia, as it is reflected in the extreme values of the warmest days. There is also a tendency to overestimate the precipitation maxima especially over southern Asia and in the tropics.

The decomposition of the monthly mean height variability is carried out in terms of a rotated empirical orthogonal function (REOF) analysis (Richman 1986). The resulting four leading REOFs (Fig. 9.3a–h) appear to provide a clean separation of the observed (MERRA) variability into patterns consisting of (in order of decreasing variance) ENSO, the NAM, the PNA, and the NAO.[3] The results from the model show a very realistic representation of those patterns although with somewhat different variance. In particular, the main deficiency of the model simulation is a too strong PNA (Fig. 9.3g) that shows up as the first mode in the simulations (it appears as the third mode in MERRA). The time series of the associated principal components or PCs (Fig. 9.3i–l) show that PC 1 (PC 3 in the model simulations, Fig. 9.3i) indeed appears to be linked to ENSO variations as well as a trend towards more positive values, with all three model runs tracking the observed behavior very closely. The PC time series for the model runs and MERRA associated with the second REOF (the NAM, Fig. 9.3j), suggest less control from the SST, with the observations showing a tendency toward negative values after the early 1990s. While the model results do not show such a tendency, the observed values do tend to fall within the spread of the model ensemble. It should be noted that our observed NAM time series is consistent with longer time series[4] of the NAM (beginning in 1950) that are characterized by predominantly negative values prior to 1985, with peak positive values occurring in the early 1990s, followed by a negative trend.

The PNA (third REOF for MERRA and first in the model runs, Fig. 9.3k) shows a trend toward more negative values. As discussed earlier, longer time series starting in 1950 show that this downward tendency follows an extended period of an upward tendency that peaked in the mid 1980s. Here, the observed and simulated behavior does show some similarity suggesting some dependence on SST or GHGs, although that appears to be stronger for the model runs (the individual ensemble members are more similar to each other than to MERRA). The time variability of the NAO (fourth REOF, Fig. 9.3l) shows little control from any external forcing, although the three model simulations again seem to be more similar to each other than to MERRA. There is no obvious trend, but there is again (like for the NAM) some tendency toward more negative values for the observations in recent years. This is part of a longer-term variability[5] that resembles the behavior of the NAM, characterized by primarily negative values from 1950 to 1980, with peak maximum values in the early 1990s, followed by the downward trend.

[3]We note that the SAM also occurs during the boreal winter, though it does not show here as a leading mode. We shall see later that during austral winter it does show up prominently as the leading mode.

[4]http://www.cpc.ncep.noaa.gov/products/precip/CWlink/daily_ao_index/JFM_season_ao_index.shtml

[5]http://www.cpc.ncep.noaa.gov/products/precip/CWlink/pna/JFM_season_nao_index.shtml

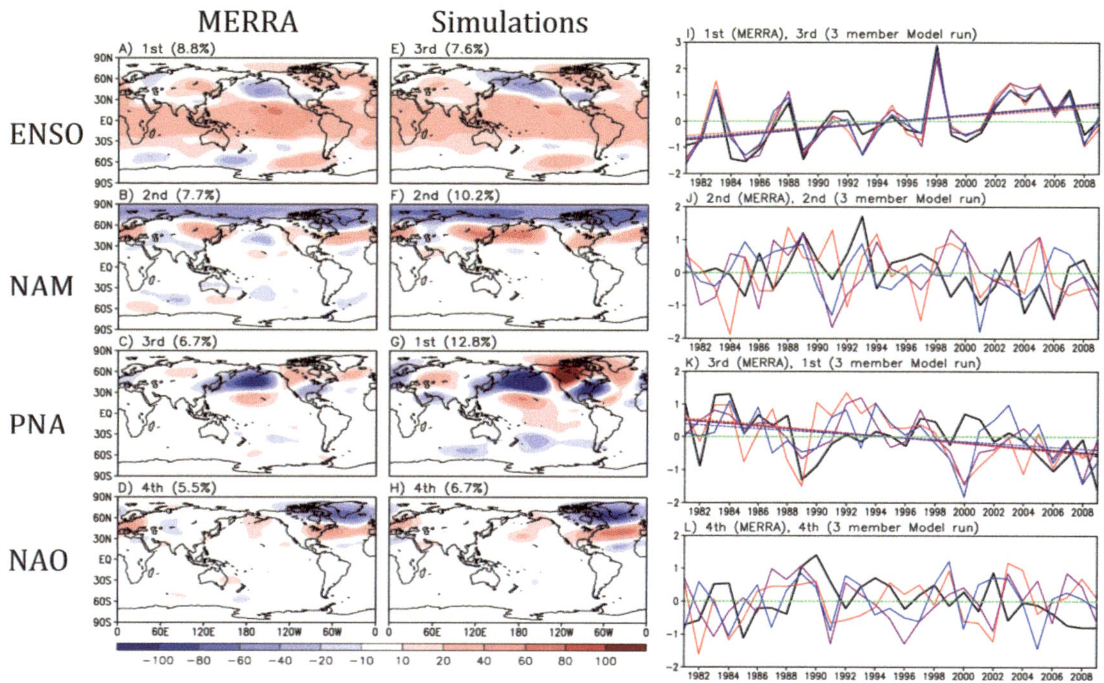

Fig. 9.3 *Left set of panels*: The four leading rotated empirical orthogonal functions (REOFs) of the monthly 250 hPa height field for JFM, 1981–2009. *Left panels* (**a–d**) are from MERRA and the *right panels* (**e–h**) are from GEOS-5 (note the reordering of the GEOS-5 REOFs to match MERRA). *Right set of panels* (**i–l**): The time series of the four leading PCs. The *black line* is from MERRA and the *colored lines* are the three model ensemble members. The monthly values are averaged to produce seasonal (JFM) means before plotting. For the first and third mode the trend lines for each ensemble member and MERRA are significant at the 10% level. The PCs are normalized to have unit variance, so amplitude information is contained in the spatial maps of the REOFs

The above results indicate that the various modes of variability consist of some patterns that are externally forced (by SST or GHGs or ozone) and others that are internal to the atmosphere, with some evidence that the latter may be impacted (modulated) by external forcing. We shall see in Sect. 9.3, that the NAM, PNA and NAO are indeed fundamentally internal (to the atmosphere) modes of variability, in that they appear as the leading REOFs of model simulations that have no interannual variability in external forcing (SST, GHGs, or ozone).

9.2.1.2 Regional Impacts of Climate Variability

We next turn to an evaluation of the relationships between the variations of the boreal winter leading large-scale modes of variability (ENSO, NAM, PNA, NAO) and regional weather including changes in storminess, near surface temperature, and precipitation. Results are presented for both MERRA (or gridded observations) and the model simulations. In the latter case, the correlations are the average of the correlations computed for each ensemble member. In all cases, a linear trend is removed before computing the correlations.

Overall, the results show that the main features of the regional impacts of the climate modes are well reproduced by the model (Fig. 9.4). This includes the enhanced storminess and precipitation along the US Gulf States during El Nino, the warmer (colder) temperatures over northern Eurasia (the Mediterranean and North African regions) during a positive NAM, the increased storminess and warming over northern Europe and the warming over the US eastern seaboard during a positive NAO, and the warming over Alaska and western Canada during a positive PNA. The main deficiency appears to be that the PNA is more strongly tied to SST variability than is observed.

Specific features of the ENSO-related correlations (*panels on the top left side of Fig.* 9.4) include a substantial El Nino warming throughout the tropics in all ocean basins, with the typical off-equatorial cooling in the Pacific. The correlations over the extratropical land areas are generally weak, with the model showing more pronounced correlations over North America than MERRA. The precipitation correlations (Fig. 9.4b, f) show the typical ENSO-related pattern in the Pacific, with positive values over the central and eastern tropical Pacific and negative values in the Pacific warm pool that extend eastward into the subtropics of both hemispheres. The sea level pressure (SLP) correlations (Fig. 9.4c, g) are consistent with the Southern Oscillation pattern. The weather variance, as estimated by the daily meridional wind variability (v'^2), shows (Fig. 9.4d, h) negative correlations in the central Pacific and positive correlations in the tropical Atlantic and the Indian Ocean and parts of the Pacific warm pool. The enhanced (reduced) storm track along the U.S. southern tier of states during El Nino (La Nina) is clearly evident in both the simulations and MERRA.

The correlations of both the MERRA and simulated temperatures 2 m above the ground (T2m) with the NAM (*top right panels of Fig.* 9.4) are characterized by negative values over the Arctic and Greenland, positive values over northern

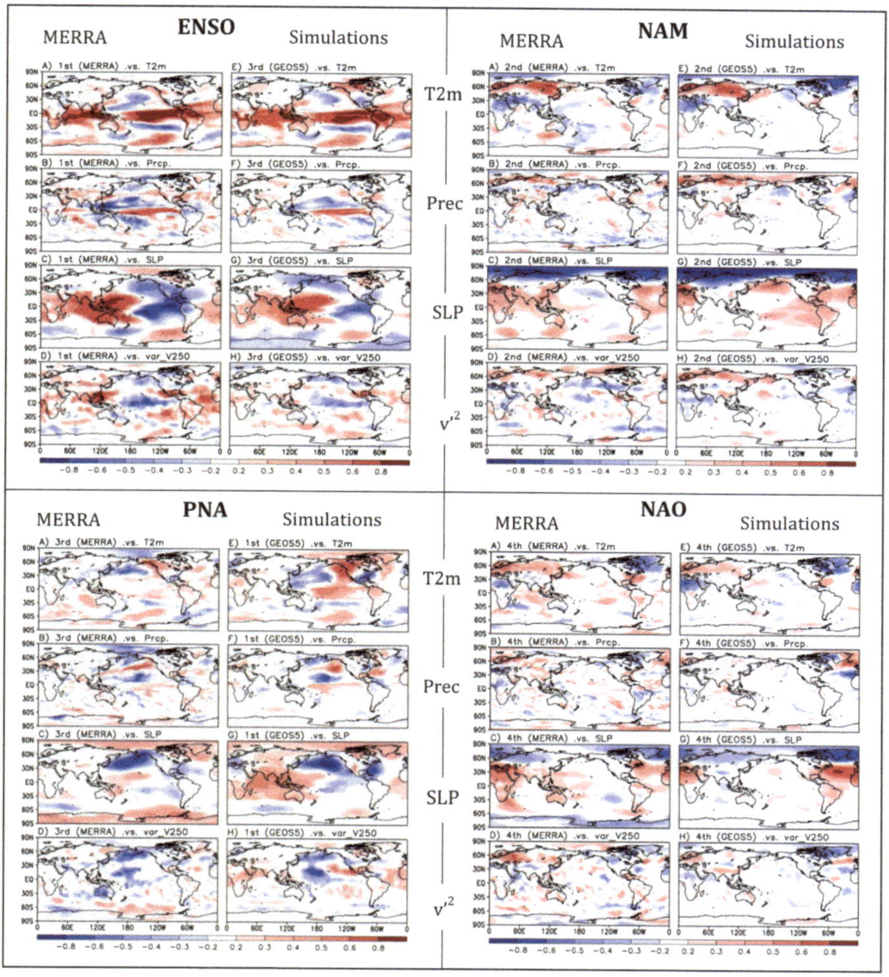

Fig. 9.4 Correlations between the leading monthly REOFs (the corresponding PCs- seasonally averaged) and seasonal mean T2m, precipitation, sea level pressure, and 250 hPa v'^2. For each set of panels, the results for MERRA are on the *left* and the results for GEOS-5 are on the *right*. *Top left set of panels*: MERRA REOF 1 and GEOS-5 REOF 3. *Top right set of panels*: MERRA REOF 2 and GEOS-5 REOF 2. *Bottom left set of panels*: MERRA REOF 3 and GEOS-5 REOF 1. *Bottom right set of panels*: MERRA REOF 4 and GEOS-5 REOF 4. Values are detrended before computing the correlations. The model results are the average of the correlations computed separately for each ensemble member

Scandinavia and Russia extending eastward to Japan, and again negative values over Northern Africa and parts of southern Asia (Fig. 9.4a, e). The precipitation correlations (Fig. 9.4b, f) are positive over much of northern Russia (with a tendency for negative values to south) and Scandinavia. The SLP exhibits negative correlations in the Arctic (north of 60°N) and primarily positive values south of that, with the model showing positive correlations in the eastern Pacific that are not in the MERRA results (Fig. 9.4c, g). The correlations with v'^2 (Fig. 9.4d, h) show positive values across northern Russia and Scandinavia, and a tendency for negative values to the south in the latitude band 30–40°N, though the observed negative correlations over the US are not reproduced in the model results.

The observed and simulated PNA correlations (*bottom left panels of Fig.* 9.4) exhibit negative values for T2m over the western North Pacific and southeastern US, with positive values over much of northwest North America (Fig. 9.4a, e). The simulations show a substantial region of positive correlations with T2m over the central and eastern tropical Pacific that indicate a too strong link to SST variability. The precipitation correlations (Fig. 9.4b, f) are positive over the North Pacific extending into the Gulf of Alaska, and these are flanked by negative correlations to the north and south. Positive correlations also occur in the tropical Pacific and in the Gulf of Mexico extending into the North Atlantic. The correlations with SLP (Fig. 9.4c, f) are consistent with the height anomalies, and are dominated by negative correlations in the North Pacific and the southeast U.S., and a tendency for positive correlations in the eastern tropical Pacific and (for the model) also in the Indian Ocean. Correlations with v'^2 (Fig. 9.4d, h) are mostly negative over the North Pacific, and the U.S., with some positive correlations (more so for the model) extending from the eastern subtropical Pacific across Mexico and into the Atlantic.

The observed and simulated NAO correlations (*bottom right panels of Fig.* 9.4) exhibit positive values for T2m (Fig. 9.4a, e) over much of Europe and Russia, and the southeast United States. Negative correlations occur over northeastern North America and Greenland, as well as northern Africa. Correlations with precipitation (Fig. 9.4b, f) show a north/south dipole over the North Atlantic and Europe (positive over Northern Europe) and that is reflected to some extent by a north/south dipole in the correlations with weather variability (positive over Northern Europe, Fig. 9.4d, h). Correlations with SLP (Fig. 9.4c, g) are primarily negative north of 60°N and positive over the North Atlantic (south of 60°N) and northern Africa.

9.2.1.3 Long-Term (Decadal Scale) Changes

In the previous sections we examined the variability on monthly time scales, and found that the leading modes of variability and their connections with weather and regional surface meteorology are generally well simulated by the model. Time series of these modes showed that they vary on both interannual and longer time scales, with a few modes showing clear evidence of trends over the three decades examined here. In this subsection we look more directly at the longer-term changes that have occurred, as they are reflected in the differences between the first and second half of the record.

Figure 9.5 shows the observed (MERRA and GPCP) and model simulated differences in the 250 hPa height, daily weather variability (v'^2), T2m, and precipitation. The most pronounced differences in the 250 hPa height field between

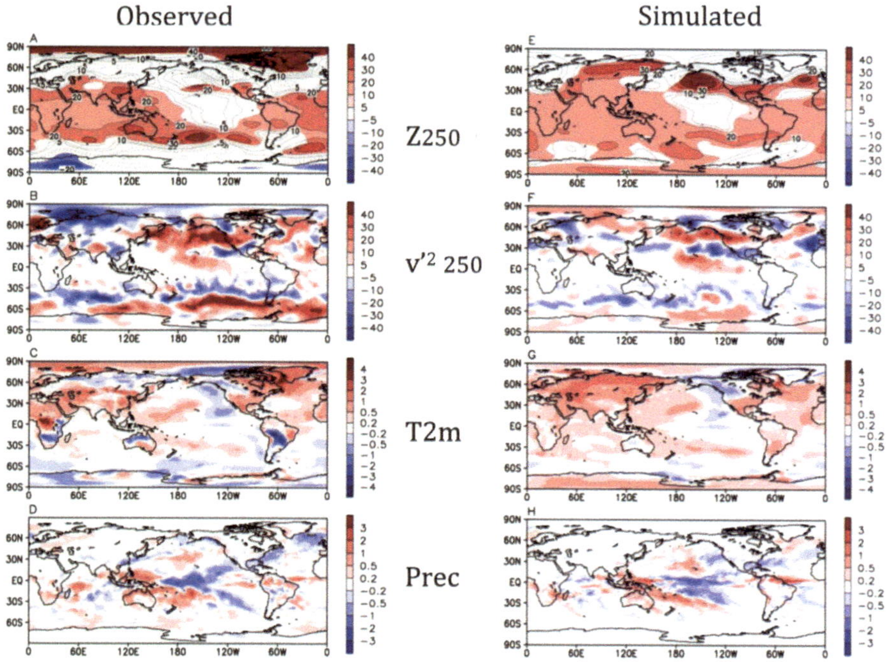

Fig. 9.5 The difference between the means: (1995–2009) minus (1981–1994) for JFM. The *left panels* are for the observations (MERRA and GPCP precipitation) and the *right panels* are for the AMIP runs. *Top panels* are the 250 hPa height differences (only values significant at the 10% level are *shaded*, units are meters), the *second row* consists of the differences in the 250mb daily V-wind variance (units are m/s squared), the *third row* shows the 2 m temperature differences (units are °C), and the *last row* shows the precipitation differences (units are mm/day)

the two periods in both MERRA and the simulations (Fig. 9.5a, e) are the overall increase in the tropics and subtropics and a decrease in the tropical eastern Pacific. The observations show a height decrease in the southern polar region, and an increase in height in the northern polar region: these are absent from the model simulations. The latter reflects the general trend towards more negative values of the NAO after 1990 in the observations (see e.g., Fig. 9.31). The weather variability (Fig. 9.5b, f), shows a distinct poleward shift in the SH in both the observations and simulations. Over North America, there is an increase in weather variability over Canada and a decrease over the US southern tier of states. There is little agreement between the observations and model results over Eurasia in terms of the changes in weather variability. The surface temperature changes (Fig. 9.5c, g) show a general warming, with cooling over western Canada, Alaska, and (for the observations) over northern Russia.

Over the ocean the temperatures have warmed over the North Atlantic, much of the western Pacific and parts of the Indian Ocean. In contrast, the eastern Pacific

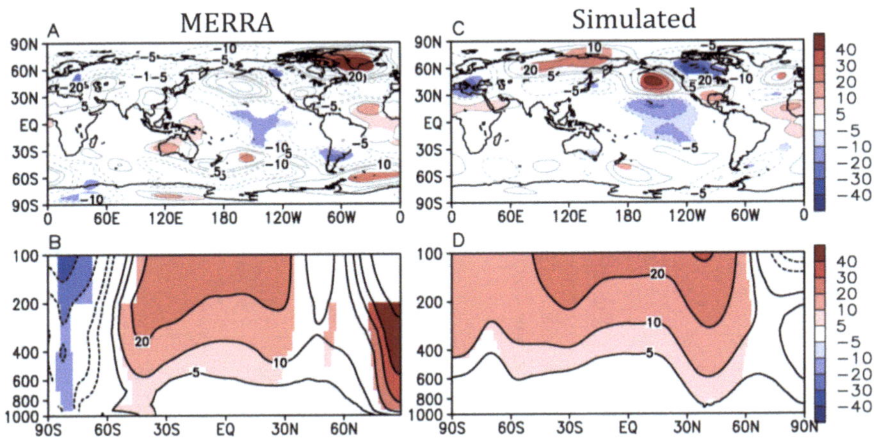

Fig. 9.6 The difference between the means: (1995–2009) minus (1981–1994) for JFM. The *left panels* are for MERRA and the *right panels* are for the AMIP runs. *Top panels* are the 250mb height differences with the zonal mean removed. The *second row* displays the zonal mean height difference (only values significant at the 10% level are *shaded*). Units are meters

shows some evidence of cooling. This appears to be part of a longer-term trend extending back at least one half century (e.g., Kumar et al. 2010). The precipitation differences (Fig. 9.5d, h) in the Pacific resemble the response to La Nina, with reduced precipitation in the tropical central and eastern Pacific extending polewards in both hemispheres (across the southern US and towards the southern tip of South America). The Pacific warm pool and the tropical Atlantic both show enhanced precipitation. The above changes in precipitation are consistent with those described in Hoerling et al. (2010).

Kumar et al. (2010) note that for the recent period (1980–2008), the global teleconnection pattern associated with La Niña has been associated with higher heights than in previous decades from the tropics to the mid-latitudes. In Fig. 9.6a, c we remove the "background" height increase by removing the zonal mean heights before computing the 250 hPac height differences. This shows more clearly the canonical La Nina response in the eastern Pacific with anomalies extending northward and eastward across North America. The cross section of the zonal mean height differences (Fig. 9.6b, d) highlights the general height increases that maximize above 200 hPa. The zonal means also emphasize the differences between the model and MERRA at high latitudes noted earlier.

The changes in the extremes between the two time periods are depicted in Fig. 9.7 in terms of the differences in the 10-year return values of the number of heavy rain days, warm days, and cold nights. Results are based on a fit of the exceedances to a Generalized Pareto Distribution (GPD, Coles 2001). This approach has the advantage over the block maxima approach applied earlier to the maxima (e.g., Fig. 9.2), in that it uses more data – this is especially important here in view of

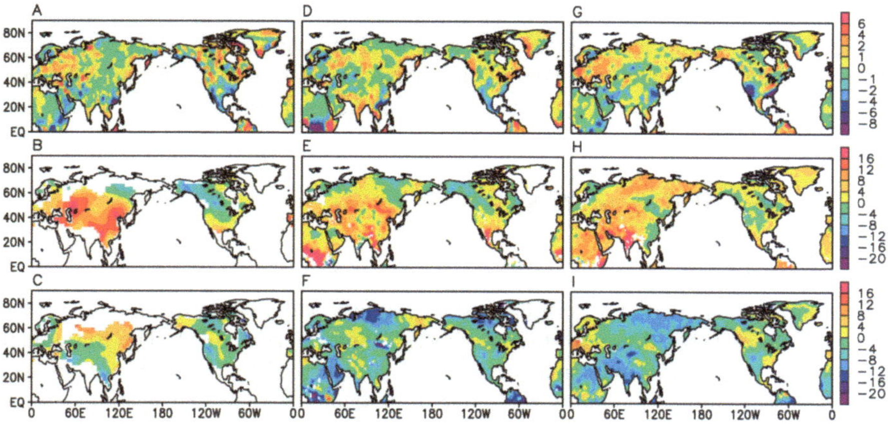

Fig. 9.7 The difference (JFM, 1995–2009 minus 1981–1994) in the 10-year return values of the number of heavy rain days (>5 mm/day), warm days (greater than the 90th percentile of the daily maximum temperatures) and cold nights (less than the 10th percentile of the daily minimum temperatures) for observations (*left panels*), MERRA (*middle panels*) and AMIP runs (*right panels*). Results are based on fits of the exceedances to a GPD

the short (15 year) records for the two time periods. The results for the changes in the number of heavy rain days (Fig. 9.7a, d, g) show general agreement between the observations, MERRA, and simulations, although there are differences in the details. The reduction in the number of heavy rain days in the southern United States is prominent in all three. So is the increase over eastern Russia, Canada, southern Asia, and northeast Brazil. The main discrepancy with the simulations occurs over northern Russia, where the model shows an increase and the observations and MERRA show a slight decrease in the number of heavy rain days.

The spatial pattern of the differences for the number of warm days (Fig. 9.7b, e, h) to a large extent reflect the distribution of the differences in the mean temperature (Fig. 9.5c, d), with an increase in the number of warm days over much of central and southern Asia, and the southern United States and Mexico. The apparent reduction in the number of warm days over northern Asia for the observations (and MERRA) versus an increase for the model is not inconsistent with the difference between the observed/MERRA and simulated mean temperature changes in that region (Fig. 9.5c, d). Over northwest North America, the observations and MERRA also show a reduction in the number of warm days while the model results are mixed.

The largest discrepancies with the observations occur for the changes in the number of cold nights (Fig. 9.7c, f, i), with the model and MERRA showing a general decrease over much of Asia, while the observations indicate an increase especially over central and northern Asia, though the northern areas have a substantial number of missing observations during the last decade. Over North America, there is more agreement with all three showing a tendency for an increase in the number of cold nights over the Pacific Northwest, and a decrease over the southwest.

Overall, the simulated (and MERRA) temperature extremes appear to reflect the basic shift to warmer conditions, with a general increase in the number of warm days and a decrease in the number of cold nights. This is less so for the observations, though that may be largely an artifact of gaps in the temperature record.

9.2.2 *Austral Winter (JAS)*

We turn next to the austral winter. In this case, we choose the Southern Hemisphere (SH) late winter 3-month period of July, August and September (JAS), as the appropriate juxtaposition with the late boreal winter period, JFM. The SH winter is also (like the boreal winter hemisphere) characterized by a number of distinct modes of variability on monthly and longer time scales. Perhaps the best know is the Southern Annular mode (SAM; Thompson and Wallace 2000) already mentioned earlier. As reported in IPCC (2007), the SAM has exhibited an upward trend over the past 30 years, corresponding to a decrease in surface pressure over the Antarctic and an increase over the southern mid-latitudes although the mean SAM index since 2000 has been below the mean of the late 1990s, but above the long term mean. An upward trend in the SAM has occurred in all seasons, but the largest trend occurs during the southern summer (Mo 2000; Thompson et al. 2000; Marshall 2003). Based on an analysis of the structure and seasonality of the observed trends in SH circulation, Thompson and Solomon (2002) suggest that the trends have been largely induced by stratospheric ozone depletion. In contrast, Ding et al. (2011) report that, while the SAM exhibits an upward trend during the summer, during winter it has exhibited a negative trend since 1979, associated with an increase in geopotential heights over high latitudes. This appears to be in part due to the nature of the SAM index, which is a superposition of both intrinsic high latitude variability and a tropically-forced component. The SAM is known to have impacts on South America (Silvestri and Vera 2003), Australia (Hendon et al. 2007), New Zealand (Kidson et al. 2009), South Africa (Reason and Rouault 2005), and the Antarctic Peninsula (Lefebvre et al. 2004).

In addition to the SAM, there are several distinct wave structures that are the counterparts to the PNA in the NH, consisting of the Pacific South American (PSA) modes 1 and 2 (Kidson 1988; Ghil and Mo 1991; Mo 2000). These zonal wave 3 structures extend from the tropical South Pacific polewards and eastward across southern South America. Both PSAs have been linked to variations in ENSO (e.g., Karoly 1989; Mo 2000).

Before examining the modes of variability, we again begin by first looking at the ability of the model to simulate the basic climatological fields and overall variability.

9.2.2.1 Climatology and Variability

The model captures that JAS stationary wave pattern quite well (Fig. 9.8a, e, i). Again, as during JFM, the correspondence in the details of the JAS troughs

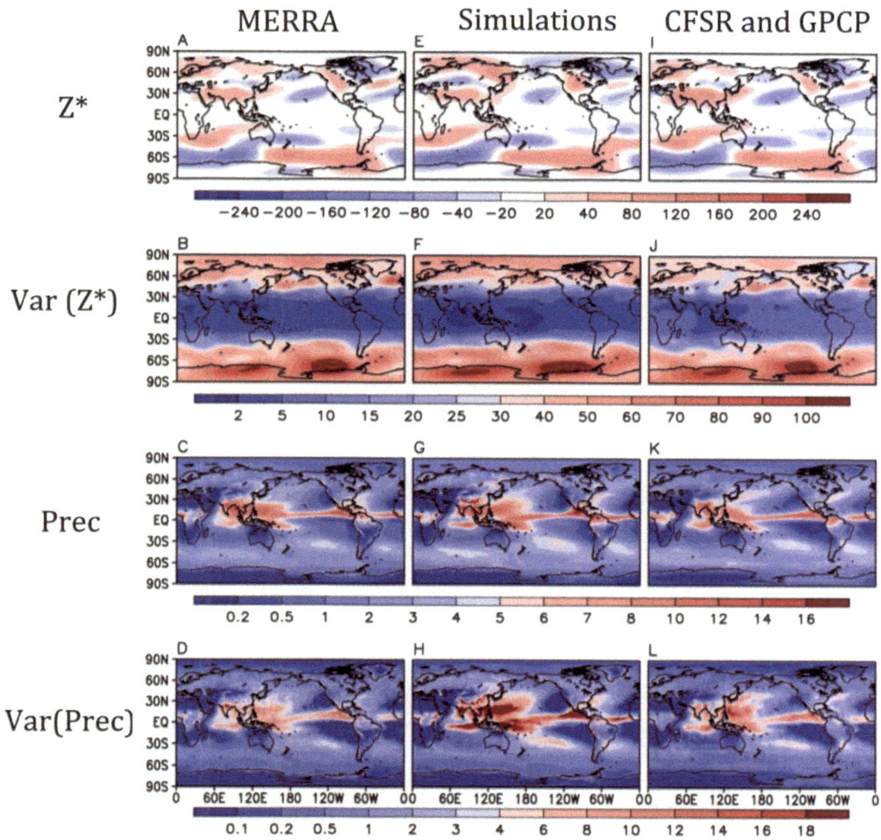

Fig. 9.8 The (JAS, 1980–2009) 250 hPa eddy height climatology (*top row*: **a**, **e**, **i**) and monthly variance (*second row*: **b**, **f**, **j**) fields from MERRA (*left panels*), GEOS-5 simulations (*middle panels*), and CFSR (*right panels*). Units are meters and meters squared times 0.01, respectively. The *third* and *fourth rows* are the same as the *first* and *second rows* except for precipitation and the *right panels* are from GPCP observations. Units are mm/day and $(mm/day)^2$, respectively

and ridges in both hemispheres is remarkable. The variance patterns (Fig. 9.8b, f, j) are also well simulated included the region of enhanced variability in the SH high latitudes near 120 W. The basic precipitation pattern is simulated well (Fig. 9.8c, g, k), although the area of Pacific warm pool precipitation extends too far north in the western Pacific, and the precipitation in the eastern end of the Pacific ITCZ is excessive. The overall pattern of precipitation variance (Fig. 9.8d, h, l) is quite realistic and reflects the spatial distribution of the mean precipitation, although the amplitude is excessive.

As an overall assessment of weather activity we again focus on the patterns of daily v'^2 at 250 hPa. This shows (Fig. 9.9a, b) that the model reproduces the basic patterns of SH weather variability quite well, matching both the spatial pattern

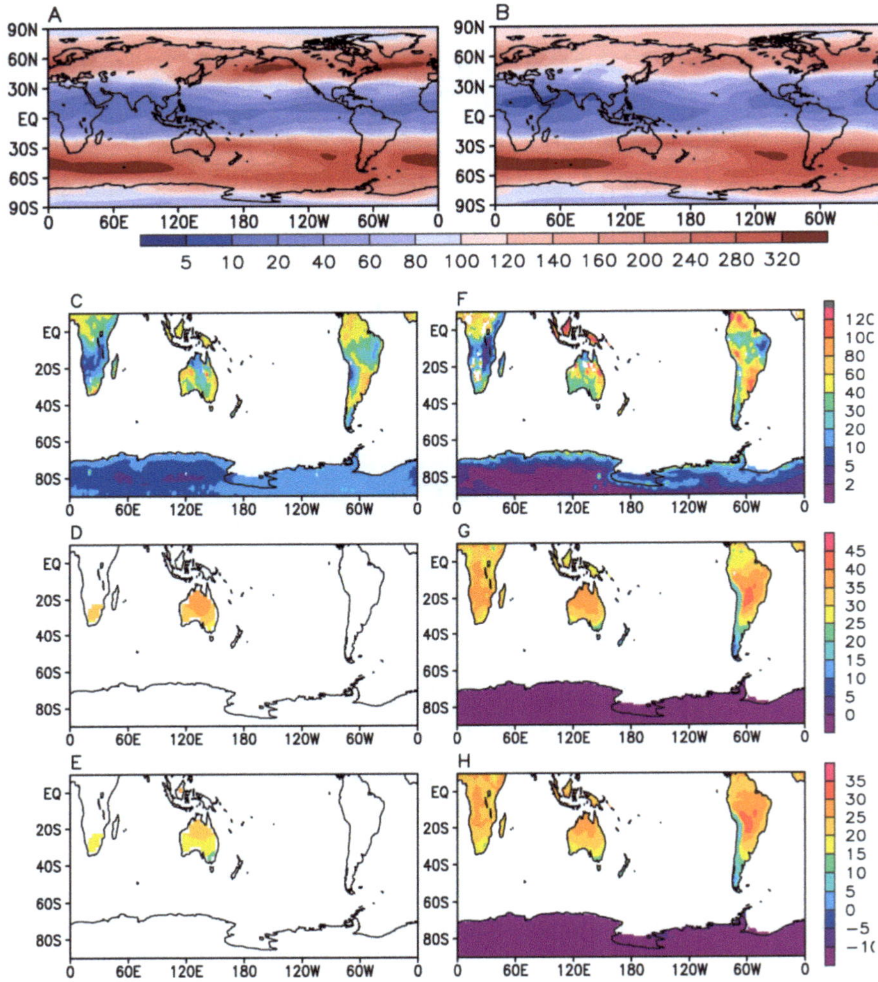

Fig. 9.9 *Upper panels*: The 250 hPa variance of the daily v-wind for JAS, 1980–2009 from MERRA (**a**) and the simulations (**b**). *Lower panels*: Maps of 10-year return values for daily maximum precipitation (**c** and **f**, units are mm/day), warmest day (**d** and **g**, units are °C) and warmest night (**e** and **h**, units are °C). *Left* is for observations (GPCP or HadGHCN), and *right* is for the AMIP runs. *White regions* indicate missing or insufficient data to produce reliable fits to the GEVD

and amplitude of the variability. We note that in the NH, the simulated weather variability is overall weaker than observed during this season, especially in both storm tracks although the regions of the maxima in variance are again quite realistic.

The 10-year return values of the precipitation maxima (Fig. 9.9c, f) are simulated reasonably well over Australia, Africa and South America during JAS, although

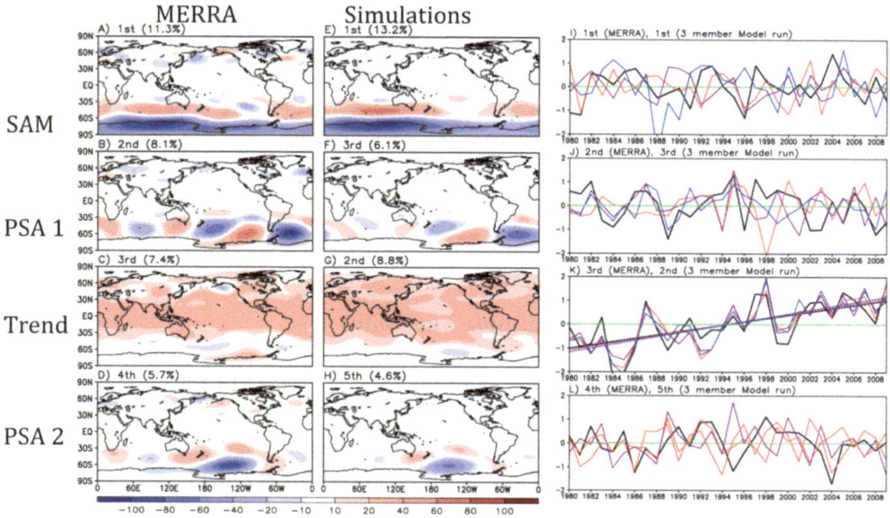

Fig. 9.10 *Left set of panels*: The four leading rotated empirical orthogonal functions (REOFs) of the monthly 250 hPa height field for JAS, 1980–2009 (note for the model the second and third REOFs are switched). *Left panels* are from MERRA and the *right panels* are from GEOS-5. *Right set of panels*: The time series of the four leading PCs of the height field REOFs. The *black line* is from MERRA and the *colored lines* are the three model ensemble members. The monthly values are averaged to produce seasonal (JAS) means before plotting. For the third mode the trend lines for each ensemble member and MERRA are significant at the 10% level. The PCs are normalized to have unit variance, so amplitude information is contained in the spatial maps of the REOFs

the amplitude tends to be too large. Also shown in Fig. 9.9 are the 10-year return values of the warmest day (d and g) and warmest night (e and h). The comparison is to a large extent limited to Australia (due to insufficient daily temperature observations over most of the rest of the SH land areas). While the spatial pattern of the temperature extremes over Australia look reasonable, the values are too large indicating a warm bias in the model.

We next turn to the REOFs of the 250 hPa height field. The four leading REOFs are shown in the left panels of Fig. 9.10a–h. The first (Fig 9.10a, e) is clearly associated with the SAM, showing a zonally symmetric structure, with however, maximum loadings in the eastern hemisphere (Indian Ocean sector), in both the observed and simulated patterns. The PCs associated with the SAM (Fig. 9.10i) show little coherence and no apparent trends. While previous studies have found an upward trend in the SAM over the last 30 years, observations show that this is largely confined to the SH summer and autumn months (Marshall 2003).

The second REOF (third in the model simulations, Fig. 9.10b, f) has a zonal wave number 3 structure extending across the middle latitudes of the Southern Hemisphere (with the largest amplitude in the South Pacific and South Atlantic). The variations do not seem to be strongly forced (little coherence among the ensemble members, Fig. 9.10j) but there is some suggestion that the waves emanate from the

Pacific warm pool. The third REOF (the second REOF in the model simulations, Fig. 9.10c, g) is associated with an almost global height increase and a very robust upward trend (Fig. 9.10k). The fourth REOF (Fig. 9.10d, h) has a wave structure confined to the South Pacific Ocean (south of 30°S), and also seems to have little if any coherence among the ensemble members or with MERRA (Fig. 9.10l). We shall see in Sect. 9.3 that for JAS, REOFs 1, 2 and 4 (as defined by MERRA) are the leading internal atmospheric modes of variability, analogous to the NAM, PNA, and NAO during JFM.

9.2.2.2 Regional Impacts of Climate Variability

We next focus on the regional impacts of the four 250 hPa REOFs (Fig. 9.11). The *top left set of panels of Fig.* 9.11 show the T2m correlations (Fig. 9.11a, e) with the SAM (the leading 250 hPa REOF). Correlations are strongest (negative values) over Antarctica in the eastern hemisphere, with weak positive values over parts of the Southern Oceans. The results for MERRA (Fig. 9.11a) show positive correlations over the eastern tropical Pacific with negative correlations in the western Pacific, suggestive of an ENSO connection (see also Ding et al. 2011). In contrast, the simulations (Fig. 9.11e) show only very weak correlations in the tropical Pacific, indicating a much weaker link to ENSO. Both model and MERRA show weak but coherence correlations with precipitation (Fig. 9.11b, f) with alternating bands of negative correlations over Antarctic, positive correlations just north of Antarctica and again negative correlations just south of Australia. The correlations with SLP (Fig. 9.11c, g) show a very clear pattern of negative correlations over Antarctic and positive correlations over the Southern Indian Ocean. The observed and simulated correlations with v'^2 (Fig. 9.11d, h) also show alternating bands of negative and positive correlations (negative over Antarctica, positive just north of that, and again negative correlations in the southern Indian Ocean) extending eastward along the southern coast of Australia, New Zealand, and further east into the south Pacific.

The correlations with the third MERRA (second model) REOF (the Trend mode) are shown in the *bottom left set of panels of Fig.* 9.11. The correlations with T2m (Fig. 9.11a, e) show that the strongest positive correlations occur over the northern Indian Ocean, parts of the Pacific warm pool, the far eastern tropical Pacific, and the Atlantic warm pool extending eastward across the Atlantic. Over the land areas the correlations tend to be positive, including parts of North America, northern South America, southern Asia, and northern Africa. Correlations with precipitation (Fig. 9.11b, f) tend to be weak and disorganized, with the most coherent correlations occurring in the Pacific ITCZ (where they are positive) and across northern South America where the correlations are negative. Correlations with SLP (Fig. 9.11c, g) are weak and show little agreement between the model and MERRA. MERRA has positive correlations over Indonesia, while the model has positive correlations over the central tropical/subtropical Pacific. Correlations with v'^2 (Fig. 9.11d, h) are weak and disorganized but tend to be positive, especially for MERRA.

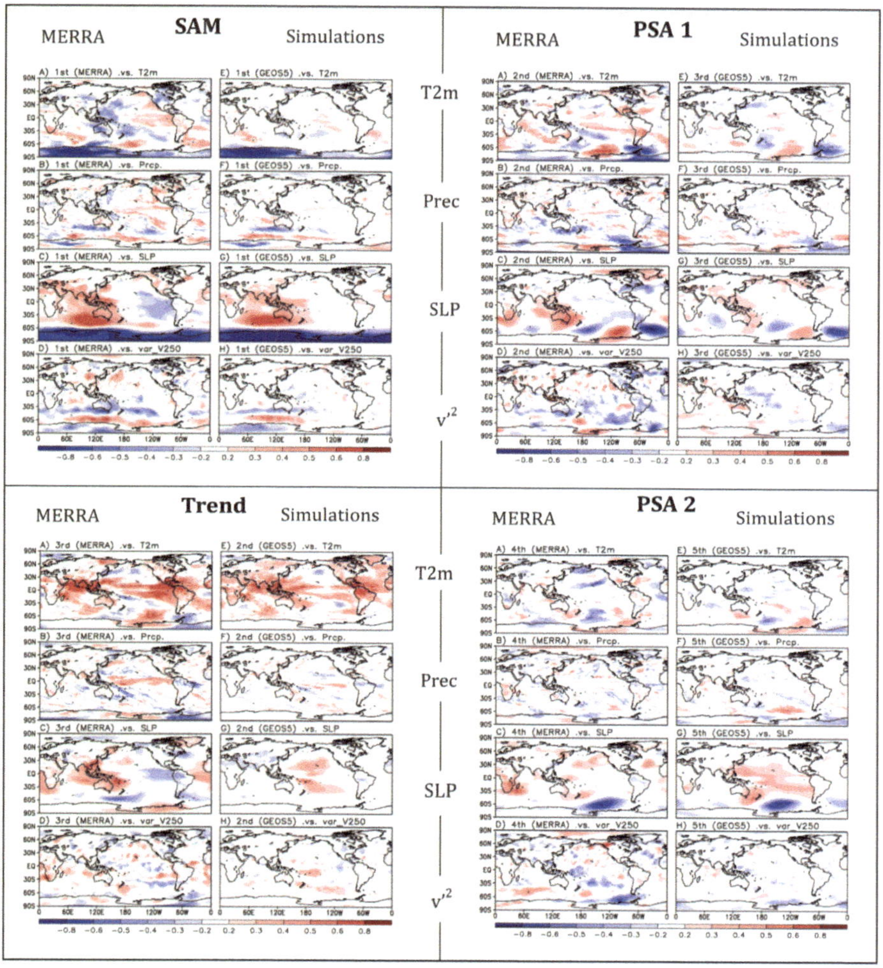

Fig. 9.11 Correlations between the leading monthly REOFs (the corresponding PCs- seasonally averaged) and seasonal mean T2m, precipitation, sea level pressure, and 250 hPa v'^2. For each set of panels, the results for MERRA are on the *left* and the results for GEOS-5 are on the *right. Top left set of panels*: MERRA REOF 1 and GEOS-5 REOF 1. *Top right set of panels*: MERRA REOF 2 and GEOS-5 REOF 3. *Bottom left set of panels*: MERRA REOF 3 and GEOS-5 REOF 2. *Bottom right set of panels*: MERRA REOF 4 and GEOS-5 REOF 5. Values are detrended before computing the correlations. The model results are the average of the correlations computed separately for each ensemble member

Finally, we also show in Fig. 9.11 the correlations with the second and fourth REOFs corresponding to PSA 1 (*upper right set of panels*) and PSA 2 (*lower right set of panels*). These both exhibit a wave structure in the high latitude South Pacific that is reflected in both the T2m (Fig. 9.11a, e) and SLP (Fig. 9.11c, g)

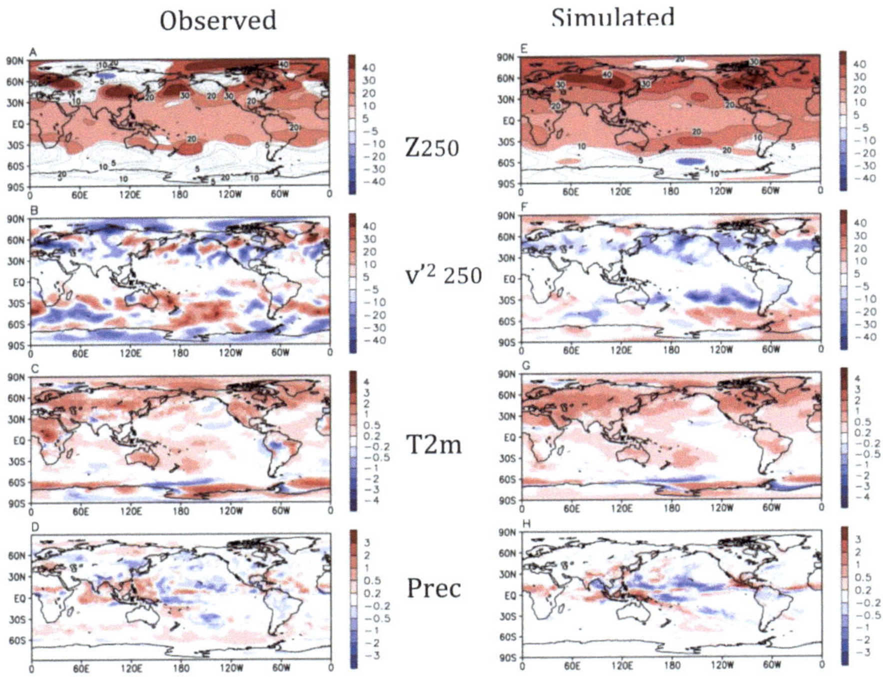

Fig. 9.12 The difference between the means of the two 15-year periods (1995–2009) and (1980–1994) for JAS. *The left panels* (**a–d**) are for the observations (MERRA and GPCP precipitation) and the *right panels* (**e–h**) are for the AMIP runs. *Top panels* are the 250 mb height differences (only values significant at the 10% level are *shaded*, units are meters), the *second row* consists of the differences in the 250 mb daily V-wind variance (units are m/s squared), the *third row* shows the 2 m temperature differences (units are °C), and the *last row* shows the precipitation differences (units are mm/day)

correlations, and that extend from Australia southward and eastward to southern South America. Both patterns also appear to impact the western hemisphere of Antarctica (see the T2m and Precipitation correlations). The MERRA-based and simulation-based correlations are generally quite similar, with neither showing a strong impact/connection with the tropics.

9.2.2.3 Long-Term (Decadal Scale) Changes

Figure 9.12 shows global maps of the differences between the means of the two 15-year periods (1995–2009) and (1980–1994) for JAS. The 250 hPa height differences (Fig. 9.12a, e) again (as during JFM) show for both MERRA and the simulations a general increase throughout the tropics and subtropics, with some hemispheric asymmetry in the sense that the increases extend further poleward to include much of the middle and high latitudes of the NH. In the SH the maximum increases occur near 30°S, while in the NH the maximum increase occurs near 60°N. In contrast to JFM, there is no clear signature of a La Nina response (presumably because of the seasonal increase in the tropical Pacific easterlies). Instead, there is some evidence of an enhanced response over the Atlantic, reflecting a response to the warming of the Atlantic Ocean during this time period. In fact, the pattern of enhanced positive anomalies extending westward and poleward from the tropical Atlantic into the North and South Pacific oceans, and over Eurasia, is reminiscent of the response to an idealized uniform warming of the Indian Ocean (simulation not shown). The main differences between the MERRA pattern and that of the simulations is the more complex wavy structure in the MERRA 250 hPa differences at middle and high latitudes, which likely reflects the larger sampling errors in the MERRA results.

There is little agreement between the model results and MERRA regarding the changes in weather statistics (Fig. 9.12b, f), with the MERRA difference patterns showing generally very little spatial coherence. The model results on the other hand do show coherence with generally reduced weather variability in the NH middle latitudes and in the SH near 30°S over the Pacific, and with increased weather variability over the high latitudes of both hemispheres. Both the model and MERRA show an increase in T2m over North America, central Asia, the Sahel and Australia (Fig. 9.12c, g). They disagree on the T2m temperature changes over central Africa and parts of South America, though it should be noted that these are regions where the MERRA reanalysis is not reliable. The precipitation differences (Fig. 9.12d, h) also show little agreement between GPCP and the model, especially in the tropical Pacific.

Figure 9.13 further highlights the areas of agreement and disagreement among the MERRA and model difference fields. After removing the zonal mean, the 250 hPa differences (Fig. 9.13a, c) are generally quite noisy, although there is some evidence of a wave response over the SH Pacific in both MERRA and the simulations. The zonal mean height differences (Fig. 9.13b, d) show height increases above about 700 mb that extend northward from about 30°S. In MERRA the differences maximize near 200 hPa and in the North Polar region, while in the simulations they continue to increase above 100 mb with maximum values occurring near 60°N.

The differences in the austral winter extremes between the two time periods are depicted in Fig. 9.14. Here again we present the differences in the number of heavy rain days (Fig. 9.14a, d), warm days (Fig. 9.14b, e) and cold nights (Fig. 9.14c, f), though only for MERRA and the simulations (the limited number of observations in the SH do not allow reliable estimates of the differences in the extremes). The changes in the number of days with heavy precipitation show substantial differences between MERRA and the simulations with, for example, an increase over northern South America in MERRA and a decrease in the simulations. Both agree that there has been an increase in the number of heavy precipitation days in east-central South America and parts of Africa. Both the MERRA and the simulated temperature extremes are characterized by an increase in the number of warm days

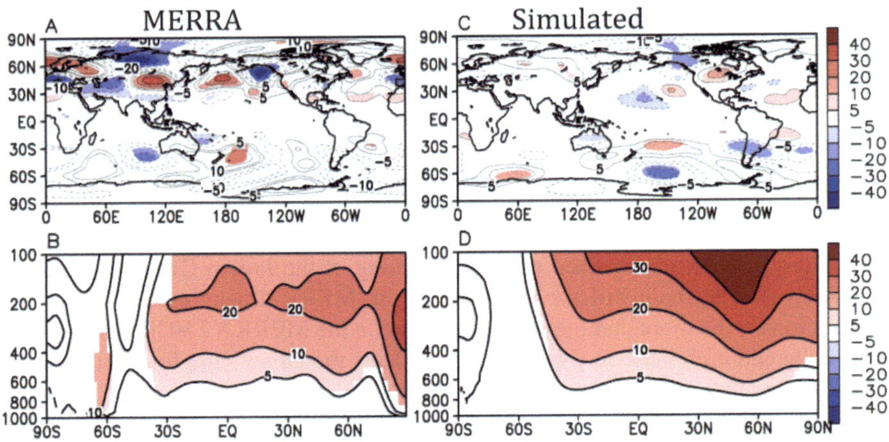

Fig. 9.13 The difference between the means of the two 15-year periods (1995–2009) and (1980–1994) for JAS. The *left panels* are for MERRA and the *right panels* are for the AMIP runs. *Top panels* are the 250 mb height differences with the zonal mean removed, while the *second row* shows the zonal mean height differences (only values significant at the 10% level are *shaded*, units are meters)

and a decrease in the number of cold days over much of the SH land masses with the largest changes occurring over northern South America, and Africa. Some of the major differences between MERRA and the simulations occur over central and southern South America, with MERRA showing an increase in the number of cold nights, while the simulations show a slight decrease.

Without observational estimates, the reliability of the simulated changes in the SH extremes is difficult to assess. While we do have the reanalysis estimates in the SH, their quality is in question, since these too suffer from insufficient observational constraints on the precipitation and surface temperature fields, as well as from potential model bias.

9.3 Impact of CO_2 Doubling and Uniform SST Increase

We next turn to the analysis of more idealized simulations in which the model is forced with uniform increases in SST or doubled CO_2. As already noted, these are not meant to provide realistic scenarios of future climate, but merely to assess sensitivities of the model climate and variability to such changes in forcing, and to help provide some insight into the changes that have already occurred. These runs of course do not have ENSO or any other anomalous SST forcing, nor do they have any variations in trace gases or aerosols. Here our assessment of changes in variability is limited to any changes in internal atmospheric/land variability.

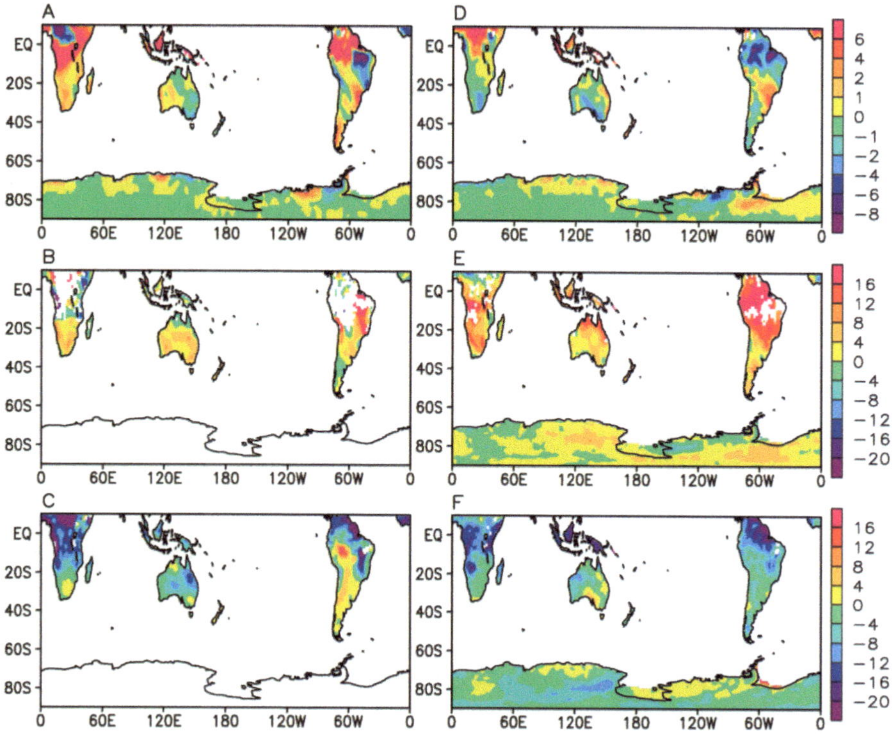

Fig. 9.14 The difference (JAS, 1995–2009 minus 1980–1994) in the 10-year return values of the number of heavy rain days (>5 mm/day), warm days (greater than the 90th percentile of the daily maximum temperatures) and cold nights (less than the 10th percentile of the daily minimum temperatures) for MERRA (*left panels***a–c**) and AMIP runs (*right panels***d–f**). Results are based on fits of the exceedances to a GPD

There are three different 20-year simulations consisting of a control run forced with a repeating annual cycle of the climatological (1981–2005 mean) SST, a run that is the same as the control but with doubled CO_2, and another run that is the same as the control but with a globally uniform 2°C increase in SST. This change is large in the sense that that it is considerably greater than what has occurred in the last three decades. In particular, we note that the actual SST changes over the last three decades are generally less than 0.5 °C and are far from uniform (see for example the T2m changes in Figs. 9.5 and 9.12; see also Deser et al. 2010). Also, the CO_2 increase over the last three decades is approximately 14%. Details of the experiments are provided in Appendix A. We begin by presenting the results for JFM. All model results are presented as differences from the control run.

9.3.1 Boreal Winter (JFM)

9.3.1.1 Impact on the Mean Climate and Weather Variability

Some care must be taken in interpreting AGCM simulations with doubled CO_2, since coupled IPCC model simulations show quite clearly that much of the global warming signal is realized as a warming of the oceans, and the atmosphere in turn responds to the associated changes in the SST. Nevertheless, it is instructive to try to isolate the fast response of the atmosphere and land to the change in radiative forcing, as discussed recently by I. Held[6] to see how much that might contribute to the continental warming.

The direct radiative impact (top left panels of Fig. 9.15) on the atmosphere of doubling CO_2 is a very modest warming of the troposphere with maximum zonal mean warming (between 0.2 and 0.5 °C) occurring at the tropopause (in the tropics and summer hemisphere) and the lower tropospheric Arctic, with cooling in the stratosphere (Fig. 9.15a). At the land surface, modest warming is largely confined to the northern middle latitudes (Fig. 9.16g). The tropospheric jets show a tendency to shift poleward (Fig. 9.15b). Changes in weather variability are small (order 10%), with most of the main storm track regions showing some reduction in storm track activity (Fig. 9.16e). The Hadley Cell shows a slight weakening (Fig. 9.15d), and the main relative humidity changes are a small reduction (<5%) in the tropics above 100mb (Fig. 9.15e). Precipitation changes are small and unorganized, with some overall preference for reduced precipitation (Fig. 9.16c).

The impact of a uniform 2°C increase in the SST of the world's oceans (Fig. 9.15f–j) consists of warming throughout the troposphere with maximum zonal mean warming of greater than 4°C occurring in the tropics just below the tropopause (Fig. 9.15f). Significant cooling is limited to the tropical and Arctic stratosphere. The relative humidity changes (Fig. 9.15j) are positive at the tropopause and in the tropical troposphere, with reduced values in the extratropical troposphere (poleward of about 30°lat) of both hemispheres. The zonal mean wind changes (Fig. 9.15g) show a poleward and upward shift of the westerly jets, with anomalous easterlies in the upper tropical troposphere and the high latitudes of the Northern Hemisphere. The upper level troposphere height differences (Fig. 9.16b) show reduced heights at high latitudes resembling an enhanced positive phase of the annular modes (NAM and SAM) in both hemispheres. Surface warming exceeds 1°C over much

[6] http://www.gfdl.noaa.gov/blog/isaac-held/2011/05/24/11-is-continental-warming-a-slave-to-warming-of-the-ocean-surface/more-1919

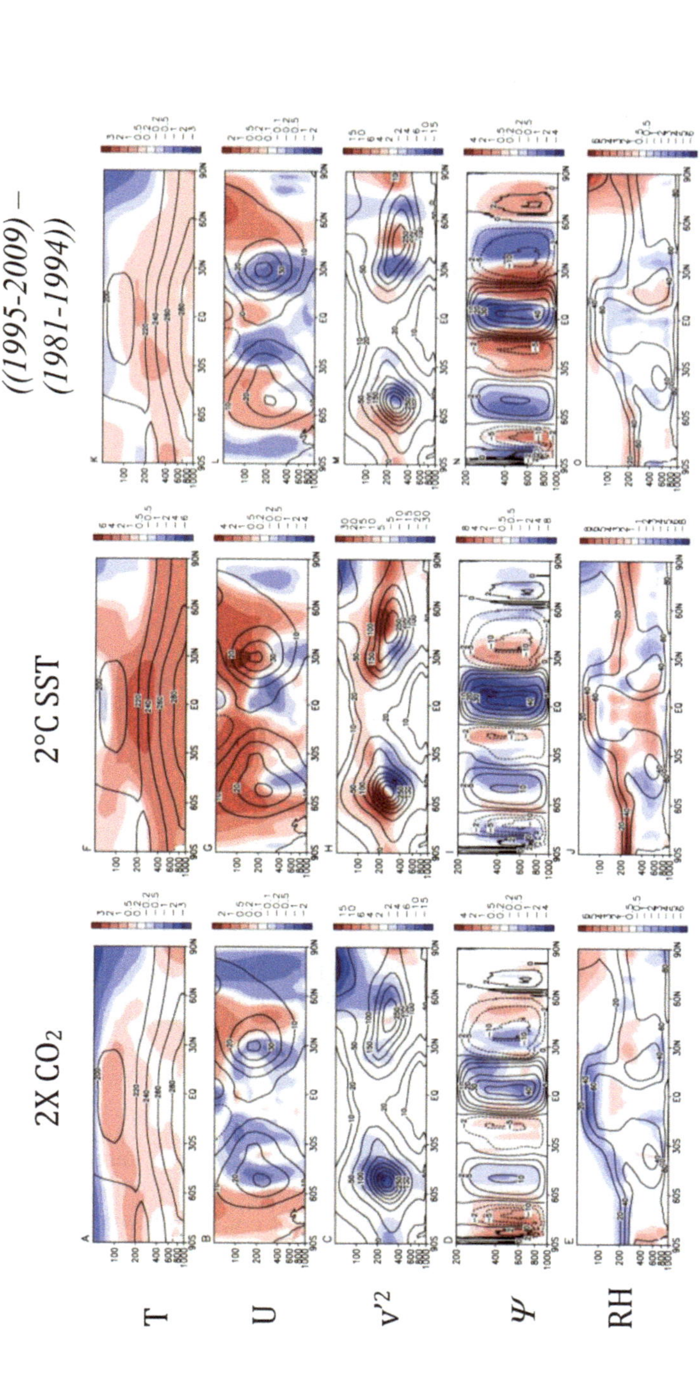

Fig. 9.15 JFM zonal mean differences (*shaded*) between experiment and control for temperature, zonal wind, variance in v (6hrly), meridional stream function (ΨX0.01), and relative humidity. *Left panels* (**a–e**): double CO_2; *middle panels* (**f–j**): global 2°C increase in SST. Contours are the mean control values. The *right panels* (**k–o**) show the ensemble mean difference ((1995–2009) – (1981–1994)) fields (*shaded*) from the three AMIP simulations. The contours are the mean of the period (1981–1994). Note the smaller contour intervals for the double CO_2 run and observed changes

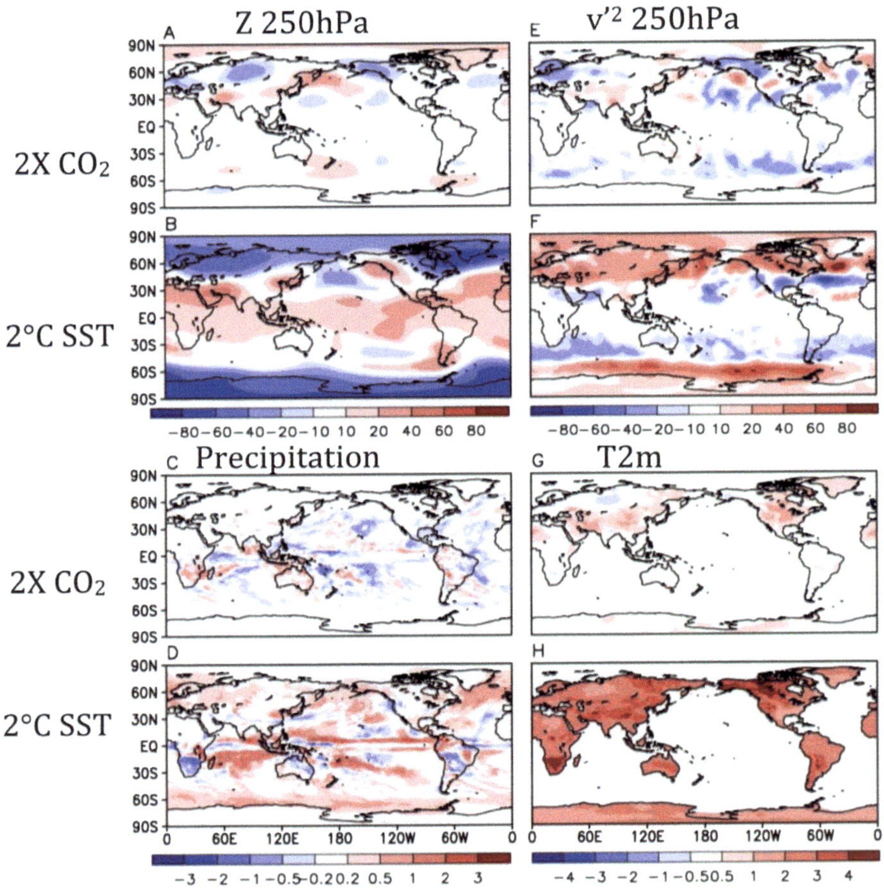

Fig. 9.16 JFM differences from the control run and the experiments with double CO_2, and 2°C SST increase. The *top left panels* (**a–b**) show the 250 hPa height (meters, with the global means removed: these are 10,532, 10,540, and 10,651 m for the control, double CO_2 and 2°C SST runs, respectively). The *top right panels* (**d–f**) show the 250mb v'^2 (m^2). The *bottom left panels* (**c–d**) show the precipitation (mm/day), and the *bottom right panels* (**g–h**) show the T2m (°C)

of the world's land areas, with some regions showing increases of more than 3°C (Fig. 9.16h). The strength of Hadley Cell is reduced (Fig. 9.15i). The strength of the adjacent Ferrel Cells is also reduced in a way that is consistent with a poleward expansion of the Hadley Cell.

Precipitation changes (Fig. 9.16d) are mostly positive with increases occurring over the SPCZ, the tropical/subtropical southern Indian Ocean, on the northern fringe of the Pacific ITCZ, and the high latitudes (generally poleward of 60°) especially in the North Atlantic. Some reduction in precipitation occurs over South

Africa, eastern Australia and central South America. The storminess (as measured by the variance of 200 hPa daily v-wind) shows a well-defined poleward shift in both hemispheres (Fig.9.16f).

The above results are compared (in Fig. 9.15k–o) with the changes that were simulated to have occurred in the last three decades during JFM. The results show a weakening and poleward expansion of the Hadley (Fig. 9.15n) that is consistent with observational studies (Fu et al. 2006). Such changes appear to be a robust response in coupled models to GHG forcing (Held and Soden 2006; Lu et al. 2007), and have been linked to an increased tendency for long-term drought in the subtropics (e.g., Seager et al. 2007). The expansion of the Hadley Cell is associated with a poleward shift (Fig. 9.15l) of the subtropical jets (e.g., Lu et al. 2007). The direct radiative impacts show similar features although these are likely too weak to contribute in a substantial way to the changes that have occurred in the last three decades. One thing of note is the substantial cooling in the NH stratosphere near the pole under CO_2 doubling (Fig. 9.15a), which may contribute to the observed changes in that region. Whether any of the observed continental interior warming is at all the result of the fast response to CO_2 changes on the land surface is not clear, though any such impact would again be small. Overall, the impacts of a CO_2 doubling presented here appear to be consistent with those presented by I. Held[8] based on the Geophysical Fluid Dynamics Laboratory (GFDL, *AM2.1/LM2.1)* AGCM. Evidence for the predominantly oceanic control of the recent world-wide warming over land can be found in Compo and Sardeshmukh (2009).

9.3.1.2 Impact on Climate Variability

In view of the considerable world-wide impact of the leading modes of variability, it is of interest to determine whether they are impacted (either as a change in amplitude or spatial structure) by changes in CO_2 or SST. Hu et al. (2011) addressed this issue in a comparison of IPCC scenario A1B and control runs with the CCSM3 model. They found that the leading seasonal mean modes of internal variability are largely unaffected by an increase in GHG and aerosol concentrations.

Our results are consistent with Hu et al. in that the leading modes of variability found in the AMIP runs and MERRA (in particular, the NAM, NAO and PNA) exist also in the control and other idealized fixed forcing runs. Of course the atmospheric ENSO mode does not appear since it is a response to SST anomalies. In fact, we have computed the REOFs separately for each experiment and found only minor differences in the patterns (differences that are likely well within the sampling errors). In order to assess any potential changes in variability, we have computed the REOFs from the combined set of 3 runs (first removing the means of the individual runs separately), to obtain a combined set of REOFs (Fig. 9.17a–d). This shows that the NAM, NAO and PNA are the three leading modes, with the fourth mode resembling the North Pacific Oscillation (NPO, also found in the AMIP runs

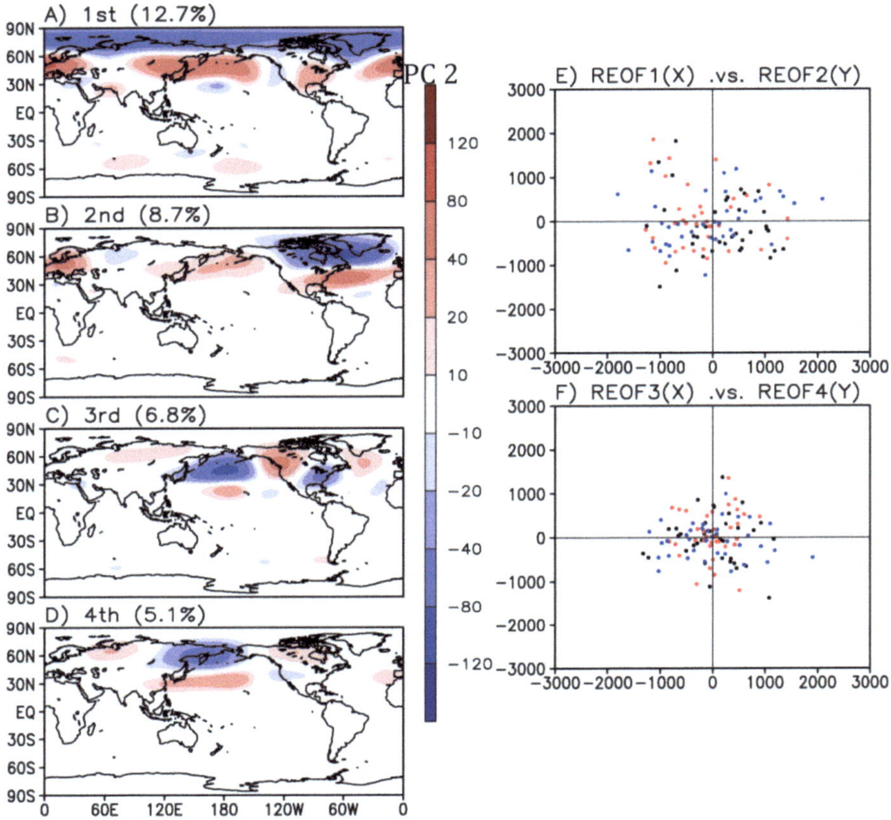

Fig. 9.17 *Left panels* (**a–d**): Leading monthly JFM 250 hPa height REOFs of the combined control, 2°C SST increase, and double CO_2 runs. The anomalies are computed with respect to the means of the individual runs. *Right panels* (**e–f**): Scatterplots of the leading REOFS (the principal components, PCs). *Black dots*: control run. *Red dots*: double CO_2 run. *Blue dots*: +2°C SST run

and MERRA as a higher mode – though not discussed in the previous sections). The NPO is well known as an important pattern of subseasonal variability (e.g., Linkin and Nigam 2008). Figure 9.17e, f show scatterplots of the amplitudes of the 4 leading PCs. These show no obvious differences in the variability in the various runs. Table 9.1, however, suggests that the variability of the NAM increased by about 2/3 and the variability of the NPO decreased by about 1/2, in the 2°C SST run. Also, it appears that the PNA variance decreased by about 1/2 in the doubled CO_2 run. It should however be emphasized that with just 20 years for each experiment the above results are only marginally significant at the 5% level (based on a F-test for the ratio of variances).

We next turn to the impacts on the JAS climate and variability.

Table 9.1 JFM variances (X 100,000) of the four leading REOFs in the idealized experiments

	Control	$2XCO_2$	SST + 2°C
MODE 1	3.00	3.05	**4.98**
MODE 2	2.77	2.93	1.99
MODE 3	1.96	**1.02**	3.10
MODE 4	1.84	1.72	**0.96**

Values in bold are significantly different from the control at 5%

9.3.2 Austral Winter (JAS)

9.3.2.1 Impact on Mean Climate and Weather Variability

The direct radiative impact on the atmosphere of doubling CO_2 during JAS (Fig.9.18a–e) is in many ways quite similar to the response during JFM, although with the hemispheres reversed. The impacts on the temperature (Fig. 9.18a) and zonal winds (Fig. 9.18b) are somewhat stronger than for JFM, with more extensive zonal mean warming in the Northern Hemisphere, greater cooling in the SH polar stratosphere, and larger changes (poleward and upward shift) in the zonal wind in the Southern Hemisphere. The Hadley cell changes (Fig. 9.18d) are small. The largest changes in RH (Fig. 9.18e) are negative in the tropical tropopause and positive in the Antarctic stratosphere (compared to the Arctic stratosphere for JFM). The 250 hPa height differences (Fig. 9.19a) are more coherent than for JFM with generally enhanced heights in the high latitudes (north of 45°N) of the NH, and reduced heights over the high latitudes (south of 60°S) of the SH. Modest surface warming (Fig. 9.19g) again occurs over much of the extratropical land area but it is more extensive over Russia. Changes in storminess (Fig. 9.19e) again show a general reduction especially in the NH middle latitudes.

The zonal mean impact of a uniform 2°C increase in the SST of the world's oceans (Fig. 9.18f–j), are quite similar to that already shown for JFM, including a general warming throughout the troposphere, an upward and equatorward expansion of the upper level westerlies (Fig. 9.18g), enhanced v'^2 (Fig. 9.18h) above the climatological maxima accompanied by a reduction equatorward and below the climatological maxima, a reduction in the strength of the Hadley Cell (Fig. 9.18i), and RH increases (Fig. 9.18j) near the tropopause. The spatial maps of the changes show an upper level height response (Fig. 9.19b) exhibiting reduced heights in both polar regions, and enhanced heights equatorward of about 30° latitude (again, having a zonally symmetric component that is reminiscent of the positive phase of the annular modes). Surface warming (Fig. 9.19h) is wide spread over all land areas with some of the largest warming occurring over the U.S. Great Lakes region. Precipitation changes (Fig. 9.19d) are generally positive in the tropical regions of the ITCZ and the Pacific warm pool extending westward across southern Asia.

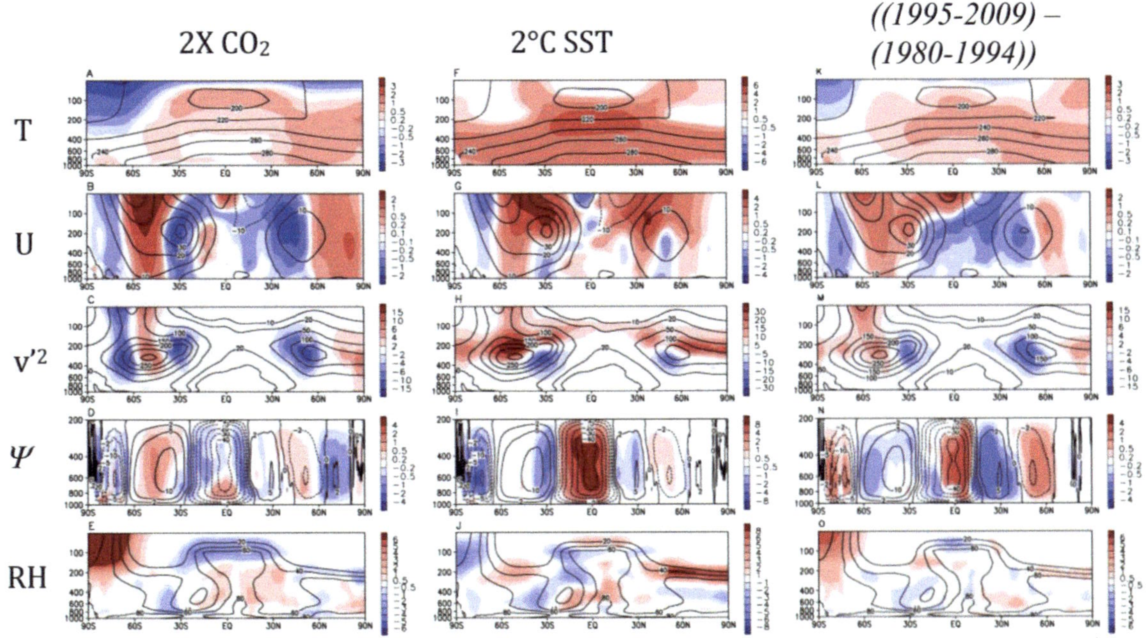

Fig. 9.18 JAS zonal mean differences (*shaded*) between experiment and control for temperature, zonal wind, variance in v (6hrly), meridional stream function(ΨX0.01), and relative humidity. *Left panels* (**a–e**): double CO_2; *middle panels* (**f–j**): global 2°C increase in SST. Contours are the mean control values. *The right panels* (**k–o**) show the ensemble mean difference ((1995–2009) – (1980–1994)) fields (*shaded*) from the three AMIP simulations. The contours are the mean of the period (1980–1994). Note the smaller contour intervals for the double CO_2 run and observed changes

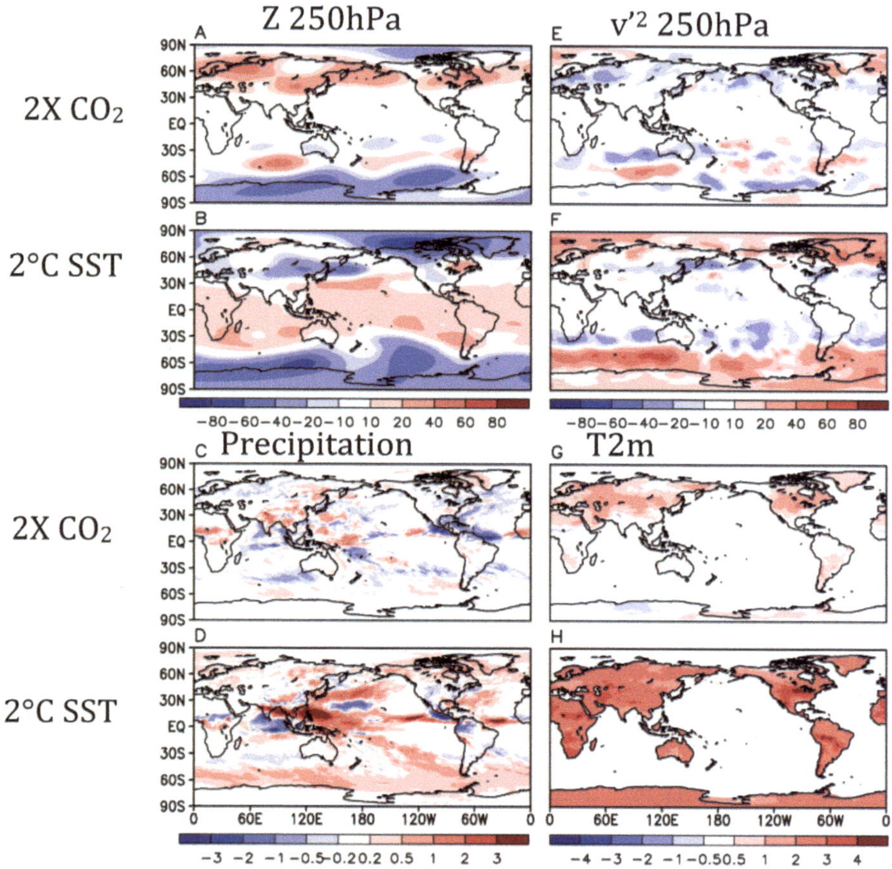

Fig. 9.19 JAS differences from the control run and the experiments with double CO_2, and 2°C SST increase. The *top left panels* (**a–b**) show the 250 hPa height (meters, with the global mean removed: these are 10,627, 10,639, and 10,748 m for the control, double CO_2 and 2°C SST runs, respectively). The *top right panels* (**e–f**) show the 250mb v'^2 (m^2). *The bottom left panels* (**c–d**) show the precipitation (mm/day), and the *bottom right panels* (**g–h**) show the T2m (°C)

The above results are compared (in Fig.9.18k–o) with the changes that were simulated to have occurred in the last three decades during JAS. The comparison indicates that the weakening of the Hadley Cell that has occurred in the last three decades during JAS (just as for JFM) is consistent with observational studies (Fu et al. 2006), and as already noted above, appears to be a robust response in coupled models to GHG forcing (Held and Soden 2006; Lu et al. 2007). Also, the expansion of the Hadley Cell is associated with a poleward shift of the subtropical jets as discussed in Lu et al. (2007). The asymmetrical warming (mostly occurring in the NH in the zonal mean) is, in contrast, more similar to what can be expected from

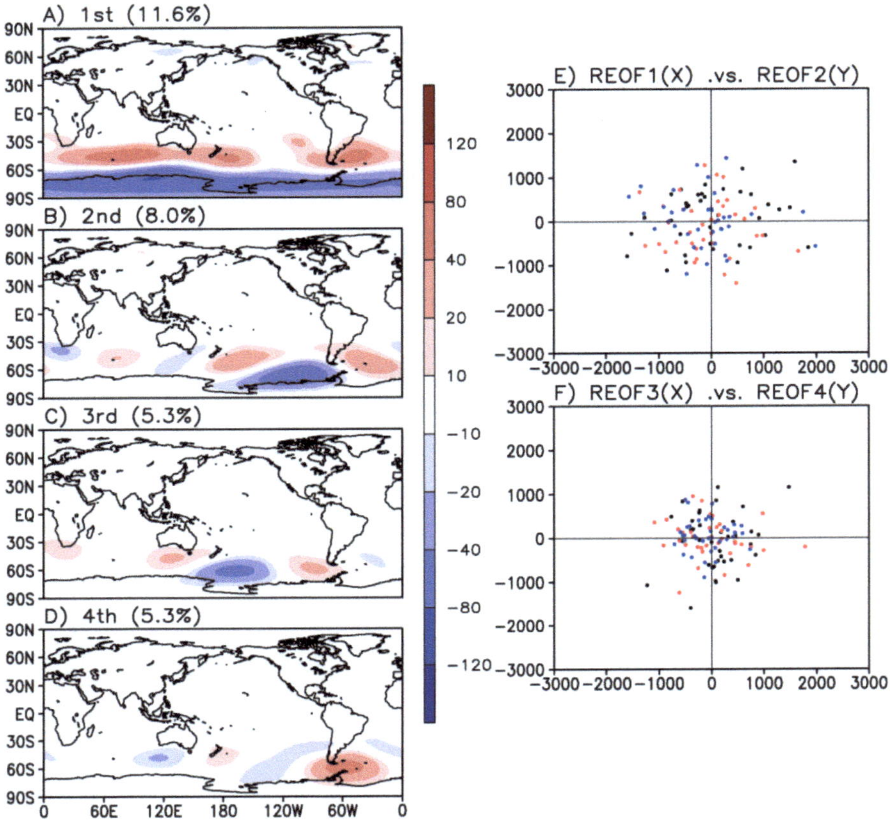

Fig. 9.20 *Left panels* (**a–d**): Leading monthly JAS 250 hPa height REOFs of the combined control, 2°C SST increase, and double CO_2 runs. The anomalies are computed with respect to the means of the individual runs. *Right panels* (**e–f**): Scatterplots of the leading REOFS (the principal components, PCs). *Black dots*: control run. *Red dots*: double CO_2 run. *Blue dots*: +2°C SST run

the direct radiative impacts of a doubling in CO_2. Also, the cooling in the SH polar stratosphere (Fig. 9.18k), the reduction in RH in the tropical tropopause (Fig. 9.18o) and the increase in the southern polar stratosphere are consistent with the direct radiative impacts of the increase in CO_2.

9.3.2.2 Impact on Climate Variability

The leading internal SH modes found in the AMIP runs and MERRA (Fig. 9.10) are largely unchanged in the control and other idealized forcing runs. This includes the SAM, and the two PSA modes (Fig. 9.20a–d). The trend mode does not appear as expected since there is no external time varying forcing. The scatterplots of the

Table 9.2 JAS variances (X 100,000) of the four leading REOFs in the idealized experiments

	Control	$2XCO_2$	SST + 2°C
MODE 1	4.27	**2.52**	3.22
MODE 2	2.39	2.39	2.09
MODE 3	1.60	2.06	**0.87**
MODE 4	2.52	**1.14**	**0.87**

Values in bold are significantly different from the control at the 5% level

PCs (Fig. 9.20e–f) indicate that there is little if any change in the variance of the internal models. There is some indication that the CO_2 increase appears to decrease the variance of the SAM by about 40%, while the SST warming acts to reduce the variability of PSA modes (Table 9.2).

9.4 Summary and Discussion

Weather extremes are inextricably linked to variability on a broad range of time scales. As such, understanding the causes of any changes in the characteristics of weather extremes requires an understanding of the connections between extremes and climate variability. This was addressed here in the framework of AMIP-style simulations in which the model was forced with observed SSTs, GHGs and ozone. In particular, we have reviewed the ability of the GEOS-5 AGCM, run at moderately high horizontal resolution (50 km), to simulate the major modes of monthly climate variability and weather extremes in the two winter hemispheres during the period 1980–2009. We also examined the nature of the longer-term (decadal) changes that occurred during this time period, including the changes in extremes. Idealized AGCM experiments were used to help guide the interpretation of the results.

Several questions were posed in the Introduction that we now attempt to answer. Specifically:

Does the GEOS-5 model reproduce the observed winter climatological fields?

In short, the results show quite clearly, that the GEOS-5 AGCM when forced by the observed SST and GHGs and ozone does reproduce the (1980–2009 mean) climatological fields quite well, including the stationary waves and precipitation fields in both winter hemispheres.

Does it have the correct low-frequency (monthly mean) modes of variability?

The GEOS-5 model appears to do a credible job in reproducing the observed variability, with several caveats. The model overestimates the monthly mean precipitation variability in the tropics. It also overestimates the upper tropospheric height variability in the North Pacific during JFM. The model does a remarkably good job in reproducing the leading patterns of monthly variability in both winter hemispheres. These include the ENSO response, the PNA, NAO, NAM, SAM and two Pacific South American (SPA) patterns. In addition, the model reproduces the key trend modes during both seasons. The main deficiency appears to be a too strong PNA variability (hence too much variability in the North Pacific) that appears to reflect a too strong sensitivity of the atmosphere to SST forcing in the central tropical Pacific.

Having multiple ensemble members also allowed an assessment of the degree to which the various modes are constrained by external forcing (SST, GHG and ozone). The results show a strong forcing for the trend and ENSO modes, while the lack of correlation between the individual ensemble members for the other modes indicates that these are largely internal to the atmosphere (unforced). This is confirmed with the idealized control simulation (having no interannual variations in external forcing) that reproduced the leading modes (PNA, NAO, NAM, SAM, and PSA). There is some evidence that the PNA is impacted/modulated by SST at longer time scales, and that appears to contribute to the tendency for a La Nina – like (negative PNA) response over the United States during the last three decades. Trends in other modes (for example those identified in previous studies of the NAO, NAM and SAM) emerge only in the context of longer records and/or during the warm season.

Does it have the correct weather variability and extremes?

The model reproduces the spatial structure and amplitude of the JFM and JAS winter storms tracks. Over land, the model also reproduces the basic patterns and amplitudes of precipitation extremes over the winter hemispheres, as reflected in the 10-year return values of the daily precipitation maximum and the warmest day and warmest night. The model does however tend to overestimate precipitation maxima at low latitudes especially during JAS, and suffers from a substantial day time cold bias in high latitude land areas during JFM. Efforts to validate the simulated daily extremes, are hampered by missing data, especially for the temperature record. The problem is especially severe in the SH where reasonably good coverage is largely confined to Australia.

Are the linkages between climate variability and regional weather simulated correctly?

The model does a remarkably good job of reproducing the predominant teleconnections associated with ENSO, the NAO, PNA, NAM, SAM and SPA modes. This includes the changes over the US associated with ENSO (e.g., enhanced storminess and precipitation along the US Gulf States), over Australia, New Zealand and Antarctica associated with the SAM, and over Eurasia associated with the NAM and NAO. It also includes warmer (colder) temperatures over northern Eurasia (the Mediterranean and North African regions) during a positive NAM, the increased storminess and warming over northern Europe and the warming over the US eastern seaboard during a positive NAO, and the warming over Alaska and western Canada during a positive PNA.

What are the longer-term changes that have occurred over the last three decades and are these reproduced in the model?

There is a very clear trend towards more positive upper tropospheric heights throughout the tropics and subtropics (extending into much of the middle latitudes) during the austral winter in both MERRA and the model simulations. During the boreal winter (JFM) the positive trend is again evident in the height field but it is intermingled with a strong ENSO signal in both the MERRA and simulations. The ENSO component of the trend appears to favor stronger warm events and contributes to an overall warming in the tropics and subtropics. As mentioned above, there is however imbedded in the warming trend, a La Nina-like (negative PNA) pattern. This is associated with a northward shift in the storm tracks and warmer temperatures over the United States. The relationship between these two distinctly different behaviors of a tendency towards both an El Nino like general atmospheric warming, and an embedded trend toward a La Nina like atmospheric response is unclear, but the latter appears to reflect the long-term response of the PNA (an intrinsically internal mode of monthly variability) to the cooling of the central and eastern tropical Pacific over the last three decades. One possibility is that this reflects the apparent shift from eastern to central Pacific ENSOs (e.g., Ashok et al. 2007), resulting in a greater sensitivity of the PNA to the cooler Pacific in recent years.

During JFM the model shows a distinct surface warming over northern Asia that, in the reanalysis, is a region of slight cooling over the last three decades. Other discrepancies between the reanalysis and the model during JFM occur in the polar regions, where the reanalysis shows a negative height change over the Southern Hemisphere and a positive change in the Northern Hemisphere, that are not reproduced in the simulations. The extent to which these represent true trends (that are not reproduced by the model) or whether they are simply statistical residuals of the variability associated with the SAM and the NAM is not clear. It is noteworthy that, despite these discrepancies, both the reanalysis and simulations show a zonally-symmetric poleward shift in the SH storm tracks during JFM.

During JAS, the model and reanalysis are consistent in showing warmer surface temperatures throughout North America, central Eurasia, Australia, and northern Africa. There is little consistency in the changes in the storm tracks, with the reanalysis showing little spatial coherence in the changes, while the model shows a poleward shift in the SH and a general reduction in storminess over the NH middle latitudes. Similarly, there is no agreement in the changes in the zonal mean jets. These results highlight the lack of dominant teleconnections and associated SST-forced changes in planetary waves during JAS, with the general continental warming presumably induced by the warmer oceans as discussed previously.

The observed and simulated changes in boreal winter extremes between the two periods reflect the continental warming and, over North America, the La Nina (negative PNA) response, including the northward shift of the storm tracks and the generally warmer conditions over the southern US. In particular, much of the southern United States shows a decrease in the number of days with heavy precipitation, while much of Europe and eastern Russia show an increase. Much of southern and central Asia, the Mediterranean region, and the southern US are characterized by an increase in the number of warm days. The major discrepancy between the simulations and observations is that the latter show a reduction in the number of warm days (and an increase in the number of cold days) in northern Russia while the simulations produce the opposite, although the substantial number of missing observations in the recent decade make the observational results unreliable in that region. The simulated austral winter temperature extremes show an overall increase in the number of warm days and a decrease in the number of cold days over much of the SH land masses with the largest changes occurring over northern South America, southern Africa and northern Australia. The simulated changes in the number of days with heavy precipitation is more complicated showing for example a reduction over the Amazon basin, and an increase over equatorial Africa. The reliability of the simulated changes in the SH extremes is however unclear, since the observationally-based estimates suffer from limited data coverage and uncertainties in the quality of the reanalysis data.

What can idealized AGCM experiments tell us about the recent, and possible future changes in climate variability and weather extremes?

During JFM, the warming of the world's oceans by 2°C leads to several key changes, including a poleward shift in the storm tracks, a poleward and upward shift in the middle latitude jets, a weakening and poleward expansion of the Hadley Cell, and a general warming over land that in some regions exceeds 3°C. The direct radiative impact on the atmosphere of a CO_2 doubling is weak, showing a maximum warming (>1°C) near the tropical and SH tropopause, a cooling above the tropopause, especially in the NH high latitudes, a weak poleward shift in the middle latitude jets (approximately 0.5 m/s change), and continental interior warming of up to 1°C in the NH. Comparisons with the simulated changes of the last three decades suggests that the expansion and weakening of the Hadley Cell is consistent with the impact of the SST warming, as is the observed SH poleward shift in the subtropical jet and storm tracks, and the continental warming.

During JAS the uniform SST increase also produces a poleward shift in storminess, an upward shift in the middle latitude jets, a weakening and poleward expansion of the Hadley Cell, and a general warming of the continents. The direct radiative impact on the atmosphere of a CO_2 doubling is again weak, showing the largest warming (>0.5°C) occurring near the tropical tropopause and in the NH middle and high latitude troposphere, together with an interior continental surface warming of up to 1°C in the NH. Cooling occurs in the stratosphere, especially in the SH high latitudes. A comparison with the simulated changes of the last three decades indicates that the weakening of the Hadley Cell is again (as during JFM) consistent with the response to the uniform SST increase. That is also true for the SH poleward shift of the subtropical jets and storm tracks, while the NH changes in the jet and storm tracks are less clearly linked to the SST changes. In fact, the asymmetrical warming (mostly occurring in the NH in the zonal mean) is similar in pattern to the direct radiative impact of increases in CO_2. Also, the reduction in RH in the tropical tropopause and the increase in the upper level south polar region is consistent with the direct radiative impacts of an increase in CO_2.

Concluding Remarks

The climate community is being challenged to provide increasingly more comprehensive societally-relevant information about the impacts of climate change that go well beyond broad statements about how much the global mean temperature will change. This in turn requires increasingly more comprehensive assessments of the quality of climate models to reproduce past regional climate impacts as well as the full spectrum of observed climate variability including those aspects (such as weather extremes) that are likely to have the greatest impact on society.

Our analysis shows that the wintertime climate and weather variability and trends of the last three decades are comprised of a rich assortment of internal (to the atmosphere) and externally forced (SST, GHGs, ozone) modes that have profound impacts on regional climates on monthly and longer time scales. Any assessment of long-term changes, including changes in extremes, must take these modes into account. In particular, we found that the long-term warming trends of the last three decades are intertwined with ENSO and the PNA. In addition, our more idealized runs highlight the important role of the oceans in warming the continents, as well as their controls on other key climate features such as the Hadley Cell, subtropical jets and storm tracks. Those runs also show that the spatial structure of the leading internal modes of variability is largely unchanged, even with substantial changes in external forcing.

Our assessment of the GEOS-5 model AMIP simulations indicates that the model is of sufficient quality to provide realistic assessments of the wintertime impacts of long-term changes in SST and GHGs on regional climates and weather extremes. Gaps in the observational coverage and the uncertain quality of the reanalysis data, however, hinder our ability to validate the model results over much of the SH and parts of Asia, especially for assessing changes in the extremes.

We did not include the warm and transition seasons in our analysis because we believe current climate models do not provide sufficiently realistic representations of such basic features as warm season continental precipitation, including the diurnal cycle and meso-scale convective systems that are critical to achieving realistic summer weather extremes in middle latitudes. Nor do they adequately resolve the most intense tropical storms that are in some regions the most important extreme events in terms of their impacts on society. In part, this is a resolution issue, although some necessary model improvements will likely still require better parameterizations, at least for the resolutions that are expected to be feasible for climate models during the next decade.

A.1 Appendices

A.1.1 Appendix A

A.1.1.1 The GEOS-5 Model and Experiments

The NASA Goddard Earth Observing System (GEOS-5) Atmospheric General Circulation Model or AGCM (Rienecker et al. 2008) employs the finite-volume dynamics of Lin (2004). This dynamical core is integrated with various physics

Table A.1 List of experiments

Experiments	Time period	Initial conditions	SST and sea ice	Radiative gases, ozone, aerosols
AMIP	1980–2009	MERRA: Jan 2, 9, and 16th of 1980	Observed (HadISST)	"Observed"– see text
Control	20 years	MERRA: 1980 Jan 2	Repeating mean annual cycle from HadISST (1981–2005)	Gases are constant IPCC 1992 values (Table A.2); ozone is climatological (1981–2005); aerosols are 2002 values
Global warming	20 years	Same as control	Same as control plus uniform 2°C increase over the global oceans	Same as control
Double CO_2	20 years	Same as control	Same as control	Same as control, but double CO_2

packages (Bacmeister et al. 2006) under the Earth System Modeling Framework (Collins et al. 2005) including the Catchment Land Surface Model (Koster et al. 2000), and a modified form of the Relaxed Arakawa-Schubert convection scheme described by Moorthi and Suarez (1992). For the experiments described here we used version 2.4 of the AGCM. The model was run with 72 hybrid-sigma vertical levels extending to 0.01 hPa, and ½° (about 50 km) horizontal resolution on a latitude/longitude grid. Atmospheric variables and surface variables were output every 6 h, and for some variables (in particular those with a strong diurnal cycle) every 3 h.

In addition to the atmospheric model inter-comparison (AMIP) style runs forced with observed SST, several other more idealized experiments were run with a repeating climatological annual cycle in the SST and sea ice. These include a control, a run with a uniform 2°C increase in the global SST, and a run with doubled CO_2 (Table A.1)

For the AMIP runs, CO_2 consists of the time varying annual global mean values provided by IPCC/CMIP5. The other greenhouse gases (GHGs: CH4, N2O, CFC-11, CFC-12, and HCFC-22), stratospheric water vapor (H2O), and ozone (O_3) are relaxed to time varying zonal averages with a 2-day e-folding time. The zonal averages of the GHGs are taken from simulations of 1950–2010 with the GEOS chemistry climate model (CCM, Pawson et al. 2008), and are calibrated (bias corrected) to the tropospheric concentrations specified by CMIP5 (Meinshausen et al. 2011). Stratospheric H_2O is also taken from the CCM. In both cases, GHGs and H_2O, 5-year running averages are first computed to reduce the influence of interannual variability in the CCM fields. Ozone is specified from AC&C/SPARC

Table A.2 Radiative gases from 1992 IPCC specifications

Radiative gas	Amount
CO_2	3.56E-04
CH_4	1.714E-06
N_2O	3.11E-07
CFC_{11}	2.68E-10
CFC_{12}	5.03E-10
CFC_{113}	8.20E-11
CFC_{22}	1.05E-10

monthly averages (ftp-esg.ucllnl.org) from 1870 to 2005, and is converted to zonal means before interpolation onto GEOS-5 layers. For all seven gases, the relaxation fields have realistic latitudinal, vertical, and seasonal variations imposed on their specified trends. Two-day e-folding times allow the species contours to sufficiently follow planetary-scale potential vorticity deformations in the stratosphere.

Aerosols are computed using the Goddard Chemistry, Aerosol, Radiation, and Transport model (GOCART, Chin et al. 2002; Colarco et al. 2009) in GEOS-5. The GOCART module is run online within the GEOS-5 AGCM; that is, the aerosols and other tracers are radiatively interactive and transported consistently with the underlying hydrodynamics and physical parameterizations (e.g., moist convection and turbulent mixing) of the model. GOCART treats the sources, sinks, and chemistry of dust, sulfate, sea salt, and black and organic carbon aerosols. Aerosol species are assumed to be external mixtures. Total mass of sulfate and hydrophobic and hydrophilic modes of carbonaceous aerosols are tracked, while for dust and sea salt the particle size distribution is explicitly resolved across five non-interacting size bins for each.

Both dust and sea salt formulations have wind-speed dependent emission functions, while sulfate and carbonaceous species have emissions principally from fossil fuel combustion, biomass burning, and biofuel consumption, with additional biogenic sources of organic carbon. Sulfate has additional chemical production from oxidation of SO2 and DMS, and we include a database of volcanic SO2 emissions and injection heights. For all aerosol species, optical properties are primarily from the commonly used OPAC data set (Hess et al. 1998). This framework also includes the representation of CO tracers, which have emissions from fossil fuel, biofuel, and biomass burning. The online CO processes in GEOS-5 derive from Bian et al. (2007), and include indirect production of CO from oxidation of natural and anthropogenic non-methane hydrocarbons, chemical production from methane (CH4) oxidation, and losses through reaction with OH.

A.1.1.2 MERRA and Other Observations

Our analysis is based in part on MERRA (Rienecker et al. 2011). MERRA is an atmospheric reanalysis that was produced with the Goddard Earth Observing

System Data Assimilation System Version 5 (GEOS-5) documented in Rienecker et al. (2008), consisting of the GEOS-5 atmospheric model and the Grid-point Statistical Interpolation (GSI) analysis system, the latter being a system jointly developed by the GMAO and NOAA's National Centers for Environmental Prediction. The GEOS-5 assimilation system includes an incremental analysis update (IAU) procedure (Bloom et al. 1996) that slowly adjusts the model states toward the observed state. This has the benefit of minimizing any unrealistic spin down (or spin-up) of the water cycle. MERRA was run at a resolution of ½° latitude × 2/3° longitude with 72-levels extending to 0.01 hPa. More information about MERRA can be found at: http://gmao.gsfc.nasa.gov/research/merra/. This study uses standard monthly mean (JJA) and hourly output that is provided on 42 pressure levels at a horizontal resolution of 1° latitude × 1.25° longitude for the period 1979–2010. A more limited comparison is made with the NOAA Climate Forecast System Reanalysis (CFSR, Saha et al. 2010).

We also make use of various observations consisting of the daily and monthly mean Global Precipitation Climatology Project (GPCP) precipitation data (Adler et al. 2003), and the HadGHCN daily and monthly temperature data (Caesar et al. 2006). Some of the results are also based on daily NOAA Climate Prediction Center precipitation data (Xie et al. 2007).

A.1.2 Appendix B

A.1.2.1 Some Examples of Fits to the GEV Distribution

In order to estimate the 10-year return values of the maximum daily precipitation and the warmest days and warmest nights, we fit the maxima to the Generalized Extreme Value (GEV, Coles 2001) distribution:

$$\frac{1}{\sigma} t(x)^{\xi+1} e^{-t(x)},$$

where

$$t(x) = \begin{cases} \left(1 + \xi \frac{x-\mu}{\sigma}\right)^{-1/\xi} & \text{if } \xi \neq 0 \\ e^{-(x-\mu)/\sigma} & \text{if } \xi = 0 \end{cases}$$

and ζ, μ and σ represent that shape, location and scale of the distribution respectively. Examples of the fits are presented in Fig. B.1.

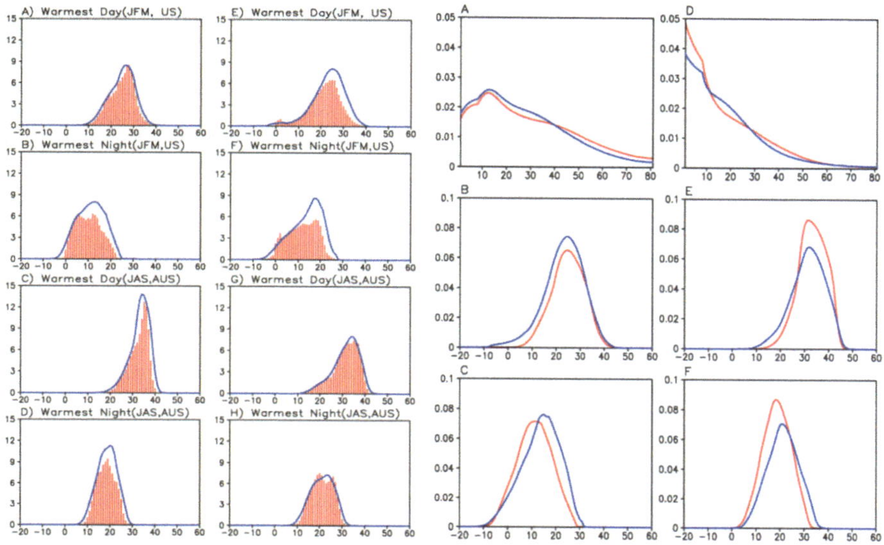

Fig. B.1 *Left set of panels* (**a–h**): Spatially averaged histograms with superimposed GEV fits from observations (*left*) and GEOS5-AMIP runs (*right*). *Top*: warmest days and nights temperature occurrences in percentage averaged over the US during JFM (*red bars*). Corresponding GEV fits are denoted by *blue curves*. *Bottom*: same but over Australia for JAS. *X-axis* denotes the actual temperature values whereas the *Y-axis* denotes the percentage values. *Right set of panels* (**a–f**): Direct comparisons of the spatially averaged GEV fits to the observations (*red*) and AMIP simulations (*blue*) for maximum daily precipitation (*top panels*), warmest day (*middle panels*), and warmest nights (*lower panels*). Spatial averaging was performed for the US (130W-70W, 25N-45N) for JFM on the *left*, and for Australia (100E-160E, 50S-10S) for JAS on the *right*

References

Adler RF, Huffman GJ, Chang A, Ferraro R, Xie P, Janowiak J, Rudolf B, Schneider U, Curtis S, Bolvin D, Gruber A, Susskind J, Arkin P (2003) The Version 2 Global Precipitation Climatology Project (GPCP) monthly precipitation analysis (1979-present). J Hydrometeorol 4:1147–1167

Ashok K, Behera SK, Rao SA, Weng H, Yamagata T (2007) El Nin˜o Modoki and its possible teleconnection. J Geophys Res 112:C11007. doi:10.1029/2006JC003798

Bacmeister JT, Suarez MJ, Robertson FR (2006) Rain re-evaporation, boundary-layer/convection interactions and Pacific rainfall patterns in an AGCM. J Atmos Sci 8, SRef-ID: 1607-7962/gra/EGU06-A-08925

Bian H, Chin M, Kawa R, Duncan B, Arellano A Jr, Kasibhatla R (2007) Uncertainty of global CO simulations constraint by biomass burning emissions. J Geophys Res 112:D23308. doi:10.1029/2006JD008376

Bloom S, Takacs L, DaSilva A, Ledvina D (1996) Data assimilation using incremental analysis updates. Mon Weather Rev 124:1256–1271

Bojariu R, Gimeno L (2003) The role of snow cover fluctuations in multiannual NAO persistence. Geophys Res Lett 30(4). doi:10.1029/2002GL015651. issn: 0094–8276

Caesar J, Alexander L, Vose R (2006) Large-scale changes in observed daily maximum and minimum temperatures: creation and analysis of a new gridded data set. J Geophys Res 111:D05101. doi:10.1029/2005JD006280

CCSP SAP 1.3, Schubert S, Arkin P, Carton J, Kalnay E, Koster R (2008) Reanalysis of historical climate data for key atmospheric features. In: Dole R, Hoerling M, Schubert M (eds) Reanalysis of historical climate data for key atmospheric features: implications for attribution of causes of observed change. A Report by the U.S. Climate Change Science Program and the Subcommittee on Global Change Research. National Oceanic and Atmospheric Administration, National Climatic Data Center, Asheville, NC, p 11–46

CCSP SAP 3.3 (2008) Weather and climate extremes in a changing climate. Regions of focus: North America, Hawaii, Caribbean, and U.S. Pacific Islands. In: Karl TR, Meehl GA, Miller CD, Hassol SJ, Waple AM, Murray WL (eds) A report by the U.S. Climate Change Science Program and the Subcommittee on Global Change Research. Department of Commerce, NOAA's National Climatic Data Center, Washington, D.C, 164 pp

Chang EKM, Fu Y (2002) Interdecadal variations in Northern Hemisphere winter storm track intensity. J Climate 15:642–658

Chang Y, Schubert S, Suarez M (2012) Attribution of the extreme U.S. east coast snowstorm activity of 2010. J Climate 25(11):3771–3791

Chin M, Ginoux P, Kinne S, Torres O, Holben BN, Duncan BN, Martin RV, Logan JA, Higurashi A, Nakajima T (2002) Tropospheric aerosol optical thickness from the GOCART model and comparisons with satellite and sun photometer measurements. J Atmos Sci 59:461–483

Colarco P, da Silva A, Chin M, Diehl T (2009) Online simulations of global aerosol distributions in the NASA GEOS-4 model and comparisons to satellite and ground-based aerosol optical depth. J Geophys Res 115:D14207. doi:10.1029/2009JD012820

Coles S (2001) An introduction to statistical modeling of extreme values. Springer-Verlag Heidelberg, 208 pp

Collins N, Theurich G, DeLuca C, Suarez M, Trayanov A, Balaji V, Li P, Yang W, Hill C, da Silva A (2005) Design and implementation of components in the earth system modeling framework. Int J High Perform Comput Appl 19:341–350. doi:10.1177/1094342005056120

Compo G, Sardeshmukh PD (2009) Oceanic influences on recent continental warming. Clim Dyn 32:333–342. doi:10.1007/s00382-008-0448-9

Deser C, Phillips AS, Alexander MA (2010) Twentieth century tropical sea surface temperature trends revisited. Geophys Res Lett 37:L10701. doi:10.1029/2010GL043321

Ding Q, Steig EJ, Battisti DS, Wallace JM (2011) Influence of the tropics on the Southern Annular Mode. Submitted to J Clim

Ding Q, Wang B, Wallace JM, Branstator G (2011) Tropical-extratropical teleconnections in boreal summer: observed interannual variability. J Climate 24:1878–1896. doi:http://dx.doi.org/10.1175/2011JCLI3621.1

Fraedrich K, Muller K (1992) Climate anomalies in Europe associated with ENSO. Int J Climatol 12:25–31

Fu Q, Johanson CM, Wallace JM, Reichler T (2006) Enhanced mid-latitude tropospheric warming in satellite measurements. Science 312:1179

Gates WL (1992) AMIP: The Atmospheric Model Intercomparison Project. Bull Am Meteorol Soc 73:1962–1970

Gershunov A, Barnett T (1998) ENSO influence on intraseasonal extreme rainfall and temperature frequencies in the contiguous US: observations and model results. J Climate 11:1575–1586

Ghil M, Mo KC (1991) Intraseasonal oscillations in the global atmosphere. Part II: Southern Hemisphere. J Atmos Sci 48:480–490

Gillett NP, Zwiers FW, Weaver AJ, Stott PA (2003) Detection of human influence on sea-level pressure. Nature 422(6929):292–294

Gong G, Entekhabi D, Cohen J (2003) Modeled Northern Hemisphere winter climate response to realistic Siberian snow anomalies. J Climate 16:3917–3931

Held IM, Soden BJ (2006) Robust responses of the hydrological cycle to global warming. J Climate 19:5686–5699

Hendon HH, Wheeler MC, Zhang C (2007) Seasonal dependence of the MJO–ENSO relationship. J Climate 20:531–543

Hess M, Koepke P, Schult I (1998) Optical properties of aerosols and clouds: the software package OPAC. Bull Am Meteorol Soc 79:831–844

Hoerling M, Eischeid J, Perlwitz J (2010) Regional precipitation trends: distinguishing natural variability from anthropogenic forcing. J Climate 23:2131–2145

Hu Z-Z, Kumar A, Jha B, Huang B (2011) An analysis of forced and internal variability in a warmer climate in CCSM3. J Climate 25(7):2356–2373

Hurrell JW (1995) Decadal trends in the North Atlantic Oscillation: regional temperatures and precipitation. Science 269:676–679

Hurrell JW (1996) Influence of variations in extratropical wintertime teleconnections on Northern Hemisphere temperature. Geophys Res Lett 23:655–668

Hurrell JW, Deser C (2009) North Atlantic climate variability: the role of the North Atlantic Oscillation. J Mar Syst 78(1):28–41. doi:10.1016/j.jmarsys.2008.11.026

Hurrell JW, Kushnir Y, Ottersen G, Visbeck M (2003a) The North Atlantic Oscillation: climatic significance and environmental impact, Geophysical monograph series. American Geophysical Union, Washington, DC

Hurrell JW et al (2003b) An overview of the North Atlantic Oscillation. In: Hurrell JW et al (eds) The North Atlantic Oscillation: climatic significance and environmental impact, Geophysical monograph, 134. American Geophysical Union, Washington, DC, pp 1–35

Hurrell JW et al (2004) Twentieth century North Atlantic climate change. Pt I: assessing determinism. Clim Dyn 23:371–389

IPCC (1992) Climate change 1992: the supplementary report to the IPCC scientific assessment. In: Houghton JT, Callander BA, Varney SK (eds). Cambridge: Cambridge University Press, 205 pp

IPCC (2007) Climate change 2007: the physical report of the Intergovernmental Panel on science basis. Contribution of Working Group I to the fourth assessment Climate Change. In: Solomon S, Qin D, Manning M, Chen Z, Marquis M, Averyt KB, Tignor M, Miller HL (eds). Cambridge University Press, Cambridge, UK/New York, 996 pp

Jones PD, Osborn TJ, Briffa KR (2003) Pressure-Based measures of the North Atlantic Oscillation (NAO): a comparison and an assessment of changes in the strength of the NAO and in its influence on surface climate parameters. Geophys Monogr 134:51–62

Kao H-Y, Yu J-Y (2009) Contrasting Eastern-Pacific and Central-Pacific types of ENSO. J Climate 22:615–632

Karoly DJ (1989) Southern Hemisphere circulation features associated with El Nin˜o–Southern Oscillation events. J Climate 2:1239–1251

Kenyon J, Hegerl GC (2008) The influence of ENSO, NAO and NPI on global temperature extremes. J Climate 21:3872–3889. doi:10.1175/2008JCLI2125

Kidson JW (1988) Interannual variations in the Southern Hemisphere circulation. J Climate 1:1177–1198

Kidson J, Renwick JA, McGregor J (2009) Hemispheric-scale seasonality of the Southern Annular Mode and impacts on the climate of New Zealand. J Climate 22:4759–4770

Koster RD, Suárez MJ, Ducharne A, Stieglitz M, Kumar P (2000) A catchment-based approach to modeling land surface processes in a GCM, part 1, model structure. J Geophys Res 105: 24809–24822

Kumar A, Jha B, L'Heureux M (2010) Are tropical SST trends changing the global teleconnection during La Niña? Geophys Res Lett 37:L12702. doi:10.1029/2010GL043394, 6pp

Lefebvre W, Goosse H, Timmermann R, Fichefet T (2004) Influence of the Southern Annular Mode on the sea ice–ocean system. J Geophys Res 109:C09005. doi:10.1029/2004JC002403

Lim Y-K, Schubert SD (2011) The impact of ENSO and the Arctic oscillation on winter temperature extremes in the Southeast United States. Geophys Res Lett 38:L15706. doi:10.1029/2011GL048283, 5 pp

Lin S-J (2004) A vertically Lagrangian finite-volume dynamical core for global models. Mon Weather Rev 132:2293–2307

Linkin ME, Nigam S (2008) The North Pacific Oscillation–West Pacific Teleconnection pattern: mature-phase structure and winter impacts. J Climate 21:1979–1997
Lu J, Vecchi GA, Reichler T (2007) Expansion of the Hadley cell under global warming. Geophys Res Lett 34:L06805. doi:10.1029/2006GL028443
Mariotti A, Dell'Aquila A (2011) Decadal climate variability in the Mediterranean region: roles of large-scale forcings and regional processes. Clim Dyn 38(5–6):1129–1145
Marshall GJ (2003) Trends in the Southern Annular Mode from observations and reanalyses. J Climate 16:4134–4143
Meinshausen M, Smith SJ, Calvin KV, Daniel JS, Kainuma MLT, Lamarque J-F, Matsumoto K, Montzka SA, Raper SCB, Riahi K, Thomson AM, Velders GJM, van Vuuren D (2011) The RCP greenhouse Gas concentrations and their extension from 1765 to 2300. Clim Change (Special Issue). doi:10.1007/s10584-011-0156-z
Mo K (2000) Relationships between low-frequency variability in the Southern Hemisphere and sea surface temperature anomalies. J Climate 13:3599–3610
Moorthi S, Suarez MJ (1992) Relaxed Arakawa-Schubert, a parameterization of moist convection for general-circulation models. Mon Weather Rev 120:978–1002
Newman M, Shin S-I, Alexander MA (2011) Natural variation in ENSO flavors. Geophys Res Lett 38:L14705. doi:10.1029/2011GL047658
Pawson S, Stolarski RS, Douglass AR, Newman PA, Nielsen JE, Frith SM, Gupta ML (2008) Goddard Earth Observing System chemistry-climate model simulations of stratospheric ozone-temperature coupling between 1950 and 2005. J Geophys Res 113:D12103. doi:10.1029/2007JD009511
Quadrelli R, Wallace JM (2004) A simplified linear framework for interpreting patterns of Northern Hemisphere wintertime climate variability. J Climate 17:3728–3744
Raible CC, Stocker TF, Yoshimori M, Renold M, Beyerle U, Casty C, Luterbacher J (2005) Northern Hemispheric trends of pressure indices and atmospheric circulation patterns in observations, reconstructions, and coupled GCM simulations. J Climate 18:3968–3982
Reason CJC, Rouault M (2005) Links between the Antarctic Oscillation and winter rainfall over western South Africa. Geophys Res Lett 32:L07705. doi:10.1029/2005GL022419
Richman MB (1986) Rotation of principal components. J Climatol 6:293–335
Rienecker MM, Suarez MJ, Todling R, Bacmeister J, Takacs L, Liu H.-C, Gu W, Sienkiewicz M, Koster RD, Gelaro R, Stajner I, Nielsen JE (2008) The GEOS-5 data assimilation system – documentation of versions 5.0.1 and 5.1.0, and 5.2.0. In: NASA technical report series on global modeling and data assimilation, NASA/TM-2008-104606, vol 27, 92 pp
Rienecker MM et al (2011) MERRA – NASA's modern-era retrospective analysis for research and applications. J Climate. doi:10.1175/JCLI-D-11-00015.1
Saha S et al (2010) The NCEP climate forecast system reanalysis. Bull Am Meteorol Soc 9: 1015–1057
Scaife AA, Folland CK, Alexander LV, Moberg A, Knight JR (2008) European climate extremes and the North Atlantic Oscillation. J Climate 21:72–83
Schubert SD, Chang Y, Suarez MJ, Pegion PJ (2008) ENSO and wintertime extreme precipitation events over the contiguous United States. J Climate 21:22–39
Seager R, Ting MF, Held IM, Kushnir Y, Lu J, Vecchi G, Huang H-P, Harnik N, Leetmaa A, Lau N-C, Li C, Velez J, Naik N (2007) Model projections of an imminent transition to a more arid climate in Southwestern North America. Science 316(5828):1181–1184. doi:10.1126/science.1139601
Silvestri GE, Vera CS (2003) Antarctic Oscillation signal on precipitation anomalies over southeastern South America. Geophys Res Lett 30:2115. doi:10.1029/2003GL018277
Thompson DWJ, Solomon S (2002) Interpretation of recent Southern Hemisphere climate change. Science 296:895–899
Thompson DWJ, Wallace JM (2000) Annular modes in the extratropical circulation. Part I: month-to-month variability. J Climate 13:1000–1016
Thompson DWJ, Wallace JM (2001) Regional climate impacts of the Northern Hemisphere annular mode. Science 293:85–89

Thompson DWJ, Wallace JM, Hegerl GC (2000) Annular modes in the extratropical circulation: part II: trends. J Climate 13:1018–1036

Trenberth KE, Stepaniak DP, Smith L (2005) Interannual variability of the patterns of atmospheric mass distribution. J Climate 18:2812–2825

Turner J (2004) The El Niño–Southern Oscillation and Antarctica. Int J Climatol 24:1–31

Xie P, Yatagai A, Chen M, Hayasaka T, Fukushima Y, Liu C, Yang S (2007) A gauge-based analysis of daily precipitation over East Asia. J Hydrometeorol 8:607–626

Yeh S-W et al (2009) El Niño in a changing climate. Nature 461:511–514

Chapter 10
Uncertainties in Observed Changes in Climate Extremes

Kenneth E. Kunkel

Abstract Some of the sources of uncertainty in evaluating observed changes in climate extremes include the highly non-linear, chaotic behavior of the dynamical climate system; inhomogeneities in the climate station records, and incomplete sampling of the climate system. There is no known theoretical basis for characterizing uncertainties arising from the chaotic nature of the climate system. Ensemble simulations as well as long control simulations from climate models offer an opportunity to characterize this source of uncertainty, realizing that climate models introduce their own layer of uncertainty. Prominent examples of inhomogeneities in climate station records include changes in instrumentation, station moves, changes in measurement methodology, and changes in time of observation. Examples of the potential effects of an instrument change in the U.S. cooperative observer network and a shift in the predominant time of observation are provided. Sampling uncertainties are greater for precipitation than temperature extremes, because precipitation extremes are smaller in scale. A Monte Carlo approach to quantify sampling uncertainty is described, based on station distribution in the U.S. cooperative observer network. Also, the use of statistical tools (such as Kendall's tau, generalized extreme value theory, and bootstrap resampling) and climate model ensemble experiments are discussed.

K.E. Kunkel (✉)
Department of Marine, Earth, and Atmospheric Sciences, North Carolina State University, Asheville, NC, USA

Cooperative Institute for Climate and Satellites, National Climatic Data Center, National Oceanic and Atmospheric Administration (NOAA), Asheville, NC, USA
e-mail: Ken.Kunkel@noaa.gov

A. AghaKouchak et al. (eds.), *Extremes in a Changing Climate*, Water Science and Technology Library 65, DOI 10.1007/978-94-007-4479-0_10,

10.1 Overview of Fundamental Issues Underlying Uncertainty

A fundamental motivation for efforts to detect changes in extremes is to provide information about the future. If there has been a real change in the characteristics of the climate that have affected extreme events, then there is a basis for anticipating characteristics in the future, perhaps as simple as a maintenance of recent levels or a continuation of recent trends. Thus, the central underlying question is whether there has been a real change in the climate system. Answering this question can be difficult because of the following considerations.

The physical laws governing the behavior of the climate system are highly non-linear. The result is a dynamical system that is chaotic, that is, the temporal evolution is highly sensitive to the initial state of the system. Extremes arise naturally out of this chaotic system. Furthermore, there is no known fundamental constraint on the frequency, intensity, or location of such extremes. It is at least conceivable that changes that appear to be real are simply a reflection of this internal chaotic dynamically-driven variability and these apparent changes carry no predictive power for the future state of the system. Thus, even if there is high confidence that a trend has been observed, this may not meet the ultimate objective of our analysis.

Since extremes are by definition rare, normal statistics that rely on a large number of samples simply do not apply. Without *a priori* knowledge of the statistical characteristics of extreme events, there is uncertainty whether the tools we apply are relevant to the specific type of event being studied. Extreme value theory may be applicable, but it has not been determined how the governing laws of the climate system lead to a particular functional distribution for extreme events. Any such relationships are purely empirical at this juncture. As such, their predictive power is unknown.

Limitations in the observational network add another layer of uncertainty. There may be changes in the network over time, such as in the spatial density and location of observing stations and in the methods of measurement, to name two examples. Real changes in the climate system may be masked, partially or wholly, by such changes which can create artifacts in the data. It can be difficult or impossible to quantify such artificial changes and separate out the real changes in the climate system.

10.2 Specific Sources of Uncertainty

10.2.1 Chaotic Climate System

The uncertainties that arise from the chaotic nature of the climate system have not been quantified in any theoretical sense. Ensemble simulations of climate models offer some hope of quantifying this, with the underlying realization that climate models introduce their own layer of uncertainty.

Under constant boundary conditions (e.g. state of the ocean and land surfaces and composition of the atmosphere), the dynamic and thermodynamic laws of the atmosphere will on occasion naturally produce extreme conditions at any specific location. However, the naturally-occurring frequency and intensity of such extremes may be quite sensitive to the boundary conditions. For example, the spatial pattern of sea surface temperatures (SSTs) is known to have effects on a wide variety of climate characteristics, including extremes. A particular SST pattern might enhance or suppress the likelihood of a specific type of extreme.

The climate system is a coupled system. Atmospheric conditions affect SST patterns and land surface characteristics. Thus, the natural climatology of extreme events at a specific location arises from complex interactions among the various components of the climate system, not just from the specific laws governing atmospheric dynamics and thermodynamics. This introduces formidable complexities in understanding and challenges for modeling of the climate system. An atmosphere-only climate model requires much less computer time than a climate model with a fully-coupled ocean and land surface. As a result, much longer simulations of atmosphere-only models are available. Such long simulations can be very valuable in understanding the natural characteristics of extreme events. However, the results may be unrealistic since coupled processes, particularly between the oceans and atmosphere, are likely to be crucially important.

10.2.2 Measurements: Climate Station Inhomogeneities

The observations that are used to investigate changes in extremes have a variety of characteristics that affect the description of extremes and there are number of sources of inhomogeneities.

Instrumental changes are common in long time series of data and often lead to artificial shifts in extreme metrics. For example, the U.S. National Weather Service Cooperative Observer Network (COOP) is the core climate network of the U.S. Established in 1890, the COOP's long records have been an indispensable source for identifying long-term trends in core climate conditions. As one example of an important instrumental change, the NWS began replacing the liquid-in-glass (LIG) thermometer with an electronic system, starting in the 1980s. Not only was the sensor changed but the radiation shield for the new sensor, a stacked-plate configuration, was completely different than the Cotton Region Shelter (CRS) used for the LIG. Inter-comparison experiments indicated that there is a systematic difference between these two instrument systems, with the newer electronic system recording lower daily maximum temperatures (T_{max}) and higher daily minimum temperatures (T_{min}) (Quayle et al. 1991; Hubbard and Lin 2006; Menne et al. 2009). Menne et al. (2009) estimate that the mean shift (going from CRS/LIG to MMTS) is -0.52 K for T_{max} and $+0.37$ K for T_{min}. As will be shown later, these magnitudes of shifts can have noticeable effects on extreme metrics time series. Adjustments for these differences can be applied to monthly mean temperature to

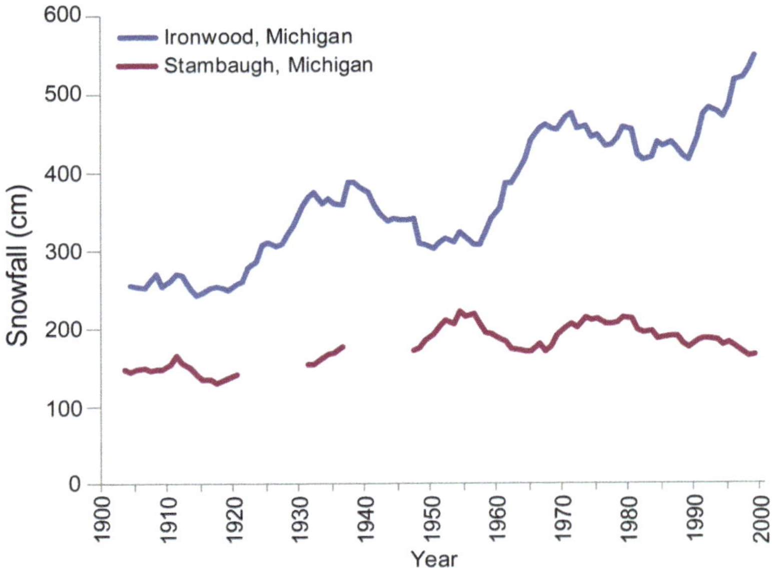

Fig. 10.1 Time series of total annual snowfall at Ironwood, Michigan, USA (*blue*) and at Stambaugh, Michigan, USA (*red*), smoothed with a 11-year moving average filter

create homogeneous time series. However, it is not known whether the differences are quantitatively the same for the small subset of extreme daily temperature values. It is important to keep in mind this potential source of uncertainty.

Changes in the characteristics of sites and/or station moves can introduce artificial shifts or trends in the data. If measurement equipment has been moved, the new site may still be considered to be the same as far as the observing network is concerned and may be identified as such with no change in name or identifier. In the COOP network, a station is generally not given a new name or identifier unless it moves at least 5 miles and/or changes elevation by at least 100 ft (National Weather Service 1993). However, changes less than this can still result in artificial shifts in the climate record that are comparable to, or larger than, real fluctuations. For example, a time series of annual snowfall for Ironwood, MI shows a large jump in totals around 1965 (Fig. 10.1). This occurred around the time of a move of the station that was too small to result in its identification as a different station (Kunkel et al. 2007a). This station is located in the lake effect snow belt of Lake Superior. Another nearby station (Stambaugh, Michigan) does not show a jump (Fig. 10.1). Based on comparisons with Stambaugh and a number of other nearby stations, the Ironwood snowfall record was judged to be inhomogeneous by Kunkel et al. (2009).

Site characteristics can change over time and affect a station's record, even if no move is involved. A common source of such changes is urbanization around the station, which will generally cause artificial warming, primarily in T_{min}

(Karl et al. 1988), the magnitude of which can be several degrees in the largest urban areas. Most research suggests that the overall effect on national and global temperature trends is rather small because of the large number of rural stations included in such analyses (Karl et al. 1988; Jones et al. 1990). Nevertheless, such effects may be important when examining local or regional trends in areas with large urban centers.

Changes in observational practices are another source of artificial shifts. The COOP stations record observations at a daily resolution. However, in the 1990s they were given the option to measure snowfall at 6-hourly intervals and sum the 6-hourly totals for a daily total. Because of compaction of snow, the application of this practice tends to inflate totals relative to daily resolution sampling.

Changes in the time that observations are taken can also introduce shifts (Karl et al. 1986). In the COOP network, typical observation times are early morning or late afternoon, near the usual times of the daily minimum and maximum temperatures, respectively. Because observations occur near the times of the daily extremes occurrences, a change in observation time can have a measurable effect on averages, irrespective of real changes. The study by Karl et al. (1986) indicates that the difference in monthly mean temperatures between early morning and late afternoon observers can be in excess of 2°C. The quantitative impact on metrics of daily temperature extremes could be substantial. An example illustrates this. Assume that a station, with an early morning observation time, experiences a severe cold wave with minimum temperatures below the 0.1% percentile threshold on two consecutive mornings, the minimum occurring at the observation time of 7 am local time. Warming begins on the afternoon of the second day and the minimum temperature during the third night is 10° warmer. On the second morning, the maximum-minimum thermometer will be reset at the time of the minimum temperature. In this case, the minimum temperature observed on the third morning will be the temperature at the time of thermometer reset on the second morning and thus will also be below the 0.1% percentile threshold. In the recorded data, it appears that there are three consecutive extremely cold nights, rather than the two that actually happened. By contrast, the record for an afternoon observer will only have two consecutive very cold mornings, reflecting the actual situation. If the station was a morning observer throughout the history of the station, this artificial bias will not affect trends. However, if that station was an afternoon observer in the early part of its history, then changed its observation time to early morning, there could be an artificial bias toward an upward trend in extremely cold days. There has been in fact a major shift from a preponderance of afternoon observers in the early and middle part of the twentieth century to a preponderance of morning observers at the present time. In the 1930s, nearly 80% of the COOP stations were afternoon observers (Karl et al. 1986). By the early 2000s, the number of early morning observers was more than double the number of late afternoon observers (Menne et al. 2009). The specific quantitative impact of this shift on trends in extreme temperatures has not been investigated and will undoubtedly be dependent on the exact metric used.

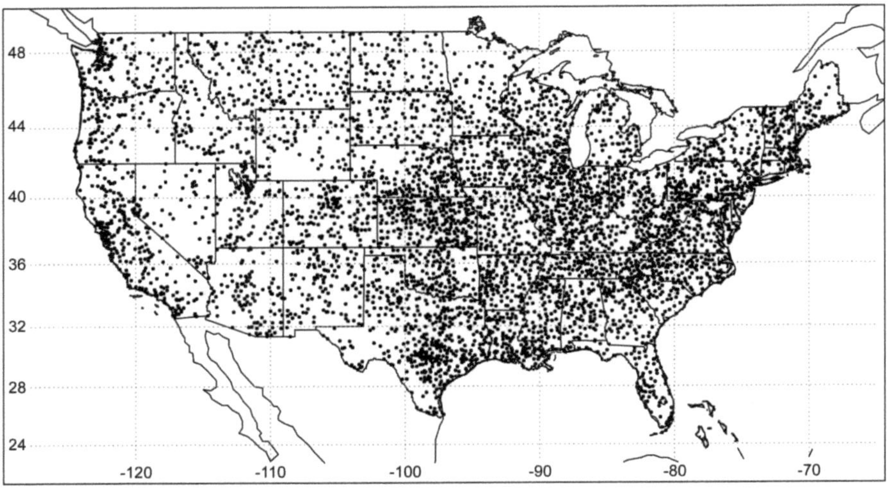

Fig. 10.2 Locations of COOP stations with less than 10% missing precipitation data for the period of 2001–2010

10.2.3 Measurements: Sampling of Physical System

The spatial and temporal sampling of the climate system is limited and does not necessarily capture all important extreme events. The current number of active COOP stations is approximately 7,000. This has declined since a peak of over 8,000 in the 1960s. If these stations were evenly distributed, this would represent about one station per 1,100 km^2, or one every 30–35 km. However, the distribution is not even, the density of stations being lower in the sparsely populated regions of the intermountain West. For example, the density of active stations in Nevada is only about one per 3,300 km^2. Figure 10.2 shows the locations of stations with less than 10% missing precipitation data for the period of 2001–2010. Most areas of the U.S. have rather uniform coverage. However, the lower density of stations in the intermountain West is apparent.

The impact of the density on measurement of extremes is dependent on the spatial coherence of the various types of extremes. Temperature extremes are coherent on fairly large spatial scales, much larger than the average spacing of COOP stations. The same is true of extreme snowstorms, although in this case there can be small scale bursts of snowfall that may be missed by the network. By contrast, many extreme precipitation events are convective in nature, leading to high variability on small spatial scales. Groisman et al. (2005) found that the correlation distance for extreme precipitation events averaged 95–250 km. Fortunately, this is larger than the average spacing of COOP stations of 30–35 km and, thus, the modern COOP network should detect the normal extreme precipitation event, even when it is convective in nature.

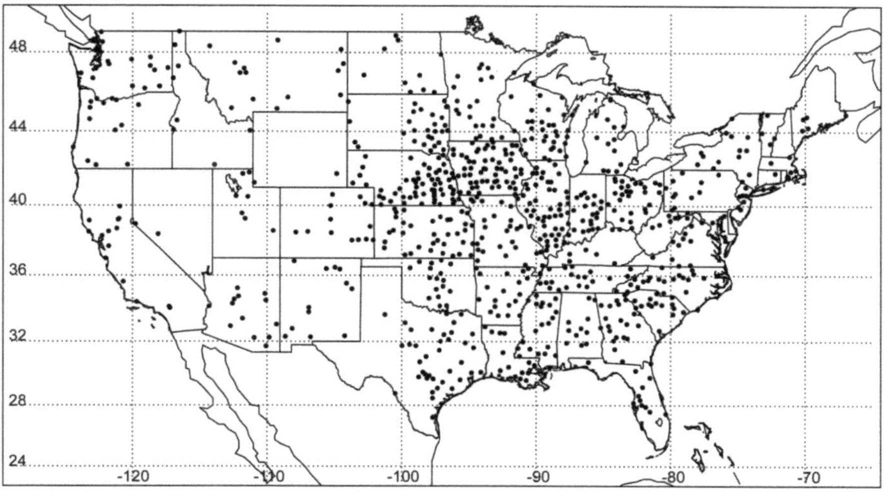

Fig. 10.3 Locations of COOP stations with less than 10% missing precipitation for 1895–2010

These constraints become more important when examining long-term trends in extremes. In this case, it is necessary to use stations with long records to avoid biases caused by stations dropping in and out of the network. The number of stations that can be used for such analyses is typically much less than the total number of current COOP stations. Kunkel et al. (2003, 2007b) examined trends for the period of 1895-present using only stations with less than 10% missing precipitation data. Updating their analysis to the present time, Fig. 10.3 shows the locations of 748 stations with less than 10% missing precipitation data for the period of 1895–2010. This translates roughly into one station every 10,000 km^2, or about one station every 100 km or so and close to the lower end of the correlation distance for extreme precipitation events. In the western US, the density of stations available was much lower than this average number (see Fig. 10.3); this is low enough that some (unknown) percentage of extremes are not sampled.

10.3 Methods for Quantification of Uncertainty

10.3.1 Monte Carlo Experiments

The Monte Carlo technique is a brute-force, but effective, approach to quantification of uncertainty. Essentially, a Monte Carlo technique uses a random number generator to sample from known theoretical or empirical distributions of some aspect(s) of the natural and observing system. This sampling is repeated many times to establish a distribution function of the outcome, from which uncertainty bounds can be estimated. The application of this approach can be computationally-demanding. In addition, a Monte Carlo experiment must be carefully designed to achieve an accurate estimation of uncertainty.

An example will illuminate this approach. Kunkel et al. (2007b) designed a Monte Carlo experiment for the purpose of quantifying the uncertainties in measuring extreme precipitation event frequencies using a 930-station network. This network is likely too sparse in some areas to sample all events, leading to some level of uncertainty in determining trends.

The experiment was designed around the denser modern network. Specifically, a total of 6,353 stations had relatively complete reporting records for the 30-year period of 1971–2000. Their first step was to create an artificial time series of length 110 years by randomly selecting years from 1971 to 2000. Of course, there were repeat years, but that was not important to the goals of this experiment, which simply required an observed, and therefore realistic, sequence of daily data that preserves spatial and temporal relationships. The next step was to create an artificial observing network by randomly choosing 930 stations from the 6,353 total. The random selection was modified to ensure that the spatial variations in station density in the real network (fewer stations in the west, more in the east) were maintained in the artificial network. An artificial daily precipitation time series for each of these stations was created by concatenating the data from the random sequence of years chosen above. All stations used the same sequence of years, thereby preserving spatial coherence. Finally, the artificial time series were analyzed to produce a national time series of the occurrence of precipitation extremes from which time-block averages and trends were computed. The selection of stations to produce an artificial network from which extreme precipitation metrics were calculated was repeated 500 times. The distribution of block averages and trends was created, from which 5 and 95 percentile uncertainties limits were determined. This analysis determined that the 95% confidence limits on the trends from the artificial time series were smaller than the magnitude of the observed trends for 1-, 5-, and 20-year return period events. The study concluded that, while the limited spatial density of long-term stations, particularly in the western U.S., does add considerable uncertainty to trend estimates, the observed trends were larger than the estimated uncertainties and thus were statistically significantly different from zero.

10.3.2 Standard Statistical Tools

Standard statistical tools are the usual method to estimate uncertainty, essentially using the properties of the time series itself. A solid foundation of statistical theory underlies this approach and is generally a powerful and effective way to estimate uncertainty. In most cases, extremes metrics data are not normally distributed. Thus, standard linear regression is not the preferred method to examine trends. A number of other approaches are typically used.

A common nonparametric approach (e.g. Alexander et al. 2006) is Kendall's tau-based slope estimator (Sen 1968). Statistical significance is assessed by looking at the sum of the signs of the differences of all possible pairs of data points. Specifically, the test statistic S is calculated as:

$$S = \sum_{j=1}^{N-1} \sum_{i=j+1}^{N} sgn\left(x_i - x_j\right)$$

where

$$\begin{aligned} sgn(y) &= 1 \quad & y > 0 \\ sgn(y) &= 0 \quad & y = 0 \\ sgn(y) &= -1 \quad & y < 0 \end{aligned}$$

The variance of S, s, is given by

$$\sigma_S^2 = \frac{N\,(N-1)\,(2N+5)}{18}$$

The z statistic is then

$$\begin{aligned} z &= \frac{S-1}{\sigma_S} \quad & S > 0 \\ z &= 0 \quad & S = 0 \\ z &= \frac{S+1}{\sigma_S} \quad & S < 0 \end{aligned}$$

The estimate of the magnitude of the trend is the median of all non-zero pairwise trends. Some level of serial correlation is commonly present in extremes time series and will artificially raise the significance level. Estimates of trends and their statistical significance should take this into account. Wang and Swail (2001) describe a pre-whitening process that can be applied to time series to minimize the effects of serial correlation.

Spatial correlation will also artificially increase the level of significance in evaluating the reality of changes over large areas. For example, a single extreme meteorological event may be manifested at several stations. Or, a weather pattern persisting for some time may cause a sequence of extremes affecting a specified area. In both cases, the time series at adjacent stations are affected by the same meteorological phenomena and are not completely independent. The question that is of relevance for establishing changes in the climate is whether the number of stations with statistically significant trends over a large area of interest is statistically significant. Using the 95% limit, just by chance we should expect 5% of the stations to exhibit either a statistically significant negative trend or a positive trend. If the time series of the stations were completely independent, then the overall field of stations would exhibit statistical significance if more than 5% of the stations exhibited a positive or negative trend. However, when considering the spatial dependence of station time series, the number of stations where statistical significance is achieved has to be higher than 5% in order to establish that there is a real change in extremes. The required level of significance can be determined by a bootstrapping resampling procedure (Livezey and Chen 1983). Assume that each station or grid point has a 100-year time series of some metric of extremes. The bootstrapping procedure consists of randomly selecting a sequence of years and constructing artificial time series based on this sequence. Then a trend is calculated. The same sequence of years is applied to all of the stations/grid points. Finally, the number of stations/grid points with statistically significant trends is counted. This process is repeated many times, perhaps 500–1,000 (Kiktev et al. 2003). The distribution of the number of stations/grid points with statistically significant trends is computed. The 95% value of that distribution then provides the threshold for field significance. For example, Kunkel et al. (1999) found that this threshold was 10% for a particular metric of extreme precipitation for the coterminous U.S., that is, stations representing at least 10% of the area had to have a statistically significant trend to consider that the field exhibited statistical significance.

In many studies, the Generalized Extreme Value (GEV) distribution has been found to be suitable as a fit to the tails of the distribution for atmospheric variables. The probability distribution of the GEV, G(y), is given by:

$$G(y) = \frac{1}{\sigma}\left(1 + \varepsilon\frac{y-\mu}{\sigma}\right)^{-1-\frac{1}{\varepsilon}} \exp\left[-\left(1+\varepsilon\frac{y-\mu}{\sigma}\right)^{-\frac{1}{\varepsilon}}\right], \varepsilon > 0, y > \mu - \frac{\sigma}{\varepsilon}$$

$$G(y) = \frac{1}{\sigma}\exp[-\frac{y-\mu}{\sigma} - \exp\left(-\frac{y-\mu}{\sigma}\right)], \varepsilon = 0$$

$$G(y) = \frac{1}{\sigma}\left(1 + \varepsilon\frac{y-\mu}{\sigma}\right)^{-1-\frac{1}{\varepsilon}} \exp\left[-\left(1+\varepsilon\frac{y-\mu}{\sigma}\right)^{-\frac{1}{\varepsilon}}\right], \varepsilon < 0, y < \mu - \frac{\sigma}{\varepsilon}$$

where ε, μ, and σ are called the shape, location, and scale parameters. A trend can be introduced as a trend in the location parameter or log of the shape parameter (Katz 2010). In Zwiers et al. (2011), trends were studied by assuming that the temporal changes can be expressed as an additive change in the location parameter with the other parameters remaining constant. The purpose of the study was to determine whether the influence of anthropogenic forcing on the climate system had a detectable influence on observed changes in extreme temperatures. The maximum likelihood method was used to find the estimates of the parameters that best fit the observed and modeled data. The bootstrapping method described above was used to ascertain the overall uncertainty in estimates. In this particular case, the data were divided into 5-year blocks and these blocks were randomly reshuffled to create artificial time series. The location parameter of the reshuffled time series was estimated using maximum likelihood. The reshuffling and estimation of the location parameter was repeated many times to determine the uncertainties in the location parameter due to natural variability. Using this approach, they found that the external forcing of the climate system was clearly detectable in the observed changes in extreme temperatures, that is, the observed changes were highly unlikely to have been due to natural variability in the climate system.

10.3.3 Climate Model Ensemble Experiments

While statistical approaches provide a well-established numerical framework for estimating uncertainties, the functional fits and other numerical aspects are in many respects an artificial framework. For example, the GEV fits the tails of the distribution of extremes well. But, there is no physics in the equation, as is also the case for all statistical functions. This may be satisfactory for mild extremes because there are a substantial number of actual samples in the observational record and a good fit by the GEV implies an adequate representation of the underlying climate system characteristics. But, the most extreme events are another matter. What is the return period of the 1930s Dust Bowl? What about the 2003 European heat wave? or the 2010 Russian heat wave? These are singular events in the observational record and, at many locations affected by them, the observed conditions are considerably more extreme than for any other similar period. Such observations can be included in a fit to the GEV and a return period, or trend, can be computed but the uncertainties in such estimates must be considered very high.

Exploring the true probability of the most extreme conditions requires another approach. Climate models offer a possible solution. They are based on the fundamental physical laws governing the climate system. They can produce the dynamic chaotic behavior that is characteristic of the system. In principle then, they can produce events like the singular ones described above and provide insights into their probability.

Two types of model simulations are most relevant. One is the long control simulation, often using pre-industrial concentrations of greenhouse gases and sometimes present-day concentrations. Some modeling groups have made very long simulations (Table 10.1), over 1,000 years, in support of the Intergovernmental Panel on Climate Change (IPCC) 4th Assessment Report. In total, the 24 models listed in Table 10.1 produced over 12,000 simulated years in these control experiments. These can be quite valuable in assessing the likelihood of very rare events. But, the estimation of such risks assumes a climate system with no change in the anthropogenic forcing influences. The other type is the simulation of the twentieth century using best estimates of the time-dependent forcing during the industrial era, of the order of 100 years in length. Many modeling groups have run multiple ensembles, with different initial conditions. The 24 models have produced 80 separate simulations of the twentieth century. Although there has been considerable research on these simulations, the use of these simulations to explore extreme events has probably not been fully exploited.

Table 10.1 Simulations by climate models used for the IPCC 4th Assessment Report

Model name	Twentieth century (# simulations)	Control simulation #years (# simulations)
BCC	4	223(1)
CCSM3	9	1,330(3)
CGCM3.1 (T47)	5	1,001(1)
CGCM3.1 (T63)	1	501(1)
CNRM-CM3	1	501(1)
CSIRO-Mk3.0	3	430(2)
CSIRO-Mk3.5	3	1,000(1)
ECHAM5/MPI-OM	3	506(1)
ECHO-G	5	652(2)
FGOALS-g1.0	3	351(1)
GFDL-CM2.0	3	500(1)
GFDL-CM2.1	3	500(1)
GISS-AOM	2	502(2)
GOISS-EH	5	400(1)
GISS-ER	9	500(1)
INM-CM3.0	1	330(1)
IPSL-CM4	1	720(1)
MIROC3.2 (hires)	1	100(1)
MIROC3.2 (medres)	3	500(1)
MRI-CGCM2.3.2	5	500(2)
PCM	4	650(2)
UKMO-HadCM3	2	422(2)
UKMO-HadGEM1	2	341(1)

This includes the number of simulations of the twentieth century and the number of model simulation years and number of individual simulations for control simulations using either pre-industrial or present-day greenhouse gas concentrations

An example of application of climate models is the study of Deser et al. (2010), who used a 10,000-year long control model simulation of the atmospheric component of the CCSM model to explore atmospheric variability for temperature, precipitation, and sea level pressure to estimate the uncertainties due to internal atmospheric variability. From the standpoint of extremes, the use of a long simulation from an atmospheric model not fully coupled to an ocean model may be potentially useful, although it may underestimate variability of those extremes that arise from the coupling of specific sea surface temperature conditions with specific atmospheric states, something that may well be very important for certain extended extremes such as severe droughts or extended wet spells.

10.4 Applications

10.4.1 Extreme Precipitation Trends in the U.S.

The spatial coherence of individual precipitation extreme events is relatively small. Thus, the uncertainty arising from the limited spatial density of the observational network may be quite large. On the other hand, the U.S. COOP network has used the same basic technology throughout its existence. Thus, uncertainty arising from instrumentation and observation practice changes is quite low. To quantify the major source of uncertainty (i.e. the limited spatial density of observing stations), the Monte Carlo approach was used to estimate uncertainties in 22-year block averages of a metric of the occurrence of extreme precipitation events in the U.S. (Kunkel et al. 2007b), using the methodology described in Sect. 10.3.1. Figure 10.4 shows a sample of the results, for events exceeding a return period of 20 years. The whiskers around each block average, representing the confidence interval, were obtained from the distribution of values from the 500 Monte Carlo simulations. Two questions were of interest, expressed as the following null hypotheses: (1) heavy precipitation event frequencies during 1895–1916 and 1917–1938 are not different,

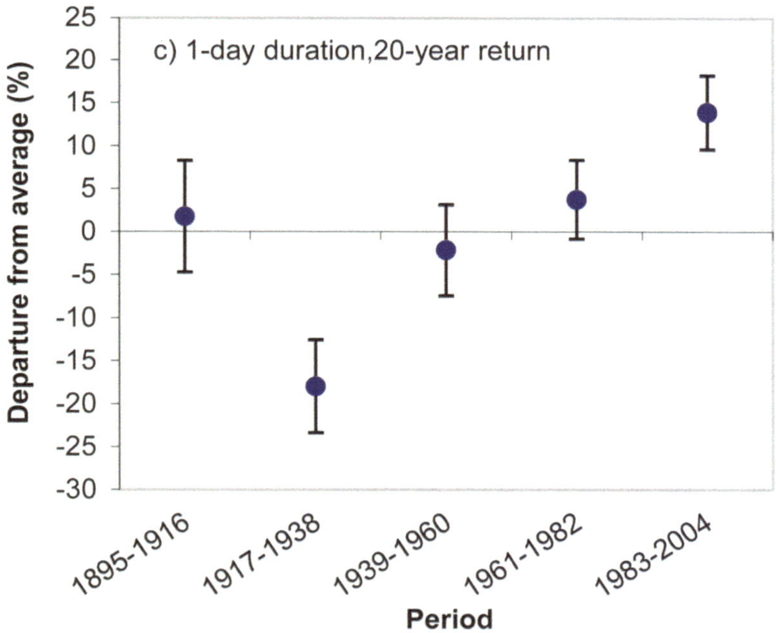

Fig. 10.4 U.S. national index of daily extreme precipitation events exceeding a 1-in-20 year threshold, averaged over 22-year consecutive blocks and expressed as deviations from the long-term average. Whiskers indicated the 95% confidence interval, based on Monte Carlo simulations (From Kunkel et al. 2007b)

and (2) heavy precipitation event frequencies during 1895–1916 and 1983–2004 are not different. To determine statistical significance of these two block average differences, two further calculations are needed. As discussed in Payton et al. (2003), the nominal value of the confidence interval (e.g., 95%) for a single period mean is not the same as the confidence level for testing differences of two period means by examining overlapping intervals. For large samples, there is a factor of $\sqrt{2}$ between the normal variate Z for the individual period means and the value of Z for testing the null hypothesis using overlapping intervals. For example, for testing the null hypotheses at the $\alpha = 0.05$ level ($Z = 1.96$), the appropriate confidence interval for the individual period means is approximately 84% ($Z = 1.96/\sqrt{2} = 1.39$); that is, if the 84% confidence intervals for each period mean do not overlap, then the null hypothesis is rejected at the 0.05 significance level. A second adjustment is needed because we are testing two differences. Since multiple comparisons are being made, an adjustment must be applied because multiple comparisons increase the probability that the null hypothesis will be rejected by chance (that is, increase the probability of a Type I error, incorrectly rejecting the null hypothesis). The Bonferroni adjustment (Snedecor and Cochran 1980) estimates this as α/m where m is the number of comparisons. For two comparisons, the probability value for testing at the $\alpha = 0.05$ significance level is 0.05/2 or 0.025, that is, the confidence interval for testing each comparison needs to be 97.5% in the present case of two comparisons when testing the null hypothesis at the $\alpha = 0.05$ significance level. The use of the Bonferroni adjustment is a conservative approach because it tends to increase the probability of a Type II error (not rejecting the null hypothesis incorrectly). Combining both adjustments for the specific example of achieving a significance level of 0.05 to reject the null hypothesis, the confidence interval for individual period means must be 89%, that is, the null hypothesis is rejected at $\alpha = 0.05$ if the 89% confidence intervals for the individual period means do not overlap. For the results in Fig. 10.4, there are a very small number of events occurring at any single station (i.e. sampling uncertainty), leading to large uncertainties at individual stations. However, on a national average basis, this analysis demonstrated that the changes over time are larger than the uncertainty and thus we can state with confidence that there have been real changes in the occurrence of these extreme events.

10.4.2 Heat and Cold Wave Trends

Temperature extremes are typically of relatively large spatial scale. Thus, the sampling uncertainty associated with limited station density is much less of a source of uncertainty than with precipitation. In the COOP network, the change of instrumentation from CRS/LIG to MMTS (mostly occurring by 1990), along with associated changes in radiation shielding, is a possible source of uncertainty. While artificial shifts have been quantified for mean values, the quantitative effect on extreme values is not known and thus the possible effects on trends have not been documented. Menne et al. (2009) estimated that the change from CRS/LIG to MMTS results in a mean shift of −0.52 K in T_{max} and +0.37 K in T_{min}. A sensitivity analysis of the potential effects of such changes was conducted, assuming that the mean changes above apply to extreme episodes as well. The test involved adjusting all post-1989 temperature observations by the mean shifts of −0.52 K for T_{max} and +0.37 K for T_{min}. A heat wave index was calculated for 4-day duration hot spells exceeding the threshold for a 1-in-5 year recurrence. Calculations were done for each station and then aggregated to a national index. The heat wave calculations were done separately first using only T_{max} and then only T_{min}, essentially representing a "hot day" index and a "warm night" index, respectively. The results are shown in Fig. 10.5. When the temperature observations are used without correction (Fig. 10.5, top), a potentially very interesting result is obtained. Starting in the 1980s, the T_{min} index is considerably greater than the T_{max} index, implying that spells of very warm nights are becoming relatively more frequent than very warm days. However, after applying the mean correction factors for all stations (Fig. 10.5, bottom), this difference between the T_{max} and T_{min} indices mostly disappears. Thus, this change in instrumentation, if not considered in analysis, introduces biases sufficiently large to affect conclusions about trends.

The quantitative effects of the time of observation bias in the COOP network on metrics of extreme temperatures could be significant, but has not been systematically investigated to date. A test of potential effects was conducted for extreme cold episodes. In the first step of the analysis, a U.S. national time series of the number of episodes of 4-day duration exceeding the 1-in-5 year threshold was calculated. These multi-day duration episodes could be affected by the shift in time of observation within the COOP networks from predominantly late afternoon to predominantly early morning. The net effect would be an artificial upward trend in occurrences, added to any real trends. In the second step of the analysis, a national time series of the number of 1-day extreme temperatures exceeding a 1-in-5 year threshold was calculated. In this latter case, it was required that there be a minimum 2-day gap between episodes. This requirement eliminates the double counting of very cold days that can occur with early morning observations and should eliminate any artificial trend bias. These two time series are compared in Fig. 10.6. The two time series are very comparable. The small observed differences between the two could be due to the different extreme event definitions, although there could be some effect due to the time of observation bias. However, linear trend lines fitted to these two time series (not shown) are virtually identical; the linear trend is −1.0% decade^{-1} for the 4-day duration metric and −1.2% decade^{-1} for the 1-day metric. There is a small positive difference for the 4-day metric compared to the 1-day metric and this small residual could be due to the time of observation bias, or could simply be due to the different definitions of the metrics.

An analysis of a heat wave index found similar results (in this case, the time of observation bias would introduce an artificial downward trend). This pilot analysis suggests that time of observation bias is unlikely to affect the overall conclusions about trends in extreme temperature.

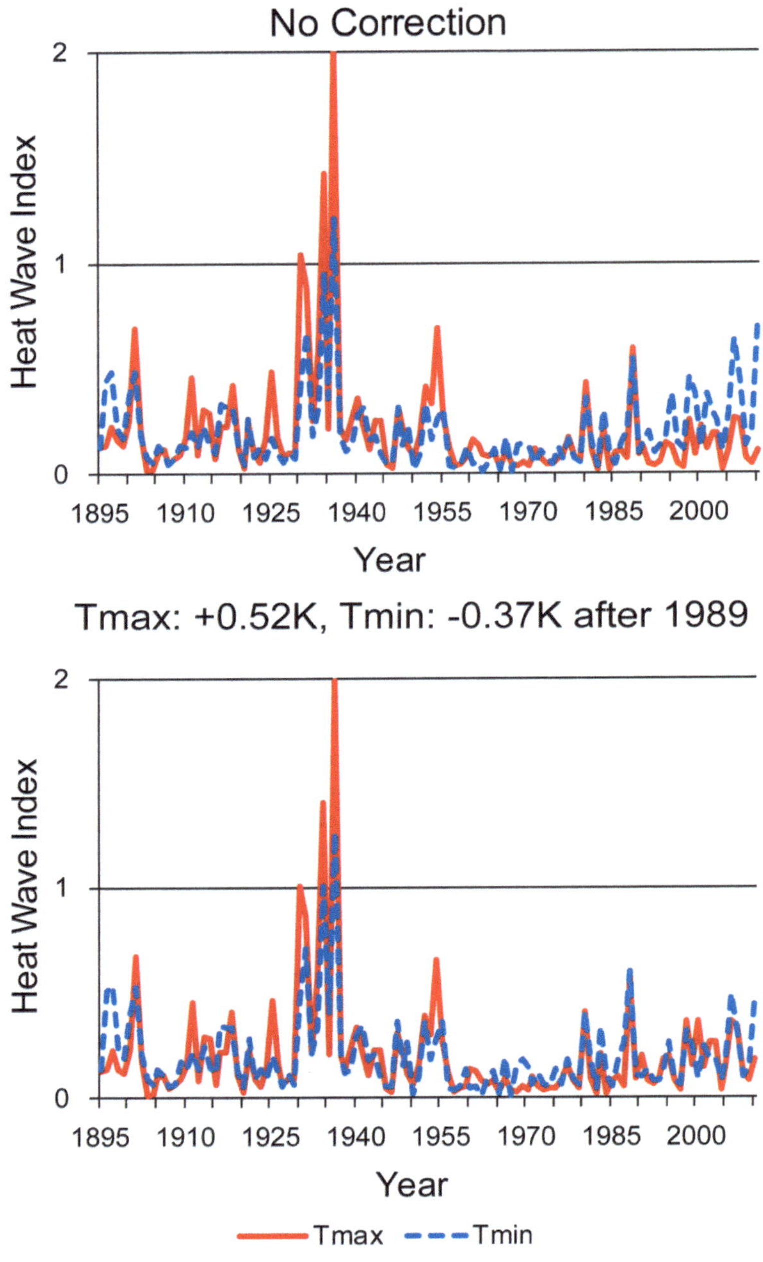

Fig. 10.5 Time series of an index for heat waves, aggregated for the U.S. based on a network of 711 long-term temperature observing stations, and calculated separated for T_{min} and T_{max}. (*Top*) No correction to observed values. (*Bottom*) Every daily value adjusted after 1989 by +0.52 K for T_{max} and −0.37 K for T_{min} before calculating index. A 5-year recurrence interval threshold is used to identify events

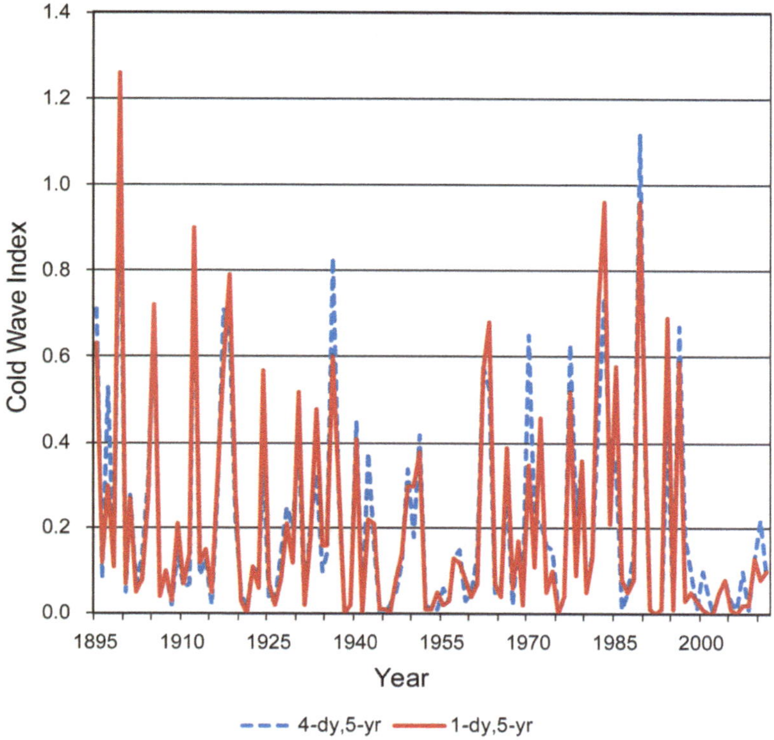

Fig. 10.6 Time series of an index for cold waves, aggregated for the U.S. based on a network of 711 long-term temperature observing stations, and calculated separated for 4-day duration episodes (*blue*) and 1-day duration (*red*). A 5-year recurrence interval threshold is used to identify events

The other likely greatest source of uncertainty in extreme temperature metrics is the sampling uncertainty associated with the rarity of extremes. Standard statistical tools that evaluate uncertainty based on the variance in the data provide the best way of quantifying this.

10.4.3 The Hurricane Problem

There is great interest in documenting any changes in the frequency and intensity of hurricanes. A number of studies have documented possible trends (see Kunkel et al. 2008 for summary). However, there has been much concern about the quality of the data and possible artificial shifts (e.g. Landsea et al. 2004). There are a number of issues with the record that arise from sampling over the oceans, changing technologies and methodology. Trend studies have look at time series extending back to the latter half of the nineteenth century. In the early part of the record, detection of hurricanes over the oceans required the occurrence of an intersection with a ship track or an island. Aircraft reconnaissance in the 1940s and satellite data in the 1960s greatly improved detection and tracking. Studies have been done to estimate the likely number of missed hurricanes in the early part of the record by superimposing hurricane tracks from the modern (since the beginning of satellite observations) era over ship tracks in the early part of the record to see how many modern hurricanes would have been missed in the early part of the record (Vecchi and Knutson 2011). This analysis indicates that in the latter part of the nineteenth century, the annual number of tropical cyclones is likely undercounted by 2–3, compared to an average annual value of about 10. This estimated undercount decreases to less than 1 per year after World War II.

Trends in the total power associated with hurricanes have even greater potential bias problems. Ascertaining the magnitudes of winds in a tropical cyclone is a more challenging problem than determining location and track. For TCs over ocean areas, the only in situ observations are from aircraft. As noted above aircraft reconnaissance began only in the 1940s. While this has continued for Atlantic TCs, this was discontinued for Pacific TCs in 1989. Inferred winds from satellites provides a uniform approach (Kossin et al. 2007) but it is not a direct approach and therefore subject to some uncertainty; however the uniformity of the data source is a big advantage and makes this the preferred approach for the satellite era.

10.4.4 The Tornado Problem

Tornadoes are characterized by their small spatial scale, relative to most other major weather hazards, and their local destructiveness. In addition, the occurrence of tornadoes over much of the historical record depended entirely on reports by weather spotters and the general public. In recent years, Doppler radar provides a supplemental record, although it is an indirect measure and any radar detection must be confirmed by ground inspection or the above-mentioned reports. A source of bias in the number of reported tornadoes is the change in population in tornado-prone regions (increasing the chances of detection and reporting) and a greater public awareness and interest in seeing and reporting events. There is a very strong upward trend in the number of weak tornadoes (Fig. 10.7). Is this real? If one examines only stronger tornadoes, there is no trend. Does this mean that there has been an increase in only weak tornadoes? Almost surely not. The expert consensus is that the upward trend in total tornadoes is a sampling artifact (Brooks and Dotzek 2008; Kunkel et al. 2012). Specifically, larger populations in tornado-prone regions and greater public interest have led to more complete reporting of the weaker tornadoes.

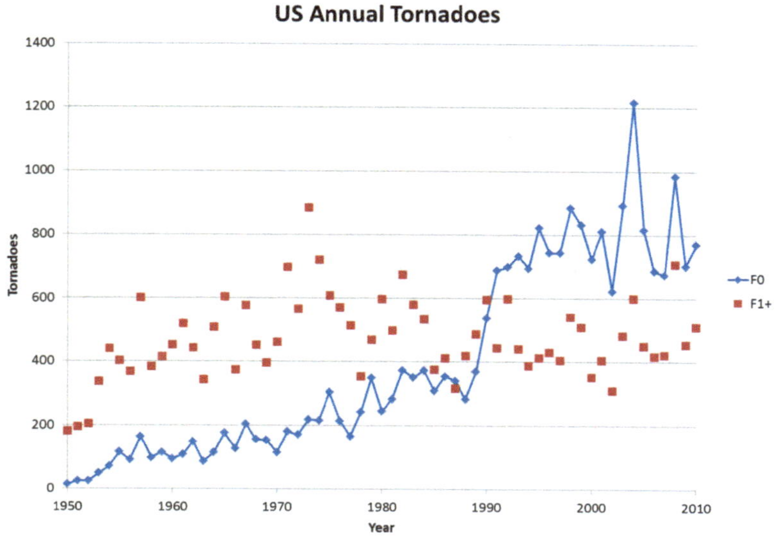

Fig. 10.7 Reported tornadoes in NWS database from 1950 to 2010. Blue line is F0 tornadoes, red dots are F1 and stronger tornadoes (From Kunkel et al. (2012), courtesy of H. Brooks)

10.5 Concluding Remarks

To reiterate the main point, a central objective of research in this arena is to determine whether there are real changes in the characteristics of extreme events. The most serious challenge to achieving this objective is not the adequacy of the statistical tools. Rather, the adequacy of the observational data and understanding of the physical climate system pose the greatest challenges. Much work has gone into data set development and improvement. Nevertheless there are opportunities to further this.

Tree ring data are the iconic example of proxy data. Many other proxies have been developed and provide unique insights into the behavior of the climate system. The further development of such data sets can provide additional insights into the behavior of extremes events, particularly in areas where proxy data have hitherto been unavailable.

The use of climate simulations to explore potential extreme behavior in the climate system is currently limited by model limitations. For example, model resolution is not sufficient to directly simulate the typical extreme precipitation event. As model resolution increases and representation of physical processes in the models improves, the ability of these models to simulate extremes should improve also. This will provide the opportunity to explore the behavior of more types of extremes.

References

Alexander LV, Zhang X, Peterson TC, Caesar J, Gleason B, Klein Tank AMG, Haylock M, Collins D, Trewin B, Rahimzadeh F, Tagipour A, Rupa Kumar K, Revadekar J, Griffiths G, Vincent L, Stephenson DB, Burn J, Aguilar E, Brunet M, Taylor M, New M, Zhai P, Rusticucci M, Vazquez-Aguirre JL (2006) Global observed changes in daily climate extremes of temperature and precipitation. J Geophys Res 111:D05109. doi:10.1029/2005JD006290

Brooks HE, Dotzek N (2008) The spatial distribution of severe convective storms and an analysis of their secular changes. In: Diaz HF, Murnane R (eds) Climate extremes and society. Cambridge University Press, Cambridge, pp 35–53, 340 pp

Deser C, Phillips A, Bourdette V, Teng H (2010) Uncertainty in climate change projections: the role of internal variability. Climate Dyn. doi:10.1007/s00382-010-0977-x

Groisman PY, Knight RW, Easterling DR, Karl TR, Hegerl GC, Razuvaev VN (2005) Trends in intense precipitation in the climate record. J Climate 18:1326–1350

Hubbard KG, Lin X (2006) Reexamination of instrument change effects in the U.S. HistoricalClimatology Network. Geophys Res Lett 33:L15710. doi:10.1029/2006GL027069

Jones PD, Ya P, Groisman M, Coughlan N, Plummer WCW, Karl TR (1990) Assessment of urbanization effects in time series of surface air temperature over land. Nature 347:169–172

Karl TR, Williams CN Jr, Young PJ, Wendland WM (1986) A model to estimate the time of observation bias associated with monthly mean maximum, minimum, and mean temperatures for the United States. J Climate Appl Meteorol 25:145–160

Karl TR, Diaz HG, Kukla G (1988) Urbanization: its detection and effect in the United States climate record. J Climate 1:1099–1123

Katz RW (2010) Statistics of extremes in climate change. Clim Chang 100:71–76

Kiktev D, Sexton DMH, Alexander L, Folland CK (2003) Comparison of modeled and observed trends in indices of daily climate extremes. J Climate 16:3560–3571

Kossin JP, Knapp KR, Vimont DJ, Murnane RJ, Harper BA (2007) A globally consistent reanalysis of hurricane variability and trends. Geophys Res Lett 34:L04815. doi:10.1029/2006GL028836

Kunkel KE, Andsager K, Easterling DR (1999) Long-term trends in extreme precipitation events over the conterminous United States and Canada. J Climate 12:2515–2527

Kunkel KE, Easterling DR, Redmond K, Hubbard K (2003) Temporal variations of extreme precipitation events in the United States: 1895–2000. Geophys Res Lett 30:1900. doi:10.1029/2003GL018052

Kunkel KE, Palecki M, Hubbard KG, Robinson D, Redmond K, Easterling D (2007a) Trend identification in 20th century U.S. Snowfall: the challenges. J Atmos Ocean Technol 24:64–73

Kunkel KE, Karl TR, Easterling DR (2007b) A Monte Carlo assessment of uncertainties in heavy precipitation frequency variations. J Hydrom 8:1152–1160

Kunkel KE, Bromirski PD, Brooks HE, Cavazos T, Douglas AV, Easterling DR, Emanuel KA, Groisman PYa, Holland GJ, Knutson TR, Kossin JP, Komar PD, Levinson DH, Smith RL (2008) Observed changes in weather and climate extremes. In: Karl TR, Meehl GA, Miller CD, Hassol SJ, Waple AM, Murray WL (eds) Weather and climate extremes in a changing climate. Regions of focus: North America, Hawaii, Caribbean, and U.S. Pacific Islands. A Report by the U.S. Climate Change Science Program and the Subcommittee on Global Change Research, Washington, DC

Kunkel KE, Karl TR, Brooks H, Kossin J, Lawrimore J, Arndt D, Bosart L, Changnon D, Cutter SL, Doesken N, Emanuel K, Groisman PYa, Katz RW, Knutson T, O'Brien J, Paciorek CJ, Peterson T, Redmond K, Robinson D, Trapp J, Vose R, Weaver S, Wehner M, Wolter K, Wuebbles D (2012) Monitoring and understanding trends in extreme storm statistics: state of knowledge. Bull Am Meteor Soc (accepted)

Kunkel KE, Palecki M, Ensor L, Hubbard KG, Robinson D, Redmond K, Easterling D (2009) Trends in twentieth-century u.s. snowfall using a quality-controlled dataset. J Atmos Oceanic Technol 26:33–44. doi:http://dx.doi.org/10.1175/2008JTECHA1138.1

Landsea CW, Anderson C, Charles N, Clark G, Dunion J, Fernandez-Partagas J, Hungerford P, Neumann C, Zimmer M (2004) The Atlantic hurricane database re-analysis project: documentation for the 1851-1910 alterations and additions to the HURDAT database. In: Murnane RJ, Liu KB (eds) Hurricanes and typhoons: past, present and future. Columbia University Press, New York, pp 177–221

Livezey RE, Chen WY (1983) Statistical field significance and its determination by Monte Carlo techniques. Mon Weather Rev 111:46–59

Menne MJ, Williams CN Jr, Vose RS (2009) The U.S. Historical Climatology Network monthly temperature data, version 2. Bull Amer Meteorol Soc 90:993–1007

National Weather Service (1993) Cooperative program operations. National Weather Service Observing Handbook No. 6, 56 pp

Payton ME, Greenstone MH, Schenker N (2003) Overlapping confidence intervals or standard error intervals: what do they mean in terms of statistical significance? J Insect Sci 3:1–6

Quayle RG, Easterling DR, Karl TR, Hughes PY (1991) Effects of recent thermometer changes in the cooperative station network. Bull Amer Meteorol Soc 72:1718–1724

Sen PK (1968) Estimates of the regression coefficient based on Kendall's Tau. J Am Stat Assoc 63:1379–1389

Snedecor GW, Cochran WG (1980) Statistical methods. The Iowa State University Press, Ames, 507 pp

Vecchi GA, Knutson TR (2011) Estimating annual numbers of Atlantic hurricanes missing from the HURDAT database (1878–1965) using ship track density. J Climate 24(6):1736–1746. doi:10.1175/2010JCLI3810.1

Wang XL, Swail VR (2001) Changes of extreme wave heights in Northern Hemisphere oceans and related atmospheric circulation regimes. J Climate 14:2204–2220

Zwiers FW, Zhang X, Feng Y (2011) Anthropogenic influence on long return period daily temperature extremes at regional scales. J Climate 24:881–892. doi:http://dx.doi.org/10.1175/5422010JCLI3908.1

Chapter 11
Uncertainties in Projections of Future Changes in Extremes

Levi D. Brekke and Joseph J. Barsugli

Abstract Water resource managers share a common challenge in understanding what climate change could mean for future hydroclimate extremes. In order to make decisions about whether to invest in adaptation measures today or to wait for more convincing information, it is critical that managers understand the uncertainties of projecting changes in extremes. Uncertainties arise from several methodological choices including criteria that drive selection of global climate projection information to frame the assessment, whether and how to bias-correct global projection information, and how to represent local controls on how to spatially downscale translations of these projections. This chapter highlights such uncertainties, focusing on projected changes in two precipitation metrics: annual total and annual maximum daily amount, and for both typical (i.e. median metrics) and extreme conditions (i.e. annual totals related to drought, having 0.01 and 0.05 cumulative probabilities; and, annual maximum daily amounts related to floods, having 0.95 and 0.99 cumulative probabilities). The evaluation is conducted on 53 daily precipitation projections over the contiguous U.S., southern Canada and northern Mexico. Focusing on two future periods, and the chapter presents: (a) assessed changes in typical metric conditions and determining their significance, (b) assessed changes in extreme metric conditions, (c) decomposition of uncertainties in both types of changes relative to three sources of global climate projection uncertainty (emissions scenario, global climate model, internal variability), and (d) characterization of how projected changes may be sensitive to spatial downscaling.

L.D. Brekke (✉)
Bureau of Reclamation (U.S. Department of the Interior), Research and Development Office, Denver, CO 303-445-2494, USA
e-mail: lbrekke@usbr.gov

J.J. Barsugli
Cooperative Institute for Research in Environmental Sciences, University of Colorado at Boulder, Boulder, CO 303-497-6042, USA
e-mail: joseph.barsugli@colorado.edu

A. AghaKouchak et al. (eds.), *Extremes in a Changing Climate*, Water Science and Technology Library 65, DOI 10.1007/978-94-007-4479-0_11,

11.1 Introduction

Water resource managers share a common challenge in understanding how to evaluate future hydroclimate possibilities and what they mean for the systems they manage. Such possibilities affect both natural resources and built infrastructure. Anticipation of these possibilities requires consideration of both typical and extreme hydroclimate conditions. Understanding the uncertainty and credibility of these projected possibilities is critical to the process of making decisions on whether to invest in adaptation measures today or delay until more certain and credible information becomes available.

This chapter focuses on the precipitation aspects of future hydroclimate possibilities, although the assessment methods presented herein are general and may be applied to assess changes in other hydroclimatic variables. It has been suggested that climate change could affect both heavy precipitation and drought-related extremes (USGCRP 2009). This premise is used as context to demonstrate how climate projections might be assessed for (1) changes in typical conditions for contrasting precipitation metrics, (2) changes in extreme occurrences of these metrics, and (3) evaluated to understand how variance in projected changes might be associated with various sources of global climate projection uncertainty. Along the way, other assessment choices are identified as contributing additional uncertainties.

Before conducting such an assessment, it is useful to first consider physical mechanisms that might establish expectations for assessment outcomes. Any such mechanism begins with recognition that future climate forcing will change as greenhouse gases accumulate in the atmosphere (Christensen et al. 2007). The task is then to relate these changes in global climate forcing to regional/local changes in precipitation. One such mechanism starts from the premise that as the atmosphere warms, its water-holding capacity increases according to the Clausius-Clapeyron relationship. In association, this is expected to lead to a small increase in global mean precipitation (Karl et al. 2008), where the increase is constrained by the global surface energy budget (Held and Soden 2006). It is also expected to affect heavy precipitation (Karl et al. 2008), for which the Clausius-Clapeyron relation only provides an approximate physical constraint (e.g., Allen and Ingram 2002), recognizing that regional to local atmospheric dynamics also influence local precipitation. So a critical question remains as to how global air temperature increases will affect atmospheric circulation and synoptic-scale weather patterns in ways that enhance or counteract the Clausius-Clapeyron effect at the local scale. One such atmospheric circulation response to global warming that has been suggested in several studies is the weakening of tropical Hadley circulation (Held and Soden 2006) and associated expansion of the tropical Hadley Cell (Lu et al. 2007; Seager et al. 2007). The result of this circulation response is an increase in atmospheric subtropical static stability, associated broadening of the atmospheric subsidence zone (i.e. "deserts" regions around the planet), and associated poleward displacement of the wetter middle-latitude storm tracks. This "Hadley Expansion" effect is seen in the majority of recent climate projections, leading to

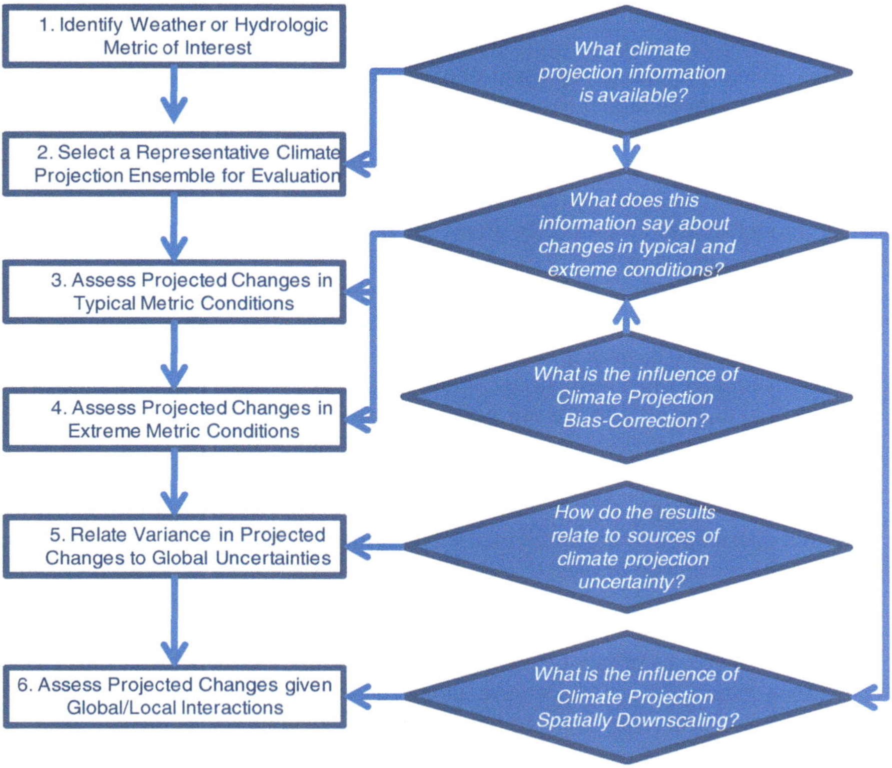

Fig. 11.1 Approach for assessing projected changes in extremes and associated uncertainties

precipitation reductions over traditionally dry sub-tropical regions (Seager et al. 2007; Christensen et al. 2007). Likewise, these same projections show increasing precipitation over the middle-to-higher latitudes, suggesting that the "Clausius-Clapeyron" relationship is leading to increased moisture content of weather systems tracking over these regions. Revisiting the suggestion of how heavier precipitation and more severe precipitation droughts could occur over a given region, the degree to which either impact occurs may depend on the interplay of these large-scale effects at a given geographic position and how these effects interact with regional to local controls on precipitation.

Noting these physical paradigms, the assessment of projected changes in precipitation metrics and associated extremes might be conducted using the following list of general steps, which might also be adapted to assessing changes in temperature, runoff and other hydroclimatic extremes (Fig. 11.1):

1. Identify Weather or Hydrologic Metric of Interest: The metric is defined by variable, statistic, duration and spatial extent (e.g., annual total precipitation over a regional watershed, annual maximum-day precipitation over a local basin).

2. Select a Representative Climate Projection Ensemble for Evaluation: The assessment considers climate projections that reflect current and best available climate science, and preferably represent choices in specifying future climate forcing, choices in climate model structure, and multiple realizations of internal climate variability given these choices. Having various options for future climate forcing and climate model structure, as well as facing the prospect of having multiple realizations of internal climate variability for given choice of climate forcing and model structure, ultimately drives the need to consider a climate projection ensemble in this step.
3. Assess Projected Changes in Typical Metric Conditions: The focus here is on gaining a sense of how gradual global warming may affect "typical" metric conditions, indicated as period-average or period-median. Such assessment provides a useful context when evaluating climate change implications for extreme or rare metric occurrences. Changes in typical conditions should also be subjected to statistical testing to determine whether they are significant.
4. Assess Projected changes in Extreme Metric Conditions: The focus here is to identify an appropriate statistical distribution of metric possibilities, and then fit that distribution to projected metric conditions from a given period. The fitted distribution can then be used to infer extreme and rare metric occurrences having N-year return periods, potentially longer than the sample period (e.g., application of extreme value theory and the metric "annual maximum-daily precipitation" to estimate frequency of rare heavy precipitation events). Again, changes in N-year extreme events for a given metric should be subjected to statistical testing to determine whether they are significant.
5. Relate Variance in Projected Changes to Global Uncertainties: Three global sources of climate projection uncertainty might be considered when trying to understand uncertainties in projected local metrics: (1) choice among scenarios for global greenhouse gas emissions, (2) choice among global climate models for simulating climate system response to a given emission scenario, and (3) choice among "initializations" that define alternative climate-system states at the beginning of climate projections. Various methods of variance decomposition might be used to allocate local change variance to these sources (e.g., Hawkins and Sutton 2009, 2010).
6. Assess Projected Changes given Global/Local Interactions: This step in the evaluation considers how local features (e.g., topography, land surface) can control variability in local surface climate, and how such local control might interact with global influences on future climate. The process of considering such local features and controls on local response to global climate change are is often referred to as "spatial downscaling" and may be conducted using methods that range from simple statistical techniques to complex and more physically based dynamic simulation.

When executing these steps, various sources of uncertainty and scoping choices become apparent. Interest in these sources of uncertainty varies depending on the purpose of the assessment. For example, for the sake of climate or impacts model development the question might be: how much gain is expected from increased resolution? For rationalizing planning investments informed by projected hydroclimate conditions, the assessment may be focused on the overall degree of uncertainty, or at least on how the decision-support information is sensitive to choices in the assessment design – e.g., is the information sensitive to choice of greenhouse gas emissions scenario, or how the climate projection information is bias-corrected and spatially downscaled?

This chapter quantifies some of the sources of uncertainty that affect projections' portrayal of future hydroclimatic extremes, with thought toward what this means for water resources planning and decision-making under uncertainty. Five such questions are common to assessments of this nature and are highlighted on Fig. 11.1:

- *Which climate projection information is available?*
- *What does this information say about changes in typical and extreme conditions?*
- *What is the influence of climate projection bias-correction?*
- *How do the results relate to sources of climate projection uncertainty?*
- *What is the influence of climate projection spatial downscaling?*

This chapter demonstrates the assessment steps of Fig. 11.1 and explores uncertainties associated with method options along the way. Note that although Fig. 11.1 may be interpreted to suggest that these questions need to be addressed in some order, it may be more appropriate to address these questions in concert, as choices on climate projection ensemble and method of climate projection bias-correction may bear influence on approach for spatial downscaling, and vice versa. Lastly, it is important to recognize that this chapter does not delve into the issue of whether GCMs adequately simulate hydroclimate extremes. Rather, the starting point for this discussion is that readers have access to an ensemble of global climate projections that report information on extremes, that this information presents a projection distribution of extremes, and that this distribution can be related to various sources of global to local uncertainty. This is not to say that adequacy issues should be dismissed.

11.2 Step 1. Identify Precipitation Metrics of Interest

In defining precipitation metrics and extremes, there are a continuum of conditions that might be considered, varying in scale and frequency of occurrence (Karl et al. 2008). Definition choices for a given assessment will vary with respect to which precipitation characteristics matter for a given resource situation. For example, local storm water management districts may be concerned with heavy precipitation extremes and the intensity of storm events having duration hours to days and over areas ranging from small catchments to larger watersheds. In contrast, local water supply managers may be concerned about "dry extremes" or sustained precipitation departures below normal, and the intensity of these departures during periods ranging from weeks to years and over areas that range from local watershed to large regional basins.

For simplicity of illustration in this chapter, focus is placed on precipitation extremes associated with two precipitation metrics:

- *Annual*: annual total precipitation and extreme departures below typical "annual total precipitation" (i.e. precipitation droughts of annual duration).
- *1daymax*: annual maximum-daily precipitation and extreme departures greater than typical "annual maximum-daily precipitation" (i.e. rare heavy precipitation events).

Although this chapter focuses on these two metrics, readers may wish to consider other metrics of hydroclimatic extremes. For example, many CMIP3 modeling groups (see below) computed and archived the so-called "Frich Indices" (Frich et al. 2002) which consist of various measures of extreme temperature and precipitation. The STARDEX project –"Statistical and Regional Dynamical Downscaling of Extremes for European regions" – (Goodess 2005) also identified a set of extremes indices for temperature and precipitation that partially overlaps with the Frich indices. Another potential starting point for exploring other metrics is the list developed by the Expert Team on Climate Change Detection and Indices (ETCCDI, Alexander et al. 2006).

11.3 Step 2. Select a Representative Climate Projections Ensemble

Recognizing the goal of representing choices in future climate forcing, climate model structure, and multiple realizations of internal variability given these choices, the best available information of projected global climate is that generated through the World Climate Research Programme (WCRP) Coupled Model Intercomparison Project (CMIP).[1] The third phase of this effort, CMIP3 (Meehl et al. 2007), includes climate simulations from 23 different coupled atmosphere-ocean general circulation models (GCMs), where each GCM has been applied to simulate future global climate response to different pathways for future greenhouse gas emissions (emissions scenario). Internal climate variability is investigated by performing multiple simulations for with a GCM starting from different initializations of the climate system. In this way, different phases of interannual and interdecadal climate variability can be sampled.

[1]We acknowledge the modeling groups, the Program for Climate Model Diagnosis and Intercomparison (PCMDI) and the WCRP's Working Group on Coupled Modeling (WGCM) for their roles in making available the WCRP CMIP3 multi-model dataset, available at: http://www-pcmdi.llnl.gov/projects/cmip/index.php. Support of this dataset is provided by the Office of Science, U.S. Department of Energy.

Recent evaluation of current climate projection information has led to recommendations that local hydroclimate impacts assessments be informed by a multi-model ensemble of climate projection information (e.g., Pierce et al. 2009, 2011; Mote et al. 2011) rather than seeking to identify a "best set of climate models" to provide projection information. While some studies have shown that ranking climate models has led to a separation of future responses (e.g., Walsh et al. 2008), others have shown that consideration in climate model skill has generally had little influence on change detection and attribution (e.g., Santer et al. 2009; Pierce et al. 2009) or characterization of future climate change distributions (e.g., Brekke et al. 2008). It has also been shown that selection of an appropriate set of performance metrics for ranking climate models can be a challenge, as all climate models appear to do something well relative to their peers and metric relevance varies with application (e.g., Gleckler et al. 2008; Brekke et al. 2008).

Given these findings, this demonstration is informed by a large and representative subset of CMIP3 projections with no efforts made to discriminate based on perceived climate model skill. However, other factors did determine model and projection membership. One factor related to time-step of temporal reporting. While CMIP3 climate simulations are executed on a sub-hourly time-step, output is generally reported in a time-aggregated form, most often as monthly or daily gridded time series information. This chapter's evaluation considers daily precipitation conditions and therefore requires daily precipitation projections. Other factors involve possible pathways of future climate forcing and internal variability of the climate system. This relates to Step 5 where the evaluation considers how global sources of climate projection uncertainty influence variance in local changes, and sets up the need to assemble a projections ensemble representing GCMs that have been applied to simulate different emissions scenarios using (ideally) multiple initializations. Given these considerations, a subset of 9 CMIP3 models were identified at the Lawrence Livermore National Laboratory's Program for Climate Model Diagnosis and Intercomparison (PCMDI) multi-model CMIP3 data repository for which there were daily precipitation projections reported for three Emissions Scenarios (B1, A1B and A2 as defined in IPCC 2000),[2] and simulated from one to five initial conditions (i.e. 53 projections listed in Table 11.1, with numbers below Emissions Scenario indicating initialization number, or "run" number). The reader may wish to review CMIP3 documentation (Meehl et al. 2007) for information about the climate models (e.g., spatial resolution, shared modules and parameterizations).

Each precipitation projection was sampled within a region bounded by −125 to −67°W and 25 to 53°N and for three climate-simulation periods: 1961–2000, 2046–2065, and 2081–2100. Ideally a single period from simulated twentieth century to projected twenty-first century would have been sampled from each daily precipitation projection. However, most of the CMIP3 modeling centers only contributed daily results to PCMDI for these three periods.

---

[2]For the ECHAM5/MPI-OM model, PCMDI did not contain projections of the A1b emissions scenario.

Table 11.1 CMIP3 projections considered in this evaluation

WCRP CMIP3 model I.D.	Reference	Emission scenario (IPCC 2000) and initial conditions (runs)		
		B1	A1b	A2
CGCM3.1 (T47)	Flato and Boer (2001)	1,2,3	1,2,3	1,2,3
CNRM-CM3	Salas-Mélia et al. (2005)	1	1	1
ECHAM5/MPI-OM	Jungclaus et al. (2006)	1		1
ECHO-G	Legutke and Voss (1999)	1,2,3	1,2,3	1,2,3
GFDL-CM2.0	Delworth et al. (2005)	1	1	1
GFDL-CM2.1	Delworth et al. (2005)	1	1	1
IPSL-CM4	IPSL (2005)	1	1	1
MIROC3.2 (medres)	K-1 model developers (2004)	1,2	1,2	1,2
MRI-CGCM2.3.2	Yukimoto et al. (2001)	1,2,3,4,5	1,2,3,4,5	1,2,3,4,5
Total		**18**	**17**	**18**

To set up an efficient multi-model evaluation of projected changes over the domain of interest (Fig. 11.1), the projections were processed in three ways:

- REGRID: Each climate projection's output was interpolated from native GCM spatial resolution to a 2 degree grid within the domain of interest using the SYMAP interpolation scheme (Shepard 1984).
- (Bias-Corrected) BC: Focusing on a common period of observation and simulation overlap (1950–1999), each REGRID projection was translated into a bias-corrected (BC) projection following the quantile-mapping methodology described in Maurer et al. (2010). Application of this method involved adopting Maurer et al. 2002 as a representation of observed historical weather conditions coarsened to REGRID resolution. The procedure results in the BC projections having simulated 1950–1999 daily precipitation conditions that are statistically consistent with observations (Maurer et al. 2002) during that period.
- (Bias Corrected Constructed Analog) BCCA: Each BC projection was spatially downscaled using a non-dynamical, constructed analogs (CA) technique (Hidalgo et al. 2008; Maurer and Hidalgo 2008), producing daily precipitation projections at 1/8 degree spatial resolution.[3]

Step 3 of this evaluation considers both REGRID and BC projections. Steps 4 and 5 then focus on BC results before some limited-area BCCA results are presented in Step 6 (Fig. 11.2).

It is foreshadowed at this point that bias-correcting climate projection information can influence the change assessment. Also, it should be understood that some type of bias-correction will be conducted in assessing projected changes in

[3]The resulting REGRID, BC and BCCA projections are available at the online Bias Correction and Downscaled WCRP CMIP3 Climate Projections archive (http://gdo-dcp.ucllnl.org/downscaled_cmip3_projections/dcpInterface.html).

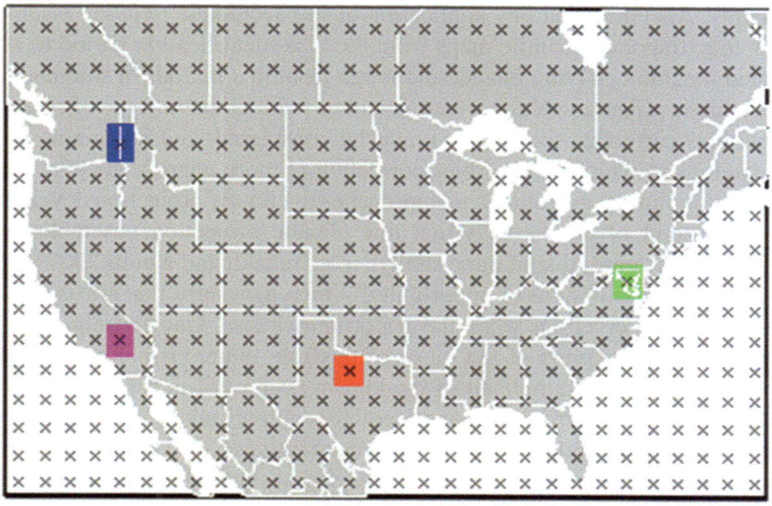

Fig. 11.2 Gridded domain of daily precipitation projections. Map shows 2-degree gridded domain of daily precipitation projections and also four 2 × 2 degree grid cells highlighting focus locations for evaluations in this chapter: Northwest (*blue*, centered on 117W and 47N), Southern Sierra (*magenta*, centered on 117 W and 35N), Southern Plains (*red*, centered on 99W and 33N) and Mid-Atlantic (*green*, centered on 77W and 39N)

extremes, and the degree of influence depends on how the bias-correction method is carried out. For example, an implicit bias-correction technique is to simply assess for changes within REGRID projections and impose those changes on an observed reference in order to imply future conditions; this might be labeled "bias-correction in the mean". Alternatively, bias-correction may be conducted to account for variable degree of bias under simulated warm versus cool or wet versus dry conditions. An example of this type of method is illustrated in Maurer et al. 2010 and is sometimes referred to as quantile mapping. It might also be labeled as "bias-correction in the distribution". A by-product of the quantile mapping approach to bias-correction is that the BC projected precipitation trends can be different than those from REGRID projections because the quantile-mapping method modifies the variance of GCM output during the period of historical bias-identification (Pierce et al. 2011). It also modifies the resultant temporal evolution of statistical properties from that period (e.g., variance, skew, and higher moments). Other applications of quantile-mapping bias-correction have involved first identifying and removing the trend in REGRID projections, applying bias-correction to residual information, and then reinserting the REGRID trend afterwards (e.g., Wood et al. 2002 and application of quantile mapping to bias-correct temperature projections, where doing so seemed logical given that the trend in projected temperature was large relative to the historical envelope of variability). Whether to apply such a trend constraint is a subjective choice, and as will be illustrated, introduces change uncertainty. Also note that any application of this technique requires assumptions on how to extrapolate quantile maps from the bias-identification period to address simulated magnitudes lying outside the range from this period. In summary, some type of climate projection bias-correction will be featured in this type of assessment, multiple bias-correction techniques might be considered, and it remains a matter of research as to which type of bias-correction is more appropriate given current limitations of climate modeling and approaches to climate projection development.

Another point of emphasis is that this climate projection ensemble may be referred to as an "ensemble of opportunity" (e.g. Tebaldi and Knutti 2007), perhaps analogous to the term "ships of opportunity" – merchant ships that are used to make oceanographic observations when the opportunity arises. One consequence is that the opportunistic sampling of the range of possible futures is neither random nor systematic, nor does it sample all of the known sources of uncertainty. Creating a probabilistic projection from such an ensemble requires making many assumptions, and the resulting probabilistic depiction of uncertainty is heavily dependent on those assumptions (Tebaldi and Knutti 2007). Therefore it should be kept in mind that the central trajectory and spread of any quantities that are derived from the "ensemble of opportunity" could be influenced by systematic biases and errors pervasive throughout the projections or in subsequent processing steps such as bias correction and downscaling. One view is that it is futile to assign weighting factors to models in order to create probability distributions from climate model ensembles, but rather to view the ensemble as the "non-discountable range of possibilities" (Stainforth et al. 2007). Annan and Hargreaves (2010), on the other hand, present evidence that the models are "statistically indistinguishable" from one another and from observations, and that an equal weighting of models is a reasonable approach. Our approach in this analysis is to acknowledge the conceptual and practical difficulties in quantifying the absolute uncertainty, while characterizing the sources of variance within the ensemble of opportunity, treating each model with equal weight.

11.4 Step 3. Assess Projected Changes in Typical Precipitation Conditions

The assessment now proceeds to evaluation of projected changes in typical *Annual* and *1daymax* precipitation. In this case "typical" is defined as period-median condition of the given metric. The assessment is conducted on both REGRID and BC projections in order to highlight how uncertainty in projected change may be introduced by the act of bias-correcting the projections. Later in this chapter (Step 6), the assessment is conducted for a limited region in the domain to illustrate how uncertainty is introduced during the process of spatial downscaling, from BC to BCCA information.

Two different views might be adopted when assessing projected changes in hydroclimate conditions:

- Transient: This view requires tracking the ensemble of projection information through time (e.g., time evolution of global average temperature conditions shown on Figure SPM-5 from IPCC 2007). Such tracking draws attention to the temporal evolution of ensemble statistics, such as the ensemble median indicating the centrally expected condition at a time stage, or the ensemble variance indicating the distribution of metric possibilities at a time stage. The transient view invites attention to the projected rate of change in the metric conditions. Applied on a projection-specific basis, the transient view also reveals presence of any low-frequency variability in the projection, which may confound interpretation of whether changes within that projection relate to human or natural forcing.
- Period-change: This view focuses attention on two climate periods, typically a reference historical period and a subsequent historical or projected future period. Changes in metric period-statistics are then assessed. One example is illustrated in IPCC (2007), which shows spatially distributed changes in global mean-seasonal precipitation between two decade periods (IPCC 2007, Figure SPM-7). In climate impacts literature, the period-change view is generally more prevalent than the transient view. Period-change assessments may be applied on a projection-specific or ensemble-pooled basis and with focus on changes in a variety of period-statistics (e.g., period mean and higher moments, minimum, maximum, etc). Interpretation of whether projection-specific period-changes relate to human or natural forcing may be challenged by the presence of low-frequency variability within in projections, particularly for precipitation projections (Hawkins and Sutton 2010).

This chapter applies a period-change view, only because the selected climate projections ensemble (Step 2) is specified for three disjoint periods. Metrics were assessed on a water year basis (i.e. October through September). The single historical period (1961–2000) was split into two historical periods of roughly the same duration as the future periods. This resulted in four periods being considered: (1) water years 1962–1981 serving as the reference historical period (i.e. October 1961 through September 1981), (2) water years 1982–2000, (3) water years 2047–2065, and (4) water years 2082–2100.

The evaluation first focuses on the ensemble-median change in period-median condition for both *Annual* (Fig. 11.3) and *1daymax* (Fig. 11.4). The goal was to understand how consensus change geographically varies through time. Attention to variance in projected changes and associated uncertainties are deferred to subsequent illustrations in Step 5. Change results for typical *Annual* conditions suggest that both "Clausius-Clapeyron" and "Hadley-Expansion" paradigms influence results in the contiguous U.S. domain (Fig. 11.3), consistent with findings from other efforts (USGCRP 2009; Seager et al. 2007). Focusing on the two historical periods, changes from 1962–1981 to 1982–2000 for both REGRID and BC results are minimal throughout the domain, with changes of plus or minus a few percent. Focusing on the future periods and changes from the 1962–1981 period, geographic patterns of change are more apparent. For both REGRID and BC conditions, the

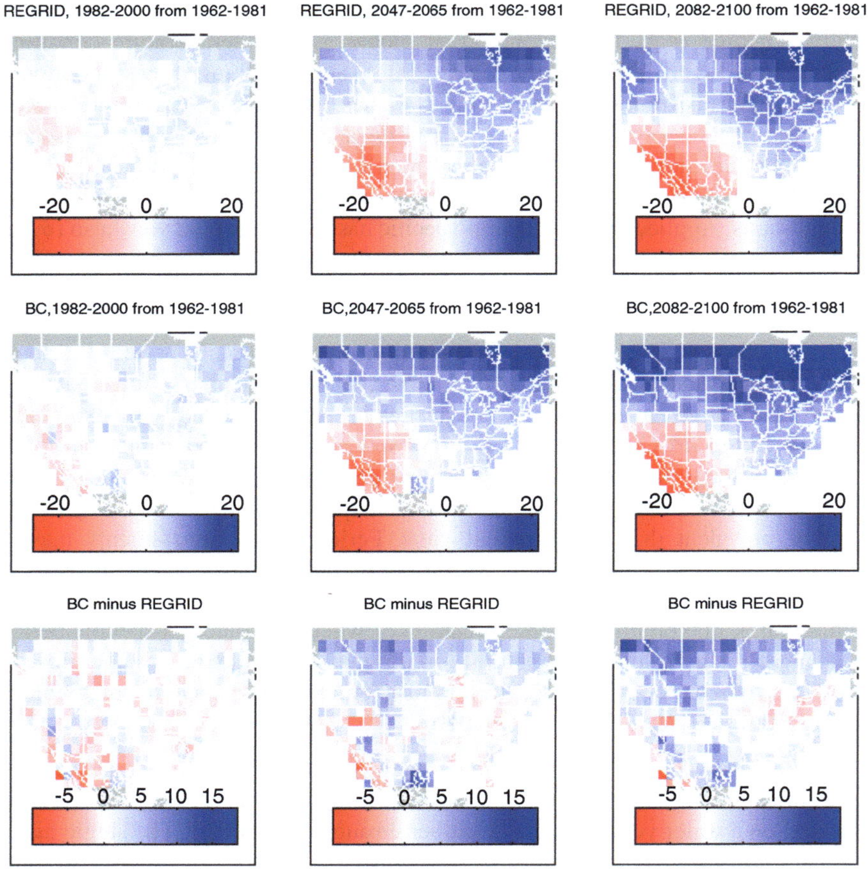

Fig. 11.3 Percentage change in period-median annual total precipitation (*Annual*). Maps show spatially distributed, ensemble-median projected change in period-median condition for REGRID and BC projections (*top* and *middle* rows, respectively) and for three period-pairs (*columns*). The *bottom row* shows the mapped differences in this change measure when using BC versus REGRID projections

northwest to northeast region shows increases in period-median *Annual* conditions, potentially due to the "Clausius-Clapeyron" effect being prominent over these regions. Switching to the southwest region, both REGRID and BC projections suggest a decrease in precipitation, which is well associated with the "Hadley Expansion" effect, and suggests that perhaps this effect is more prominent than any "Clausius-Clapeyron" affecting precipitation-bearing weather over this region.

Comparison of changes in REGRID and BC *Annual* (Fig. 11.3, bottom row) reveals how the act of bias-correction can introduce uncertainty. Throughout much of the domain, and notably over southern Canada and the U.S. Great Plains, the BC changes are wetter than those from REGRID, and for some locations by more

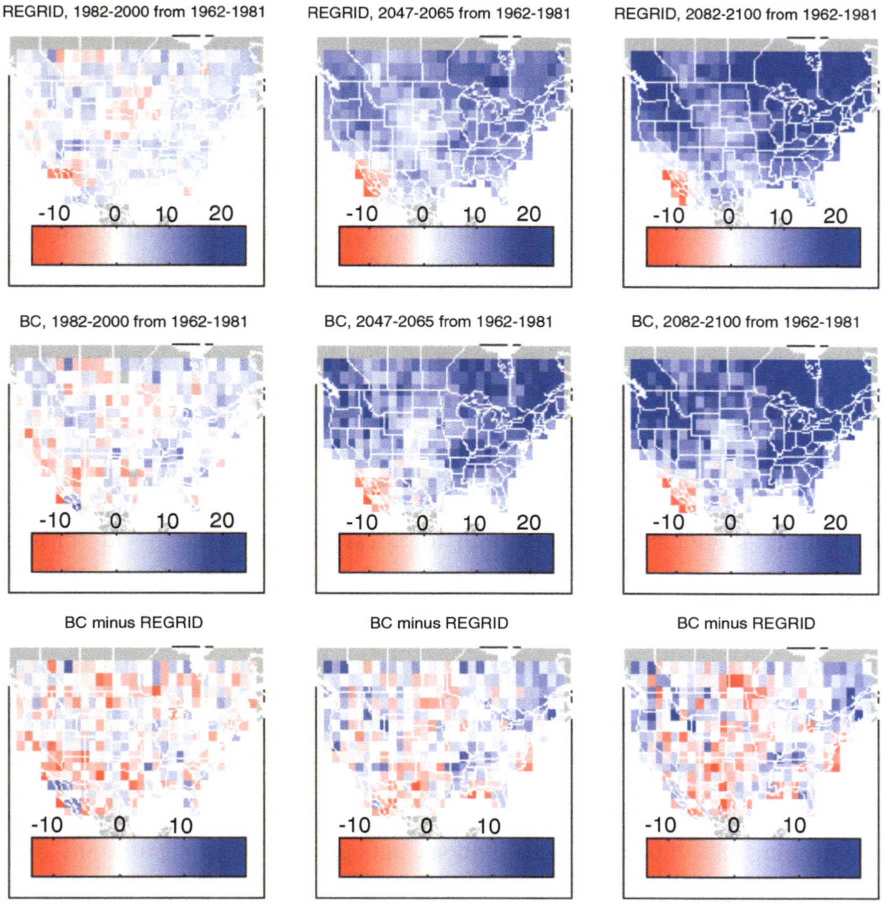

Fig. 11.4 Percentage change in period-median annual maximum-day precipitation (1daymax). Same as Fig. 11.3, but for the precipitation metric 1daymax

than 10%. However, some regions show BC changes that are drier than those from REGRID, including Ohio River Valley and areas within the Southwestern U.S. This illustrates how quantile-mapping bias-correction modifies the variance of REGRID output (Pierce et al. 2011) and, without the constraint of trend preservation (Wood et al. 2002), can lead to a different evolution of statistical properties in the BC changes compared to REGRID changes.

Geographic change patterns for typical *1daymax* conditions (Fig. 11.4) are similar to those found for *Annual*. Changes from the reference 1962–1981 to 1982–2000 show little to no change, and no apparent geographic pattern of change. For the two future periods, the same regions emerge as areas of increased (Northwest to Northeast) and decreased metric conditions (Southwest). However, there are three noteworthy differences between projected changes in typical *Annual* and *1daymax*. First, the areas of positive change are generally greater for *1daymax* than for *Annual*, and encroach on the the Annual areas of neutral to negative change. This result perhaps highlights areas (e.g., southern Plains, southern Intermountain West) that may experience both minor decreases in annual total precipitation and increases in heavy daily precipitation events when they occur (USGCRP 2009). Second, the spatial change patterns have less noise or more spatial fidelity for the *Annual* condition compared to the *1daymax* condition, suggesting that interpreting *1daymax* change in a single grid-cell should be informed by consideration of surrouding grid-cell changes within the greater region. Third, the comparison of BC and REGRID results shows that quantile-mapping bias-correction can affect locally expressed *1daymax* changes, but with less spatial fidelity than how it affected *Annual* changes. In sum, these differences in spatial change patterns for *Annual* and *1daymax* suggest that projected changes in *1daymax* may not linearly trend with changes in *Annual*, especially at the cell resolution of REGRID and BC information.

Even as the preceding evaluation illustrates projected change in typical metric conditions, a complementary question has yet to be addressed: are these changes statistically significant? One approach to assessing significance is to consider projection-specific period-distributions of metric conditions, compare distributions between two periods using a two-sample Komolgorov-Smirnov test (Massey 1951), and then determine whether to reject a null hypothesis that the two period distributions originate from a common underlying distribution (e.g., at a specified significance level). This type of test considers the potential influence of change in median and variance and was applied using a 0.10 significance level to compare period-distributions of metric conditions. Figure 11.5 indicates the count of BC projections where the null hypothesis was rejected when applied to each of the three periods following the reference period. Results show that most projections do not contain 1982–2000 metric distributions that appear to differ significantly from 1962 to 1981 distributions. For the two future periods, geographic patterns emerge where a majority of projections appear to express different period-distributions, with greater frequency of null rejection occurring over the Northwest and Northeast regions. It's notable that the frequency of null rejections is greater for the *Annual* metric than for *1daymax*. So although the Northwest and Northeast regions of positive change were arguably greater for *1daymax*, the ability to characterize such changes as being significant was confined to the interior areas of both regions and/or supported by fewer projections. This result seems to be due to test results for *1daymax* being more influenced by *1daymax* period variance compared to associated tests for *Annual*. In relation to this, a Wilcoxon Rank Sum test (Gibbons 1985) was also applied to evaluate for apparent changes in period-median conditions. Geographic patterns of null rejection frequency were found similar to those from Komolgorov-Smirnov tests (results not shown), implying that the latter may be primarily influenced by apparent change in median rather than change in distributional variance and shape properties.

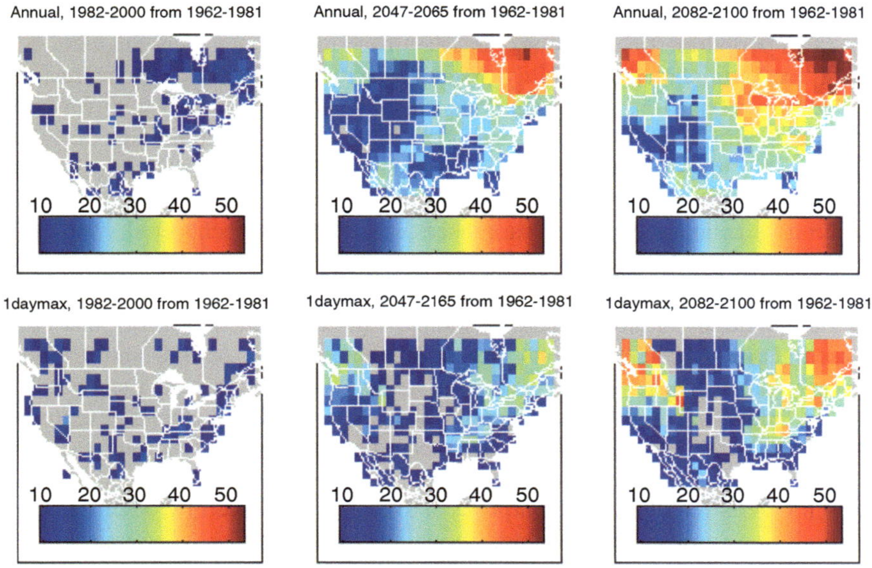

Fig. 11.5 Count of projections showing significant differences in period-distributions of annual and 1daymax. Plots show spatially distributed count of BC projections where, when comparing the distribution of 1962–1981 Annual (*top row*) or 1daymax values (*bottom row*) to the associated distribution of values from a subsequent period (1982–2000, 2047–2065, or 2082–2100 for the columns, *left to right*, respectively), a Komolgorov-Smirnov two-sample test (alpha 0.10) resulted in a rejection of the null hypothesis that the two period distributions come from a common underlying distribution. Only locations with counts of at least nine (which would occur with probability less than 0.1, assuming a binomial distribution) are shown

11.5 Step 4. Assess Projected Changes in Extreme Precipitation Conditions

The assessment now proceeds to evaluation of projected changes in extreme *Annual* and *1daymax* precipitation. For *Annual*, an extreme occurrence would be an atypical amount of annual total precipitation, whether it's a large or small total. In this case, focus is placed on the latter as such annual "droughts" can test capabilities in managing water supply systems to satisfy various demands (i.e. *Annual* events expected to occur once in every 20–100 years). For *1daymax*, an extreme occurrence would be an atypical amount of annual maximum-day precipitation. In this case, focus is placed on large and somewhat rare *1daymax* events that are relevant to the design and planning of various water resources systems (i.e. *1daymax* events that are expected to occur once in every 20–100 years). From this step forward, evaluations are conducted only on BC projections, having already noted in Step 3 how climate projection bias-correction can introduce uncertainty about assessed changes.

In order to estimate rare metric occurrences, cumulative probability distributions are fit to period-metric conditions, and then used to estimate event magnitudes of a specified cumulative probability associated with return-periods of interest (i.e. $p = 0.05$ and 0.01 for small-magnitude *Annual*, and $p = 0.95$ and 0.99 for large-magnitude *1daymax*). For *Annual* samples, a gamma distribution with maximum likelihood estimated parameters was used (Hahn and Shapiro 1994). For *1daymax* samples, a three-parameter generalized extreme value (GEV) distribution with maximum likelihood estimated parameters was used (Embrechts et al. 1997; Kotz and Nadarajah 2000). Both distributions were fit in two ways: (1) to "projection-specific" conditions (i.e. projection-, period-, and grid cell-specific samples of metric conditions, where N years of sample values informed each fit), and (2) to "ensemble-informed" conditions (i.e. projections pooled-, period- and grid cell-specific samples, where $53 \times N$ years of sample values informed each fit). It is acknowledged that other statistical methods might be used to characterize extremes, and that this chapter simply adopts one method common to engineering practice as a means to evaluate uncertainty in projections of extremes. In principle, the method chosen here is for illustration purposes and could be replaced by other candidate methods (e.g., peaks over thresholds, fitting other distributions to extremes).

The first part of this evaluation step focuses on uncertainty introduced from the choice of fitting projection-specific or ensemble-informed distributions. Focusing on *1daymax* and the four grid cell locations highlighted on Fig. 11.2, results show that the central tendency and shape of projection-specific distribution varies considerably (Fig. 11.6). Results also show that the ensemble-informed distribution lies somewhat centrally within the spread of projection-specific fits. When the evaluation focuses on ensemble-median changes in extremes, results were found to be insensitive to whether the result was sampled from the collective of projection-specific fits or sampled from the single ensemble-informed fit.

Broadening to changes in extremes throughout the domain, only the results from ensemble-informed fits are shown on Figs. 11.7 and 11.8, which respectively contain results for small-magnitude *Annual* and large-magnitude *1daymax* extremes for three return periods: 2-, 20- and 200-year. The 2-year return period represents sampling the distributions at $p=0.50$ and should show change geographic patterns similar to the period-median changes already discussed in Step 3. Focusing first on *Annual* extremes and the two future periods, it's apparent that changes in extreme dry years (20- and 100-year return periods) do not necessarily follow changes in typical years (2-year return periods). Changes in typical annual precipitation suggested the Northwest and Northeast as regions of positive change and the Southwest as a region of negative change. When focus is switched from typical to extreme years, the Southwest region expands to include much of the southern tier of the domain, and the regions of positive change are confined to the more northwest and northeastern areas. Perhaps most striking is that the projected change in 100-year "annual precipitation droughts" is expected to be neutral to negative over nearly all of the contiguous U.S. even though sizeable portions of the northwest to northeast are expected to receive an increase in typical year precipitation (Fig. 11.3).

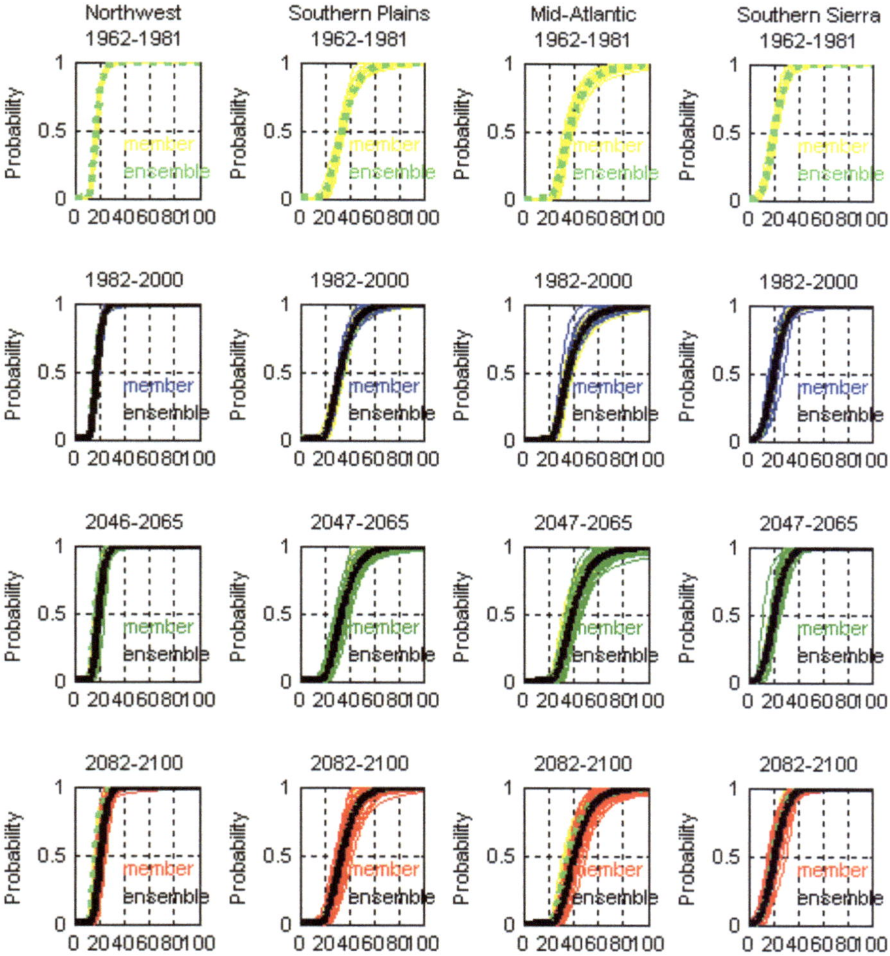

Fig. 11.6 Estimated generalized extreme value distributions fit to period samples of 1daymax at four grid cells. Distributions were fit to data from BC projections (x-axis) from a given period (*row*) and grid cell (*column*, Fig. 11.2). Distributions were fit using two types of samples: projection-specific samples resulting in 53 distributions associated with 53 projections for a given period and grid cell (lines colored other than *black*); and, an ensemble-pooled sample resulting in a single distribution representing all 53 projections for a given period and location (*heavy black line* except for 1962–1981 period, where it's a *dashed green line*). For periods following 1962–1981, the distributions from 1962 to 1981 are also shown, plotted under the given period's distributions (*yellow and green lines* from *top row* of plots)

Switching focus to *1daymax* extremes and the two future periods, results show that changes in more common *1daymax* extremes (20-year return periods) generally follow changes in typical years (2-year return periods), but not so for the more uncommon *1daymax* extremes (100-year return periods). In addition, as the

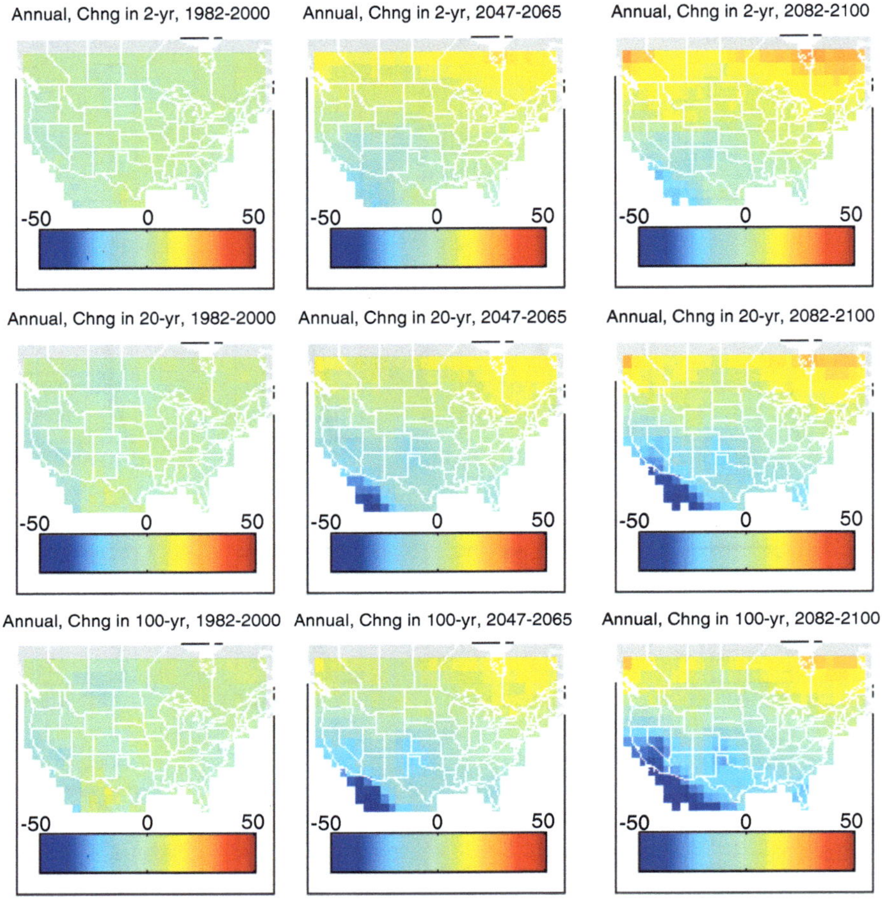

Fig. 11.7 Percentage change in estimated annual values having N-year return periods. Maps show percentage change in estimated Annual values based on (1) fitting gamma distributions to ensemble-pooled BC projected conditions during 1962–1981, 1982–2000, 2047–2065, and 2082–2100 periods, (2) estimating events having 2-, 20- and 100-year return periods (*rows*) as the period-specific gamma distributions sampled at 0.50, 0.05 and 0.01 cumulative probabilities, and (3) assessing changes in these return period values for the three latter periods relative to the 1962–1981 period (*columns*)

distributions are used to infer the more uncommon extremes, the spatial fidelity of changes tends to breakdown. This suggests that at this grid-cell resolution there may be greater uncertainty in estimating changes in 100-year return period values than changes in 20- and 2-year return period values.

Before concluding this section, it is noted that uncertainties in estimating projected changes in extremes are introduced through the choice of distribution form and method of distribution fitting. For example, although the gamma distribution

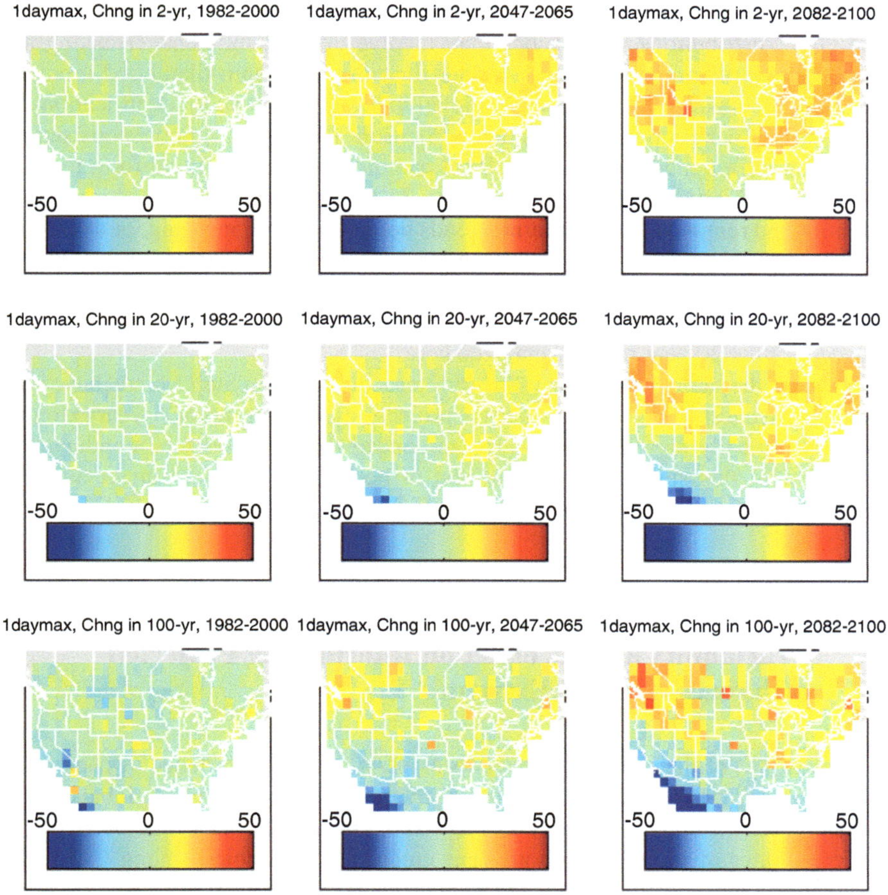

Fig. 11.8 Percentage change in estimated 1daymax values having N-year return periods. Maps show percentage change in estimated 1daymax values based on (1) fitting generalized extreme value (GEV) distributions to ensemble-pooled BC projected conditions during 1962–1981, 1982–2000, 2047–2065, and 2082–2100 periods, (2) estimating events having 2-, 20- and 100-year return periods (*rows*) as the period-specific GEV distributions sampled at 0.50, 0.95 and 0.99 cumulative probabilities, and (3) assessing changes in these return period values for the three latter periods relative to the 1962–1981 period (*columns*)

has been suggested as an appropriate distribution to describe *Annual* precipitation possibilities (Wilks 2006) and changes due to global warming (Watterson and Dix 2003), it may not be the ideal distribution to describe conditions for some U.S. locations (or grid cells in this case). Likewise, the GEV distribution may be commonly featured in engineering practice to describe heavy precipitation intensity-duration-frequency relationships (e.g., NOAA Atlas 14, Volume 6, Version 2.0: California (Perica et al. 2011)), but may not be the ideal distribution for some locations. Also, estimating parameters for a given distribution can be done multiple ways (e.g., for the GEV distributions, approaches may involve using L-moments (Hosking and Wallis 1997) or product moments rather than the maximum-likelihood moments used in this illustration). Parameter estimation is also sensitive to identification and treatment of data outliers, efforts made to ensure temporally homogenous sample data, and efforts made to constrain local parameter estimates to conform to some larger spatial pattern of parameter values. In this case, no efforts were made to identify and removed period-internal trends before distribution fitting, or to identify regional (multi grid-cell) constraints on local (grid-cell) parameter values. Wehner (2010) discusses several practical issues in applying GEV analysis to climate data, including trend and bias removal. He notes that trend removal is not generally needed for the 20 year periods analyzed here.

Finally, the same Komolgorov-Smirnov tests discussed in Step 3 could have been performed to test whether the changes in projection-distributions of N-year extremes were statistically significant. It is recommended that such tests be conducted to complement this step if there is interest in being able to comment on whether assessed changes in N-year metrics are significant.

11.6 Preliminary for Step 5: Low-Frequency Climate Variability and Its effect on Interpreting Projected Changes in Local Extremes

The next part of the evaluation involves assessing how the variance of changes in *Annual* or *1daymax* extremes decomposes relative to three sources of global climate projection uncertainty: (1) choice among future climate forcing scenarios involving different rates of global greenhouse gas emissions, (2) choice among GCMs for simulating climate system response to a given emission scenario, and (3) choice among "initializations" that define alternative climate-system states at the beginning of climate projections. The existence of multiple emissions scenarios (IPCC 2000) and multiple GCMs is well recognized among the practitioner community. However, the third source of global uncertainty, also referred to as "internal" climate variability because it is not the result of an imposed climate forcing, can bear significant implications for interpreting regional to local climate changes within global simulations decades into the future (Hawkins and Sutton 2009), particularly for precipitation and in the mid-latitudes (Hawkins and Sutton 2010).

Some of the uncertainty associated with internal variability stems from decadal to multidecadal modes of the climate system. Other aspects arise from finite sampling of shorter-term climate and weather variability that would exist even in the absence of low-frequency climate modes. Considering our climate projections ensemble, it is clear that low-frequency variability is expressed in the climate projections, but to varying degree depending on climate model and geographic location. This low-frequency variability can confound interpretation of projected extremes. For example, Kendon et al. (2008) indicated that a single 30-year simulation is not adequate to infer changes in extreme precipitation in regions where there is large multi-decadal variability. One also needs to recognize that GCMs should be scrutinized for more than just their expression of low-frequency climate system variability but also for whether this simulated variability seems reasonable or plausible relative to observations.

Internal variability, if not recognized as such, can give the erroneous appearance of a large change in precipitation when a period-change view is applied. To illustrate, consider projection-specific changes in period-median metrics at the four locations introduced in Step 4 (Fig. 11.1) from 1962–1981 to 2047–2065 (futr1) and 2082–2100 (futr2). Recognizing that the projections share twenty-first century climate forcing scenarios that have a somewhat monotonic increase in climate forcing, it's reasonable to expect a somewhat monotonic increase in global air temperature. It might also be reasonable to expect that this implies a monotonic change in local precipitation if it's sensitive primarily to changes in global climate forcing. However, as shown for the four locations on Fig. 11.9, many projections do not express monotonic change from the reference period *through* the future periods. Some projections show more positive change by futr1 transitioning to less positive change by futr2 (green projections on Fig. 11.9); others show the opposite (red projections). For these projections, it appears that there is low-frequency precipitation variability that is significantly influencing period-change in precipitation at these grid cells.

When this analysis is broadened to consider presence of this type of low-frequency variability at all grid cells (Fig. 11.10), regions emerge where it appears that projected changes may be more sensitive to low-frequency variability. Focusing on the right-column maps of Fig. 11.10, it appears that changes in typical *1daymax* conditions over the high-plains and southern portions of the contiguous U.S. are sensitive to projected low-frequency climate variations. For *Annual*, these same regions are identified, although the frequency of projections exhibiting low-frequency variations is generally reduced except for over the southern Intermountain West. Comparing these results to those from Step 3, it is evident that the geographic pattern of projections having changes affected by low-frequency variability seems to generally align with the geographic pattern of locations where the period-changes in typical metric conditions did not appear to be statistically significant (Fig. 11.5).

11.7 Step 5. Relate Variance in Projected Changes to Global Uncertainties

The evaluation now considers how uncertainties in projected local changes relate to three sources of uncertainty in global climate projection uncertainty: future emission path, climate model sensitivity, and internal climate variability (as simulated by the models). Hawkins and Sutton (2009, 2010) apply variance decomposition as

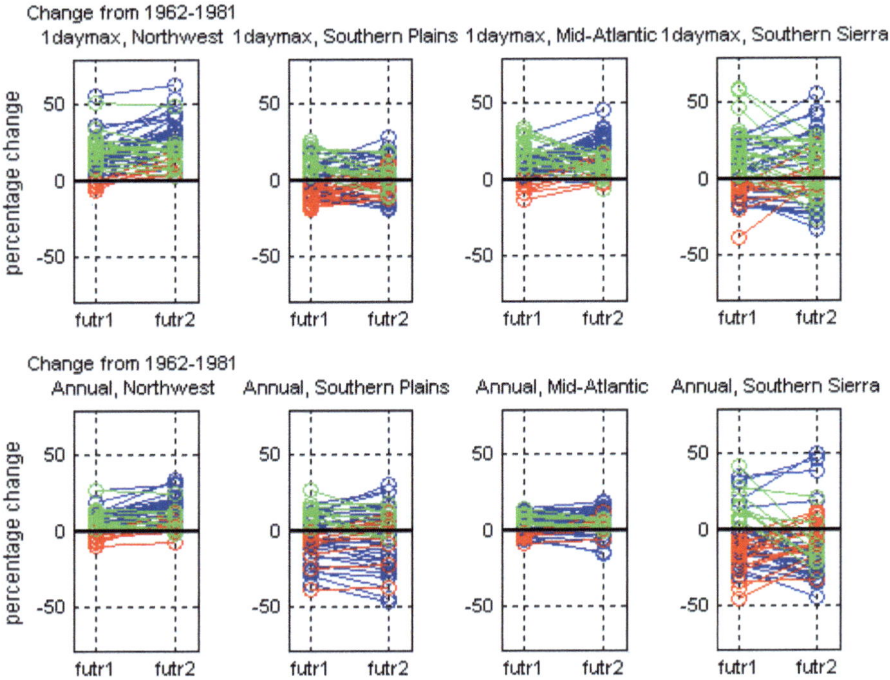

Fig. 11.9 Period-to-period stability of projected changes in extremes – four grid cells. Figures show ensemble of BC projected changes in period-median values for 2047–2065 (futr1) and 2082–2100 (futr2) relative to relative to 1962–1981. All projections are initially shown as *blue circles* connected by *blue lines*. Projections are then color-recoded if they have change values that do not monotonically increase or decrease through time, either from more negative to more positive change (*red*), or from more positive to more negative change (*green*)

a way to relate the uncertainty of annual temperature and precipitation at global and regional scales to these three sources of global projection uncertainty (i.e. HS2009 and HS2010, respectively). Here we provide a similar analysis of variance, focusing on the *1daymax* metric of heavy precipitation focusing on the ensemble of BC precipitation projections.

The ensemble of 53 available projections poses a challenge for this purpose as it contains an unequal number of runs for each model (Table 11.1). Most models have only a single run during the past, and only three runs (one for each emissions scenario) into the future. The small sample size makes it difficult to separate the uncertainty introduced by internal variability from the other sources. In order to gain as much statistical power as possible for the variance decomposition all changes are computed with respect to 1962–2000. This choice differs from the rest of the paper where changes are with respect to 1962–1981.

Following Kendon et al. (2008) the precipitation conditions in each projection are regionalized by pooling the daily values within a 3×3 array of adjacent cells centered on each grid-cell. For coastal grid-cells, only the data overlying terrestrial

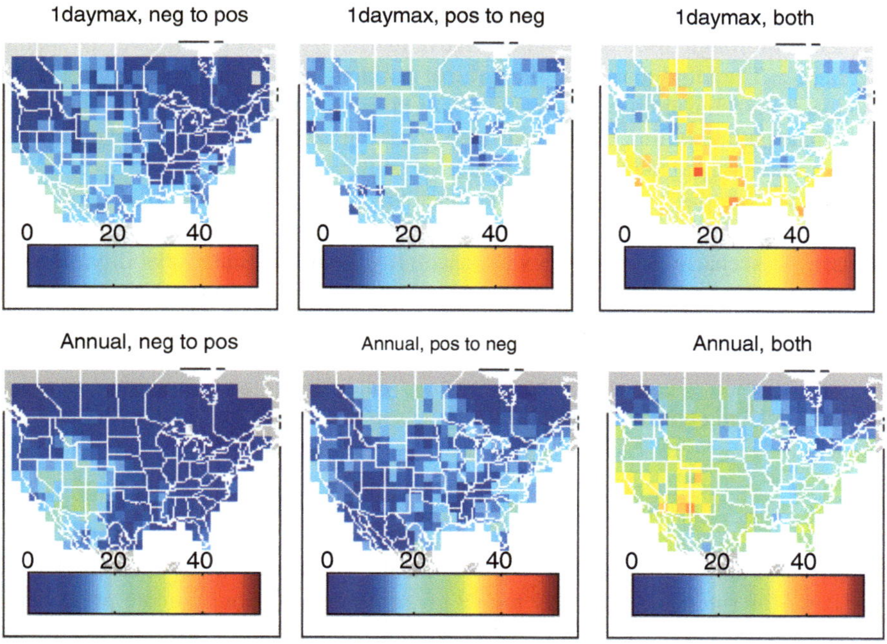

Fig. 11.10 Period-to-period stability of projected changes in extremes – full domain. Following Fig. 11.9, maps show the count of BC projections expressing projected changes in period-median conditions (1daymax on the top row, Annual on the *bottom* row) that do not monotonically increase or decrease through futr1 and futr2 (Fig. 11.9), transitioning from more to less negative (*neg to pos*, *left column*), more to less positive (*pos to neg*, *middle column*) or either (*right column*)

areas can be pooled given the limitations of BC projections generation that relies on a land-based observational dataset for bias correction. Unlike Kendon et al. (2008), we rescale the data before pooling, dividing by the projection-specific 1961–2000 mean *1daymax*:

$$q_{s,m,i}(x, y, t) = \frac{\tilde{q}_{s,m,i}(x, y, t)}{\langle \tilde{q}_{s,m,i}(x, y, t) \rangle_{t=1961..2000}},$$

where $\tilde{q}_{s,m,i}(x, y, t)$ is the *1daymax* value at each grid-cell (x = longitude, y = latitude) for each year (t) and the brackets denote the average over the period shown. The subscripts denote emissions scenario s, model m, and run i. As described in Step 4, a GEV distribution is fit to the pooled data to obtain the 2-, 20- and 100-year return values for each of the two time periods in the future at each grid-cell. The only difference from Step 4 is that cell-specific GEV distributions are fit to the pool of values from a 3 × 3 array of grid cells centered on the given cell rather than to only the values from the given cell. Though the individual grids in the pooled sample are not strictly independent, a spot check of several grid-cells shows that the GEV distribution provides a good fit to the pooled data. Kendon et al. (2008) present a thorough discussion of the relative merits of spatial pooling versus spatial averaging. They find that a small degree of pooling, such as the 3×3 pooling, can increase the robustness of estimated changes in extreme precipitation. They also note that spatial pooling is no substitute for having an ensemble of simulations that can sample many phases of multidecadal variability.

To preview the approach, the variance in a grid-cell's projected change in *1daymax* will be decomposed and assigned to the following order of uncertainties: emissions scenario, climate model sensitivity, and internal variability. Before presenting this analysis of variance, emissions scenario- and model-specific mean changes are first inspected. First consider the variation among the emissions scenarios. For a given *1daymax* statistic $z_{s,m,i}$, such as the N-year return value at a given location, we compute the mean fractional change over individual runs for each model and emissions scenario,

$$Z_{s,m} = \frac{\langle z_{s,m,i} \rangle_{i=1,I(m),future}}{\langle z_{m,i} \rangle_{t=1962-2000}} - 1,$$

where the number of runs, *I(m)*, would ideally be the same for each model. In our case, the run count varies by model given the limits of our climate projections "ensemble of opportunity." In this way, no individual model is given more weight simply because there are more runs in the dataset. We then average $z_{s,m}$ over the models to get the mean response for each emissions scenario at each future time period, z_s, having an N-year return period. The emissions-specific change patterns, averaged over all models (not shown) are quite similar to one another for either period and in general show larger amplitude changes in the later time period.

Now consider the variance associated with climate models and the same statistic. The challenge of isolating the model-specific changes is highlighted in Fig. 11.9, which shows the temporal instability of the sign of change when considering individual model runs. In most locations, one would expect that the model-specific changes would scale monotonically with the increased emissions, with deviations from monotonicity due to internal variability. Given the small number of runs for most of the available models, one way to increase the sampling of internal variability is to note that the simulations using the three emissions scenarios diverge from one another due to the chaotic nature of the climate system. As a result, by mid-century the three emissions scenarios that were run for each initial condition are effectively independent samples of each model's internal variability.

The average of $z_{s,m}$ over emissions scenarios yields an estimate of the spatial pattern of change for each model for 2047–2065 and 2082–2100. For example, the resulting patterns reveal striking differences among model-specific results when considering 2082–2100 change from 1962 to 2000 in the 2-year return value of *1daymax* (Fig. 11.12). When the model-specific patterns for the two time periods were computed independently, the average pattern correlation across all models (excluding the mpi model) was 0.86 (2-year), 0.80 (20-year), 0.77 (100-year). This agreement is a good indication that the sampling is sufficient to discriminate

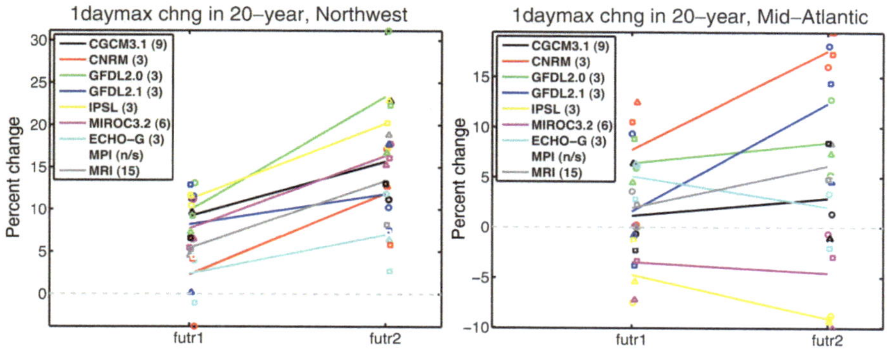

Fig. 11.11 Percent change in 20-year return value for 1daymax for two regions. The panels show change for the Northwest (*left*) and Mid-Atlantic (*right*) locations (Fig. 11.2) from 1962–2000 to 2047-2065 (futr1) and 2082-2100 (futr2) for each of the models noted. The total number of runs for each model, including all emissions scenarios, is shown in parentheses. The model-colored lines indicate the mean over all emissions scenarios and runs for a given model; the model-colored circles (a1b), triangles (a2), and squares (b1) show the average of all runs for each separate emissions scenario. The ECHAM5 (MPI) model is not shown because only two emissions scenarios were available

among the broad patterns of change in each model. The model-specific change at two locations for the 20-year return value of 1daymax precipitation is shown in Fig. 11.11. While not perfect, we see that there is a tendency for each model-specific average to show monotonic change in time. For the mid-Atlantic location we note a consistent model-specific behavior for models that show both increasing and decreasing extremes.

The model-specific and emissions-specific results motivate an analysis of variance approach that differs from the method of HS2009. HS2009 used a polynomial fit in time to $z_{s,m,i}$ at each location in order to isolate the underlying model sensitivity from the low-frequency internal variability. The lack of continuous time periods in our projections ensemble and the similarity in the spatial patterns motivate a different approach based on the concept of pattern scaling (Mitchell et al. 1999). Pattern scaling assumes that the change in a variable of interest can be approximated by a model-specific spatial pattern multiplied by a scale factor α that represents the overall magnitude of the climate forcing in a given emissions scenario. Formally,

$$Z_{s,m,i}\,(x,y,\tau) = \alpha\,(s,\tau)\cdot P_m\,(x,y) + \varepsilon_i\,(x,y,s,\tau)$$

where P_m is the spatial pattern for model m determined as the change from the mean of 1962–2000 to the mean of the two future period means (2047–2065 and 2082–2100). The scaling factor, α, is associated with emissions scenario s and future period τ. Unlike Mitchell et al. (1999) we do not assume a linear relationship of the scale factor to the global mean temperature. Instead, the scaling factors are determined by linear regression of the projection-specific changes for a given model,

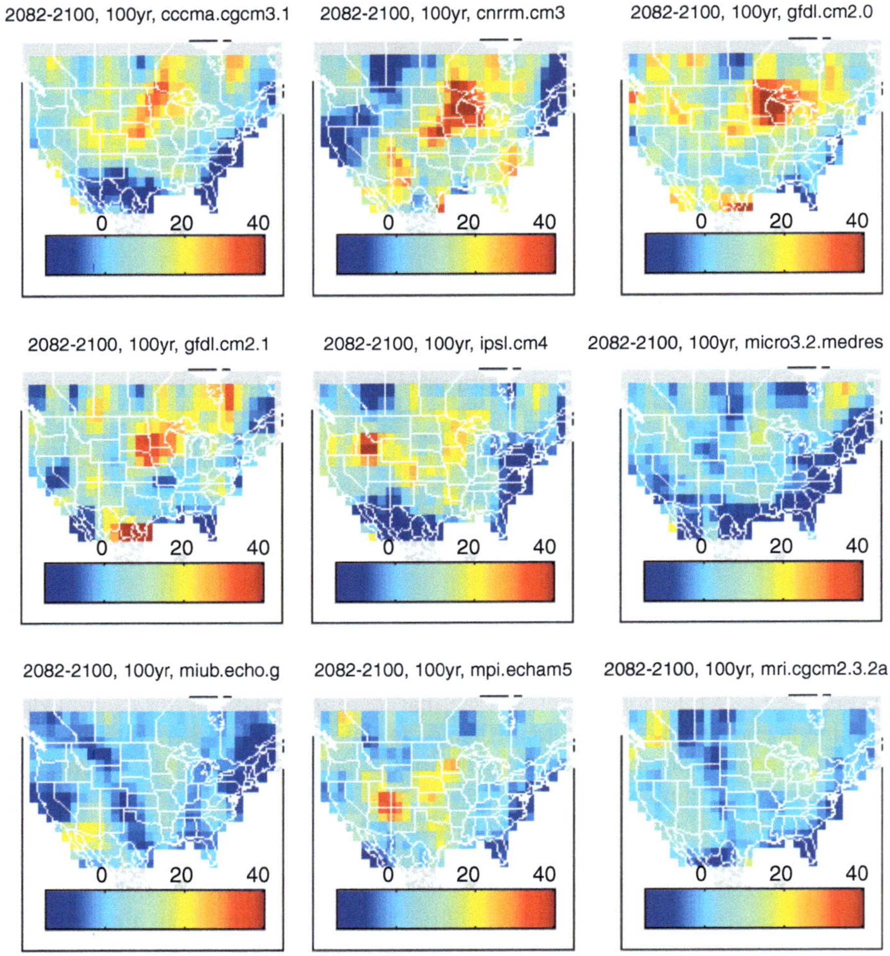

Fig. 11.12 Percent change in 1daymax having a 100-year return value for nine climate models from 1962–2000 to 2082–2100. Changes are computed from residual BC projections of 1daymax (having period- and emissions scenario-specific means removed) and reflect average change over all runs simulated by the given model (**Table**11.1). Note that data for the mpi model was available only from the B1 and A2 emissions scenarios, so its estimates are not as robust as for the other models

$Z_{s,m}$, onto the pattern P_m for that model. The model-specific scale factors are then averaged over all models to obtain a scale factor for each emissions scenario and future period. Considering only the first term on the right hand side of the above equation, $\widehat{z}_{s,m}(x, y, \tau) = \alpha(s, \tau) \cdot P_m(x, y)$ represents the combined effects of emissions scenario and model choice (Fig. 11.12).

Proceeding to the analysis of variance, we provide a linear decomposition of the sources of variance as in HS2009, but with the variance contributed by emissions

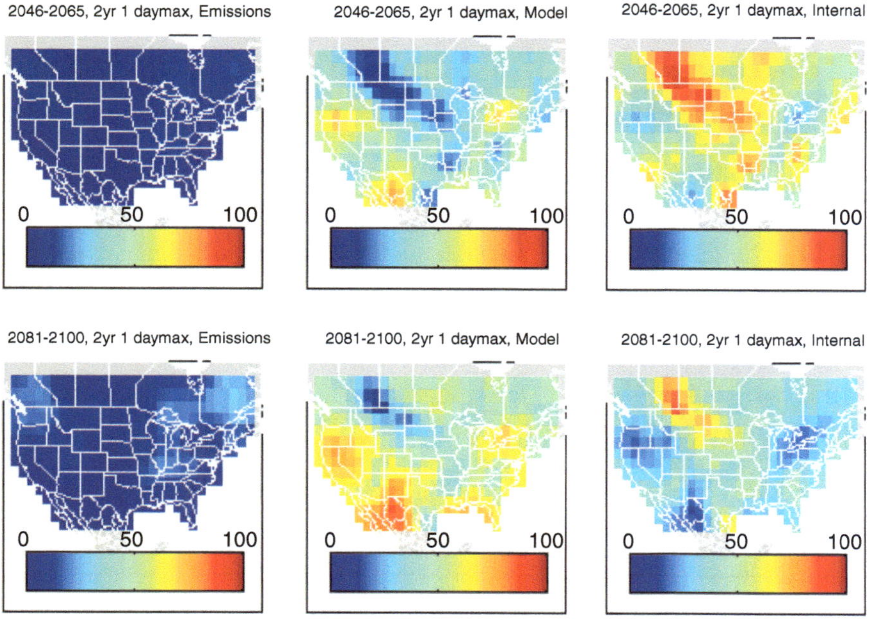

Fig. 11.13 Percentage of variance in projected change in 2-year 1daymax explained by three sources of global climate projection uncertainty. Percentage variance is allocated to choice among emissions scenarios (*left*), climate models (*middle*), and internal variability (*right*) for change from 1962–2000 to 2047–2055 (*top row*) and 2082–2100 (*bottom row*)

scenario and model computed from $\widehat{z}_{s,m}$ rather than from their polynomial curves. The emissions scenario variance is computed first for each model, and then averaged over all models. This differs from HS2009, who compute the mean over all models first, then compute the variance. In the case that the mean over all models is zero, as in the right panel of Fig. 11.11, this might lead to an erroneous result. The model variance is computed on the model-specific changes averaged over all emission scenarios. The internal variability is then determined from the variance of the residual $\varepsilon_{i.}$ for each model run, averaged over all models and emissions scenarios. As in HS2009 we assume that there are no interactions among the sources of variability so that the total variability follows from $\sigma^2_{total} = \sigma^2_{model} + \sigma^2_{emissions} + \sigma^2_{internal}$.

This variant on the HS2009 method is also very simple, and no doubt a more sophisticated analysis could be attempted. For reasons stated in HS2009 this method probably yields underestimates of the overall uncertainty in these projections. In addition, the estimates of the relative sources of variance will depend on the assumptions implicit in the use of pattern scaling as well as on the choice of baseline period, and on whether or not to regionalize the data.

Figures 11.13 and 11.14 respectively show the percentage of the variance accounted for by the three factors for the 2-year and 100-year return values. Emissions scenario uncertainty is only significant in the latter period, as one would

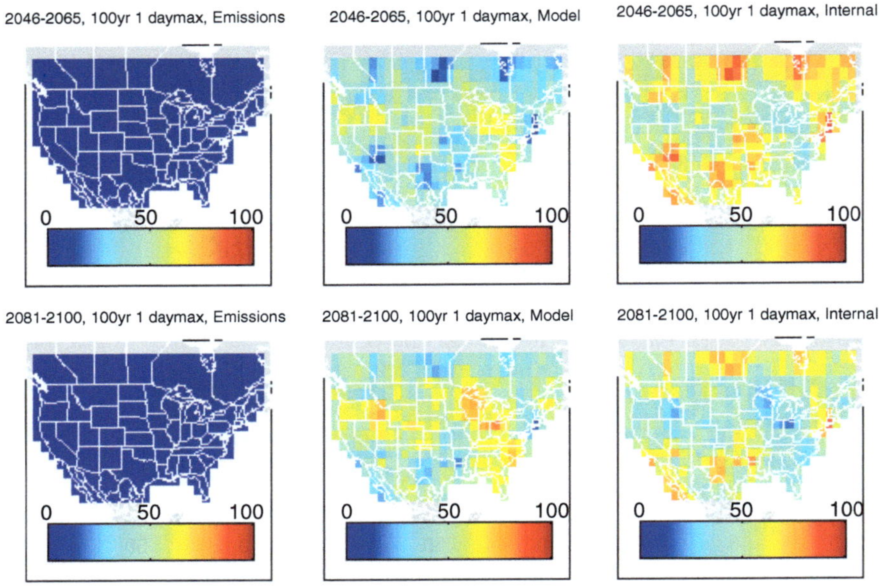

Fig. 11.14 Percentage of variance in projected change in 100-year 1daymax explained by three sources of global climate projection uncertainty. Similar to Fig. 11.13, but for the 100-year return period

expect from the relatively small difference in the associated climate forcing at mid-century (IPCC 2000), and even then only for the 2-year return values. Even in the period 2082–2100 it generally accounts for less than 20% of the variance in much of the domain. As noted in HS2010, the emissions scenarios generally play a larger role at larger spatial scales. Much of the model and internal uncertainty has large amplitude at sub-continental scales, whereas emissions uncertainty acts to alter the pace and ultimate magnitude of change globally.

The uncertainty from internal variability is generally largest for the period 2047–2065 for both return periods. For the period 2082–2100, both model uncertainty and internal variability play a very large role over much of the domain. For the 2-year values, model uncertainty dominates on the west coast with values up to 80% and is in the range or 30–50% over much of the domain. Even for the 100-year return value, model uncertainty is a significant factor.

This leads to several issues when interpreting consensus and spread among ensemble changes. First of all, there are regions where the model signals are not in the same direction (e.g., Fig. 11.11 highlighting changes for two of the grid-cell "regions" shown on Fig. 11.2). For such regions, the multi-model mean signal may be much smaller than the signal in most of the individual models. This leads to an adaptation dilemma – the uncertainty can increase into the future – reflecting potential for negative and positive changes– even though the mean signal is small.

To test the robustness of our results, we performed the decomposition of variance using alternative methodological choices. We performed three-way analysis of variance (MATLAB *anovan* function) on $z_{s,m,i}(x,y,\ \tau)$ with and without allowing a statistical interaction term between the model and emissions scenario factors. The results were broadly similar in pattern for the 2- and 20-year values, with the interaction term contribution comparable to the emissions scenario contribution. The 100-year values were substantially different, with different model patterns for the two time periods, and much more variance attributed to internal variability. Concern over small sample size and the resulting instability of model patterns motivated us to adopt the pattern scaling approach that assumes the same underlying model patterns of change for both periods, thus effectively doubling the sample size. Another choice related to applying the model REGRID versus BC projections, leading to qualitatively similar results in the patterns and amount of variance explained by each factor at least when the scaling step was done. Lastly, the analysis was conducted before scaling and pooling. The resulting analysis of variance was similar in the broad patterns, but much noisier on the grid scale with a much greater dominance of internal variability when the data were not pooled. Scaling (which accounts for the different model's biases in simulating extremes during the historic period), does not change the broad patterns, but does reduce the apparent contribution of the emissions scenario. Given the typically small ensemble sizes for climate model projections, regionalization of extreme values can reduce the influence of grid-scale internal variability.

Finally, the role of internal variability should not be discounted. In this paper we have not addressed how well the models reproduce the observed internal variability in extremes. The large scale of the climate model grid-cells may not be able to resolve processes such as land-atmosphere feedback that may enhance the variability of local extremes. Because of the ad-hoc design of the CMIP3 experiment, it is not possible to cleanly separate the inter-model variability from the internal variability for our purposes. Larger ensemble sizes for each model would be desirable. We have used the available daily projections from the CMIP3 archive at PCMDI, and combined the analysis for two future time periods to gain what statistical power we could. Nonetheless, our method may be underestimating the contribution of multi-decadal internal variability.

11.8 Step 6: Assess Changes Given Global/Local Interactions

Up to this point the demonstrations in this paper have analyzed changes in daily precipitation at the spatial scale of GCM output. In contrast, many applications of climate data to water resources management first involve "downscaling", that is the translation of global climate projection to a finer spatial scale, ideally recognizing local climate forcing and local controls on how global climate change might influence local conditions. Choosing among candidate downscaling methods introduces additional uncertainties in the evaluation of extremes beyond those already originating from choices in global climate projection (Step 5). Some studies have quantified relative amount of uncertainty introduced by GCMs versus subsequent downscaling when assessing local hydroclimate impacts (e.g., Wilby and Harris 2006; Crosbie et al. 2011).

Quantifying the uncertainties introduced by choice of downscaling method is difficult to assess without conducting the idealized experiment where GCM projections are subjected to both multiple statistical downscaling methods and dynamical downscaling using multiple RCMs. Nonetheless, the effect of simple downscaling methods on extremes may be transparent. For example, an annual change-factor ("delta") method based on historical observations would simply multiply the *annual* and *1daymax* values by that change factor. A simple spatial disaggregation method such as was used in Wood et al. (2002) would, at least approximately, transfer the percentage changes in extremes seen at the GCM scale to the finer scale.

More involved statistical and dynamical methods need to be evaluated explicitly for their effect on extremes. As an example, consider a more involved statistical method where the ensemble of BC daily precipitation projections has been downscaled to a 1/8 degree spatial grid using the constructed analogues technique (Hidalgo et al. 2008; Maurer and Hidalgo 2008). The BCCA technique creates daily fields of precipitation through linear regression on the 30 nearest historical analogs to the projected patterns as measured by a domain-wide root-mean-square distance. Because the method is applied on the domain-wide pattern, it is not constrained to reproduce the same local extremes seen in the GCM data. Nonetheless, constructed analogues do well at reproducing the upper end of the precipitation distribution when tested on historical data (Hidalgo et al. 2008). As an illustration, Fig. 11.15 shows the *Annual* and *1daymax* values for the BCCA dataset for part of the western U.S. These values may be compared against values for the BC dataset shown in the middle rows of Figs. 11.3 and 11.4. While the BCCA data expresses an overall north-south gradient in the change field similar to that in the BC results, the percentage change in the BCCA dataset is enhanced over the higher topography of the Sierra Nevada demonstrating that portrayed changes can be affected by this type of downscaling.

Given that GCMs are capable of modeling the changes in synoptic-scale weather patterns, statistical downscaling methods are meant to capture the changes in regional and local conditions associated with the large-scale changes, under the assumption that the (often very simple) statistical relationship used in the downscaling does not change into the future. Different statistical downscaling methods give different projections of extremes, even when applied to the same GCM data (Haylock et al. 2006; Fowler et al. 2007). The STARDEX project compared many downscaling methods, and noted that no single method stood out as clearly superior, and recommended that multiple downscaling methods be considered (Goodess 2005).

In many locations, regional and mesoscale atmospheric dynamics are responsible for extremes and in particular for extremes of precipitation at daily and shorter time scales, and for regions of high topographic relief. To capture these with more confidence, it may be desirable to downscale the GCM using a regional climate model (RCM), a process known as dynamical downscaling. The North American Regional Climate Change Assessment Program (NARCCAP) coordinated a

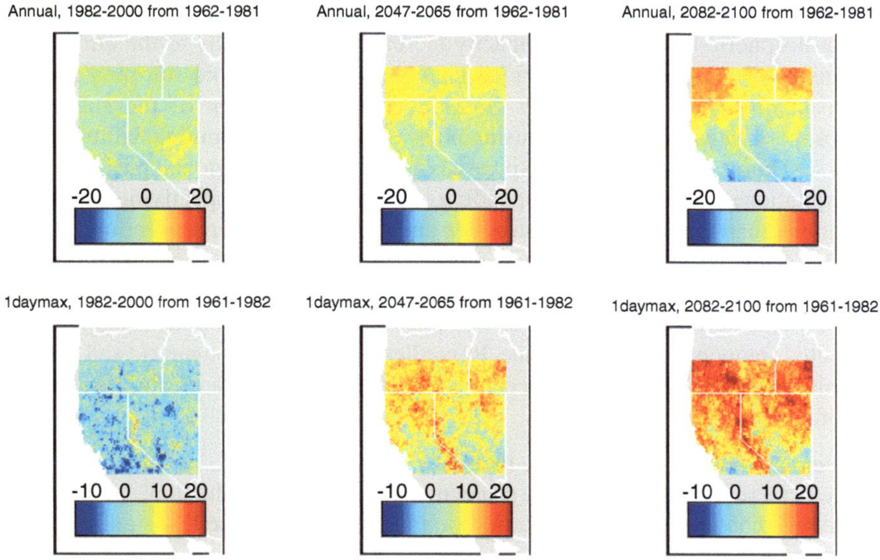

Fig. 11.15 Percentage change in downscaled period-median annual and 1daymax precipitation. Maps show spatially distributed, ensemble-median projected change in period median condition for BC projections that have been spatially downscaled using Constructed Analogues (i.e. BCCA projections)

multi-model dynamically downscaled ensemble at approximately 50 km resolution covering most of North America. For the European and West African domains, the ENSEMBLES project produced a multi-model ensemble at approximately 25 km resolution. RCMs, like GCMs have biases in simulating the climate, and even at 25 km do not resolve many processes that are important for precipitation extremes. Also as noted earlier, the need to consider uncertainty in a resource assessment motivates consideration of climate projections from a large multi-GCM ensemble. The typically smaller ensemble size featured in multi-RCM experiments – and in particular the small number of driving GCM runs that have been downscaled – can be a hindrance to their use. However, the increased resolution can aid in studies of the processes that relate to extremes. In addition, RCMs can be used to investigate the impact of local and regional climate forcing such as changes in land cover and land use or regional aerosol and dust emissions.

11.9 Summary

This chapter explores various uncertainties associated with projecting changes in extreme precipitation conditions, associated with both daily heavy precipitation events and annual droughts. Physical paradigms for changes in extreme precipitation relate to two global scale drivers– the “Clausius-Clapeyron” relationship where the atmospheric water-holding capacity increases exponentially as the atmosphere warms, and the “Hadley Expansion” phenomena where global warming leads to a weakening of tropical Hadley circulation and broadening of the subtropical atmospheric subsidence zones causing poleward migration of middle latitude storm tracks. These paradigms are expressed in contemporary global projections of precipitation to varying degree depending on choices made in specifying the climate forcing scenario, climate model structure and climate system initialization.

The scope of relevant projected precipitation extremes is very broad, varying from localized and acute heavy precipitation events to regional sustained precipitation deficits. Further, for any metric, extremes may be defined by their magnitude or expected frequency of reoccurrence. Regardless of precipitation metric and extreme definition, the evaluation of projected changes in extremes can feature a common set of steps:

1. Identify weather or hydrologic metric of interest.
2. Select a representative climate projection ensemble for evaluation.
3. Assess changes in typical metric conditions, and evaluate the significance of these changes.
4. Assess changes in extreme metric conditions, and evaluate the significance of these changes.
5. Relate variance in projected changes to global uncertainties.
6. Assess changes given global/local interactions.

In Step 1, focus may be placed on many types of weather or hydrologic variables, statistical metrics of these variables. For this demonstration, focus was placed on two annual precipitation metrics: total annual precipitation (*Annual*) and maximum-day precipitation (*1daymax*). In Step 2, ensemble consideration is necessary in order to support exploring dimensions of uncertainty, regardless of whether model validation is conducted beforehand in a way that bears influence on inform ensemble selection. For this demonstration, changes in the two metrics were evaluated within a 53-member ensemble of daily CMIP3 projections over the contiguous United States.

In Step 3, the demonstration initially focused on consensus projected change (ensemble-median) in typical metric conditions (period-median) from a reference period of water years 1962–1981 to three subsequent periods: water years 1982–2000, 2047–2065 and 2082–2100. Results revealed that typical *Annual* conditions were expected to increase in the northwest and northeast regions of the domain and decrease in the southwest reaches of the domain, consistent in their broad spatial patterns and magnitude with earlier findings based on the CMIP3 multi-model ensemble (USGCRP 2009). Similar geographic patterns of change were found for typical *1daymax* conditions, although the region of increased metric conditions was somewhat expanded to the south. Analyses were conducted on two versions of the climate projections: one using regridded results from GCM simulations (REGRID) and another using bias-corrected GCM results (BC) where “bias-correction in the distribution” was conducted using a quantile mapping technique. Comparison of results showed that changes varied between REGRID and BC projections, sometimes by greater than 10%, highlighting how the act of bias-correcting climate projections introduces uncertainty into the change assessment. Finally, statistical testing showed that BC projections most frequently expressed significant change in the northwest and northeast regions of the domain for both metrics, and also the southwest region for *Annual.*

In Step 4, the demonstration switched focus from changes in typical to changes in extreme *1daymax* and *Annual* events. Extremes values for *1daymax* and *Annual* were respectively estimated from generalized extreme value and gamma probability distributions fit to metric period-samples at each grid cell. Three return periods were considered: 2-year (akin to period-median), 20- and 100-year. Results revealed that geographic patterns of changes in extreme conditions (20- and 100-year return) do not necessarily align with the patterns of typical conditions (2-year). For *Annual* extremes, the region of negative change expanded from the southwest to the east and north. For *1daymax* extremes, the region of positive change largely covered the same extent as that associated with typical *1daymax.* Combining these two results, it's apparent that some middle domain regions are projected to experience a combination of more severe annual droughts and more extreme heavy precipitation events, which is also consistent with earlier findings (USGCRP 2009). Although not demonstrated, the same tests applied to evaluate the significance of typical changes could also be applied to evaluate the significance of extreme changes.

Having characterized changes in typical and extreme conditions, the demonstration moved onto Step 5 and considered how variance in these projected changes partitions to three sources of global climate projection uncertainty: climate forcing scenario (Emissions), climate model sensitivity (Model), and internal climate variability (Internal). Following the variance decomposition approach of Hawkins and Sutton (2009, 2010), and adapting it from their transient application to the period-change application in this study, results showed the Emissions choices account for only a small fraction of change variance for most of the domain and for both projected twenty-first century periods. Exceptions to this finding are the northwest and northeast regions, which showed significant sensitivity to emissions path by 2082–2100. The relative influence of Model and Internal sources of uncertainty varied regionally, with Model uncertainty found to account for a generally greater share of variance for the typical *1daymax* (2-year return) than for the extreme *1daymax* (100-year return). Reasons for why the Internal source of uncertainty receives considerable variance allocation include the presence of low-frequency (i.e. decadal to multidecadal) modes of variability in the climate system and simulated within the GCMs to varying degree (rightly or wrongly), and sampling uncertainty in determining extremes from relatively short climate records. These findings were found to be generally robust when the variance decomposition was implemented using different choices in the analytical design (e.g., whether to scale metric conditions, whether to apply to REGRID or BC projections, whether to decompose using an additive model following HS2009 or a statistical model that permits Emissions-Model interaction). However, estimates of the 100-year values of *1daymax* were more sensitive to the choice of method. The dominance of model variance for 20-year values is in general agreement with Wehner (2010), who notes a similar result for 20-year temperature extremes.

In summary consideration of Steps 3– 5, a fair amount of caution should be used in interpreting the GCM-scale precipitation extremes, particularly in any single grid-cell. There are many uncertainties introduced by the parameterization of convective storms and the coarse representation of orographic precipitation and mesoscale atmospheric circulations. It is known that tropical storms are not well simulated by the CMIP3 models. Yet the GCMs demonstrate that there are large scale patterns in how the water budget may change, and that there are large-scale sources of uncertainty in these projections.

Finally, the demonstration moved to the final step, considering how regional atmospheric circulation and local climate controls might interact with global climate projections to determine local precipitation change. Such consideration invites the task of spatially downscaling climate projections to more local resolution. Schemes for doing so may be simple to complex statistical methods or complex dynamical methods. The dynamical approach is preferable if the purpose is to characterize how regional/local factors might modulate the change signal from global projections. However, dynamical downscaling remains sensitive to regional climate modeling approach (similar to how global climate projection remains sensitive to GCM choice), and such techniques have yet to be applied to the projections and period scope considered in this analysis. Alternatively, a daily non-dynamical technique has been applied to these projections (constructed analogues) and evaluation of these results over a limited west-coast region reveals that the spatial pattern of changes can be modulated even by such non-dynamical techniques, introducing further uncertainty into the assessment.

In conclusion, this chapter demonstrates the following:

- The reader may evaluate projections for relevant precipitation metrics and then apply traditional probabilistic methods to these projected metrics in order to characterize extreme metric events.
- Characterization of projected extreme events is confounded by climate model simulation biases, and choice of whether to characterize extremes with or without climate model bias-correction. Although a quantile mapping bias-correction without trend constraint was implemented in this demonstration, there are multiple types of bias-correction techniques available and it remains a matter of research to identify an appropriate method. Given the current situation, it may be preferable to consider multiple bias-correction techniques if the interest is in better disclosure of projected impacts uncertainty.
- Projected extremes and changes in projected extremes vary across the projection ensemble, and in different ways depending on extreme metric and geographic location.
- Statistical testing can reveal the statistical significance of apparent changes in the projection-distribution of typical and extreme metric conditions, and is recommended as a complement to any such change assessment.

- Low-frequency variability exists in climate projections and to varying degree depending on source climate model and geographic location. Its occurrence in projections confounds interpretation of projected changes in precipitation extremes.
- Using variance decomposition, variance in projected changes in extremes may be associated with three sources of global climate projection uncertainty: climate forcing (emissions), climate model sensitivity (models), and internal climate system variability (model initialization). For the continental U.S. and precipitation metrics considered, much of the variance is attributable to climate model sensitivity and internal climate system variability.
- Uncertainties also arise from changes in regional circulation and local controls on climate, which are partially revealed through downscaling of GCM results to local scale. Statistical methods are efficient in accomplishing this task. However, the effects of regional circulation and local controls on climate are likely to be best resolved through dynamical downscaling.

Acknowledgments We acknowledge the contribution of the anonymous reviewers and discussions with Dr. Gilbert Compo in improving the manuscript. Dr. Barsugli was supported by the Western Water Assessment, a NOAA Regional Integrated Sciences and Assessments program at the University of Colorado and the NOAA Earth System Research Laboratory. Dr. Brekke was supported by the Reclamation Research and Development Office and Science and Technology Program. We also acknowledge the WCRP Working Group on Coupled Modeling as well as the archive hosts at Lawrence Livermore National Laboratory's PCMDI and Green Data Oasis for their roles in respectively making available the WCRP CMIP3 multi-model dataset and the Bias Corrected and Downscaled WCRP CMIP3 Climate Projections dataset (containing REGRID, BC and BCCA data discussed herein).

References

Alexander LV et al (2006) Global observed changes in daily climate extremes of temperature and precipitation. J Geophys Res 111:D05109. doi:10.1029/2005JD006290

Allen MR, Ingram WJ (2002) Constraints on future changes in climate and the hydrological cycle. Nature 419(6903):224–232

Annan JD, Hargreaves JC (2010) Reliability of the CMIP3 Ensemble. Geophys Res Lett 37:L02703. doi:10.1029/2009GL041994

Brekke LD, Dettinger MD, Maurer EP, Anderson M (2008) Significance of model credibility in estimating climate projection distributions for regional hydroclimatological risk assessments. Clim Chang 89(3–4):371–394

Christensen JH, Hewitson B, Busuioc A, Chen A, Gao X, Held I, Jones R, Kolli RK, Kwon W-T, Laprise R, Magaña Rueda V, Mearns L, Menéndez CG, Räisänen J, Rinke A, Sarr A, Whetton P (2007) Regional climate projections. In: Solomon S, Qin D, Manning M, Chen Z, Marquis M, Averyt KB, Tignor M, Miller HL (eds) Climate change 2007: the physical science basis, contribution of working group I to the fourth assessment report of the intergovernmental panel on climate change. Cambridge University Press, Cambridge/New York, 94pp

Crosbie RS, Dawes WR, Charles SP, Mpelasoka FS, Aryal S, Barron O, Summerell GK (2011) Differences in future recharge estimates due to GCMs, downscaling methods and hydrological models. Geophys Res Lett 38:L11406. doi:10.1029/2011GL047657

Delworth TL et al (2005) GFDL's CM2 global coupled climate models part 1: formulation and simulation characteristics. J Climate 19:643–674

Embrechts P, Klüppelberg C, Mikosch T (1997) Modelling extremal events for insurance and finance. Springer, New York

Flato GM, Boer GJ (2001) Warming asymmetry in climate change simulations. Geophys Res Lett 28:195–198

Fowler HJ, Blenkinsop S, Tebaldi C (2007) Linking climate change modelling to impact studies: recent advances in downscaling techniques for hydrological modelling. Int J Climatol 27: 1547–1578

Frich P, Alexander LV, Della-Mara P, Gleason B, Haylock M, Klein Tank AMG, Peterson T (2002) Observed coherent changes in climatic extremes during the second half of the twentieth century. Clim Res 19:193–212

Gibbons JD (1985) Nonparametric statistical inference. Marcel Dekker, New York

Gleckler PJ, Taylor KE, Doutriaux C (2008) Performance metrics for climate models. J Geophys Res 113(D06104)

Goodess C (ed) (2005) STARDEX: statistical and regional dynamically downscaling of extremes for European regions. Detailed project report: January 2002 to July 2005. Section 6: detailed report, related to overall project duration. Climatic Research Unit, University of East Anglia, Norwich, UK, 49 pp

Hahn GJ, Shapiro SS (1994) Statistical models in engineering. Wiley, Hoboken

Hawkins E, Sutton RT (2009) The potential to narrow uncertainty in regional climate predictions. Bull Am Meteorol Soc 90(8):1095–1107

Hawkins E, Sutton RT (2010) The potential to narrow uncertainty in projections of regional precipitation change. Climate Dyn. doi:10.1007/s00382-010-0810-6

Haylock MR, Cawley GC, Harpham C, Wilby RL, Goodess CM (2006) Downscaling heavy precipitation over the United Kingdom: a comparison of dynamical and statistical methods and their future scenarios. Int J Climatol 26(10):1397–1415

Held IM, Soden BJ (2006) Robust responses of the hydrological cycle to global warming. J Climate 19(21):5686–5699

Hidalgo HG, Dettinger MD, Cayan DR (2008) Downscaling with constructed analogues: daily precipitation and temperature fields over the United States. Report CEC-500-2007-123 from the California Energy Commission, Public Interest Energy Research Program, Sacramento, CA, 62 pp

Hosking JRM, Wallis JR (1997) Regional frequency analysis, an approach based on L-moments. Cambridge University Press, Cambridge

IPCC (2000) In: Nakicenovic N, Swart R (eds) Emissions scenarios. Special report of the intergovernmental panel on climate change. Cambridge University Press, Cambridge/New York, 570 pp

IPCC (2007) In: Alley R et al (eds) Climate change 2007: the physical science basis – summary for policymakers, contribution of working group I to the fourth assessment report of the Intergovernmental Panel on Climate Change. Cambridge University Press, Cambridge/New York, 18 pp

IPSL (2005) The new IPSL climate system model: IPSL-CM4. Institut Pierre Simon Laplace des Sciences de l'Environnement Global, Paris, p 73

Jungclaus JH, Botzet M, Haak H, Keenlyside N, Luo J-J, Latif M, Marotzke J, Mikolajewicz U, Roeckner E (2006) Ocean circulation and tropical variability in the AOGCM ECHAM5/MPI-OM. J Climate 19:3952–3972

K-1 model developers (2004) In: Hasumi H, Emori S (eds) K-1 coupled model (MIROC) description, K-1 technical report, 1. Center for Climate System Research, University of Tokyo, Tokyo, 34 pp

Karl TR, Meehl GA, Miller CD, Hassol SJ, Waple AM, Murray WL (2008) Weather and climate extremes in a changing climate – regions of focus: North America, Hawaii, Caribbean, and U.S. Pacific Islands, Synthesis and Assessment Product 3.3 Report by the U.S. Climate Change Science Program and the Subcommittee on Global Change Research, 180 pp

Kendon EJ, Rowell DP, Jones RG, Buonomo E (2008) Robustness of future changes in precipitation extremes. J Climate 21:4280–4297

Kotz S, Nadarajah S (2000) Extreme value distributions: theory and applications. Imperial College Press, London

Legutke S, Voss, R (1999) The Hamburg atmosphere-ocean coupled circulation model ECHO-G. Technical report, No. 18, German Climate Computer Centre (DKRZ), Hamburg, 62 pp

Lu J, Vecchi GA, Reichler T (2007) Expansion of the Hadley cell under global warming. Geophys Res Lett 34(L06805)

Massey FJ (1951) The Kolmogorov-Smirnov test for goodness of fit. J Am Stat Assoc 46(253): 68–78

Maurer EP, Hidalgo HG (2008) Utility of daily vs. monthly large-scale climate data: an intercomparison of two statistical downscaling methods. Hydrol Earth Syst Sci 12:551–563

Maurer EP, Wood AW, Adam JC, Lettenmaier DP, Nijssen B (2002) A long-term hydrologically-based data set of land surface fluxes and states for the conterminous United States. J Climate 15(22):3237–3251

Maurer EP, Hidalgo HG, Das T, Dettinger MD, Cayan DR (2010) The utility of daily large-scale climate data in the assessment of climate change impacts on daily streamflow in California. Hydrol Earth Syst Sci 14:1125–1138. doi:10.5194/hess-14-1125-2010

Meehl GA, Covey C, Delworth T, Latif M, Mcavaney B, Mitchell JFB, Stouffer RJ, Taylor KE (2007) The WCRP CMIP3 multimodel dataset – a new era in climate change research. Bull Am Meteorol Soc 88(9):1383–1394

Mitchell JFB, Johns TC, Eagles M, Ingram WJ, Davis RA (1999) Towards the construction of climate change scenarios. Clim Change 41:547–581

Mote P, Brekke L, Duffy PB, Maurer E (2011) Guidelines for constructing climate scenarios. Eos Trans AGU 92(31). doi:10.1029/2011EO310001

Perica S, Dietz S, Heim S, Hiner L, Maitaria K, Martin D, Pavlovic S, Roy I, Trypaluk C, Unruh D, Yan F, Yekta M, Zhao T, Bonnin G, Brewer D, Chen L-C, Parzybok T, Yarchoan J (2011) NOAA Atlas 14 precipitation-frequency Atlas of the United States volume 6 version 2.0. California, U.S. Department of Commerce, National Oceanic and Atmospheric Administration, National Weather Service, Silver Spring, Maryland, 239 pp

Pierce DW, Barnett TP, Santer BD, Gleckler PJ (2009) Selecting global climate models for regional climate change studies. Proc Natl Acad Sci U S A 106(21):8441–8446

Pierce DW, Das T, Cayan DR, Maurer EP, Miller N, Bao Y, Kanamitsu M, Yoshimura K, Snyder MA, Sloan LC, Franco G, Tyree M (2011) Probabilistic estimates of California climate change by the 2060s using statistical and dynamical downscaling. Report CEC-XXX-XXXX by the California Energy Commission, Public Interest Energy Research Program, Sacramento, CA, 67 pp (in review)

Salas-Mélia D, Chauvin F, Déqué M, Douville H, Gueremy JF, Marquet P, Planton S, Royer JF, Tyteca S (2005) Description and validation of the CNRM-CM3 global coupled model. CNRM working note 103, 36 pp

Santer BD, Taylor KE, Gleckler PJ, Bonfils C, Barnett TP, Pierce DW, Wigley TML, Mears C, Wentz FJ, Bruggemann W, Gillett NP, Klein SA, Solomon S, Stott PA, Wehner MF (2009) Incorporating model quality information in climate change detection and attribution studies. Proc Natl Acad Sci U S A 106(35):14778–14783

Seager R, Ting M, Held I, Kushnir Y, Lu J, Vecchi G, Huang H-P, Harnik N, Leetmaa A, Lau N-C, Li C, Velez J, Naik N (2007) Model projections of an imminent transition to a more arid climate in southwestern North America. Science 316(5828):1181–1184

Shepard DS (1984) Computer mapping: the SYMAP interpolation algorithm. In: Gaile GL, Willmott CJ (eds) Spatial statistics and models. Reidel, Norwell, pp 133–145

Stainforth DA, Allen MR, Tredger ER, Smith LA (2007) Confidence, uncertainty and decision-support relevance in climate predictions. Philos Trans R Soc A 365:2145–2161. doi:10.1098/rsta.2007.2074

Tebaldi C, Knutti R (2007) The use of the multi-model ensemble in probabilistic climate projections. Philos Trans R Soc A 365(1857):2053

U.S. Global Climate Research Program (USGCRP) (2009) In: Karl TR, Melillo JM, Peterson TC (eds) Global climate change impacts in the United States. Cambridge University Press, Cambridge, 196 pp

Walsh JE, Chapman WL, Romanovsky V, Christensen JH, Stendel M (2008) Global climate model performance over Alaska and Greenland. J Climate 21:6156–6174

Watterson IG, Dix MR (2003) Simulated changes due to global warming in daily precipitation means and extremes and their interpretation using the gamma distribution. J Geophys Res 108(D13):4379. doi:10.1029/2002JD002928

Wehner M (2010) Sources of uncertainty in the extreme value statistics of climate data. Extremes 13:205–217

Wilby RL, Harris I (2006) A framework for assessing uncertainties in climate change impacts: low-flow scenarios for the River Thames. UK. Water Resour Res 42:W02419. doi:10.1029/2005WR004065

Wilks DS (2006) Statistical methods in the atmospheric sciences, vol 59, 2nd edn, International geophysics series. Academic, London, 627 pp

Wood AW, Maurer EP, Kumar A, Lettenmaier DP (2002) Long-range experimental hydrologic forecasting for the Eastern United States. J Geophys Res Atmos 107(D20):4429. doi:10.1029/2001JD000659

Yukimoto S, Noda A, Kitoh A, Sugi M, Kitamura Y, Hosaka M, Shibata K, Maeda S, Uchiyama T (2001) The new Meteorological Research Institute coupled GCM (MRI-CGCM2) – model climate and variability. Pap Meteor Geophys 51:47–88

Chapter 12
Global Data Sets for Analysis of Climate Extremes

David R. Easterling

Abstract Anthropogenic forced climate change is expected to result in changes in both the mean climate state and climate extremes. Documenting changes in extremes, such as drought, heavy precipitation, heat waves or tropical storms requires long-term observations of climate at sufficiently high temporal resolution to identify individual extreme events and is spatially complete enough to use at both global and regional scales. This chapter discusses various data sets that are suitable for examining changes in extremes in the observed record. Issues discussed include problems with data quality, such as data homogeneity and data availability. Data sets discussed include global sets for monthly and daily temperature and precipitation, tropical cyclones, and reanalyses. In addition, a brief section on climate observing networks is included outlining new networks in the U.S. and Canada that have been designed to alleviate many of the problems discussed here.

12.1 Introduction

Changes in climate are expected to lead to changes in the occurrence of extreme events (Easterling et al. 2000). Climate extremes can result from natural variability, forcing due to increasing greenhouse gases, or more likely some combination of the two. For example, some of the more robust climate change signals related to extremes in both the observed record and in model simulations for the future are increases in the number of unusually warm days and nights and, decrease in the number of unusually cold days and nights and an increase in heavy precipitation events (Seneviratne et al. 2012). Other changes include a likely increase in the incidence of hurricanes in the north Atlantic since about 1970 (Kunkel et al. 2008).

D.R. Easterling (✉)
National Climatic Data Center, National Oceanic and Atmospheric Administration (NOAA), Asheville, NC, USA
e-mail: David.Easterling@noaa.gov

A. AghaKouchak et al. (eds.), *Extremes in a Changing Climate*, Water Science and Technology Library 65, DOI 10.1007/978-94-007-4479-0_12,

On the other hand, although drought has increased in many regions, and is expected to increase in many regions, the global signal on drought showing some increase is not as strong as other extremes (Seneviratne et al. 2012).

Documenting observed changes in extremes is often difficult owing to the scarcity of long-term data suitable for analyzing extremes in many regions of the globe. This chapter discusses data sets and issues with data that are available to examine climate extremes, and associated elements of the climate system such as storms and drought.

12.2 Data Issues Impacting the Analysis of Extremes

By definition climate extremes are rare events, thus there is a lack of observed climate data documenting these events. The data often have problems such as missing data, a lack of precision, or just are not available for a given region. If data are available they may not have the proper sampling resolution. For instance monthly averaged temperature data is readily available for most parts of the world for relatively long periods of time (e.g. since 1900) and are suitable for studying extremes such as drought, however to examine most extremes daily or even hourly data are necessary. Long-term data with this kind of temporal resolution is often not readily available even for regions with monthly averaged data.

Since most true extremes are rare, many researchers have defined "extremes" in such a way as to increase the number of observations for more robust statistical analysis, but this often results in events that may not be truly extreme. For example, studies examining the observed record have defined unusually warm or unusually cold days using the 90th percentile or 10th percentile resulting in 10% of the observations being unusually warm and 10% being unusually cold (Alexander et al. 2006). Similarly for precipitation heavy and extreme daily rainfall events have been defined using the 90th and 99th percentiles (Groisman et al. 2005). Return period thresholds have also been used (e.g. 1, 5, or 20 year return period amounts) as an alternative way of defining an extreme threshold amount (Kunkel et al. 1999, 2003).

The assessment of observed changes in most climate extremes requires the use of longer-term observations of the climate elements and phenomena examined in this section. Here issues are discussed related to data and observations used to examine observed changes in extremes. This will allow the reader to place the results in other chapters and their uncertainties in context with the data used to derive these results.

Studies examining observed changes in climate elements such as temperature and precipitation over the past few decades have traditionally focused on changes in mean and variance statistics (Easterling et al. 2000). Although studies using mean and variance statistics remain important, particularly for climate change detection and attribution studies, documenting changes in extreme events is crucial since these events have the most immediate and greatest impacts on society and the natural environment. Although the situation is changing, many countries still consider higher temporal resolution climate data as being monetarily valuable and do not freely distribute their data which can be problematic. Lastly, long-term observations of climate are often available only at weather stations, such as at airports, that were designed to take observations in support of developing weather forecasts, and not for climate purposes, which can result in lower quality data.

Often the analysis of extremes in weather and climate elements involves examining the tails of a statistical distribution. This involves setting a threshold value that must be exceeded to define an extreme event. Daily temperature or precipitation extremes can be defined by establishing either an absolute threshold, such as exceeding 35°C for daily maximum temperature, or 50 mm for daily total precipitation, or a value derived based on a statistical distribution, such as the 90th percentile value for daily temperature. Fitting extreme value distributions, such as the Generalized Extreme Value (GEV) distribution or a generalized Pareto distribution, to either an annual maximum or partial duration series is another method used to examine changes in extremes (Brown et al. 2008). This can be used to define return period thresholds, such as the 10 year, 24 h precipitation amount, which is the amount that has a 10% chance of occurring in any given year. This threshold amount can then be used to define a time series of exceedances by counting how often the amount is equaled or exceeded in each year. To determine if changes are occurring a linear trend analysis is performed as discussed in Smith (2008). Another approach to determining changes is to fit the GEV to successive overlapping time periods and examine the distribution parameters for trends (DeGaetano 2009).

In any case, the results using a threshold value can be impacted by the quality of the data, and there are a number of data problems that can affect values that exceed thresholds. These problems include missing data, particularly if those values that are missing would exceed a threshold, or those missing values could affect the value of the threshold by changing the statistical results used to establish the threshold value. For instance, Kunkel et al. (2003) determined various return period thresholds for daily precipitation empirically by rank-ordering daily precipitation observations then defining a threshold for, say, the 20 year return period for 100 years of data as being the fifth highest value (one would expect 5 daily rainfall events to equal or exceed that value over 100 years). However, if a given year of data had more than 60 days missing, that year was excluded, potentially eliminating one or more extreme values from the analysis.

Other data problems include the Time-of-observation Bias (TOB) for daily temperatures discussed in detail in Karl et al. (1986). This is an issue in the U.S. Cooperative Observer Network, where observers take an observation only once per day using daily maximum and minimum thermometers. A general switch in the observing time in the network from afternoon (typically 5 pm LST) to morning (typically 7 am LST) resulted in a cool bias due to the occasional double counting of daily low temperature values. In particular, Easterling (2002) and Meehl et al. (2009) include discussions on how the TOB issue impacts extremes of temperature.

Quality control procedures designed to flag a value suspected of being erroneous can impact the research results by flagging extreme values that are truly correct, or by not flagging a truly incorrect value. This has been an issue, particularly with large daily precipitation totals due to convective storms or isolated extreme temperature events. Quality assurance checks are typically implemented to examine the data on a station by station basis. These employ both internal checks, such as climatological bounds checks (e.g. is the value reasonable for the location and season), and spatial checks using comparison with nearby climate stations. An isolated but intense thunderstorm may result in an extreme daily precipitation total at one station, but not impact any surrounding stations and result in a flagged value. In recent years particular care has been given to develop automated quality assurance procedures that minimize the flagging of valid observations (false positives), yet does remove the truly incorrect values (Durre et al. 2008).

Climate data homogeneity is another issue, and data are defined as homogeneous when the variations and trends in a climate time series are due solely to variability and changes in the climate system. Whether or not climate data are homogeneous can impact the results of an analysis of extremes. Inhomogeneities occur in a climate time series due to a variety of reasons. These reasons include changes in the location of an observing station, changes in instrumentation, the installation or removal of a wind shield on a precipitation gauge, land use/land cover changes, or changes in the daily observing time. When a change occurs it can result in either a discontinuity in the time series (slight jump) or a more gradual change that can manifest itself as a false trend (Menne and Williams 2009), both of which can impact whether a particular observation exceeds a threshold (see Fig. 12.1). Homogeneity detection and data adjustments have been implemented for longer averaging periods (e.g. monthly, seasonal, annual), and homogeneity detection and adjustments for daily and sub-daily data have recently been developed (e.g. Della-Marta and Wanner 2006), but have not been widely implemented. Even considering the issues of quality control and homogeneity, results of recent global analyses of extremes (e.g. Alexander et al. 2006) are consistent with what would be expected owing to analyses of mean values using homogeneity adjusted data (e.g. Vose et al. 2005).

Precipitation observations from rain gauges are particularly susceptible to measurement problems that can impact investigations into extremes. The two main problems with gauge data are the use of gauges with different measurement mechanisms and the under-catch problem associated with windy conditions, particularly whether the gauge has a wind shield or not (Easterling et al. 1999a). High temporal resolution precipitation data are mainly observed using gauges with a more automated mechanism such as a tipping bucket or weighing mechanism. Tipping buckets in particular have been shown to under-catch precipitation amounts in high rain rate situations (Sevruk 1982). The use of wind shields helps alleviate the issue of under-catch in high wind situations, where under-catch can be 50% or more with snow or light rain. However, the installation or removal of a wind shield at an observing site creates a discontinuity in the time series that must be addressed. Since high daily precipitation totals can be either due to a short duration, high rain rate event, or longer duration (many hours) lighter event, under-catch due to either problem can be significant.

Figure 12.1 shows the impacts of a variety of changes in precipitation measurements in the climate records of ten countries. These include changes in gauge

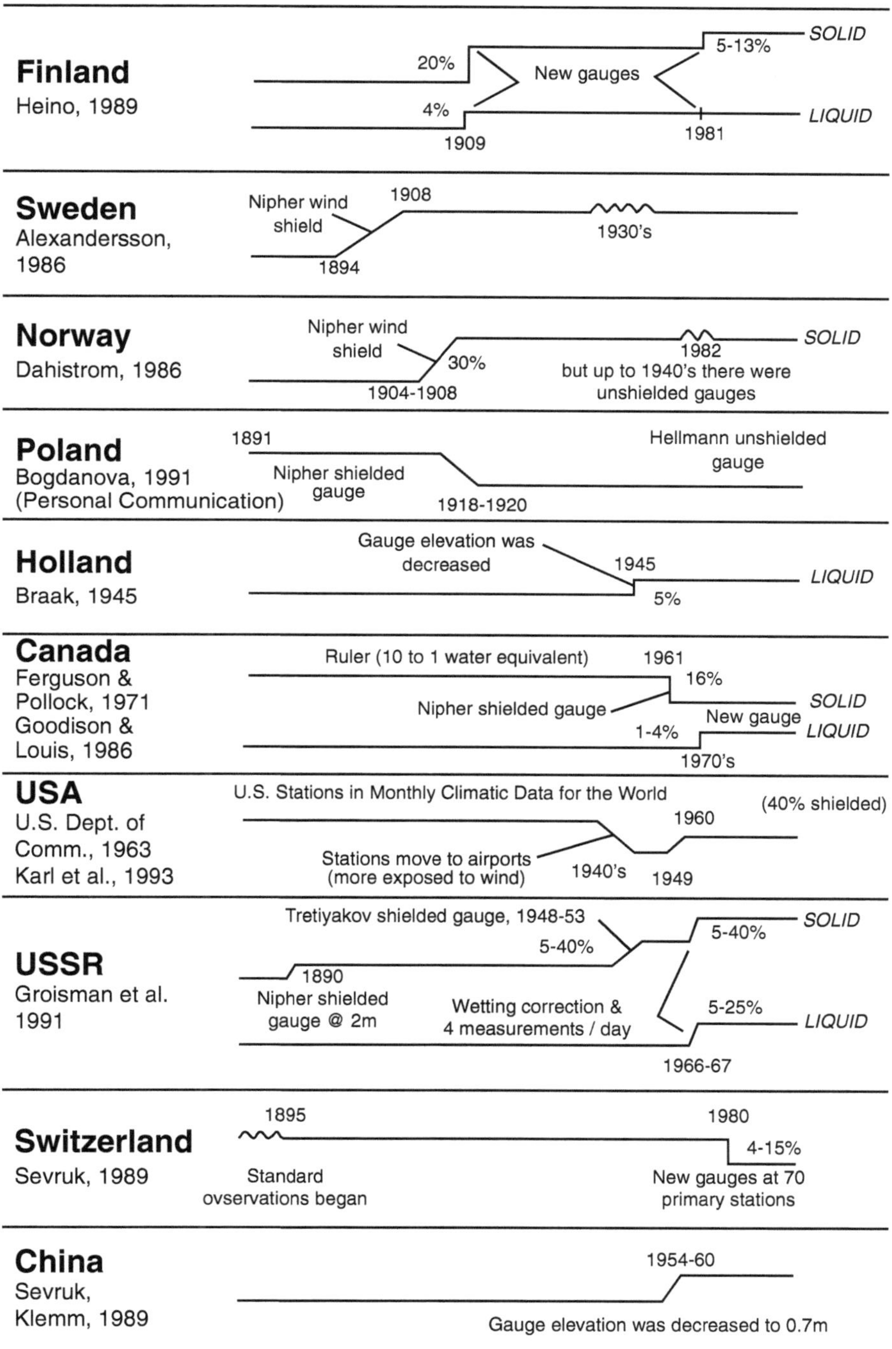

Fig. 12.1 Precipitation time series discontinuities in various countries and causes (From Karl et al. 1995)

type, gauge height, installation of wind shields and measurement practices. The discontinuity in the time series that results from one of these kinds of changes may result in an increase or decrease in the amount of precipitation measured. In some instances, such as for the U.S. precipitation record, discontinuities resulted in a decrease due to station moves to locations with windier conditions early in the record, then an increase later with the installation of wind shields at many locations.

Providing high quality observations of frozen precipitation presents special challenges. Observations of snow can be the liquid water equivalent (as from gauges), snow depth (total depth of snow on the ground), or snow fall (new snow falling since last observation). Snow depth observations are typically made using a snow board and measurement stick with the new snow fall measurement insured by clearing the snow board after each measurement (Doesken 2003). The timing (e.g. every 4 or every 6 h) of the regular measurements or failure to clear the snow board can impact the snow total since snow on the ground compacts over time (Kunkel et al. 2007).

The implementation of quality control procedures can also impact extremes in data sets. In particular, outlier checks designed to detect and evaluate the legitimacy of extreme outliers can flag as erroneous extreme values that are, indeed, valid (Easterling et al. 1999a, b). This is particularly a problem in warm season precipitation where isolated convective storms may produce extreme rainfall at an observing station, but not at near-by stations used for comparison in quality control. Outlier checks typically compare an observed value with observed values for the same time period at some number of nearest neighbor stations (Durre et al. 2008). This can be done using z-scores and comparing with z-scores from surrounding stations then flagging the observed value if it is more than some number (e.g. 3.5) of standard deviation values different. Another approach is to convert an observed time series to anomalies from the series mean and compare with anomalies from nearby highly-correlated stations, flagging the observation if the observed anomaly is different from the neighbor anomaly by some pre-determined value (e.g. 4°C). With spatially conservative variables, such as temperature, these kinds of outlier checks work quite well. However, as noted previously, for other variables that have high spatial variability on short time scales (e.g. precipitation) outlier checks may exclude valid values that are truly extreme.

Data for weather and climate phenomena, such as tropical and extra-tropical cyclones or thunderstorms can be much more problematic than for most weather and climate elements. Thunderstorms and tornadoes are not well observed and in the United States the time series of tornado occurrence since 1950 displays a false increasing trend that is mainly due to increased population density and increased numbers of people to observe them in remote areas (Kunkel et al. 2008). A similar problem occurs with thunderstorms, most thunderstorm observations come from major weather stations such as those at airports. Changes in reporting practices, increased population density and even changes in the ambient noise level at an observing station all have led to inconsistencies in the observed record of thunderstorms.

Studies examining changes in extra-tropical cyclones (ETCs) focus on changes in storm track location, intensities and frequency. Most studies have relied on model-based reanalyses that also incorporate observations into a hybrid model-observational data set. However, reanalyses can have homogeneity problems due to changes such as the introduction of satellite data in the late 1970s and other observing system changes (Trenberth et al. 2001). This homogeneity problem has been addressed in a recent reanalysis that has a much longer (130 year) period of record by assimilating only surface air pressure observations and sea-surface temperatures (Compo et al. 2011). This produces a very long and presumably homogeneous reanalysis, however it still remains to be seen whether the reanalysis produces robust depictions of the climate system in times of sparse observations.

Analyses of observed changes in tropical cyclones have been impacted by a number of issues with the historical record. Changes in reporting methods and in technology have led to major issues with the homogeneity of observed record of tropical cyclones (e.g., Landsea et al. 2004). Since data quality and reporting protocols vary between scientific agencies dealing with tropical cyclones heterogeneities are introduced when records from multiple ocean basins are combined to explore global trends (Knapp and Kruk 2010). Similar to other weather and climate observations, tropical cyclone observations are taken to support short-term forecasting needs. Changes in observing techniques are typically implemented without any overlap or calibration against existing methods. These changes often impact homogeneity of the climate record. The introduction of geostationary satellite imagery in the 1970s, and with it the introduction of tropical cyclone analysis methods, such as the Dvorak technique for estimating storm intensity, further compromises the homogeneity of tropical cyclone observations.

12.3 Climate Observing Networks

Most high temporal resolution data that are necessary for documenting changes in most extremes are taken in support of weather forecasting. Weather observing networks are typically sited in locations, such as airports, that are sufficient for observing the weather, but not ideal for providing high quality climate observations necessary for climate monitoring and research. Some networks have been established specifically to provide climate observations. For example, the U.S Cooperative Observer Network was established in the 1890s to provide temperature and precipitation data to describe the climate of the United States. The primary observations consist of daily maximum and minimum temperature, and 24 h total precipitation. Currently there are about 8,000 sites, with observations taken by volunteers. Instrumentation and training is provided by the U.S. National Weather Service, with data archived at the U.S. National Climatic Data Center. Long-term hourly precipitation data in the U.S. are provided by approximately 2,200 Hourly Precipitation Data (HPD) stations throughout the 50 states. Most of these stations have a period of record starting in the middle of the twentieth century (Doesken 2003).

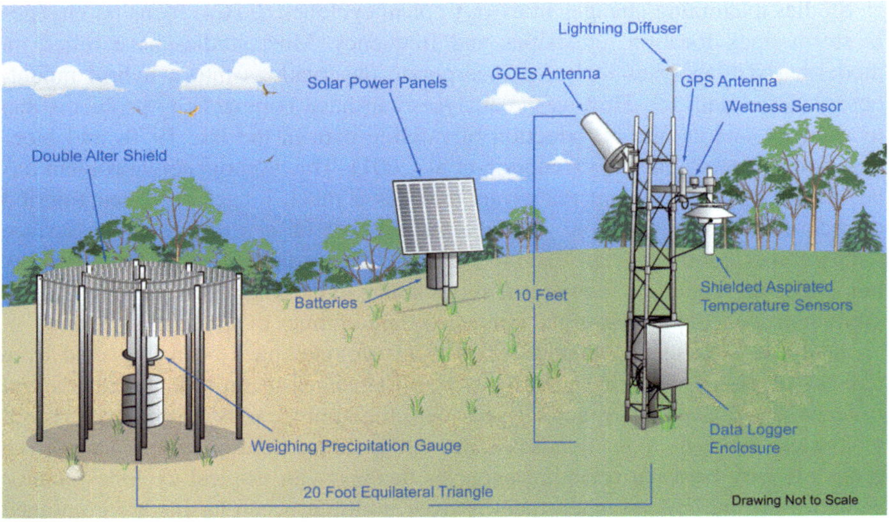

Fig. 12.2 Schematic of U.S. Regional Climate Reference Network station. The rain gauge is equipped with a double Alter wind shield

Other networks include the recently established U.S. Climate Reference Network and U.S. Regional Climate Reference Network (see Fig. 12.2 for schematic of a station) that has stations in all 50 states, and the Canadian Reference Climate Station network with 303 stations. These networks share similar instrument configuration and have been designed and sited to provide high quality climate observations. In particular, both networks employ the same weighing-type of rain gauge, with wind shields that are sited in as open an environment as possible. They provide high temporal resolution rainfall amounts (5 min). These networks also maintain triple redundancy in the observations of both temperature and precipitation. Triple redundancy means both temperature and precipitation have three independent sensors providing three observations at each observing time. This allows the detection of instrument problems, such as one sensor going out of calibration, and enables network operators to quickly replace malfunctioning equipment. Other countries are establishing similar networks, including China. These networks have been designed specifically for monitoring the climate and adhere to the climate monitoring principals originally articulated by Karl et al. (1995) and adopted by the Global Climate Observing System (GCOS). See http://www.ncdc.noaa.gov/crn/.

12.4 Data Sets for Examining Climate Extremes

There are a limited number of comprehensive data sets of temperature and precipitation with global coverage in existence (see Table 12.1). These data sets contain either daily, or monthly averaged maximum, minimum and mean temperature and

Table 12.1 List of global datasets suitable for use in extremes analyses

Data set (see text)	Temporal resolution	Elements	Observation platform/unit	Website
GHCN-M	Monthly	Temperature and precipitation, land only	Station	www.ncdc.noaa.gov
CRUTEM3 and 4	Monthly	Temperature, land only	Gridded	www.metoffice.gov.uk/hadobs
HadCRUT3 and 4	Monthly	Temperature, land and ocean	Gridded	www.metoffice.gov.uk/hadobs
GPCC	Monthly	Precipitation	Station	Gpcc.dwd.de
HadEX	Annual, Monthly	Extremes indices	Gridded and station	www.metoffice.gov.uk/hadobs
GHCN-D	Daily	Temperature, precipitation, snow	Station	www.ncdc.noaa.gov
HadGHCND	Daily	Temperature	Gridded	www.metoffice.gov/uk/hadobs
Various reanalyses	Sub-daily	Temperature, precipitation and much more	Gridded	www.reanalysis.org
IBTrACS	Sub-daily	Tropical storm location, winds, air pressure and more.	Storm	www.ncdc.noaa.gov

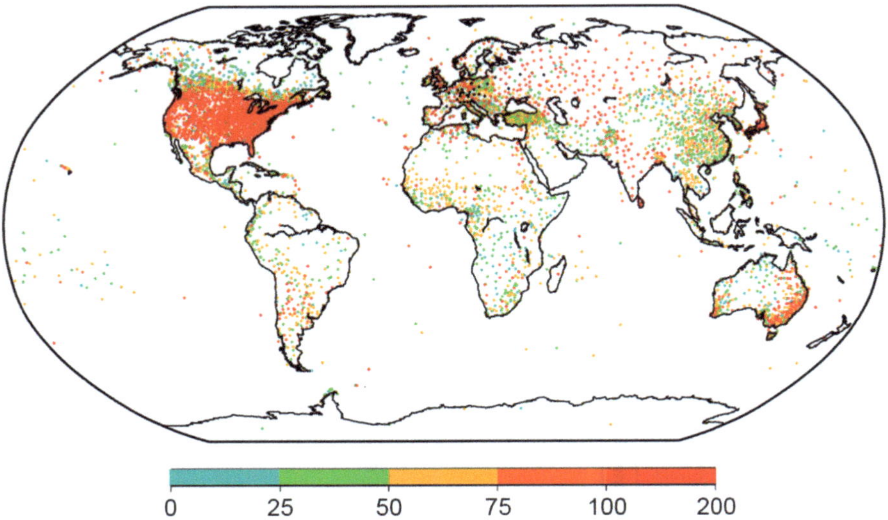

Fig. 12.3 Locations of the 7,280 observing stations in the GHCN-M inventory (From Lawrimore et al. 2012)

total precipitation at observing stations. In the context of extremes, monthly versions are suitable mainly for looking at large area, longer period extremes, such as extreme wet periods or droughts of a month or more. One such data set is the Global Historical Climatology Network monthly (GHCN-M) data set (Vose et al. 1992; Peterson and Vose 1997; Lawrimore et al. 2011). The GHCN-M contains monthly temperature for approximately 7,000 stations and monthly total precipitation for over 20,000 stations (see Fig. 12.3). Earliest records for temperature are for 1,701 and for precipitation 1,697. Examining the map in Fig. 12.3, the regions with the most stations, and longest periods of record are in North America, Europe, Australia and parts of Asia. Similar data sets for temperature for land only (CRUTEM3/4) and land/ocean (HadCRUT3/4) are compiled and maintained by the Climatic Research Unit of the University of East Anglia, jointly with the Hadley Center of the British Meteorological Office (Jones et al. 2012; Harris et al. 2012; Morrice et al. 2012) and the Global Precipitation Climatology Center Full Data Reanalysis Version 5 (GPCC) of the German Weather Service (Rudolf et al. 2011).

The Global Historical Climatology Network Daily (GHCN-D) data set contains daily maximum and minimum temperature and 24 h total precipitation, snow fall and snow depth at approximately 75,000 stations from 180 countries from around the globe, although most (~65%) report only precipitation (Menne et al. 2012). The GHCN-D provides data for maximum and minimum temperature, total daily precipitation, snowfall and snow depth. However, most of the stations (66%) only report total daily precipitation. One major issue with developing and maintaining a daily data set, such as the GHCN-D is that daily data are much less accessible than monthly data. This is partly due to the reluctance of many countries to share daily data, and partly due to a lack of these data having been digitized. GHCN-M and GHCN-D are available from the National Climatic Data Center (see Table 12.1).

The Global Precipitation Climatology Project (GPCP) provides merged satellite-rain gauge data on a global grid for the period 1979-present. There is a monthly data set on a 2.5° grid available for the period 1979-present (Adler et al. 2003) and a daily version on a 1° grid for the period 1996-present (Huffman et al. 2001). These data sets have been used to examine numerous features of the hydrologic cycle, including drought (Chen et al. 2009) and in comparisons between satellite estimates and high temporal resolution rain gauge data (Sapiano and Arkin 2009).

Although many countries are reluctant to share high temporal resolution data, they have been willing to allow value-added products based on their data to be made freely available. One such data set is a global gridded version of climate extremes indices first described in Frich et al. (2002). The data set is described in Alexander et al. (2006) and is comprised of 27 indices of temperature and precipitation computed using software developed by the Climate Research Branch of the Meteorological Service of Canada. Each of the indices are calculated using data for 2,500 stations (temperature) and 6,000 stations (precipitation) and are provided on a $2.5° \times 3.75°$ latitude/longitude grid for about 70% of the global land surface for the period 1951–2003. Indices include a number for heavy precipitation events such as annual contributions from very wet days, and a daily intensity index. Temperature indices include unusually cold (<10th percentile) days and nights and unusually warm (>90th percentile) days and nights, warm-spell duration and frost-days. The indices description and software are available at cccma.seos.uvic.ca/ETCCDI/and data are available at www.metoffice.gov.uk/hadobs/hadex/.

On a regional basis there are a number of data sets that are suitable for examining extremes. Haylock et al. (2008) describe a gridded daily high-resolution data set of temperature and precipitation for the period 1950–2006 for Europe. The data set was developed with a number of spatial resolutions to facilitate comparison with climate model simulations. The U.S. Historical Climatology Network data sets of monthly and daily temperature and precipitation for the 48 contiguous states were specifically developed for documenting century scale climate variability and change (Easterling et al. 1996, 1999a, b). Haylock and Nichols (2000) describe a high-quality daily precipitation data set for Australia and document observed changes in extreme rainfall indices.

The International Best Track Archive for Climate Stewardship (IBTrACS, Knapp et al. 2010) data set is the most complete global data set of tropical cyclone observations available. It combines data from a number of regional tropical cyclone data sets (e.g. HURDAT, Landsea et al. 2004) and consists of observations of tropical cyclone position, winds, and pressure. The driver for development of the IBTrACS was to collect all available data from the Regional Specialized Meteorological Centers, plus any other agencies that has tropical cyclone data, and provide the data in one product with a common data format. In addition to the basic track and meteorological data, summary statistics for each storm are also provided.

Reanalyses are another potential source of data for examining hydrological extremes especially large-scale storms and associated circulation patterns (see

Sect. 12.2 regarding ETCs). A reanalysis is a hybrid model-observational data set where observations are assimilated into a numerical climate model to provide physically consistent 3-dimensional depictions of the climate at a given time step (e.g. Kalnay et al. 1996). Values of temperature, precipitation, winds, pressure and all other atmospheric variables are provided for the entire globe, or in the instance of regional reanalyses, for the region of interest. Most current reanalysis data sets extend back to 1979 when appropriate satellite data became available or the middle of the twentieth century due to constraints on the availability of many of the observations included, especially upper air observations from radiosondes which only begin in large scale in the mid twentieth century. However, one reanalysis, the twentieth century reanalysis described in Compo et al. (2011), extends back into the late nineteenth century. This is possible because the twentieth century reanalysis only assimilates surface air pressure and sea-surface temperatures, both of which have enough observations back into the late nineteenth century to allow the creation of a long-term reanalysis data set. Additionally, land-surface models associated with reanalysis systems can be used to examine soil moisture and drought (e.g. Sheffield and Wood 2008). More information is available at www.reanalysis.org. Additionally Table 12.1 contains a listing of most data sets suitable for analysis of extremes.

Lastly, metadata describing observing stations are important, but detailed metadata for stations from most countries are difficult if not impossible to obtain. Metadata for long-term observing stations essentially falls into two categories: (1) Historical metadata that documents changes such as instrument changes or observing practices through time, and (2) Station locations and information about their surrounding environment. Historical metadata documenting changes at stations for most countries is impossible to collect and therefore not available for these data sets. Since these kinds of changes can impact the homogeneity of the station data this has necessitated the development of statistical methods for homogeneity evaluation as described in Sect. 12.2.

12.5 Concluding Remarks

The availability of temperature and precipitation data with sufficient spatial and temporal coverage to be useful in analysis of climate extremes remains problematic for many parts of the globe. Satellite data provide virtually complete spatial coverage, but suffer from a lack of long time series necessary for documenting longer-term (50 years +) changes in climate extremes. On the other hand, station-based *in situ* measurements of temperature and precipitation, particularly higher temporal resolution data, are often not available for large parts of the world due to a reluctance of individual countries to release them. Furthermore, obtaining long time series of *in situ* measurements, particularly in the early twentieth century is also an issue owing to the lack of digital data from many parts of the globe. In many cases these data exist in manuscript form, but have not been digitized. Even when available, in situ observations are often not homogeneous due to changes in instrumentation, observing practices, and station locations.

Other kinds of observations relevant to climate extremes, such as storms, have problems similar to those for temperature and precipitation. In particular, the tropical cyclone record suffers from inconsistencies owing to changes in technology and reporting practices, although this is less of an issue for land-falling tropical cyclones that result in many climate extremes.

In spite of the data issues discussed above, concerted efforts to rehabilitate data and provide comprehensive global data sets for the scientific community have been underway for many years (e.g. Lawrimore et al. 2011). The GCOS was established in 1992 specifically to coordinate and provide global observations of the climate system available from a number of national and international climate observing efforts (Karl et al. 2010). Additionally, new observing networks specifically designed to minimize data issues are being established and already provide high-quality, homogeneous observations suitable to address questions on climate extremes.

References

Adler RF, Huffman GJ, Chang A, Ferraro R, Xie P, Janowiak J, Rudolf B, Schneider U, Curtis S, Bolvin D, Gruber A, Susskind J, Arkin P (2003) The Version 2 Global Precipitation Climatology Project (GPCP) monthly precipitation analysis (1979-present). J Hydrometeorol 4:1147–1167

Alexander LV et al (2006) Global observed changes in daily climate extremes of temperature and precipitation. J Geophys Res Atmos 111:D05109. doi:10.1029/2005JD006290

Brown SJ, Caesar J, Ferro CAT (2008) Global changes in extreme daily temperature since 1950. J Geophys Res Atmos 113:D05115

Chen JL, Wilson CR, Tapley BD, Yang ZL, Niu GY (2009) 2005 drought event in the Amazon River basin as measured by GRACE and estimated by climate models. J Geophys Res Solid Earth 114, Art. No.B05404

Compo G et al (2011) The twentieth-century reanalysis project. Q J R Meteorol Soc 137. doi:10.112/qj.776

DeGaetano AT (2009) Time-dependent changes in extreme-precipitation return-period amounts in the Continental United States. J Appl Meteorol Climatol 48:2086–2099

Della-Marta PM, Wanner H (2006) A method of homogenizing the extremes and mean of daily temperature measurements. J Climate 19:4179–4197

Doesken NJ (2003) Challenge of snow measurements. Chapter 50. In: Potter TD, Colman BR (eds) Handbook of weather, climate, and water: dynamics, climate, physical meteorology, weather systems, and measurements. Wiley, Hoboken, pp 927–949

Durre I, Menne MJ, Vose RS (2008) Strategies for evaluating quality-control procedures. J Clim Appl Meteorol 47:1785–1791

Easterling DR (2002) Recent changes in frost days and the frost-free season in the United States. Bull Am Meteorol Soc 83:1327–1332

Easterling DR, Karl TR, Mason EH, Hughes PY, Bowman DP (1996) United States Historical Climatology Network (U.S. HCN) monthly temperature and precipitation data. ORNL/CDIAC-87. Carbon Dioxide Information Analysis Center, Oak Ridge National Laboratory, U.S. Department of Energy, Oak Ridge, Tennessee

Easterling DR, Karl TR, Lawrimore JH, Del Greco SA (1999a) United States Historical Climatology Network daily temperature, precipitation, and snow data for 1871–1997. ORNL/CDIAC-118. Carbon Dioxide Information Analysis Center, Oak Ridge National Laboratory, U.S. Department of Energy, Oak Ridge, Tennessee

Easterling DR, Diaz HF, Douglas AV, Hogg WD, Kunkel KE, Rogers JC, Wilkinson JF (1999b) Long-term observations for monitoring climate extremes in the Americas. Clim Chang 42: 285–308

Easterling DR, Meehl G, Changnon S, Parmesan C, Karl TR, Mearns LO (2000) Climate extremes: observations, modeling, and impacts. Science 289:2068–2074

Frich P, Alexander LV, Della-Marta P, Gleason B, Haylock M, Klein Tank AMG, Peterson T (2002) Observed coherent changes in climatic extremes during the second half of the twentieth century. Clim Res 19:193–212

Groisman PY, Knight RW, Easterling DR, Karl TR, Hegerl GC (2005) Trends in intense precipitation in the climate record. J Climate 18:1326–1350

Harris I, Jones PD, Osborn TJ, Lister DH (2012) Updated high-resolution grids of monthly climatic observations – the CRU TS 3.1 Dataset. Int J Climatol, in review

Haylock M, Nicholls N (2000) Trends in extreme rainfall indices for an updated high quality data set for Australia, 1910–1998. Int J Climatol 20:1533–1541

Haylock MR, Hofstra N, Klein Tank AMG, Klok EJ, Jones PD, New M (2008) A European daily high-resolution gridded data set of surface temperature and precipitation for 1950–2006. J Geophys Res 113:D20119. doi:10.1029/2008JD010201

Huffman GJ, Adler RF, Morrissey M, Bolvin DT, Curtis S, Joyce R, McGavock B, Susskind J (2001) Global precipitation at one-degree daily resolution from multi-satellite observations. J Hydrometeorol 2:36–50

Jones PD, Lister DH, Osborn TJ, Harpham C, Salmon M, Morice CP (2012) Hemispheric and large-scale land surface air temperature variations: an extensive revision and an update to 2010. J Geophys Res. doi:10.1029/2011JK017139

Kalnay E et al (1996) The NCEP/NCAR 40-year reanalysis project. Bull Am Meteorol Soc 77:437–471

Karl TR, Williams CN, Young PJ (1986) A model to estimate the time of observation bias associated with monthly mean maximum, minimum, and mean temperatures for the United States. J Clim Appl Meteorol 25:145–160

Karl TR, Derr V, Easterling DR, Folland CK, Hofmann D, Levitus S, Nicholls N, Parker D, Withee G (1995) Critical issues for long-term climate monitoring. Clim Chang 31:185–221

Karl TR et al (2010) Observation needs for climate information, prediction and application: capabilities of existing and future observing systems. Proc Environ Sci 1. doi:10.1016/j.proenv.2010.09.013

Knapp KR, Kruk MC (2010) Quantifying inter-agency differences in tropical cyclone best track wind speed estimates. Mon Weather Rev 138(4):1459–1473

Knapp KR, Kruk M, Levinson D, Diamond H, Neumann C (2010) The International Best Track Archive for Climate Stewardship (IBTrACS): unifying tropical cyclone best track data. Bull Am Meteorol Soc 91:363–376. doi:10.1175/2010BAMS2930.1

Kunkel KE, Andsager K, Easterling DR (1999) Trends in heavy precipitation events over the continental U.S. J Climate 12:2515–2527

Kunkel KE, Easterling DR, Redmond K, Hubbard K (2003) Temporal variations of extreme precipitation events: 1895–2000. Geophys Res Lett 30(17):1900. doi:10.1029/2003GL018052,2003, 4 pp

Kunkel KE, Palecki M, Hubbard K, Robinson D, Redmond K, Easterling DR (2007) Trend identification in twentieth-century U.S. snowfall: the challenges. J Atmos Ocean Technol 24:64–73

Kunkel KE et al. (2008) Observed changes in weather and climate extremes. In: Karl TR, Meehl GA, Miller CD, Hassol SJ, Waple AM, Murray WL (eds) Weather and climate extremes in a changing climate. Regions of focus: North America, Hawaii, Caribbean, and U.S. Pacific Islands. A report by the U.S. Climate Change Science Program and the Subcommittee on Global Change Research. Washington, DC

Landsea CW, Anderson C, Charles N, Clark G, Dunion J, Fernandez-Partagas J, Hungerford P, Neumann C, Zimmer M (2004) The Atlantic hurricane database re-analysis project: documentation for the 1851–1910 alterations and additions to the HURDAT database. In: Murnane RJ, Liu KB (eds) Hurricanes and typhoons: past, present and future. Columbia University Press, New York, pp 177–221

Lawrimore JH, Menne MJ, Gleason BE, Williams CN, Wuertz DB, Vose RS, Rennie J (2011) An overview of the global historical climatology network monthly mean temperature data set, version 3. J Geophys Res 116:D19121. doi:10.1029/2011JD016187

Meehl GA, Tebaldi C, Walton G, Easterling D, McDaniel L (2009) Relative increase of record high maximum temperatures compared to record low minimum temperatures in the US. Geophys Res Lett 36. doi:10.1029/2009GL040736

Menne MJ, Durre I, Vose RS, Gleason BE, Houston TG (2012) An overview of the global historical climatology network-daily database. J Atmos Oceanic Technol 29:897–910. doi: http://dx.doi.org/10.1175/JTECH-D-11-00103.1

Menne MJ, Williams CN Jr (2009) Homogenization of temperature series via pairwise comparisons. J Climate 22(7):1700–1717

Morrice CP, Kennedy JJ, Rayner NA, Jones PD (2012) Quanitifying uncertainties in global and regional temperature change using an ensemble of observational estimates: the HadCRUT4 data set. J Geophys Res. doi:10.1029/2011JK017187

Peterson TC, Vose RS (1997) An overview of the global historical climatology network temperature database. Bull Am Meteorol Soc 78:2837–2848

Rudolf B, Becker A, Schneider U, Meyer-Christoffer A, Ziese M (2011) New GPCC full data reanalysis version 5 provides high-quality gridded monthly precipitation data. GEWEX News 21(2):4–5

Sapiano MRP, Arkin PA (2009) An intercomparison and validation of high-resolution satellite precipitation estimates with 3-hourly gauge data. J Hydrometeorol 10(1):149–166

Seneviratne S, Nichols N, Easterling D, Goodess C, Kanae S, Kossin J, Luo Y, Marengo J, McInnes K, Rahimi M, Reichstein M, Sorteberg A, Vera C, Zhang X (2012) Changes in climate extremes and their impacts on the natural physical environment. Chapter 3. In: IPCC special report on managing the risks of extreme events and disasters to advance climate change adaptation. Cambridge University Press, Cambridge/New York

Sevruk B (1982) Methods of correcting for systematic error in point precipitation measurements for operational use. Hydrol Rep 21, World Meteorological Organization 589

Sheffield J, Wood EF (2008) Global trends and variability in soil moisture and drought characteristics, 1950–2000, from observation-driven simulations of the terrestrial hydrologic cycle. J Climate 21(3):432–458

Smith RL (2008) Statistical trend analysis. In: Karl TR, Meehl GA, Miller CD, Hassol SJ, Waple AM, Murray WL (eds) Weather and climate extremes in a changing climate. Regions of focus: North America, Hawaii, Caribbean, and U.S. Pacific Islands. A report by the U.S. Climate Change Science Program and the Subcommittee on Global Change Research, Washington, DC

Trenberth KE, Stepaniak D, Hurrell J, Fiorino M (2001) Quality of reanalyses in the tropics. J Climate 14(7):1499–1510

Vose RS et al. (1992) The global historical climatology network: long-term monthly temperature, precipitation, sea level pressure, and station pressure data. ORNL/CDIAC-53, NDP-041, Carbon Dioxide Information and Analysis Center, Oak Ridge National Laboratory, Oak Ridge, TN, 325 pp

Vose RS, Easterling DR, Gleason B (2005) Maximum and minimum temperature trends for the globe: an update through 2004 [Online]. Geophys Res Lett 32:L23822. doi:10.1029/2005GL024379 [in print 32(23):1–5]

Chapter 13
Nonstationarity in Extremes and Engineering Design

Dörte Jakob

Abstract Dealing with nonstationarity in hydrological extremes in the design of structures is a truly multidisciplinary undertaking; requiring expertise in hydrology, statistics, engineering and decision-making. This chapter gives a broad overview over relevant key aspects in these areas including definitions of key words like 'extremes' and 'stationarity'. We briefly cover current knowledge of both climate variability and climate change and effects on hydrological extremes with particular emphasis on precipitation and floods. This is followed by a brief discussion on impacts of hydrological extremes, risk assessments and options for adaptation as well as hurdles. A large part of this chapter is dedicated to new statistical techniques (or extensions of existing techniques) to address nonstationarity in hydrological extremes through the use of time-varying parameters, moments, quantile estimates and the use of covariates. A changing climate may prove impetus to change some of the existing paradigms and explore new avenues. The need to reduce uncertainty, or alternatively derive more reliable uncertainty estimates, is exacerbated in a changing climate. One of the key strategies should be a move from deterministic to probabilistic approaches. Bayesian techniques are a promising framework in this context.

13.1 Introduction

Dealing with nonstationarity in hydrological extremes in the design of structures is a truly multidisciplinary undertaking; requiring expertise in hydrology, statistics, engineering and decision-making. This chapter aims to give a broad overview over

D. Jakob (✉)
School of Earth Sciences, University of Melbourne, Melbourne, Australia

Bureau of Meteorology, Climate and Water Division, Melbourne, Australia
e-mail: D.Jakob@bom.gov.au

A. AghaKouchak et al. (eds.), *Extremes in a Changing Climate*, Water Science and Technology Library 65, DOI 10.1007/978-94-007-4479-0_13,

relevant key aspects in these areas. In Sect. 13.1 we will be setting the scene and provide examples of recent hydrological events, their impacts and questions they raised. This section will conclude with presenting definitions of 'extremes' and 'stationarity', which are fundamental to the remainder of this chapter. Section 13.2 covers current knowledge of both climate variability and climate change and effects on hydrological extremes with particular emphasis on precipitation and floods. A brief discussion on impacts of hydrological extremes, risk assessments and options for adaptation as well as hurdles will be presented in Sect. 13.3. The heart of this chapter though is Sect. 13.4, providing details about state-of-practice and state-of-the-art approaches for deriving estimates of design rainfall and floods.

13.1.1 Setting the Scene

> *... Extreme hydrological events destroy human heritage and undermine development by breaking continuity. People have interacted with these features with varying degrees of success since history began. Sometimes it has been a failure, as floods and droughts (and desertification) have wiped out whole civilizations.* (Kundzewicz and Kaczmarek 2000)

An impressive illustration for failure to interact successfully with hydrological extremes is the fall of Angkor, the capitol of the Khmer empire. While other societal stressors came into play, it appears that decade-long droughts during the mid to late fourteenth century and the early fifteenth century interspersed with intense monsoons led to Angkor's decline and fall (Buckley et al. 2011). Recent tree-ring based reconstructions of hydroclimatic variations based on the Palmer Drought Severity Index (PDSI) are giving new insights into the demise of Angkor. Regional medieval climate variability was identified to be linked to the El Niño-Southern Oscillation (ENSO), with El Niño conditions leading to droughts and La Niña events being linked to above average rainfall over much of southeast Asia. While drought episodes affected the city's water supply and agricultural productivity, heavy monsoons damaged 'water control infrastructure'. Climate model simulations support the hypothesis that the southward shift of the Intertropical Convergence Zone (ITCZ), which occurred at roughly the same time as the Angkor droughts, may have contributed to a decline in rainfall over South and Southeast Asia.

13.1.2 Recent Extremes, Their Impact and Questions They Raised

> *Variables characterizing processes of the hydrological cycle are subject to variability. From time to time, they take values in their low or high ranges, including hydrological extremes. When there is too little or too much water, the problem becomes spectacular and of concern to the general public.* (Kundzewicz and Kaczmarek 2000)

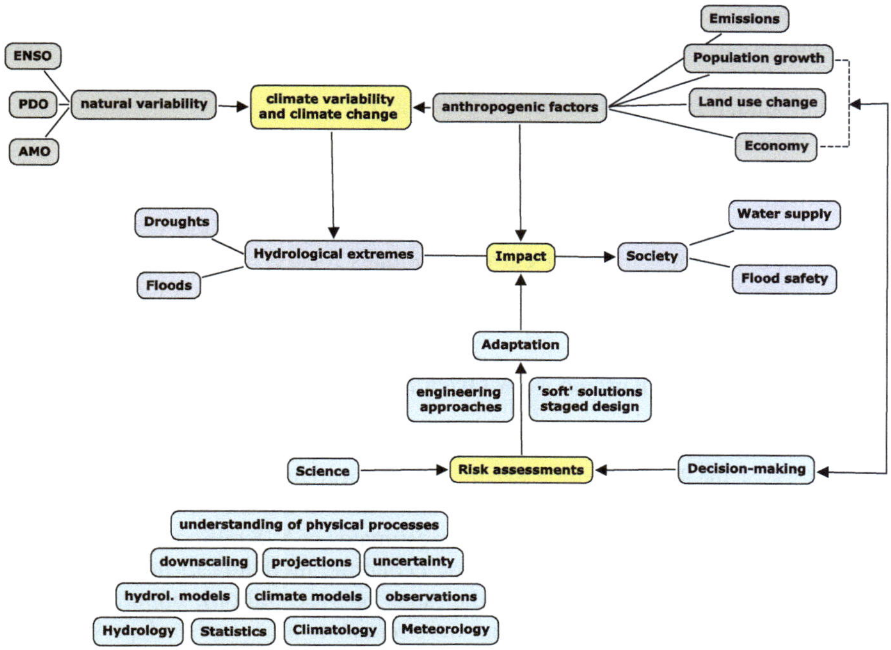

Fig. 13.1 Schematic overview over key components relevant to engineering design under nonstationarity

Some of the problems that we face today with regards to hydrological extremes will be highlighted using recent examples. These examples illustrate the breadth of challenges faced by meteorologists and climatologists in understanding the underlying physical processes and potential effects of climate change. For hydrologists, engineers and decision-makers the challenge is not only in incorporating this information in design of structures but also in developing operational guidelines. Figure 13.1 summarizes some of the key concepts relevant to engineering design under nonstationarity.

13.1.2.1 Boscastle Flood, Cornwall, UK – 16 August 2004

From a floods perspective, the year 2004 was characterized by a significant mesoscale convective event on 16 August, which caused severe flooding in the town of Boscastle. Nearly 200 mm of rainfall fell over the catchment upstream of Boscastle over a period of only 4.5 h (Dale 2005). In contrast, widespread and severe flooding across much of the UK in autumn 2000 (Marsh and Dale 2002) was due to an amalgamation of a series of rainfall events over a period of about 3 months. Clearly, the two events were due to different flood producing mechanisms. Relevant rainfall events are either frontal or convective in nature (or a combination of the two) and may be orographically enhanced. Apart from rainfall, flooding itself may be affected by a range of meteorological/hydrological parameters like snowmelt, frozen ground, tidal effects and groundwater levels.

The Boscastle event raises the question how flood-producing mechanisms might change under climate change, and in particular how frequency and intensity of the most extreme short duration rainfall events are likely to change. The estimated return period for rainfall leading to the Boscastle flood exceeds 2,000 years, sparking discussion about the validity of current estimates for the rarest rainfall and flood events (Probable Maximum Precipitation and Probable Maximum Floods).

13.1.2.2 Hurricane Katrina, New Orleans, USA – August 2005

Hurricane Katrina was one of the most expensive and deadliest natural disasters in the history of the USA. Despite excellent meteorological forecasts and warnings provided by the US National Hurricane Center well in advance of the event, about 1,300 lives were lost and total damage was estimated as $75 billion (McCallum and Heming 2006). Other estimates based on economic modeling to include indirect costs put the figure much higher (in the range of $130 billion). Important factors for the genesis of this tropical cyclone were remnants of an earlier depression and an upper tropospheric trough, together with high sea surface temperatures (above 27°C), weak vertical wind shear and high mid-level humidity. Due to areas of deep warm water (depth of water with temperature of at least 26°C) the hurricane was able to intensify to a Category 5 event. Katrina's impact was due to a combination of factors: powerful winds, heavy rainfalls and storm surges up to 10 m.

From an engineering perspective, this event raised questions about the state of flood defenses for the city of New Orleans. For climate scientists, it renewed the dispute about the impact of climate change on frequency and intensity of tropical cyclones. From a management perspective, this event raises the question of how we make best use of available information and resources.

13.1.2.3 Brisbane Flood, Queensland, Australia – 13 January 2011

Prior to the autumn of 2010, much of south-east Australia had been in drought conditions since the late 1990s, while the southwest of the country had received significantly reduced annual (and especially winter) rainfalls since the 1970s. After the breakdown of the 2009/2010 El Niño a strong La Niña event developed in autumn 2010. La Niña events are typically related to higher than average rainfall over most of Australia. The 2010 event was one of the strongest events and led to 2010 becoming the third-wettest year on record, despite prolonged dry conditions in the southwest. Unusually heavy falls were experienced in a number of states: Queensland, New South Wales, the Northern Territory and South Australia (Bureau of Meteorology 2011). Preliminary estimates of damage for the state of Queensland are in the order of one billion Australian Dollars. Much media coverage was given to the flooding of Brisbane (Queensland) on the 13 January 2011. Flood events of comparable magnitude had occurred previously under La Niña conditions in February 1893 and January 1974. The latter of these resulted in 14 deaths and $200 million damage. As a consequence of the damage sustained during the 2011 event, operation of Wivenhoe Dam came under scrutiny. Challenges in operating dams to ensure both water supply and protection were already covered in a report on the 1974 flood event (Bureau of Meteorology 1974):

> *In situations where the major flood contribution occurs in catchments below Somerset Dam and the proposed Wivenhoe Dam, there are considerable problems in deciding when to empty the flood storage. If floodwaters were retained by the dam for too long not only would there be major and prolonged flooding upstream from the storage but the dam would become virtually useless for flood mitigation downstream in the event of a repetition of excessive rainfall. Meteorologically such a situation has already occurred (in 1893 when there were three floods within a month) and a recurrence appears inevitable.*

The Brisbane floods highlight the necessity of addressing nonstationarity in hydrological extremes associated with the El Niño-Southern Oscillation against a potential long-term underlying drying in some regions of Australia in the operation of large dams.

Vulnerabilities in developing countries are much greater, as the 2010 Pakistan floods caused by heavy monsoonal rainfall show. Estimates of impact vary depending on source. According to Alertnet (http://www.trust.org/alertnet/crisis-centre/crisis/pakistan-floods-2010) this was the worst flood in Pakistan in living memory, costing 1,750 lives and affecting more than a tenth of the population.

13.1.3 Definitions

This section will cover definitions of four key terms used in the remainder of the chapter: engineering design, extremes, return period and stationarity.

13.1.3.1 Engineering Design

For the purposes of this chapter, we will take ‘engineering design’ to relate to design of structures such as roads, bridges, dams and sewers. Closely related activities include flood risk mapping, flood warnings and emergency management as well as the operation of dams and regulated river systems and mining operations. For most of this chapter, the focus will be firmly on high extremes (floods) but we note that while low extremes in precipitation and streamflows (low flows) may be considered less of a challenge for design, they can have large effect on society (e.g. water supply, hydropower production, natural environment).

Table 13.1 Extreme rainfall indices as recommended by the ETCCDI

Index	Name	Definition
RX1day	Max 1-day rainfall total	Seasonal maximum 1-day rainfall
RX5day	Max 5-day rainfall total	Seasonal maximum 5-day rainfall
R10mm	Heavy rainfall days	Annual count of days where rainfall > 10 mm
R95pT	Proportion of annual rainfall from very wet days	Total annual rainfall from wet days (≥ 1 mm) with rainfall above the 95th percentile for wet days in the 1961–1990 period, divided by the annual rainfall

13.1.3.2 Extremes

Differences in semantics between different scientific communities can lead to misinterpretations. These differences have evolved over decades and are motivated by different paradigms. While it may be desirable to bring these definitions into alignment, this may not always be feasible. Certainly there should be an emphasis on stating clearly which definition is used.

Climate Change

For the analysis of trends in intense daily precipitation Groisman et al. (2005) defined *heavy*, *very heavy* and *extreme* precipitation as the 90th, 99.7th and 99.9th percentiles respectively and this definition has become somewhat of a standard. The Expert Team on Climate Change Detection and Indices (ETCCDI) defined a suite of indices and indicators for use with both observations and climate model output. Table 13.1 gives some examples for indices defined for assessing rainfall extremes, note that *'heavy rainfall days'* refers to the frequency of exceedance of a predefined, constant threshold (10 mm).

Extreme Value Analysis

From a statistical perspective, extremes are defined in the sense of extreme value series. For annual maximum series of *daily data* this implies the 99.7th percentile (matching the definition by Groisman et al. for very heavy precipitation) while for series of annual maxima of *hourly data* the 99.99th percentile would be characteristic and for a peaks-over-threshold approach with on average three peaks per year this would equate to the 99.2nd percentile.

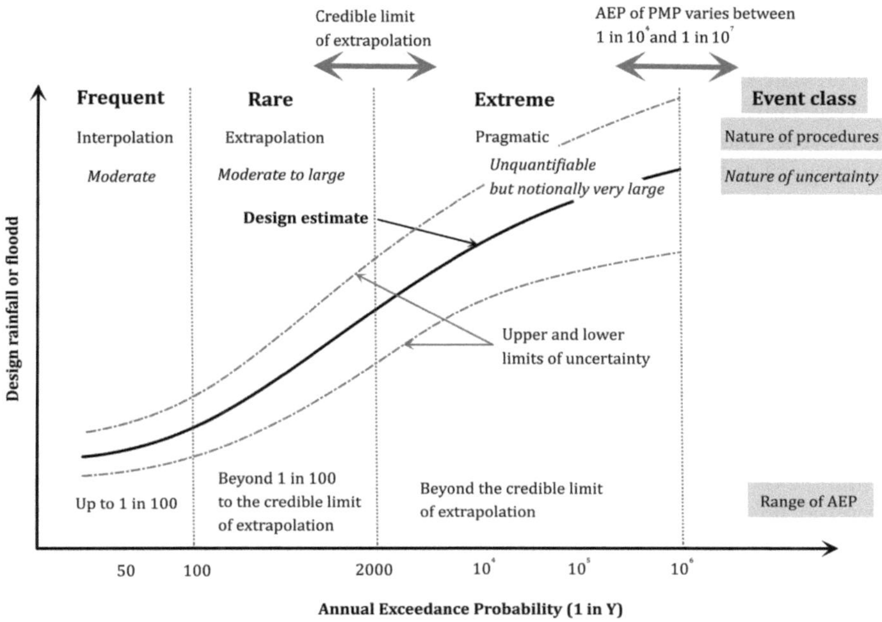

Fig. 13.2 Design characteristics of notional design event classes (Redrawn after Nathan and Weinmann 2001)

Hydrology and Engineering

Hydrologists might classify events on the basis of their frequency of occurrence as *'frequent'*, *'rare'* or *'extreme'* (Fig. 13.2). Events labeled as 'extreme' would be expected to occur on average only once every 10,000 or 100,000 years – well outside the range typically considered in climate change research or extreme value analysis.

Impact Assessments

Lastly, the term 'extreme' may be attached to an event because of the impact it has on society (damages/lives lost as in the examples provided above) rather than magnitude or rarity of its meteorological or hydrological components.

13.1.3.3 Return Period

Return period or *recurrence interval* is often used as a way of referring to concepts which more accurately should be described in terms of *Annual Exceedance Probability (AEP)* or *Average Recurrence Interval (ARI)*. The Annual Exceedance

Probability (AEP) is the probability that a given rainfall total (or flow) for a given duration will be exceeded in any 1 year. An Average Recurrence Interval (ARI) can be calculated as the reciprocal of the AEP.

Generally, and when applied to peaks-over-threshold series (series of independent events above a preselected threshold), ARI is the average, or expected, value of the *periods* between exceedances of a given rainfall total for a given duration. When specifically applied to the annual maximum series (a series that only contains the event with the largest magnitude that occurred in each year) it is the average, or expected, *number of years* between years in which there are one or more exceedances of a given rainfall total for a given duration.

Conventional definitions of ARI and AEP refer to long-term averages and are based on the notion that the probability of an event of given magnitude being equalled or exceeded does not change over the lifetime of a structure. The risk R of failure of a structure within a life-time of n years for an event magnitude with an average recurrence interval T can be calculated using Eq. 13.1. For a structure with a lifetime of 100 years designed for a '1 in 100 years' – event, the risk of failure is therefore 63.4%.

$$R = 1 - \left[1 - \left(\frac{1}{T}\right)\right]^n \tag{13.1}$$

13.1.3.4 Stationarity

As with the term 'extremes' there are different perceptions about what stationarity (or nonstationarity) implies and indeed for practical application, interpretation of the meaning of stationarity may be somewhat relative and depending on the period under consideration (year, decade, century).

Coles (2001) provides the following definition of a stationary process:

> *A random process $X_1, X_2, \ldots$ is said to be stationary if, given any set of integers $\{i_1, \ldots, i_k\}$ and any integer m, the joint distributions of $\{X_{i_1}, \ldots, X_{i_k}\}$ and $\{X_{i_1+m}, \ldots, X_{i_k+m}\}$ are identical.*

Where a 'random process' is defined as a sequence of independent and identically distributed random variables. According to Coles, *'trends, seasonality and other deterministic cycles are excluded by an assumption of stationarity.'* Often though reference is given to 'stationary time series' (rather than stochastic processes) and stationary series are considered those for which the distribution is invariant under translation in time. For most practical applications the definition of stationarity is relaxed to what is referred to as 'wide-sense stationarity', which is satisfied if neither mean nor autocorrelation change with time.

Whether we treat a series as stationary or nonstationary depends to some degree on our perception about the underlying processes: do we assume there is an underlying deterministic function defining changes or are they considered to have come about as part of a random process? This question is far from purely philosophical and in practice may boil down to answering the question whether we assume (natural) climate variability or climate change are driving observed changes. The validity of assumptions about random processes (independent and identically distributed observations) as relevant for frequency analysis is discussed in Sect. 13.4.

13.2 Climate Variability and Climate Change

With reference to 'hydrologic extremes in a changing climate' this section aims to provide some of the required background information. This short section therefore describes the respective effects of natural (or inherent) climate variability and anthropogenic climate change on hydrological extremes. Following on from the discussion above, it may prove useful for our understanding and for practical applications to distinguish between the two. In this discussion we will cover neither nonstationarity in hydrological extremes due to factors other than climate nor will we expand on abrupt climate shifts.

13.2.1 Climate Variability

Large-scale modes of atmosphere-ocean interactions are known to affect precipitation (rainfall as well as snow accumulation) and therefore the occurrence and severity of droughts and floods. Well-known examples are the El Niño-Southern Oscillation (ENSO) and the North Atlantic Oscillation (NAO). The word 'oscillation' hints at the fact that these tend to be see-saws between high/low, positive/negative or warm/cold phases generally driven by anomalously warm/cool sea surface temperatures and related changes in pressure patterns with associated changes in wind direction and speed, and cloudiness. These oscillations can be characterized using indices defined for a grid box(es) of certain latitude and longitude range, and may be based on a single parameter (e.g. the Southern Oscillation Index is calculated as the pressure difference between Darwin and Tahiti) or a combination of parameters like the multivariate ENSO Index.

Most severe droughts for the conterminous USA tend to occur for the negative (cool) phase of the Pacific Decadal Oscillation (PDO) and the positive (warm) phase of the Atlantic Multidecadal Oscillation (AMO). McCabe et al. (2004) found that more than half of the temporal and spatial variance in drought frequency can be explained by the state of AMO and PDO.

The effects of the ENSO had already been briefly touched on in the discussion of the demise of Angkor and recent flooding in Brisbane. This oscillation is one of the main drivers behind rainfall variability over large parts of Australia but its relationship with rainfall is determined by the interaction with other drivers like the Indian Ocean Dipole (IOD). The combined effect of La Niña and negative phase of the IOD typically leads to wet conditions over eastern Australia while the combination of El Niño and positive IOD on average leads to particularly dry conditions in this region. The Southern Annular Mode (SAM) affects mainly southern parts of the country where the positive phase of SAM is associated with reduced rainfall due to the poleward contraction of the westerly wind belt. On shorter time-scales, the Madden-Julian Oscillation (MJO) comes into play and is important especially for monsoonal rainfall (in the north of the continent). The patterns of rainfall relationship to each of the drivers exhibit substantial decadal variability (Risbey et al. 2009) and studies have shown that drivers themselves may be affected by climate change. For example, a recent study showed that observed positive summertime trends in SAM driven by increasing greenhouse gases are likely to reverse due to recovery of the ozone hole (Arblaster et al. 2011).

Much attention has been given to studying the large-scale oscillations discussed in this section and we now know that their effect varies with region and season, and that climate change may affect both the frequency with which these oscillations enter their positive/negative phase and the strength of these events. While the relationship of indices with rainfall is not stable over time, knowledge of these drivers should still prove useful in attempting decadal and multidecadal prediction. From a design perspective, it is desirable to include this information in the estimation of design estimates. How this can be achieved is discussed in Sect. 13.4. Knowledge of these drivers is already successfully being exploited in providing seasonal streamflow forecasts (Wang et al. 2009). Predictions at seasonal time scales could be considered a 'test bed' vis à vis longer time scales with regards to choice of suitable statistical techniques as well approaches to presenting the skill of these predictions.

13.2.2 Climate Change

Temperature records indicate an increase in global temperature of about 0.7°C over the last 100 years (IPCC 2007). Most of this warming has been attributed to greenhouse gas emissions from human activity (e.g. burning of fossil fuels).

13.2.2.1 Rainfall Extremes

According to the Clausius-Clapeyron relation, for each 1°C increase in global temperature, the precipitable water increases by about 7% but models show that the associated increase in average rainfall is typically considerably lower (about 2% per 1°C). However, Lenderink and van Meijgaard (2008) show that for a site in the Netherlands (de Bilt) for temperatures above 12°C the increase in magnitude for intense rainfall events (99th percentile) at the 1-h duration could be as high as 14% per 1°C temperature increase. O'Gorman and Schneider (2009) found that the effects of climate variations are most readily detected for the highest thresholds.

According to IPCC (2011)

> *There have been statistically significant trends in the number of heavy precipitation events in some regions. It is likely that more of these regions have experienced increases than decreases, although there are strong regional and subregional variations in these trends.*
>
> *It is likely that the frequency of heavy precipitation or the proportion of total rainfall from heavy falls will increase in the 21st century over many areas of the globe. This is particularly the case in the high latitudes and tropical regions, and in winter in the northern mid-latitudes. Heavy rainfalls associated with tropical cyclones are likely to increase with continued warming. There is medium confidence that, in some regions, increases in heavy precipitation will occur despite projected decreases of total precipitation in those regions. Based on a range of emissions scenarios (B1, A1B, A2), a 1-in-20 year annual maximum daily precipitation amount is likely to become a 1-in-5 to 1-in-15 year event by the end of the 21st century in many regions, and in most regions the higher emissions scenarios (A1B and A2) lead to a stronger projected decrease in return period.*

Detecting and attributing trends in historical rainfall extremes is complicated by the large degree of variability. Min et al. (2011) are considered the first to have succeeded in formally attributing the anthropogenic contribution to intensification of extreme precipitation over large parts of the Northern Hemisphere. However, because of the low signal-to-noise ratio, attribution fails when moving to smaller subcontinental scales.

Using downscaled climate model runs, Christensen and Christensen (2002) show that even as summers become drier over parts of northern Europe, severe 5-day precipitation (99th percentile) might increase. For Canada, Mailhot et al. (2007) found that precipitation changes at short durations exceeded those at longer durations. Based on comparison of control run (1961–1990) and future climate simulation (2041–2070) from the Canadian Regional Climate Model (CRCM), return periods were found to have halved for the 2- and 6-h durations, and to have decreased by one third for the 12- and 24-h durations. For late twenty-first century warm season precipitation (May–August), average precipitation is projected to decrease over the Central United States but at the same time an increase is projected in the intensity of both heavy precipitation events and rain in general when it does fall (Bukovsky and Karoly 2011). The same study indicates an increase in the frequency of very heavy to extreme 6-h average events, but a decrease in the frequency of all events lighter than those.

13.2.2.2 Floods

There is, however, less certainty about changes in floods (IPCC 2011):

> *There is limited to medium evidence available to assess climate-driven observed changes in the magnitude and frequency of floods at regional scales because the available instrumental records of floods at gauge stations are limited in space and time, and because of confounding effects of changes in land use and engineering. Furthermore, there is low agreement in this evidence, and thus overall low confidence at the global scale regarding even the sign of these changes.*
>
> *Projected precipitation and temperature changes imply possible changes in floods, although overall there is low confidence in projections of changes in fluvial floods.*

Changes in rainfall are amplified in streamflow, for Australia typically by a factor of 2–3 (Chiew 2006). Analogous to changes in rainfall extremes, Boé et al. (2009) found based on downscaled projections from 14 CMIP3 (Climate Model Intercomparison Project) models that an increase in high-flow magnitudes could occur in parallel with an increase in low-flow frequency and a decrease in mean discharge. However, the authors argue that such changes 'are strongly linked to changes in extreme precipitation, whose evolution at the regional scale is hard to predict'.

13.2.2.3 Sea Level Rise

One of the areas where climate change is already being addressed, is in the design of coastal structures (see also Sect. 13.3.2. Adaptation) because there is a high degree of certainty that climate change is affecting sea levels (IPCC 2011):

> *It is very likely that mean sea level rise will contribute to upward trends in extreme coastal high water levels in the future.*

13.2.2.4 Additional Considerations

This brief overview over studies into how climate change is likely to affect hydrological extremes in the future highlights that such changes are expected to be far from uniform. While there is little doubt that climate change will affect (and may already be affecting) hydrological extremes, statements like 'a warming world will lead to an intensification of the hydrological cycle' while true do not do the complexity of the processes justice and by themselves are not helpful in a decision-making context. In-depth assessments need to include geographical location, season, duration and rarity of extremes.

Geographical location Changes at regional and local scales as required in decision-making are difficult to predict.

Season Changes will not only affect the frequency or magnitude of events but seasonality of flood events is likely to be affected too through changes in antecedent conditions (snowpack, state of soil – frozen ground, catchment wetness, ground water levels). Because potential effects of climate change may vary with season, this fact should be addressed in the analysis. Clear differences in seasonal trends have already been identified. For example, for four Irish rivers Bastola et al. (2011) found increases in streamflow during winter accompanied by progressive decrease in summer. Extrapolating from known seasonal rainfall changes to changes in streamflow is complicated by the fact that seasonality of rainfall and streamflow may not match, as illustrated in Fig. 13.3.

Duration There are indications that extremes at shorter durations might be disproportionally affected (Mailhot et al. 2007; Jakob et al. 2011a, b), which is important in an urban context.

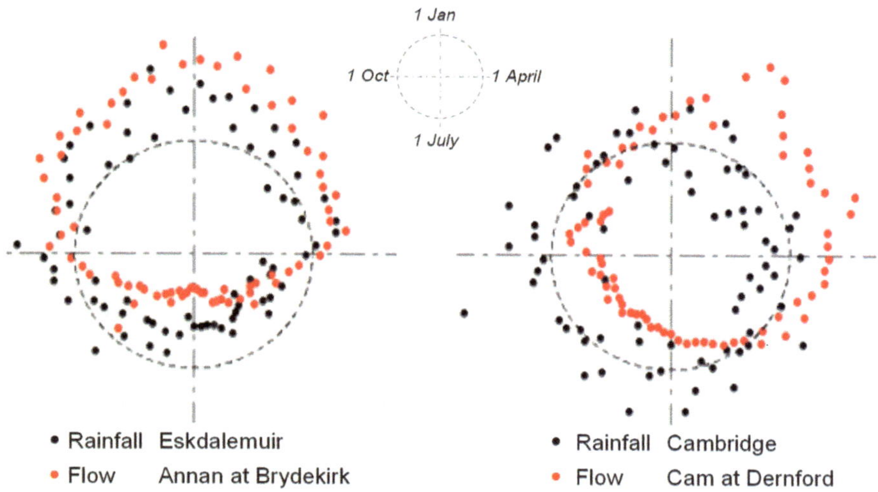

Fig. 13.3 Circular diagrams highlighting differences in seasonality in rainfall (*black dots*) and flows (*red dots*) for two catchments. Shown are standardized mean pentad values for 30 years commencing 1 March 1968. The angular plotting position is defined by the date, distance from the origin indicates magnitude. The shaley Annan is located in southwest Scotland, the chalky Cam in the southeast of England (Source: Reed 2011)

Rarity of extreme Effects of climate change could be more pronounced for more intense rainfall events (O'Gorman and Schneider 2009). Also, low and high extremes could be affected differently.

13.3 Impacts – Hydrological Extremes and Society

A special report on 'Climate Change and Water' discusses the challenges we are facing (Bates et al. 2008). Expected changes in hydrological extremes can briefly be summarized as

> *Increased precipitation intensity and variability is projected to increase the risks of flooding and drought in many areas. The frequency of heavy precipitation events (or proportion of total rainfall from heavy falls) will very likely increase over most areas during the 21st century, with consequences to the risk of rain-generated floods. At the same time, the proportion of land surface in extreme drought at any one time is projected to increase (likely), in addition to a tendency for drying in continental interiors during summer, especially in the subtropics, low and mid-latitudes.*

The report also provides an overview over main water-related projected impacts by region. Some examples have been selected to give an impression of the range of impacts with relevance for design issues.

Asia
Coastal areas, especially heavily populated Asian megadelta regions, will be at greatest risk due to increased flooding from the sea and, in some megadeltas, flooding from the rivers.

Australia and New Zealand
Risks to major infrastructure are likely to increase due to climate change. Design criteria for extreme events are very likely to be exceeded more frequently by 2030. Risks include failure of floodplain levees and urban drainage systems, and flooding of coastal towns near rivers.

Europe
By the 2070s, hydropower potential for the whole of Europe is expected to decline by 6%, with strong regional variations from a 20–50% decrease in the Mediterranean region to a 15–30% increase in northern and eastern Europe.

Latin America
Any future reductions in rainfall in arid and semi-arid regions of Argentina, Chile and Brazil are likely to lead to severe water shortages.

Other not directly design-relevant impacts relate to agriculture (damage to crops, death of livestock, land degradation) and human health (diseases, malnutrition). After Schneider and Kuntz-Duriseti (2000) the significance of climate impacts can be judged using five criteria: monetary loss, loss of life, biodiversity loss, distribution and equity, and quality of life. Most assessments in the published literature, however, express adaptation costs and benefits (avoided climate impacts) in purely monetary terms.

13.3.1 Impacts on Cities

Currently half the world's population is living in cities and this proportion is projected to rise to about 60% by 2030 with most of that growth occurring in developing countries (OECD 2008). At the same time, cities and especially coastal cities or those located at major waterways, are likely to be strongly affected by changes in hydrological extremes. An estimated 5% of the world's Gross Domestic Product (GDP) was exposed to 100-year floods in 2005. This number is expected to rise to about 9% of world GDP by the 2070s. While two thirds of that increase is expected due to socio-economic changes, another third of this projected increase is likely due to climate change and artificial subsidence (Hanson et al. 2010).

Ntelekos et al. (2010) provide an overview of the flood history of the USA and how it was shaped by major flood events. They add another dimension for assessment of impacts of changes in hydrological extremes in an urban environment: Flood frequency estimation in an urban setting is complicated due to the uncertain nature of the relationship between rainfall and flows for application in hydrological modeling. One of the exacerbating factors for the uncertainty in this relationship is nonstationarity due to changing infrastructure and climate. To add to the problem, urban streams are typically more sensitive to short rainfall durations (around 1 h) and it is 'very likely' that an increasing fraction of annual rainfall will be delivered in short duration high intensity events (IPCC 2007). An example to illustrate the magnitude of the challenge: Total loss from river flooding for metropolitan Boston (USA) is projected to exceed $57 billion by 2100, almost half of which is attributed to climate change (Kirshen et al. 2008). Suitable adaptation measures could reduce these costs by about 80%, implying that there is huge potential to reduce those impacts through suitable adaptation strategies.

13.3.2 Adaptation

For decision-making purposes *vulnerability* could be classified by degree of severity as either *'acceptable'*, *'significant negative impacts but survivable'* or *'intolerable without policy changes'* (Plate 2002). In assessing the vulnerability of a system to climate change (and 'other unfamiliar conditions') one would typically consider *reliability* (frequency of likely failure), *resilience* (time required to recover from failure), *vulnerability* (severity of consequences of failure) and a range of conditions that stress the system but don't lead to failure (robustness, brittleness).

Ziervogel et al. (2010) discuss how identifying key vulnerabilities to climate variability (both with regards to droughts and floods) may help us develop appropriate responses to climate change. As discussed earlier, developing approaches to make best use of seasonal forecasts could be considered a 'test bed' when it comes to adaptation to climate change. One of the important factors identified as hindering the uptake of seasonal forecasts by water managers was risk perception. So even the combined knowledge of the causes of climate change, likely impacts and possible solutions does not necessarily translate into action. Perception of risk depends on personal experience (or lack thereof) of climate-related events and competing risks. Communication therefore needs to address this potential barrier by conveying a consistent message on climate change and providing advice that is tailored for the target audience, making use of visualisation where possible (Adger et al. 2007).

Adger et al. (2007) provide a detailed discussion on adaptation to climate change and provide a range of examples across different sectors. They define adaptation to climate change as 'adjustments to reduce vulnerability or enhance resilience in response to *observed* or *expected* changes in climate and associated extreme weather events.' There are barriers and limits to adaptation which fall into five broad categories (IPCC 2007):

- physical or ecological
- technical, political or social
- economic
- cultural and institutional
- cognitive and informational

Two contrasting examples (coastal structures, fluvial flood risk management) are presented below to illustrate this point.

13.3.2.1 Coastal Structures

For long-lived infrastructure there are already cases where scenarios of future climate change and associated impacts have been taken into account, to date most of these address anticipated sea-level rise because there is high confidence that climate change is leading to sea level rise. Analysis for Cuxhaven, Germany (Mudersbach and Jensen 2010) suggests an increase in annual maximum water levels of about 50 cm by the year 2085. The Dutch Committee on Water Defence already recommends taking 50 cm sea-level rise into account when designing storm surge barriers and dams.

13.3.2.2 Flood Risk

There is, however, less certainty in observed changes in magnitude and frequency of floods at regional scales, and low confidence in projections of changes in fluvial floods (IPCC 2011). Fluvial flood risk can be represented as a combination of hydrological hazard and vulnerability (Merz et al. 2010). According to Merz et al. this relationship can be expressed as

$$RI(t) = \int_{h_D(t)}^{\infty} f_h(h,t)\, D(h,t)\, dh \tag{13.2}$$

where

- h is the flood water level,
- D is the damage,
- h_D is the level above which damage occurs, and
- $f_h(h,t)$ is the probability density function for the flood level h.

A large number of studies have been undertaken to assess the relative contributions of climatic and non-climatic factors to changes in flood risk, and it appears likely that non-climatic factors dominate (IPCC 2011):

> *Increasing exposure of people and economic assets has been the major cause of the long-term increases in economic losses from weather- and climate-related disasters (high confidence). Long-term trends in economic disaster losses adjusted for wealth and population increases have not been attributed to climate change, but a role for climate change has not been excluded (medium evidence, high agreement).*

The increase in economic losses together with a high uncertainty about flood risk in a changing climate have already led to a shift away from flood protection to flood risk management. In other words, a move away from structural measures to prevent flooding based on a design flood event to 'living with floods' and measures to reduce damage. Merz et al. (2010) quote examples like floating amphibious buildings.

13.3.2.3 Combined Effects of Sea Level Rise and Increased Intensities in Design Rainfall

Under the title 'Practical consideration of climate change', the Department of Environment and Climate Change, New South Wales, Australia (2007) sets out advice to be used in conjunction with existing floodplain risk management guidelines.

As a first step, it is recommended to undertake *sensitivity analyses* (in addition to those usually undertaken). Recommended timeframes are 2090–2100 for sea level rise and 2070 for rainfall. Three scenarios (low, medium, high) are suggested for sea level rise, ranging from 0.18 to 0.91 m. Likewise, three scenarios are recommended to assess the sensitivity to changes in rainfall intensities, ranging from 10–30%. Where applicable, the combined effects of sea level rise and changes in rainfall intensity should be assessed using joint probability approaches.

Secondly, and based on the results from the sensitivity analyses, the *significance of impacts from climate change for the location in question* needs to be assessed to decide whether impacts are 'minor' or 'significant'. The following questions will need to be addressed:

- Will climate change significantly affect flood hazard, flood damage, frequency of exposure, frequency of inundation or could it result in new floodways?
- What are the key sources of changes in vulnerability (sea level rise, increased rainfall intensity, a combination of the two)?
- What are the management options regarding existing or proposed development?

As part of the review of existing Flood Risk Management Plans (to be undertaken at least every 5 years) it needs to be ensured that relevant climate change impacts are adequately understood and considered in informed decision making.

Finally, *suitable management strategies for existing and proposed development* depend on the significance of ramifications. Where these are considered minor for existing development, one option might be to do nothing now but to allow for necessary upgrades in the future. Where ramifications are considered significant, other options will have to be considered:

- With regards to future development, consideration may be given to development types that allow planned retreat from affected land (e.g. caravan parks).
- Consideration could be given to setting land aside to allow for construction of a levee to manage future climate change impacts (i.e. reduce the frequency of inundation).
- Damage potential could be reduced by a requirement for two-storey housing with flood compatible structural materials in the bottom storey.

13.3.3 Decision Making Under Uncertainty

13.3.3.1 Uncertainty in Projected Changes

The majority of decision-making approaches make use of output from Global Climate Models (GCMs), either directly or indirectly since they are the only credible tool for projecting future climate. However, while there have been major advances in this area, climate models are by their very nature a simplification of highly nonlinear processes and even the latest generation of coupled GCMs does not adequately model large-scale teleconnections known to significantly affect hydrological extremes. The prime example here is the El Niño-Southern Oscillation (ENSO). Coupled climate models generally struggle to correctly model both the location of the Pacific cold tongue in Sea Surface Temperatures (generally overpredicting the westward extension) and ENSO seasonality/variability (IPCC 2007). Based on GCM output we cannot yet conclusively say whether ENSO is likely to strengthen or weaken under climate change. Because of their coarse resolution, GCMs are not capable of accurately modeling characteristics of key rainfall-producing mechanisms like the frequency and intensity of tropical cyclones or the seasonality/extent of the monsoon and associated wind patterns.

The performance of GCMs is evaluated on the basis of how well they are able to replicate current climate conditions. While adequate performance with regards to current climate would be considered a necessary condition, it does not imply reliability of climate projections. There is little agreement on how one should choose suitable models. Generally ensemble approaches are preferred. The choice of models to include in such ensembles is based on their perceived performance against a set of metrics or variables of interest and results can be used to develop weighting schemes for use with multiple models/model runs. A detailed and well-written discussion of the various components of uncertainty is presented in Stainforth et al. (2007). The authors suggest reframing the question by moving away from attempting to select the best models to using the available information to establish a 'non-discountable envelope of possibilities'. Another key reference with regards to uncertainty in (temperature) projections is Knutti et al. (2008).

Statistical assessments have been undertaken to assess and decompose uncertainty in GCM projections (Hawkins and Sutton 2009, 2010). Three important components are: (1) choice of emission scenario, (2) model structure and (3) modeled climate variability. Alternative statistical techniques like ANOVA (Yip et al. 2011) allow assessing the interaction terms (e.g. interaction between model choice and scenario). Other uncertainties in GCMs – for instance those due to 'knowledge gaps' (including uncertainty about key feedbacks like the combined lapse-rate/water vapour feedback) – are more difficult to assess quantitatively. It might be possible, however, to make use of expert knowledge with Bayesian techniques (Thyer et al. 2009). For most practical applications, GCM projections have to be processed further and it is informative to assess the relative contributions of GCM uncertainty, downscaling uncertainty (Crosbie et al. 2011), hydrological modeling (Bastola et al. 2011) and decision making-tools.

13.3.3.2 Methods for Decision Making Under Uncertainty

The approach described above has a number of drawbacks: firstly, at each of these steps uncertainties are introduced or propagated (uncertainty cascade) and secondly, it is computationally highly expensive.

An inverse approach referred to as ‘decision scaling’ (Brown 2009) was implemented in the International Upper Great Lakes Study (IUGLS). Rather than climate output, the starting point in decision scaling is ‘*asking first to identify the futures we are not prepared for and then asking if those futures are plausible*’. Based on vulnerability assessments, risk scenarios can be developed and the plausibility of these scenarios can be evaluated using a ‘climate response function’ to identify potential climate change that causes an unacceptable decrease in reliability and therefore requires some adaptation.

A white paper by the Water Utility Climate Alliance (2010) presents five decision support planning methods (DCSMs) together with a discussion of their key elements, case studies and gap analysis. These five methods are

- Classic decision analysis,
- Traditional scenario planning,
- Robust decision making,
- Real options, and
- Portfolio planning.

Important criteria for the selection of appropriate method(s) are available resources (computational demands, level of expertise required), planning horizons and ease of use of results (communicability, transparency).

Wilby et al. (2009) present an overview of tools for constructing climate change risk information with an emphasis on developing regions because of their greater vulnerability and lower capacity to adapt. The authors advocate marrying two different approaches to decision-making: evaluating the sensitivity of infrastructure to observed climate variability to answer the question: *‘How much climate change has to happen to be of practical significance?’* which also allows identifying key vulnerabilities; and scenario-led adaptation and development planning to test effectiveness of adaptation measures from a long-term perspective. If we accept the inherent uncertainty and work under the assumption that projections of future climate multiple decades ahead are not certain enough to provide us with the information required for an optimum design solution, an adaptive approach to design may be prudent (Westra et al. 2010). Under this approach, *‘solutions are implemented in a staged fashion and the need for each successive step is evaluated periodically’*.

In some cases it may be desirable to provide specific guidance despite high uncertainties, although this is rarely done. Design rainfall estimates for New Zealand can be derived using the High Intensity Rainfall System (HIRDS). In its latest version (version 3, accessed 22 May 2011, http://hirds.niwa.co.nz/) users can enter projected temperature changes to assess potential effects of climate change. The basis of this assessment is a percentage adjustment per degree warming, although users are cautioned that *‘increases might in some cases be considerably higher’*.

13.4 Deriving Design Estimates

The selection of suitable data from instrumental records and making use of non-systematic information is touched on briefly in the first part of this section while the remainder of the section presents techniques for deriving design estimates and how to address nonstationarity.

13.4.1 Data

13.4.1.1 Testing and Selecting Data Under Nonstationarity

Nonstationarity raises questions in the process of selecting data, namely selection of measurements that are representative of a 'current climate'. For a long time, the paradigm in frequency analysis has been to use long records wherever possible to increase the chance of sampling some of the largest events, and to cover multidecadal climate variability (known wet and dry periods). Changes over time in series of precipitation and streamflow may present themselves in three different ways:

- Slow, more or less monotonic trends
- Fluctuations on shorter time scales (modulating underlying trends)
- Sudden changes or breakpoints

Guidance on change detection in hydrological records and on adequate tests is given in Kundzewicz and Robson (2004). The authors advocate the use of visually exploring the data to identify patterns and features before applying formal statistical tests to assess the significance of these features. Given the nature of hydrological data (often skewed and serially correlated), the use of non-parametric tests and resampling approaches is recommended. The selection of suitable statistical tests should be based on the nature of the change to be tested for, i.e. gradual change (trend) or step change. Tests suitable for the detection of step change include Pettitt's test (Pettitt 1979), the distribution-free CUSUM test and Student's t test (for normally distributed data). Testing for trends could be based on Spearmans' rho, the Mann-Kendall test (addressing autocorrelation where required, Hamed and Rao 1998) or linear regression (assuming data are normally distributed).

The length of record required to detect a trend depends (among other factors) on the magnitude of the trend and the desired probability of detecting the trend. Longer records allow detecting smaller trends and nearly twice the record length may be required for a 99% probability of detecting a trend than for a 50% probability (Westra and Sisson 2011). Statistical testing to identify whether long-term trends in series can be detected would ideally be undertaken in a regional context because this increases the signal-to-noise ratio and therefore the likelihood of trend detection. In assessing the significance of such trends both serial correlation and cross-correlation need to be accounted for (Ishak et al. 2011). Simulations show that spatial information remains a poor substitute for increasing record length where that is possible (Westra and Sisson 2011).

An approach for the detection of network-wide trends in rainfall extremes is presented in Burauskaite-Harju et al. (2010). Analyses for 12 sites in the Netherlands were based on empirically derived percentiles (90th, 95th, 98th and 99th percentile) and percentiles estimated from fitted GPD (Generalized Pareto Distribution). Trend testing was undertaken using a multivariate Mann-Kendall test. The study identified that results from trend testing were robust to misspecification of the underlying distribution. Especially at higher percentiles, the power of trend testing was enhanced when GPD based percentile estimates were used because the sample variance was smaller than for empirical percentile estimates.

Where feasible, it might also be informative to undertake sensitivity testing to explore how design estimates are affected by choice of base period. As previously highlighted, such assessments should be undertaken on a seasonal basis and for a range of durations. In trying to select data representative of current or baseline conditions, we are faced with the conundrum that it suddenly becomes attractive to consider shorter periods of say 30-year periods but during which we may not have adequately sampled extreme events or the full range of multidecadal variability. Clearly, for structures with long life-times the use of baseline estimates will not be appropriate in a changing climate. Following this thought pattern though, the derivation of design estimates is essentially being split into two steps:

1. Derive estimates characterising 'current' climate conditions based on the analysis of historical data and
2. Develop guidance on how baseline estimates have to be adjusted based on projections of future climate.

The attractiveness of this two-step approach stems from its straightforward extension of existing techniques and paradigms. But what it does mean too is that hydrologists continue to undertake conventional analyses, only incorporating projected changes after preliminary estimates have been derived.

In the following we provide two examples of how the problem of data selection is being tackled in revisions of design estimates that have either recently commenced or are imminent. Haddad et al. (2010) describe the selection of streamflow data as part of the regional flood frequency analysis for the revision of flood frequency estimation for Australia (Haddad et al. 2010). Two trend tests (Mann-Kendall test and CUSUM test) were applied to series of annual maximum flood series. Between 10 and 20% of series were identified as having decreasing trends. This finding is likely due to a severe drought that affected southeast Australia since the 1990s. It has not yet been resolved whether the detected decrease in annual maximum flood series data is an expression of long-term climate variability or is due to climate change. A pragmatic way was chosen to deal with nonstationarity in the process of data selection: The minimum record length was increased from 20 to 25 years. While this led to a reduction of the number of candidate stations to about a quarter, it was deemed that the remaining series yield a more representative assessment of long-term flood risk.

Currently valid design rainfall estimates for Germany had been derived for the period 1951–2000 (Malitz 2005). Estimates for subdaily durations will be revised in the near future to include observations from the most recent decade (2001–2010). It has yet to be decided whether to choose a 30-year reference period (from 1981 to 2010) or a 60-year reference period (1951–2010) for the planned revision. A number of studies has already been undertaken to assess potential effects of climate change on daily rainfall totals for Germany. Preliminary investigations based on annual maxima at subdaily durations identified significant changes for only a small number of regions. The absence of strong indication for an increase in annual maximum rainfall at subdaily durations was ascribed to the fact that Germany experienced climate variability that led to exceptional magnitudes in subdaily rainfall totals in the 1950s and 1960s (Gabriele Malitz, personal communication). The two examples above highlight the difficulty in separating climate variability from the potential influence of climate change.

13.4.1.2 Instrumental Records

Design estimates will rely directly or indirectly on measurements and could therefore be affected by errors introduced either in the measurement itself (e.g. instrumentation failure), transmission of measurement, and ingestion into data bases (e.g. keying in of paper charts). Significant errors can also be introduced in the conversion of measurement to a hydrologically relevant variable like in the conversion from snow depths to snow water equivalent. Likewise, flows are often not directly measured but derived as estimates from stage discharge relationships (rating curves).

Estimates of potential errors introduced through that conversion vary widely. Cook (1987) quotes about 20–30%, another source puts the errors in the range of 10–100% (Benito et al. 2004). Certainly, in practice it may be difficult to derive suitable rating curves for the most extreme events even with the aid of hydrologic models, simply because those rare events have not been measured. Apart from precipitation and streamflow other climate variables like wind speed/direction, solar radiation, humidity and soil moisture may be required in deriving design estimates.

Other common challenges include incomplete records, artefacts in measurements (e.g. due to changes in observing practices or changes in station location) as well as insufficient network density, record length and temporal resolution. Statistical techniques have been developed to address most of these issues such as spatial techniques like kriging, procedures for infilling and disaggregation of data and homogeneity tests. Homogeneity testing based on standard normal homogeneity tests (SNHT) or two-phase regression tests (TPR) should precede testing for change in hydrological series to avoid detection of spurious trends (Pettitt 1979; Reeves et al. 2007; Toreti et al. 2011).

13.4.1.3 Historical Floods and Palaeofloods

Systematically measured flood data are used in the estimation of return periods up to about 100 years. For very large and rare events, this information can be supplemented by data from other sources, namely historical information (from documents and chronicles, e.g. Bayliss and Reed 2001) and flood information obtained through palaeoflood hydrology. *Historical floods* are extremes that have not been measured but have been documented and are accessible in public archives and local chronicles. Palaeoflood hydrology concerns itself with *'lasting effects of floods on natural indicators'* such as slackwater deposits or scour lines and has been used to lengthen flood records beyond that of instrumental records (Benito et al. 2004; Benito and Thorndycraft 2005). This information is valuable because it will provide exactly what is lacking from instrumental records due to a combination of logistic and technical challenges in measuring the most extreme events and limiting effects of short records. From historical archives, information about the precise dates and prevailing meteorological conditions could be retrieved together with some reference to flood peak levels (marked on buildings or bridges). Estimates of exact discharge level, minimum or maximum flood stage can be derived from palaeostage indicators (elevation reached by flood, flooded/non-flooded areas). Numerical dating of these events is undertaken using radiocarbon dating, luminescence methods and optical stimulates methods (Benito et al. 2004).

Non-systematical information can be incorporated into flood frequency analysis (Fig. 13.4). From a statistical perspective, this requires addressing two issues: dealing with censored data (only the major floods are known) and dealing with nonstationarity (due to climate variability), further details are given in Sect. 13.4.4.3. One major obstacle to the more widespread use of non-systematical flood data is the highly complex nature of the reconstruction of past flood events; requiring interdisciplinary cooperation between palaeohydrologists, historians, climatologists, meteorologists and statisticians.

13.4.2 Choice of Techniques

One of the key aspects of prescribing design standards is to ensure that the level of protection provided and the costs to achieve that level of protection are optimally selected. Some examples for different levels of protection are given below (Institution of Engineers Australia 1987):

- For urban drainage, standards are set to reduce inconvenience and disruptions.
- For the construction of bridges and roads, standards depend on the importance of roads and the availability of alternative routes.
- Flood mitigation schemes are typically based on 1 in 100 year ARI.

Performance standards for dams depend on height and storage capacity. For large dams engineers make use of standards-based approaches rather than statistical

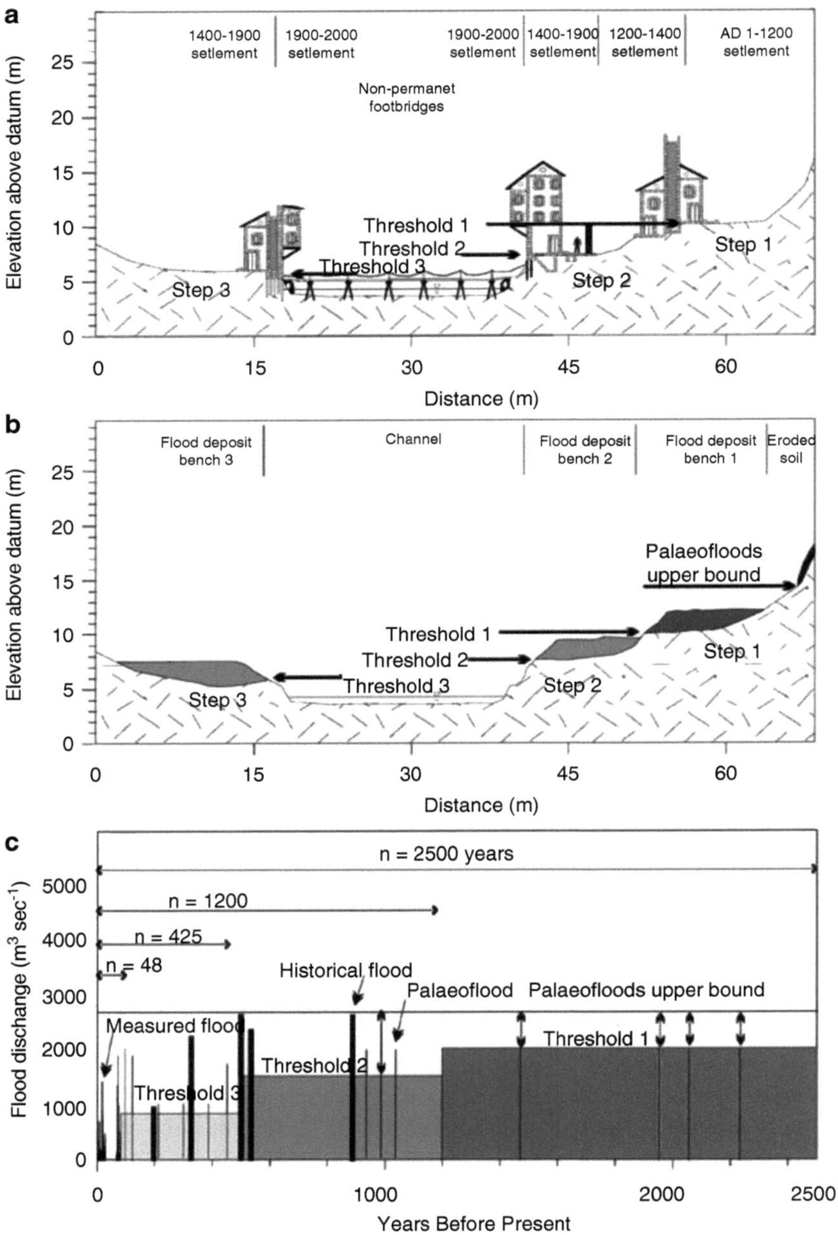

Fig. 13.4 Combining systematic and non-systematic flood information (source: Benito et al. 2004) – (**a**) Schematic diagram illustrating the changing level of flood perception through time according to progressive human settlement towards the river. This means that the threshold of flood discharges recorded in the historical documents decreases with time. (**b**) Geomorphic surfaces covered with slack-water flood deposits showing different thresholds of flood discharges. (**c**) Organisation of historical, palaeoflood and systematic data, using the described discharge thresholds, for flood frequency analysis

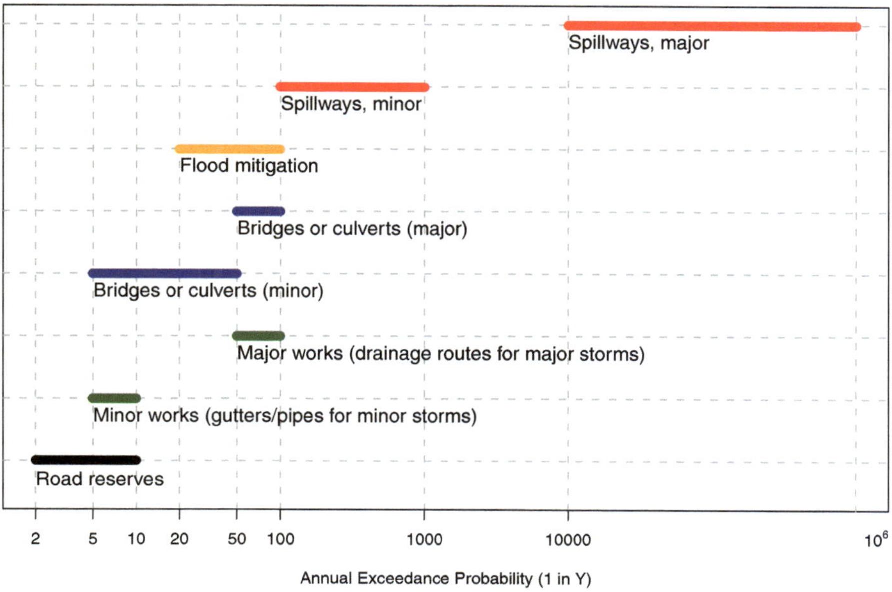

Fig. 13.5 Typical range of AEPs adopted as design standards (Redrawn after Ladson 2008)

techniques. These are applied where one has to follow a precautionary principle and failure has to be avoided regardless of cost (where loss of life is concerned). Such standards would be the Probable Maximum Flood and the Probable Maximum Precipitation. Typical ranges of Annual Exceedance Probability (AEP) for important design standards are summarised in Fig. 13.5 based on information provided in Ladson (2008).

No one technique can be used to derive estimates for all applications and AEP. Swain et al. (1998) present expert estimates of credible limits of extrapolation depending on data type. Where multiple data sources and techniques are available, these should be used to (a) derive independent estimates for validation and (b) to consolidate. For example, estimates derived through flood frequency analysis could be validated with results from rainfall-runoff modelling (Table 13.2).

As discussed above, the choice of technique will depend on both the application and data availability. With regards to data types there are basically two different approaches, a direct approach based on streamflow data and an indirect approach based on precipitation data. Frequency analysis (of either rainfall or flood data) can be used to derive estimates for the range up to about 1 in 100 years. This approach will be discussed in some detail in the following sections. Estimation of Probable Maximum Precipitation (PMP) and Probable Maximum Flood (PMF) is a somewhat separate topic and relevant procedures are presented in Sect. 13.4.7.

Table 13.2 Hydrometeorological data types and extrapolation limits for flood frequency analysis (After Swain et al. 1998)

Type of data used for flood frequency analysis	Limit of credible extrapolation for annual exceedance probability	
	Typical	*Optimal*
At-site streamflow data	1 in 100	1 in 200
Regional streamflow data	1 in 750	1 in 1,000
At-site streamflow and at-site palaeoflood data	1 in 4,000	1 in 10,000
Regional precipitation data	1 in 2,000	1 in 10,000
Regional streamflow data and regional palaeoflood data	1 in 15,000	1 in 40,000
Combinations of regional data sets and extrapolation	1 in 40,000	1 in 100,000

13.4.3 Frequency Analysis

The statistical techniques used in rainfall and flood frequency analysis respectively are based on the same fundamental concepts. An overview of techniques used for point rainfall frequency estimation in Europe (United Kingdom, Sweden, France, Germany), North America (Canada, USA) and the Southern Hemisphere (South Africa, New Zealand, Australia) is given in Svensson and Jones (2010).

13.4.3.1 Extracting the Extremes Series

Where frequency analysis has been identified as the appropriate technique and once data have been collected, quality controlled and missing data infilled as appropriate; series of extremes can be extracted. Extreme series consist either of block maxima (annual, seasonal or monthly maxima, r-largest) or exceedances above a threshold (referred to as 'peaks-over-threshold' or 'partial durations series'). Based on theoretical considerations and for sufficiently long records, estimates derived from an annual maximum approach and a peaks-over-threshold approach would be expected to be in good agreement.

For the annual maximum series, the highest-ranking event in a given calendar or water year is selected. The extraction of annual maxima is therefore simple to implement, apart from the fact that a suitable set of criteria is required to decide whether to include a year depending on the amount of missing data. The extraction of peaks-over-threshold series requires the definition of criteria to ensure the independence of subsequent events. This can be achieved by requiring a minimum drop in magnitude (for streamflow) and a minimum separation in time between selected peaks. A number of measures have been developed to assist in the selection of a suitable threshold. Thresholds can be defined either in terms of magnitude or with respect to an average number of peaks per year. Typical choices are in the range between one to three events per year. Lang et al. (1999) discuss three tests (based on the mean number of events, the mean threshold exceedance and the dispersion index) to decide on appropriate thresholds while Svensson and Jones (2010) describe threshold selection based on goodness-of-fit tests. Thresholds need to be selected high enough to satisfy the independence assumption but low enough to guarantee the sample is still large enough for parameter estimation. The additional effort in extracting peaks-over-threshold series is particularly worthwhile for short records.

13.4.3.2 Frequency Distribution

Frequency distributions are used to model the relationship between magnitude of extremes (rainfall or streamflow) and Annual Exceedance Probability (AEP). For a given AEP one can derive the corresponding magnitude which is relevant for design applications, and vice versa for a given magnitude the corresponding AEP can be estimated and the rarity of observed events can be assessed. Common choices are three-parameter distributions like the Generalised Extreme Value distribution (GEV) for annual maxima and the Generalised Pareto distribution (GPD) for peaks-over-threshold series. The three parameters here are the location, scale and shape parameters. A range of goodness-of-fit tests has been developed to test (a) whether a distribution delivers an acceptable fit and (b) to decide which of the candidate distributions results in the best fit (e.g. Hosking and Wallis 1997). The choice of distribution may also reflect physical considerations, for example whether the distribution should have an upper (or lower) bound.

13.4.3.3 Underlying Assumptions

The underlying assumption in fitting a frequency distribution to series of rainfall or streamflow extremes is that of a stationary process (see Sect. 13.1.3.4). Observations therefore should be independent and identically distributed.

Independence with respect to a time series at just one location requires that the events are not serially correlated. This assumption is likely to be satisfied for block maxima where the blocks are sufficiently long (say years) and the extraction process of peaks-over-threshold series is aimed at satisfying this assumption (for instance by requiring a minimum separation of events in time). Khaliq et al. (2006) discuss approaches to fitting distributions to dependent data.

The assumption that events have been sampled from the same population and therefore are identically distributed may not be valid. For instance where different mechanisms lead to precipitation extremes (frontal systems, convective events) or flooding (precipitation, snowmelt) this assumption may be violated, an example is shown in Fig. 13.6. In particular, very large events may be due to rare events like tropical cyclones and this could become a serious issue for deriving estimates at low AEP especially where the parameter estimates are derived using techniques

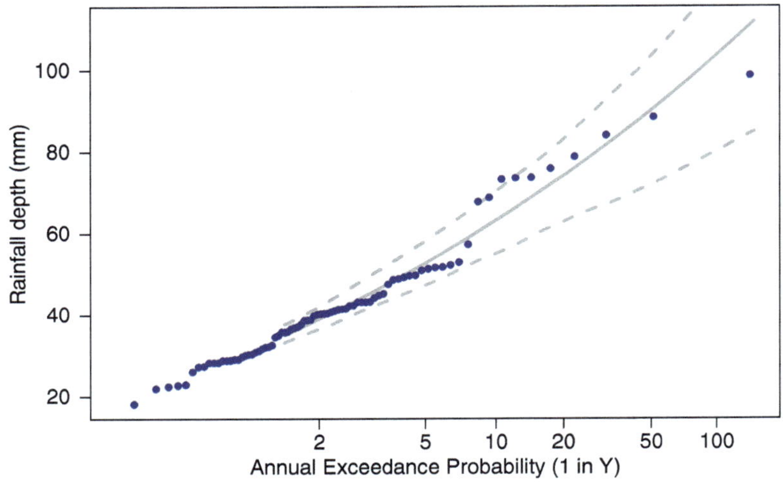

Fig. 13.6 Annual maximum series of 2 h rainfall depths at Brisbane (Queensland, Australia) based on period 1912–1993 (*dots*), GEV fitted using L-moments (*solid line*) and 95% confidence interval

that give high leverage to the most extreme events (e.g. LH-moments, Wang 1997). Approaches to address this issue include use of seasonal maxima (Svensson and Jones 2010). Seasonal maxima have been shown to be more homogeneous in respect to both distribution and trend over time (Strupczewski et al. 2001 for Polish rivers). This may indicate a need to give more emphasis to considerations of seasonality of rainfall extremes and flooding events. Van den Brink and Können (2008) suggest a tool to verify whether the largest events in a record are adequately represented by the fitted distribution. Willems (2000) presents an approach to derive compound IFD relationships for two seasons and two storm types.

13.4.4 Regional Frequency Analysis

13.4.4.1 Defining Regions

Considering the limited record lengths available and the need to extrapolate to very low AEP, optimal use should be made of available information. This is achieved through regional analysis; the underlying concept is to 'trade space for time'. In other words, data from a number of 'hydrologically similar' sites are pooled together to extend the length of record. Selection of sites to include for the purposes of regionalisation (sometimes referred to as pooling) can be based on geographical location, catchment characteristics, statistical measures characterising the extreme series or a combination of these approaches. There has been a tendency to move away from a small number of large geographically contiguous regions that were used predominantly in the past to a 'region-of-influence' approach (Hosking and Wallis 1997), that is to regions tailored to a 'site of interest'. Commonly used measures in this context are based on L-moment statistics to (a) identify homogeneous regions (i.e. sites in a region are judged to be hydrologically similar) and (b) to identify discordant sites (sites that are outliers with respect to other sites in a region).

So, what length of record is required to reliably identify the correct underlying frequency distribution? And how many years are required to derive reliable estimates for AEP 1 in 100 years?

- Generally, regional approaches are preferred over at-site assessments with the possible exception of cases where long at-site records exist and less extreme quantiles (say AEP 1 in 10 years) are required. The UK Flood Estimation Handbook (Robson and Reed 1999) recommends as a rule of thumb in flood frequency estimation to use five times as many station years as the target return period in question (5T rule). Following this rule, one would aim to include stations with a combined record length of approximately 500 years to estimate the 100-year flood.
- If cyclical behaviour rather than long-term trends are a concern, consideration should be given to the period of record as well as the record length.
- Bayesian techniques are especially useful with short records and allow deriving estimates of uncertainty. Resampling approaches are useful in constructing confidence intervals around frequency curves (Fig. 13.6). Information on uncertainty of estimates can be used to decide whether an estimate is deemed reliable or whether further investigations are required.
- Assuming the underlying assumptions are not violated, misspecification of distribution is probably important only for events rarer than 1 in 100 years (Hosking and Wallis 1997).

13.4.4.2 Index Flood Approach

This approach was pioneered by Dalrymple (1960) for flood frequency estimation and is widely used in regional frequency analysis of floods and rainfall. Fundamentally, this approach builds on the idea that sites in a region are hydrologically similar and the frequency curves for sites in the region can therefore be split into two components: a scaling factor – a site-specific index flood – and a scaled version of the frequency curve – the growth curve. The index flood is often chosen as the mean or median of the at-site annual maxima (or peaks-over-threshold). One critical aspect in this approach – apart from the choice of sites to include in the region and the size of the region – is the choice of suitable weights for combining regional and at-site information when deriving estimates at a site of interest. (Robson and Reed 1999)

13.4.4.3 Parameter Estimation

The choice of method for estimating parameters is a much-studied topic. Factors to consider in the choice of approach include the available record length and the parameters themselves. Commonly used techniques include method of moments, Probability Weighted Moments (PWM), Maximum likelihood estimators (MLE) and L-moments (Hosking and Wallis 1997) as well as modifications of the L-moment approach (LH-moments and LL-moments, Wang 1997). Hybrid techniques have also been suggested (Ailliot et al. 2011).

It is desirable to derive estimates of parameter uncertainty and a Maximum likelihood approach can be used to calculate standard errors (Hosking and Wallis 1987), which in turn can be used to construct confidence intervals. However, for small samples, maximum likelihood estimators may have very large variance and numerical solutions of maximum likelihood equations can lead to convergence problems or may result in solutions that are not physically acceptable. Alternatives are bootstrapping techniques (Faulkner 1999) and Bayesian techniques. Approaches to reduce uncertainty in parameter estimation can be based on regional information and such approaches have become increasingly popular. Examples include the use of regional shape parameters (Stedinger and Lu 1995), hierarchical approaches (Alila 1999) and regional regression (Swain et al. 1998).

Frequency analysis of palaeoflood data is based on the hypothesis that for a specified period all exceedances above a threshold have been captured. Estimates for distributional parameters are derived under the assumption of stationarity. The validity of this assumption for censored data can be checked using Lang's test (Lang et al. 1999). Maximum likelihood estimators, the method of expected moments, partial probability weighted moments or non-parametric techniques could be considered when working with censored data. Bayesian approaches appear particularly promising in this context. A review of suitable techniques is given in Ouarda et al. (1998).

13.4.4.4 Bayesian Techniques

Regional estimation may be used to mitigate lack of data but local information is then ignored. Bayesian approaches have a number of advantages in such situations: they are conceptually simpler than the classical approach to frequency estimation, they allow incorporating additional information in the inference process and the precision of parameter estimates can be assessed. A literature review on Bayesian approaches for combining local and regional information is provided in Seidou et al. (2006). In their discussion they distinguish between 'mixed approaches' where some parameters are estimated with local data and others with regional data, and 'simultaneous estimation' where both information sources are used together to estimate all parameters and quantiles. For the latter, either empirical or parametric (full) Bayesian inference can be employed.

Linear empirical Bayesian estimation combines at-site and regional information assuming that both local and regional flood quantile estimators are normally distributed. Weights are chosen to be inversely proportional to the estimator's variance. An alternative Bayesian approach is developed in Seidou et al. (2006). Prior distributions are obtained using a log-linear regression model on quantiles and quantile differences. Local observations are then used (within a Markov Chain Monte Carlo, MCMC algorithm) to infer posterior distributions of parameters and quantiles. Advantages of this approach are two-fold: it relaxes the assumption of normality of local quantiles' probability distribution and significantly improves the estimation of parameters and quantiles when short records are used.

13.4.5 Addressing Nonstationarity in Frequency Analysis

The classical notions of 'probability of exceedence' and 'return period' are no longer valid under nonstationarity. (Khaliq et al. 2006)

There are a number of confounding problems that need to be considered in choosing and developing suitable approaches to address nonstationarity in rainfall and flood frequency analysis:

- Nonstationarity in hydrological extremes may be due to climate or a range of other factors, including changes in land use/land cover. In this context, testing for break points should be undertaken prior to trend analysis. While land use/land cover changes may affect hydrological extremes directly, there are also interactions between such changes and climate change. Through these interactions it is possible that land use/land cover changes could magnify climate-driven changes in extremes (Vogel 2010).
- Since nonstationarity may be expressed in higher variability, both mean and variance should be tested. Techniques have been developed to decompose fluctuations and trends in time series; these include spectral analysis, wavelet analysis and Empirical Mode Decomposition (EMD). The latter has successfully been applied to series of hydrological time series (McMahon et al. 2008; Jakob 2011). Given that nonstationarity may come about due to a combination of factors, such analyses could be useful in gaining an understanding of the underlying mechanisms.
- Trends - especially were they are estimated on the basis of short records – may be misleading. In a similar vein, trends established at single sites may not be characteristic of larger regions. Field significance tests could be applied (Wilks 2006). Spatial pooling has been used to detect meaningful changes in extremes, for instance by using spatial averages of quantile estimates (Frei et al. 2006; Burauskaite-Harju et al. 2010).

Nonstationarity may affect series at different time scales. How this is addressed in the frequency analysis depends on the relationship between frequencies of nonstationarity and quantile estimates required. Where nonstationarity needs to be accounted for this can be achieved in a number of different ways:

- time-varying parameters
- time-varying moments
- use of covariates
- time-varying quantile estimates

Local likelihood approaches have been developed and proven to be useful exploratory tools. For these, models are fitted to extremes series using local fitting, i.e. parameter estimates can vary over time. This approach can be extended to include covariates in the local estimation. While there is merit in tackling the problem from a mathematical perspective for initial assessments, the aim should be to understand the physical mechanisms, and potentially include them as covariates. These covariates may be climate-related but other sources of nonstationarity need to be taken into account. For example, flood frequency analysis in urban watersheds is complicated by nonstationarity associated with changes in land use and stormwater infrastructure. In this situation suitable choices of covariates (in addition to rainfall and climate indices) may include indicators of the degree of urbanisation, like population. Villarini et al. (2009) undertook flood frequency analysis in such a setting using a Generalised Additive Model for location, scale and shape (GAMLSS). They found that peak discharge did not increase linearly but exhibited time trends which were most pronounced during a period of accelerated urban growth (Fig. 13.7). This example highlights the need to take more than just climate into account when addressing nonstationarity in frequency analysis. A promising technique that is not yet widely used is quantile regression. In quantile regression (and censored quantile regression) quantiles are directly related to climate indices through linear or non-linear functions.

13.4.5.1 Use of Covariates for Nonstationary GEV Parameters Fitted Using GML

Generalised Maximum Likelihood (GML) is based on the same principles as Maximum Likelihood – model parameters are estimated by maximising a log likelihood function (Coles 2001) – but with one additional constraint on the shape parameter. The GML approach can therefore be viewed as a special case of a Bayesian approach where the prior distribution is only specified for the shape parameter. A distinct advantage of GML is that both historical and regional information can be included in the formulation of the prior distribution. The parameter estimates can be derived using Monte Carlo Markov Chain (MCMC) methods. The concept of GML estimators can be extended to nonstationary models (El Adlouni et al. 2007). In the following we provide an outline of an approach to modelling under nonstationarity based on the use of a climate index as covariate.

El Adlouni et al. (2007) discuss the modelling of annual maximum precipitation for a station in California with 51 years of record where annual maxima exhibit

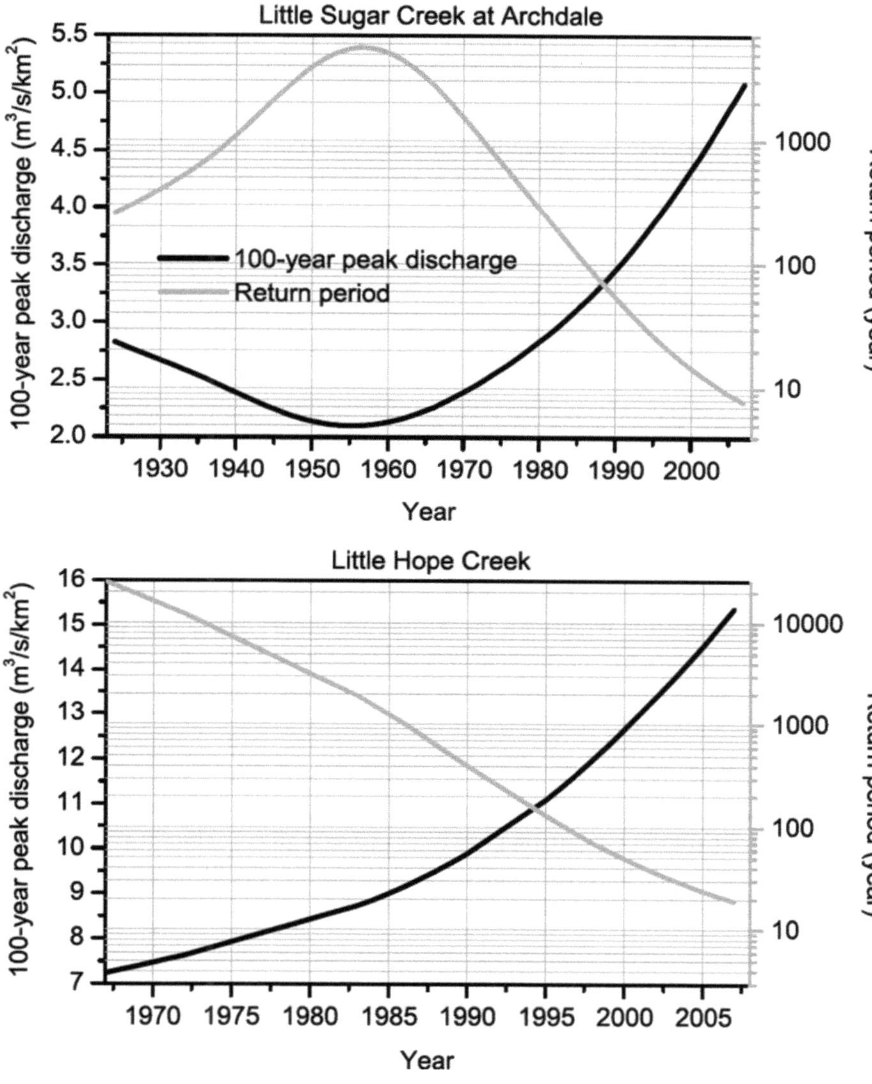

Fig. 13.7 Time trends in the value of peak discharge with 0.01 annual exceedance probability (*black line*) and in the estimated return period of the 0.01 annual exceedance flood peak derived from the 83-year gage record (*gray line*). These results are based on the GAMLSS model for Little Sugar Creek at Archdale (*top panel*) and Little Hope Creek (*bottom panel*). For reference, the Bulletin 17B 100-year peak estimates are 4.3 and 15.0 $m^3\ s^{-1}\ km^{-2}$, respectively, for Little Sugar Creek and Little Hope Creek, based on the 20-year record from 1987 through 2006 (Source: Villarini et al. 2009)

a correlation with the Southern Oscillation Index (SOI) of −0.6. Using a GML approach to estimate conditional medians ($Q_{0.5}$) and 95% confidence intervals from both the classical stationary model and nonstationary models it was found that:

- The nonstationary models represent the data variance more adequately.
- Differences in conditional quantile estimates ($Q_{0.5}$) are especially high for low SOI values, where estimates from the more complex nonstationary model (location as quadratic function of covariate) can be three times as high as estimates from the stationary model but for high SOI values the classical approach leads to underestimation.
- Precision of estimates depends on information available to establish dependence between parameters and covariates, e.g. few cases of low value SOI are available and this is borne out by wider confidence intervals at low SOI values.

Covariates can be introduced in two ways: If time itself is selected as the covariate then quantiles of the process of interest can be calculated and changes over time can be examined. Where the covariate is a time-varying stochastic process, one can compute conditional quantile estimates (given the value of the covariate). In addition to the classical model with time-constant parameters, three models were developed in Sugahara et al. (2009) (location parameter as linear and quadratic function of covariate respectively, both location and scale parameter as linear functions of covariate). Whether a more complex model explains a significantly larger part of the variance in the data was assessed based on a deviance statistics D (calculated as the difference in the maximised log likelihood function of the two models, see Coles 2001).

How the results from nonstationary frequency estimation are used in hydrological design depends on the time scales over which the covariate changes. Flood risk levels can be re-estimated on a seasonal or annual basis taking into account the current state of relevant predictors, such as the SOI index while for the design of infrastructure with long life spans risk assessment should be carried out by integrating the risk level throughout the lifetime of the structure.

13.4.5.2 Addressing Nonstationarity in Regional Frequency Analysis

The number of publications discussing nonstationary models in relation to the widely used regional flood frequency analysis (index flood approach) is still very limited. The first to develop such an approach were Cunderlik and Burn (2003) assuming temporal and spatial variation in location and scale parameter. However, this technique is mathematically complex and computationally intensive. Cunderlik and Ouarda (2006) refined this approach by assuming that the most uncertain parameter – the shape parameter – was still time-varying but spatially constant while the regional scale parameter was estimated as the weighted average of at-site scale parameters. Regional frequency analysis approaches can be used in exploring climate model projections. As discussed in the example below, for these studies nonstationarity needs to be explicitly accounted for. Studies indicate the need to (a) consider seasonal models and (b) assess changes across a range of quantiles.

Hanel et al. (2009) assume for their work on precipitation extremes that all three GEV parameters vary with time while only the location parameter varies both in time and space, and with the additional constraint that relative changes in quantiles are constant over the region. The temporal trend in the location parameter is assumed to be constant over the region of interest. This is motivated by the assumption that changes in extreme precipitation are mainly associated with large-scale changes in the atmospheric conditions. While short period convective storms may cause local flooding in summer, in winter multiday events may lead to flooding with larger spatial extent. Separate analyses were therefore undertaken for summer (1-day totals) and winter (5-day totals). Data from a regional climate model (RACMO) were compared to gridded observations for the Rhine basin. Hanel et al. chose exponential dependence on a time-dependent covariate for both the dispersion coefficient and the location parameter. Seasonal global temperature anomalies from the driving climate model (ECHAM5) were chosen as the time-dependent covariate. Changes in 1-day summer precipitation maxima were mainly found for large quantiles and these were primarily related to positive trends in the dispersion coefficient while there were smaller trends of opposite sign in quantiles close to the median. For the 5-day maximum winter on the other hand, low quantiles were increasing due to increase in the location parameter. For higher return periods this effect was offset by decrease in the shape parameter, resulting in small increase or even decrease.

An alternative to the 'spatial' GEV approaches described above is the max-stable process model. The important difference between the two is that max-stable processes also account for dependence between (point-based) records. This technique was applied to data at 30 subdaily gauges in east Australia for the period 1965–2005 and a statistically significant increase in annual maxima of 6-min rainfall by 18% was detected, with smaller increases at longer durations and an absence of significant trends at the daily duration (Westra and Sisson 2011). Global sea surface temperature, Australian annual average temperature and Southern Oscillation Index were considered as covariates. The increase in subdaily extreme precipitaton shows strongest relationship with global SST. However, this relationship is not sufficient to fully explain the strong increase in subdaily rainfall extremes and its attribution therefore requires further investigation.

13.4.6 Modelling Approaches

For estimation of flood magnitudes, Nathan and Weinmann (1995) distinguish between three different ranges of AEP:

- frequent events (AEP above 1%)
- extreme events (beyond the credible limits of extrapolation) and
- 'intermediate' events (the range between the first two categories).

For the first category, estimates could be derived directly, that is by extrapolation from flood records. Due to a lack of suitably long records this is, however, rarely the solution. More commonly an indirect approach has to be used: design rainfall estimates are derived and these are used to derive flood hydrographs. Sources of uncertainty are assumptions about spatial and temporal rainfall patterns as well as initial conditions (soil moisture and state of surface/subsurface storages). Nathan and Weinmann therefore recommend comparison of model results with at-site or regional flood frequency estimates in an attempt to reduce model uncertainty.

Estimates for the most extreme events (PMP/PMF) are dealt with using a similar technique: PMP estimates are fed into runoff-routing models to derive estimates of PMF. While estimates are generally deemed to exhibit little sensitivity to antecedent conditions like soil moisture (due to the magnitude of the PMP), they can be extremely sensitive to assumptions about initial storage conditions (reservoir may or may not be drawn down prior to the onset of a major storm). Due to a lack of suitable observations, model calibration for extreme events is especially challenging. Nathan and Weinmann (1995) recommend two strategies to explore and reduce the effects of uncertainties discussed above: (a) the use of joint probabilities (discussed below) and (b) comparing resulting estimates with envelope curves of observed extreme events.

13.4.6.1 Monte Carlo/Joint Probability Approaches

By creating a large number of replicates or scenarios rather than single estimates it is possible not only to derive a 'best estimate' of the expected value but also to provide an estimate of the inherent uncertainty. This approach can be extended to allow for observations to be treated as random variables, taking observed dependencies into account (stochastic modelling) and to make use of atmospheric models in deriving point and areal precipitation.

In recognition of the fact that design flood characteristics may result from a variety of combinations of different flood producing factors, Nathan et al. (2002) suggest use of a Monte Carlo framework to characterize hydrologic risk. With the aim of a more realistic presentation of flood generation processes, inputs are considered in terms of their joint probability distributions. This approach is an extension of current state-of-practice approaches where – apart from rainfall – input parameters are generally treated as fixed.

For flood risk analysis the complete 'flood disaster chain' from the triggering event through to its consequences needs to be assessed. However, taking into account all relevant flooding scenarios, their associated probabilities and consequences/damages in a Monte Carlo Framework leads to a highly complex model and it may therefore be desirable to simplify this approach (Apel et al. 2006). Advantages of a pared-down version of this approach include faster computation, robustness (due to fewer parameters) and a model that is easier to understand. For example, stochastic rainfall simulations and spatially distributed catchment models can be used to model runoff processes. Alternatively the approach can be simplified by using flood frequency curves and correlations between catchments instead. The authors present a Monte Carlo framework that includes the following components: hydrological load, flood routing, levee failure and outflow through levee breach, damage estimation. Similar tools could be useful in decision-making required during the design-stage and in an operational environment.

This approach provides a suitable framework for assessing hydrological risk in the presence of nonstationarity. With regards to nonstationarity in a changing climate, and where the required information is available (e.g from climate modelling) one can account for variability and change through distributional parameters, and describe changed relationships through joint probability.

13.4.6.2 Continuous Simulation

An alternative to event-based approaches used in design flood estimation is continuous simulation. An extensive review of this approach, including its origin as a tool in flood forecasting, is provided in Boughton and Droop (2003). An overwhelming array of models of different levels of complexity (e.g. lumped vs distributed models) has been developed for continuous simulation, to allow accounting for different hydrological processes in different regions and different applications but the main components tend to be loss model and flood hydrograph models. Calibration is typically based on recorded streamflows. For ungauged catchments model parameters may instead be found indirectly, through relationship with catchment characteristics. Advantages over design-based approaches are due to the fact that a number of subjective assumptions and choices are not required in continuous simulation:

- assumptions about losses,
- separation of surface runoff and baseflow and
- selection of a critical storm.

Stochastic rainfall generators can be used to produce very long series of rainfall. This allows simulation of long periods of streamflow from which to derive required estimates. While it may not be possible to evaluate the relative performance of event-based approaches and continuous simulation on the basis of estimates derived, such comparisons should prove useful in assessing uncertainty.

Continuous simulation has potential for exploring effects of climate change on design floods because simulation can be driven by suitably downscaled and corrected rainfall series derived from GCMs (Prudhomme et al. 2002). However, one of the assumptions behind this approach is that the calibration is relevant not only for the current climate but also for future climate conditions. The validity of such an assumption would have to be assessed on a case-by-case basis.

13.4.6.3 Prediction in Ungauged Climates

Modelling in a changing world can be considered analogous to the problem of ‘prediction in ungauged basins’ (PUB) as ‘prediction in ungauged climates’ (PUC), requiring extrapolation in time rather than (or in addition to) extrapolation in space. Impact assessments often implicitly assume that parameters calibrated under current climate conditions will remain valid under future conditions. When considering modelling approaches to address this problem, one might initially want to focus on developing techniques that will provide robust parameter estimates under current conditions, for instance deriving a priori estimates based on physical relationships, regionalisation of calibrated model parameters, use of multi-objective and regional calibration or alternatively use ensemble modelling and model output averaging to reduce uncertainty (Peel and Blöschl 2011). Under changing conditions, either due to a changing climate or other changes affecting hydrological processes (like land use change) the following approaches should be considered:

- *Scenario approaches* – running precalibrated models with data representative of changed conditions (either using downscaled and bias corrected GCM output or by appropriately scaling observations to reflect projected change).
- *Sensitivity methods* (either model or data-based) – using the concept of elasticity, e.g. the magnitude of changes in streamflow is quantified based on the magnitude of changes in temperature and precipitation.
- *Spatial gradients* – similar to the method of analogues in that one assumes that under a changed climate conditions hydrological processes for a catchment might resemble those for another catchment under current climate conditions.

The challenge here is in combining techniques to ensure robust parameter estimates discussed above with modelling approaches suitable for use under changing conditions. Promising in this context may be the use of 'hydrological signature modelling': Plausible parameter sets and model structures are those that reproduce hydrological signatures of interest like runoff ratio or baseflow index across timescales.

13.4.6.4 Uncertainty and Impact Studies

A fundamental problem in using hydrological models calibrated using historical data in studying the response of runoff to climate change is assessing the validity of calibration parameters. An Australian case study (Vaze et al. 2010) indicates that such models if calibrated using more than 20 years of data can be used for climate impact studies where the future mean annual rainfall does not change by more than 15–20%. It was found that it is more difficult for a model calibrated under wet conditions to predict dry conditions than vice versa. Given that projections for southeast Australia indicate a drier future, there is good reason to use models calibrated using records representative of recent drought conditions.

The role of hydrological modelling uncertainty in impact studies can be remarkably high. Three techniques that allow assessing hydrological modelling uncertainty due to both parameter and structural uncertainty are Generalised Likelihood Uncertainty Estimates (GLUE), Bayesian Total Error Analysis (BATEA,

Thyer et al. 2009) and Bayesian Model Averaging (BMA). The GLUE scheme is widely used to account for parameter uncertainty. Its strength lies in the fact that it can handle parametrically and structurally different models. However, uncertainty estimates derived using this technique are highly sensitive to a number of subjective decisions (number of models, range of model parameters, selection of threshold to differentiate between acceptable and unacceptable solutions). BMA not only accounts for uncertainty about model forms or assumptions but also propagates this uncertainty to inferences about quantities of interest.

While the GLUE and BMA techniques are conceptually different, they yielded similar results when applied to assess climate change impacts over the next century for four Irish river catchments on the basis of three GCMs, two emission scenarios and four conceptual rainfall-runoff models: progressive increase in streamflow in winter and progressive decrease in summer (Bastola et al. 2011). From a decision-making perspective, both BMA and GLUE can add value to predictions by allowing a truer assessment of the uncertainties in future projections.

13.4.6.5 Downscaling Under Consideration of Circulation Patterns

Output from GCM generally has to be pre-processed prior to use for hydrological applications. Statistical or dynamical downscaling techniques are required to increase resolution for use with hydrological models and validation of downscaled data against historical data (for instance using Taylor diagrams) tends to indicate the need for additional corrections.

For the analysis of hydrological extremes so-called ‘delta’ methods (percentage scaling) is inappropriate and quantile-quantile (Q-Q) matching approaches are preferred instead. Bárdossy and Pegram (2011) present an extension to this approach: based on sea level pressure from reanalysis data they define a set of 20 circulation patterns (CP) for use in the downscaling procedure. Their analyses suggest that the use of CP is particularly relevant where precipitation under CP differs strongly from climatology. A ‘double Q-Q transform’ is used because rainfall distributions differ between CPs (as expected) but are also found to differ between the observed period and future scenarios. Extremes relevant for hydrological applications were assessed for the Rhine basin (annual maxima of daily rainfall for flash flooding and an antecedent precipitation index related to drought conditions) and results indicate that the basin might become wetter than it has been in the past.

Non-homogeneous hidden Markov models (NHMM) have been used for some time to relate the occurrence of precipitation at multiple sites to broad-scale atmospheric circulation patterns (Hughes et al. 1999). This technique can be considered a stochastical downscaling approach: atmospheric predictors are downscaled to multisite daily precipitation *occurrence* while the precipitation *amounts* are downscaled based on conditional multiple linear regression (Charles et al. 2007).

13.4.7 *Probable Maximum Precipitation (PMP) and Probable Maximum Flood (PMF)*

PMF estimates are used in dam safety assessments. Whether one could or should attach an AEP to the PMF is a contested issue. However, most would agree that the frequency-based approaches described in Sects. 13.4.3 and 13.4.4 are not suited to estimation of the PMF.

Estimates of Probable Maximum Flood (PMF) are used in the design of a project at a particular geographical location in a given watershed (WMO 2009). PMF estimates may also assist in designing the size (dam height, reservoir storage capacity) of a structure and dimensioning of its flood carrying components (spillway and reservoir storage capacity). In particular, PMF estimates can be used to evaluate risks associated with the hazard of dam failure or flooding of nuclear power plants (Fernandes et al. 2010) and hydropower stations (Thompson 2003).

Estimation of a theoretical storm of critical duration and depth (Probable Maximum Precipitation, PMP) is undertaken to determine the Probable Maximum Flood (PMF) for a particular watershed on the basis of current knowledge of the hydrometeorological and hydrological processes under extreme conditions. The definition of PMF (after WMO 2009) addresses three key points:

- PMF is the theoretical maximum flood in a design watershed;
- poses extremely serious threats to the flood control of a given project;
- could plausibly occur in a locality at a particular time of year under modern meteorological conditions.

The estimation of PMF is undertaken outside the framework of conventional flood frequency analysis, for three reasons (Fernandes et al. 2010):

- Firstly (and perhaps most obvious) is the fact that extreme events like the PMF are considered beyond the limit of credible extrapolation based on typically available observed flood records.
- Secondly, distributions used in flood frequency analysis (e.g. Generalised Extreme Value distribution), are not generally unbounded whereas the idea of PMF implies an upper limit/bound.
- Thirdly, current deterministic techniques for estimating PMF do not allow deriving estimates of associated uncertainty.

Given the potentially disastrous implications of underdesign and the economical implications of overdesign, it appears prudent to invest substantial effort into (a) deriving reliable estimates and (b) providing guidance on the uncertainty attached to estimates of PMP/PMF in a changing climate.

13.4.7.1 Estimation of Probable Maximum Precipitation (PMP)

PMP estimates should be considered approximations and where feasible, validation of these estimates should be undertaken. Assumptions required in deriving PMF estimates on the basis of PMP estimates include antecedent moisture and unit hydrograph shapes. This means even where the PMP estimate is being held constant; the PMF estimate is not unique (Vogel et al. 2007).

Much of the following discussion on techniques for PMP estimation is based on the material provided in a WMO manual. This manual (WMO 2009) provides:

- a classification of methods together with a discussion of their suitability depending on availability of data and hydrometeorological expertise; watershed, climate and orography, and
- illustrates the use of procedures using examples from actual studies undertaken by the National Weather Service (formerly United States Weather Bureau), National Oceanic and Atmospheric Administration, United States Department of Commerce, the Australian Bureau of Meteorology, and water and power authorities in China and India.

Fundamentally, approaches will either focus on the watershed in question (direct approach) or on the meteorological event (indirect approach). The indirect approach is based on storm area and estimates are then converted into PMP for the collecting area of a particular project in the design watershed. WMO (2009) distinguishes between the following 8 methods:

(a) The local method (local storm maximization or local model)
(b) The transposition method (storm transposition or transposition model)
(c) The combination method (temporal/spatial maximization of storm or storm combination)
(d) The inferential method (theoretical model or rationality model)
(e) The generalized method (generalized estimation)
(f) The statistical method

and for extremely large watersheds:

1. The major temporal and spatial combination method;
2. The storm simulation method based on historical floods.

The statistical technique is based on the idea of the transposition of storms. On the basis of annual maximum rainfall series for a given duration, mean annual maxima and standard deviations are calculated and using the 'general frequency equation' a scaling parameter K_m is derived which can then be used to estimate PMP for the site of interest (Hershfield 1977).

The major advantage of this technique is its simplicity. Only precipitation data are required, neither dew point nor wind data have to be available. This approach is therefore suited to speedy assessments like those required for feasibility studies. Assessments will result in point estimates, and catchment averages are derived using area reduction curves. The technique is recommended only for small and intermediate catchment sizes. Estimates derived using the statistical technique may also provide an independent estimate against which to evaluate estimates derived on the basis of hydrometeorological techniques.

More commonly, techniques for estimating PMP are based on hydrometeorological assessments and maximisation of historical events. A key assumption made in deriving PMP estimates is the simultaneous occurrence of maximum moisture availability and maximum storm efficiency. Moisture maximisation is used to scale up the observed moisture availability to the maximum possible (for given location and season). Moisture availability is expressed as the moisture content of a column of air, referred to as precipitable water (PW, in mm). Estimates of this moisture content can be derived in two different ways: from upper-air soundings (using weather balloons) and from surface observations. Due to a lack of suitable upper-air data, estimates of moisture availability are usually derived from surface data under the assumption of a saturated atmosphere (pseudo-adiabatic lapse rate). For the central United States, this assumption was shown to lead to an overestimate of PMP by on average 6.9% (Chen and Bradley 2006). Upper-air soundings suggest that conditions conducive to maximum atmospheric moisture availability do not favour deep convection and heavy rainfall, raising doubt about the validity of the assumption of concurrent maximum moisture and maximum storm efficiency.

13.4.7.2 Uncertainties in Deriving PMF Estimates

Deriving estimates of PMF from PMP is based on the same concepts as deriving design flood estimates from design rainfall estimates: estimates of PMF are derived by running hydrological models where one of the key inputs is PMP. For this exercise it is generally assumed that a large storm has occurred, resulting in high antecedent moisture conditions. While PMF estimates are not very sensitive to variations in initial loss and continuing loss, estimates will depend significantly on assumptions made about the initial conditions of storage. One way to address the issue is through the use of a joint probability approach (Nathan and Weinmann 1995). Two other major sources of uncertainty in the estimation of PMF are spatial and temporal patterns of design rainfall. Together these factors may lead to uncertainties in the order of 50% of the PMF estimate.

13.4.7.3 Recent Developments

Rezacova et al. (2005) describe the development of the first set of PMP estimates for the Czech Republic, a study prompted by severe floods in July 1997 and August 2002. Two techniques (statistical technique, storm model) were used to allow comparison of PMP values derived. This study is noteworthy because radar data were used to develop maximum areal reduction factors, required in the conversion of point PMP estimates to catchment averages. Likewise, information from numerical weather prediction models was found to be useful for assessing spatially averaged rainfall for extreme events despite the fact that they might not actually match the local extremes due to resolution (Thompson 2003).

Current state-of-practice techniques for estimation of Probable Maximum Flood (PMF) result in valuable but uncertain estimates not least because the use of models for deriving PMF estimates implies using these models well outside the range of calibration. Validation of estimates using suitable benchmarks is therefore highly desirable (Nathan and Weinmann 1995). Palaeoflood and historical data may be used as a 'yardstick' against which PMF estimates can be compared because the largest floods are not necessarily captured in the instrumental record. An extreme flood event in the Gardon River (France) in September 2002 was considered the largest flood on record. Subsequent investigations found sedimentary evidence of at least five floods that were larger than the event in September 2002 (Sheffer et al. 2003). Palaeoflood data can therefore prove a useful tool used in validating estimates of the Probable Maximum Flood. Results from studies undertaken for a small number of catchments in the USA and Spain showed that existing PMF estimates may be very large overestimates (Enzel et al. 1993; Levish et al. 1996, 1997). Non-systematic information may also be useful in selecting appropriate parameters for hydrological modeling; and by including this information in flood frequency analysis, estimates of probabilities of rare events can be improved.

While the concept of PMF is essentially deterministic, estimates of exceedance probabilities are required to incorporate such events into quantitative risk assessment studies. There are different views on whether meaningful probabilities could be assigned to PMF estimates. Nathan and Weinmann (2001) argue that operational estimates might conceivably be exceeded and that conceptual foundations for estimating the AEP are unclear. To reflect the considerable uncertainty in estimating AEP for PMP, they suggest as notional upper and lower limits for the AEP plus/minus two orders of magnitude. Vogel et al. (2007) present an approach based on envelope curves for flood of record (FOR) and PMF to estimate exceedance probabilities. Based on theoretical considerations and a case study for 226 sites located across the USA they found exceedance probabilities of the order of 10^{-4}.

Fernandes et al. (2010) present a Bayesian approach to combine two apparently incompatible approaches: deterministic PMF estimates and frequency analysis of maximum flows. Key ingredients in this approach are:

- The assumption that there is a natural upper bound to the magnitude of floods (the at-site PMF) in a given region and therefore use of distributions which have an upper bound;
- Incorporating systematic (annual flood peaks) and non-systematic (historical and palaeoflood data); and
- Including information related to PMF in the formulation of an informative prior distribution.

Advantages of this approach include consistent and coherent results, and the opportunity to include information from different data sources, not least that of the PMF estimate as a reference for the upper bound of the frequency distribution. The underlying assumption of homogeneity and stationarity (especially with regards to non-systematic data) is briefly discussed and the authors argue for including these data because they provide information that is lacking from short samples: information about the largest floods.

In relation to estimation of PMP it might be worthwhile exploring the concept of atmospheric rivers and its association with extreme rainfall events. About 90% of poleward water vapour transport in the mid-latitudes occurs within narrow bands called 'atmospheric rivers' (Zhu and Newell 1998). Atmospheric rivers (AR) can lead to intense precipitation. Using daily GCM outputs of vertically integrated water vapour (IWV) and near surface (925 hPa) wind speed and direction, Dettinger et al. (2009) analyzed projected changes in frequency of such events for the winter season in California. While weakening of mid-latitude westerly winds partially compensated for a tendency to higher IWV, there is potential for increased flooding before and after the historical bounds of the flood season in California. Although the average intensity of storms is not projected to increase, occasionally 'much-larger-than-historical range storm intensities' are projected to occur under warming scenarios.

Use of physically based models in deriving PMP estimates for practical applications is still the exception although in the long term and especially with a view to assessing the potential effects of a changing climate, modelling approaches and Bayesian techniques appear the most promising recent developments. A physically based regional atmospheric model (MM5) was calibrated using NCAR (National Centre for Atmospheric Research) reanalysis data and a historical storm event was maximised using three different methods: (a) by maximising atmospheric moisture to 100% relative humidity, (b) by maintaining atmospheric conditions corresponding to the heaviest precipitation and (c) by spatially shifting atmospheric conditions to hit the watershed (Ohara et al. 2011). Maximum Precipitation (MP) estimates derived for the American River watershed in California for the 72-h duration were very similar across the three methods. Assuming initial and boundary conditions can be derived from reanalysis data for historical events and from GCM output for climate projections, this may prove a tool for assessing the potential effects of climate change on PMP estimates.

13.4.7.4 PMP and Potential Effects of Climate Change

Nonstationarity with regards to estimation of PMP due to changes in sea surface temperatures (SST) has been discussed decades ago (Schwarz 1972; Pyke 1975). Where extreme events are related to tropical storms it was recommended to make use of SST in the process of moisture maximisation. But while the question of how nonstationarity might affect hydrological extremes and design estimates is an area of active research, potential effects on estimates of PMP and PMF have so far not received much attention. One of these studies (Clark et al. 2002) argues that while global temperature is known to have increased, temperature changes for India (the study area) are less clear and while maximum persisting dew points may increase in some regions, for the region studied (India) there is no convincing evidence that this is the case.

In the second edition of the manual on PMP estimation (WMO 1986) the definition of PMP reads:

> *The greatest depth of precipitation for a given duration meteorologically possible for a design watershed or a given storm area at a particular location at a particular time of year, **with no allowance made for long-term climatic trends**.*

In the recent edition WMO (2009), PMP is defined as:

> *Theoretically, the greatest precipitation for a given duration that is physically possible over a given watershed area or size of storm area at a particular geographic location at a certain time of the year, **under modern meteorological conditions**.*

This subtle shift hints at the need to explore how PMP estimates might be changing in a changing climate. Few such studies have been undertaken, the following is based on a case study for Australia (Jakob et al. 2009).

The method applicable for most of mainland Australia is based on generalizing historical storms from a database containing 122 storms and dating back to the end of the nineteenth century. In generalising storms, features specific to a location that influence rainfall depth are removed. Steps in the generalization process include:

- deriving depth-duration-area-curves;
- defining zones depending on storm types (e.g. monsoonal trough, tropical cyclone);
- removing portion of rainfall due to orographic enhancement (remainder is labeled 'synoptic' component);
- maximizing local moisture availability; and
- adjusting for decay of storms (as storms move away from the coast and south/polewards).

In deriving PMP estimates for catchments, these local features are reconstructed. As a result, PMP estimates for a set of durations (1–7 days) are derived. As discussed above, spatial and temporal distributions are often required for further hydrological analyses.

To investigate potential effects of climate change on PMP estimates, the factors used in deriving these estimates were initially assessed separately. In addition, changes in generalized rainfall depths were assessed. These changes reflect the combined effects of changes in relative storm efficiency and moisture maximization. Some significant increases in moisture availability were found for coastal Australia, and climate models project further general increases, although with some regions of decrease. Very few significant changes in storm efficiency were found, although there is a tendency to a reduction in storm efficiency for coastal parts of eastern Australia. Typically, no significant changes were found in generalised rainfall depths, but a recent event was record breaking (both in terms of storm efficiency and generalised rainfall depth) if only for the season during which the event occurred (winter).

Since the PMP method is related to very large rainfall events, changes in both observed and projected extreme rainfall were also assessed using indices defined by the Expert Team on Climate Change Detection and Indices (ETCCDI):

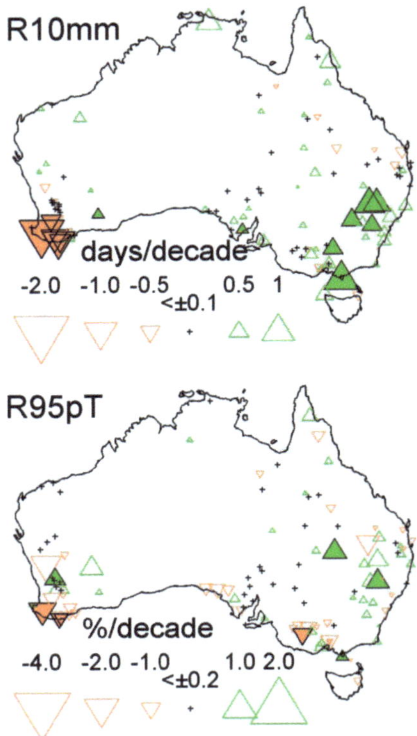

Fig. 13.8 Trends in extreme rainfall indices and annual rainfall for the period 1910–2005. Trends significant at the 0.05 level are shown as solid triangles with a black outline. Trends with a magnitude of less than 5% are shown as a black + (Source: Jakob et al. 2009)

- R10mm: Annual count of days where rainfall exceeds 10 mm
- RX5day: Seasonal maximum 5-day rainfall total
- R95pT: Proportion of annual rainfall from very wet days (total annual rainfall from wet days (≥1 mm) with rainfall above the 95th percentile for wet days in the 1961–1990 period, divided by the annual rainfall)

Long-term trends in rainfall extremes were found for only two regions (Fig. 13.8). This implies that for most of Australia current generalized estimates are representative of current climate conditions. Global climate models do not accurately model the trends of late twentieth century Australian rainfall. However, there is an indication that due to the overall increase in moisture availability in a warming climate the most extreme rainfall is likely to increase in the twenty-first century.

While this study did not confirm that PMP estimates are definitely increasing under a changing climate, confidence in the reliability of this assessment is limited due to the deterministic nature of the methodology applied to derive estimates of

PMP and the fact that it is based on historical events. While it is important to develop an understanding of how climate change might affect PMP estimates, this shouldn't remain the sole focus for further work in this area. A number of related issues remain to be addressed:

- Feasibility of using additional data sources for practical applications (e.g. radar data for spatial patterns; nonsystematic data);
- Validation of PMP estimates (through independent data and techniques);
- Quantification of uncertainty associated with operational estimates of PMP;
- Validity of current techniques (e.g. generalization) used to estimate PMP in a changing climate;
- Validity of assumption of simultaneous occurrence of maximum moisture availability and maximum storm efficiency; and
- Feasibility of using physical models in deriving PMP estimates operationally (Hardaker 1996; Leslie and Leplastrier 2008; Ohara et al. 2010).

13.5 Conclusions

This chapter started with a series of examples to set the scene. In a similar vein, an analogue will be used here for the conclusions. Michael Pollan (2009) in his book 'Food Rules: An Eaters Manual' sets out 64 rules on what one should eat, what kind of food one should eat and how one should eat. He summarises the essence of his extensive research in the following brief statement: 'Eat food. Not too much. Mostly Plants.' With regards to hydrological extremes and engineering design in a changing climate, here is an attempt to distil findings in a similarly short sentiment:

Design with change in mind. Not just climate. Think across disciplines.

While the examples presented in this chapter highlight the effect of climate on hydrological extremes, the discussion also showed that although climate is an important stressor, it would be simplistic and damaging to focus only on potential effects of a changing climate, neglecting other factors, like increasing urbanization and population growth. Economical impact assessments teach us that projections need to take into account climate change as well as other factors, namely socioeconomic changes. While a significant proportion of losses due to hydrological extremes may be attributable to climate change, suitable adaptation strategies can drastically reduce that component.

Nonstationarity has always been an issue to be dealt with in the design of structures; severe penalties have arisen in the past where this rule was not adhered to. Climate change adds a new dimension to this challenge. Beyond doubt are these facts: Humans are contributing to climate change, climate change is affecting hydrological extremes and this needs to be taken into consideration in the design of structures as well as in the assessment of existing structures.

The question is no longer whether climate models are useful but how to best use the information they provide. It may be a misguided effort to try and derive exact quantitative estimates of anticipated change. Instead the emphasis should shift to understanding the underlying processes. This is perhaps especially true for the estimation of Probable Maximum Precipitation and Probable Maximum Flood.

New and innovative approaches to decision-making are required. Considering the uncertainty in climate projections and the propagation of this uncertainty through the chain of models, from downscaling via hydrological models to decision-making tools, alternatives become attractive. The starting point could therefore be to identify thresholds, which if they were exceeded, we couldn't cope with and then with the use of climate models to assess the likelihood of such thresholds being exceeded.

Design of infrastructure requires information that is derived across a range of disciplines: meteorology, climatology, hydrology, statistics, engineering and decision-making. Addressing common misconceptions in communication between these communities should be one focal point, because even as they are becoming more engaged with each other, there is still ample room for misinterpretation because key terminology is used with different meanings (e.g. 'extreme').

A large part of this chapter covers new statistical techniques (or extensions of existing techniques) to address nonstationarity in hydrological extremes, through the use of time-varying parameters, moments, quantile estimates and the use of covariates. A changing climate may prove impetus to change some of the existing paradigms and explore new avenues. Bayesian techniques are a promising framework in this context because (a) they allow incorporating almost any type of relevant additional information (including non-systematic) into the analysis and (b) because of their superior suitability for assessing uncertainty. The need to reduce uncertainty, or alternatively derive more reliable uncertainty estimates, is exacerbated in a changing climate. This is a recurring theme throughout the discussion in this chapter and one of the key strategies should be moving from deterministic to probabilistic approaches.

Given the nature of processes leading to nonstationarity in hydrological extremes, exact estimates of how these extremes change over the next few decades will not be available. But this inherent uncertainty can be addressed in the way we plan the design of structures. It may no longer be appropriate to work towards an 'optimum' solution, instead the idea of sustainability and building with future generations in mind should be given more emphasis: Design should allow some degree of flexibility, allowing for and actually planning for future upgrades as they are required.

Acknowledgement The importance of the subject matter is evident from the number of recent journal publications and workshops on the topic. For readers looking for a small number of informative discussions and reviews focusing on particular aspects, the following is a shortlist of recommended reading. Extensive use has been made of this material in the preparation of this book chapter.

Hydrological extremes in a changing world
Bates et al. (2008)

IPCC (2011)
Kundzewicz and Kaczmarek (2000)

Rainfall frequency analysis
Svensson and Jones (2010)

Frequency analysis under nonstationarity
Khaliq et al. (2006)

Bayesian techniques
Seidou et al. (2006)

Decision making
Water Utility Climate Alliance (2010)

Crosscutting
Workshop on Nonstationarity, Hydrologic Frequency Analysis, and Water Management, 13–15 January 2010, Boulder, Colorado, Colorado Water Institute Information Series No. 109, available online: http://www.cwi.colostate.edu/NonstationarityWorkshop/proceedings.shtml.

I am grateful to all of those who have supported me in writing this chapter. In particular I would like to thank Sri Srikanthan and Karin Xuereb at the Australian Bureau of Meteorology and Amir AghaKouchak at the Department of Civil and Environmental Engineering University of California for their input during the review process.

References

Adger WN, Agrawala S, Mirza MMQ, Conde C, O'Brien K, Pulhin J, Pulwarty R, Smit B, Takahashi K (2007) Assessment of adaptation practices, options, constraints and capacity. Climate change 2007: impacts, adaptation and vulnerability. Contribution of Working Group II to the fourth assessment report of the Intergovernmental Panel on Climate Change, Cambridge University Press, Cambridge, UK, pp 717–743

Ailliot P, Thompson C, Thomson P (2011) Mixed methods for fitting the GEV distribution. Water Resour Res 47:W05551. doi:10.1029/2010WR009417

Alila Y (1999) A hierarchical approach for the regionalization of precipitation annual maxima in Canada. J Geophys Res 104(36):31645–31655

Apel H, Thieken AH, Merz B, Bloeschl G, Apel H, Thieken AH, Merz B, Bloeschl G (2006) A probabilistic modelling system for assessing flood risks. Nat Hazards 38:79–100. doi:10.1007/s11069-005-8603-7

Arblaster JM, Meehl GA, Karoly DJ (2011) Future climate change in the Southern Hemisphere: competing effects of ozone and greenhouse gases. Geophys Res Lett 38(2):1–6. doi:10.1029/2010GL045384

Bárdossy A, Pegram G (2011) Downscaling precipitation using regional climate models and circulation patterns toward hydrology. Water Resour Res 47(4):1–18. doi:10.1029/2010WR009689

Bastola S, Murphy C, Sweeney J (2011) The role of hydrological modelling uncertainties in climate change impact assessments of Irish river catchments. Adv Water Resour 34(5):562–576. doi:10.1016/j.advwatres.2011.01.008, Elsevier Ltd

Bates BC, Kundzewicz ZW, Wu S, Palutikof JP (eds) (2008) Climate change and water. Technical paper of the Intergovernmental Panel on Climate Change, IPCC Secretariat, Geneva, 210 pp

Bayliss AC, Reed DW (2001) The use of historical information in flood frequency estimation, Report to MAFF, CEH Wallingford

Benito G, Thorndycraft VR (2005) Palaeoflood hydrology and its role in applied hydrological sciences. J Hydrol 313:3–15

Benito G, Lang M, Barriendos M, Llasat MC, Francés F, Ouarda T et al (2004) Use of systematic, palaeoflood and historical data for the improvement of flood risk estimation. Review of scientific methods. Nat Hazards 31(3):623–643. doi:10.1023/B:NHAZ.0000024895.48463.eb

Boé J, Terray L, Martin E, Habets F (2009) Projected changes in components of the hydrological cycle in French river basins during the 21st century. Water Resour Res 45:W08426

Boughton W, Droop O (2003) Continuous simulation for design flood estimation – a review. Environ Model Software 18(4):309–318. doi:10.1016/S1364-8152(03)00004-5

Brown C (2009) Decision-scaling for robust planning and policy under climate uncertainty. World resources report, Washington, DC. Available online at http://www.worldresourcesreport.org

Buckley BM, Anchukaitis KJ, Penny D, Fletcher R, Cook ER, Sanod M, Nam LC, Wichienkeeof A, Minh TT, Mai Hong TM (2011) Climate as a contributing factor in the demise of Angkor, Cambodia. Proceedings of the National Academy of Science, www.pnas.org/cgi/doi/10.1073/pnas.0910827107

Bukovsky M, Karoly DJ (2011) A regional study of climate change impacts on warm-season precipitation in the Central U.S. J Climate. doi:10.1175/2010JCLI3447.1

Burauskaite-Harju A, Grimvall A, von Brömssen C (2010) A test for network-wide trends in rainfall extremes. Int J Climatol. doi:10.1002/joc.2263

Bureau of Meteorology (1974) Brisbane floods. Department of Science, Canberra

Bureau of Meteorology (2011) Special climate statement 24 (1st issued 7th January 2011, updated 24th February 2011), http://www.bom.gov.au/climate/current/special-statements.shtml. Accessed 30 Apr 2011

Charles SP, Bari MA, Kitsios A, Bates BC (2007) Effect of GCM bias on downscaled precipitation and runoff projections for the Serpentine catchment Western Australia. Int J Climatol 1690:1673–1690. doi:10.1002/joc

Chen L-C, Bradley AA (2006) Adequacy of using surface humidity to estimate atmospheric moisture availability for probable maximum precipitation. Water Resour Res 42(9):1–17. doi:10.1029/2005WR004469

Chiew FHS (2006) Estimation of rainfall elasticity of streamflow in Australia. Hydrol Sci J 51: 613–625

Christensen JH, Christensen OB (2002) Severe summertime flooding in Europe. Nature 421: 805–806

Clark C, Rakhecha PR, Hill C, Lane S (2002) Areal PMP distribution of one-day to three-day duration over India. Meteorol Appl 9:399–406. doi:10.1017/S1350482702004024

Coles S (2001) An introduction to the statistical modelling of extreme values, Springer Series in Statistics. Springer, London

Cook JL (1987) Quantifying peak discharges for historical floods. J Hydrol 96:29–40

Crosbie RS, Dawes WR, Charles SP, Mpelasoka FS, Aryal S, Barron O, Summerell GK (2011) Differences in future recharge estimates due to GCMs, downscaling methods and hydrological models. Geophys Res Lett 38(11):1–5. doi:10.1029/2011GL047657.

Cunderlik JM, Burn DH (2003) Non-stationary pooled flood frequency analysis. J Hydrol Amst 276:210–223. doi:10.1016/S0022-1694(03)00062-3

Cunderlik JM, Ouarda TBMJ (2006) Regional flood-duration- frequency modeling in the changing environment. J Hydrol Amst 318:276–291. doi:10.1016/j.jhydrol.2005.06.020

Dale M (2005) Impact of climate change on UK flooding and future predictions. Proc ICE Water Manage 158(4):135–140. doi:10.1680/wama.2005.158.4.135

Dalrymple T (1960) Flood frequency analysis, U.S. Geological Survey, Water Supply Paper 1543-A, Reston, VA

Department of Environment and Climate Change, New South Wales (Australia) (2007) Floodplain risk management guideline – practical consideration of climate change, 14 pp

Dettinger MD, Hidalgo H, Das T, Cayan D, Knowles N (2009) Projections of potential flood regime changes in California: California Energy Commission Report CEC-500-2009-050-D, 68 pp

El Adlouni S, Ouarda TBMJ, Zhang X, Roy R, Bobée B (2007) Generalized maximum likelihood estimators of the non-stationary GEV model parameters. Water Resour Res 43:W03410. doi:10.1029/2005WR004545

Enzel Y, Ely LL, House PK, Baker VR (1993) Paleoflood evidence for a natural upper bound to flood magnitudes in the Colorado river basin. Water Resour Res 29:2287–2297

Faulkner D (1999) Flood estimation handbook, vol 2: Rainfall frequency estimation. Institute of Hydrology, Wallingford, 110 pp

Fernandes W, Naghettini M, Loschi R (2010) A Bayesian approach for estimating extreme flood probabilities with upper-bounded distribution functions. Stoch Environ Res Risk Assess 24(8):1127–1143. doi:10.1007/s00477-010-0365-4

Frei C, Schöll R, Fukutome S, Schmidli J, Vidale PL (2006) Future change of precipitation extremes in Europe: intercomparison of scenarios from regional climate models. J Geophys Res 111:D06105. doi:10.1029/2005JD005965

Groisman PY, Knight RW, Easterling DR, Karl TR, Hegerl GC, Razuvaev VN (2005) Trends in intense precipitation. J Climate 18:1326–1350

Haddad K, Rahman A, Weinmann E, Kuczera G, Ball J (2010) Streamflow data preparation for regional flood frequency analysis: lessons from southeast Australia. Aust J Water Resour 14(1):17–32.

Hamed KH, Rao AR (1998) A modified Mann-Kendall trend test for autocorrelated data. J Hydrol 204:182–196

Hanel M, Buishand TA, Ferro CAT (2009) A nonstationary index flood model for precipitation extremes in transient regional climate model simulations. J Geophys Res Atmos 114:D15107

Hanson S, Nicholls R, Ranger N, Hallegatte S, Corfee-Morlot J, Herweijer C et al (2010) A global ranking of port cities with high exposure to climate extremes. Clim Chang 104:89–111. doi:10.1007/s10584-010-9977-4

Hardaker PJ (1996) Estimation of Probable Maximum Precipitation (PMP) for the Evinos catchment in Greece using a storm model. Meteorol Appl 3:137–145

Hawkins E, Sutton R (2009) The potential to narrow uncertainty in regional climate predictions. Bull Am Meteorol Soc 90(8):1095–107. doi:10.1175/2009BAMS2607.1.

Hawkins E, Sutton R (2010) The potential to narrow uncertainty in projections of regional precipitation change. Clim Dyn 37(1–2):407–18. doi:10.1007/s00382-010-0810-6.

Hershfield DM (1977) Some tools for hydrometeorologists. Preprints, 2nd conference hydrometeorology, American Meteorological Society, Boston, MA, pp 79–82

Hosking JR, Wallis JR (1987) Parameter and quantile estimation for the generalized Pareto distribution. Technometrics 29:339–49.

Hosking JRM, Wallis JR (1997) Regional frequency analysis: an approach based on L-moments. Cambridge University Press, Cambridge

Hughes JP, Guttorp P, Charles SP (1999) A non-homogeneous hidden Markov model for precipitation occurrence. Appl Statist 48:15–30

Institution of Engineers (1987) Australia Australian rainfall and runoff: a guide to flood estimation, vol 1, Editor-in-chief DH Pilgrim, Rev Edn 1987 (Reprinted edn 1998), Barton, ACT

IPCC (2007) Climate Change 2007: the physical science basis, contribution of Working Group I to the fourth assessment report of the Intergovernmental Panel on Climate Change, Solomon S, Qin D, Manning M, Chen Z, Marquis M, Averyt KB, Tignor M, Miller HL (eds). Cambridge/New York, 996 pp

IPCC (2011) Summary for policymakers. In: Field CB, Barros V, Stocker TF, Qin D, Dokken D, Ebi KL, Mastrandrea MD, Mach KJ, Plattner G-K, Allen SK, Tignor M, Midgley PM (eds) Intergovernmental panel on climate change special report on managing the risks of extreme events and disasters to advance climate change adaptation. Cambridge University Press, Cambridge/New York

Ishak EH, Rahman A, Westra S, Sharama A, Kuzcera G (2011) Trends in peak steamflow data in Australia: impacts of serial and cross-correlation. In: Proceedings of the 34th IAHR world congress, 26 June – 1 July 2011, Brisbane, Australia, pp 1–8

Jakob D (2011) Exploring links between effects of climate variations on average rainfall and rainfall extremes. In: Proceedings of the 34th IAHR world congress, 26 June–1 July 2011, Brisbane, Australia, pp 1–8
Jakob D, Smalley R, Meighen J, Xuereb K, Taylor B (2009) Climate change and probable maximum precipitation. HRS report no. 12. Australian Bureau of Meteorology, Melbourne
Jakob D, Karoly DJ, Seed A (2011a) Non-stationarity in daily and sub-daily intense rainfall – part 1: Sydney, Australia. Nat Hazards Earth Syst Sci 11(8):2263–2271. doi:10.5194/nhess-11-2263-2011
Jakob D, Karoly DJ, Seed A (2011b) Non-stationarity in daily and sub-daily intense rainfall – part 2: Regional assessment for sites in south-east Australia. Nat Hazards Earth Syst Sci 11(8):2273–2284. doi:10.5194/nhess-11-2273-2011
Khaliq MN, Ouarda TBMJ, Ondo J-C, Gachon P, Bobée B (2006) Frequency analysis of a sequence of dependent and/or non-stationarity hydro-meteorological observations: a review. J Hydrol 329:534–552
Kirshen P, Knee K, Ruth M (2008) Climate change and coastal flooding in Metro Boston: impacts and adaptation strategies. Clim Change 90(4):453–73. doi:10.1007/s10584-008-9398-9.
Knutti R, Allen MR, Friedlingstein P, Gregory JM, Hegerl GC, Meehl GA, Meinshausen M (2008) A review of uncertainties in global temperature projections over the twenty-first century. J Clim 21(11):2651–63. doi:10.1175/2007JCLI2119.1.
Kundzewicz Z, Kaczmarek Z (2000) Coping with hydrological extremes. Water Int 25(1):66–75. doi:10.1080/02508060008686798
Kundzewicz Z, Robson A (2004) Change detection in hydrological recordsa review of the methodology/Revue methodologique de la detection de changements dans les chroniques hydrologiques. Hydrol Sci J 49(1):1–19. doi:10.1623/hysj.49.1.7.53993
Ladson A (2008) Hydrology: an Australian introduction. Melbourne: Oxford University Press.
Lang M, Ouarda T, Bobée B (1999) Towards operational guidelines for over-threshold modeling. J Hydrol 225:103–1171
Lenderink G, van Meijgaard E (2008) Increase in hourly precipitation extremes beyond expectations from temperature changes. Nature 1:511–514
Leslie LM, Leplastrier M (2008) Estimating future trends in severe hailstorms over the Sydney Basin: a climate modelling study. Atmos Res 87:37–51
Levish D, Ostenaa D, O'Connell D (1996) Paleohydrologic bounds and the frequency of extreme floods on the Santa Ynez River, California, 1996, California weather symposium a prehistoric look at California rainfall and floods 1996, 19 pp
Levish D, Ostenaa D, O'Connell D (1997) Paleoflood hydrology and dam safety. In: Mahoney DJ (ed) Waterpower 97: proceedings of the international conference on hydropower. American Society of Civil Engineers, New York, pp 2205–2214
Mailhot A, Duchesne S, Caya D, Talbot G (2007) Assessment of future change in intensity-duration-frequency (IDF) curves for Southern Quebec using the Canadian Regional Climate Model (CRCM). J Hydrol 347:197–210
Malitz G (2005) Grundlagenbericht über Starkniederschlagshöhen in Deutschland, (Grundlagenbericht KOSTRA-DWD-2000), Deutscher Wetterdienst – Hydrometeorologie
Marsh TJ, Dale M (2002) The UK floods of 2000–2001: a hydrometeorological appraisal. Water Environ J 16:180–188. doi:10.1111/j.1747-6593.2002.tb00392.x
McCabe GJ, Palecki MA, Betancourt JL (2004) Pacific and Atlantic Ocean influences on multidecadal drought frequency in the United States. Proc Natl Acad Sci U S A 101(12): 4136–4141
McCallum E, Heming J (2006) Hurricane Katrina: an environmental perspective. Philos Transact A Math Phys Eng Sci 364(1845):2099–2115. doi:10.1098/rsta.2006.1815
McMahon TA, Kiem AS, Peel MC, Jordan PW, Pegram GS (2008) A new approach to stochastically generating six-monthly rainfall sequences based on Empirical Mode Decomposition. J Hydrometeorol 9:1377–1389
Merz B, Hall J, Disse M, Schumann A (2010) Fluvial flood risk management in a changing world. Nat Hazards Earth Syst Sci 10:509–527

Min S-K, Zhang X, Zwiers FW, Hegerl GC (2011) Human contribution to more intense precipitation extremes. Nature 470:378–381

Mudersbach C, Jensen J. Nonstationary extreme value analysis of annual maximum water levels for designing coastal structures on the German North Sea coastline. J Flood Risk Manage. 2010;3(1):52–62. doi:10.1111/j.1753-318X.2009.01054.x.

Nathan RJ, Weinmann PE (1995) The estimation of extreme floods – the need and scope for revision of our national guidelines. Aust J Water Resour 1:1

Nathan RJ, Weinmann PE (2001) Estimation of large and extreme floods for medium and large catchments: book VI. Australian rainfall and runoff – a guide to flood estimation, 4th edn. Engineers Australia, National Committee for Water Engineering, Australia

Nathan RJ, Weinmann PE, Hill PI (2002) Use of a Monte Carlo framework to characterise hydrological risk. ANCOLD Bull (122):55–64 Dec 2002

Ntelekos AA, Oppenheimer M, Smith JA, Miller AJ (2010) Urbanization, climate change and flood policy in the United States. Clim Change 103(3–4):597–616. doi:10.1007/s10584-009-9789-6.

OECD (2008) OECD environmental outlook to 2030. OECD Publishing, doi:10.1787/9789264040519-en.

O'Gorman P, Schneider T (2009) Scaling of precipitation extremes over a wide range of climates simulated with an idealized GCM. J Climate 22(21):5676–5685. doi:10.1175/2009JCLI2701.1

Ohara N, Kavvas ML, Kure S, Chen ZQ, Jang S, Tan E (2011) Physically based estimation of maximum precipitation over American River Watershed. California J Hydrol Eng 351–361. doi:10.1061/(ASCE)HE.1943-5584.0000324.

Ouarda TBMJ, Rasmussen PF, Bobée B, Bernier J (1998) Use of historical information in hydrologic frequency analysis. Water Sci J/Revue des Sciences de l'Eau 11:41–49

Peel MC, Blöschl G (2011) Hydrological modelling in a changing world. Prog Phys Geogr 35(2):249–261. doi:10.1177/0309133311402550

Pettitt AN (1979) A non-parametric approach to the change-point problem. Appl Statist 28(2): 126–135

Plate EJ (2002) Flood risk and flood management. J Hydrol 267:2–11

Pollan M (2009) Food rules: an eaters manual. Penguin Press, New York, 112 pp

Prudhomme C, Reynard N, Crooks S (2002) Downscaling of global climate models for flood frequency analysis: where are we now? Hydrol Process 16(6):1137–1150. doi:10.1002/hyp.1054

Pyke CB (1975) Some aspects of the influence of abnormal eastern equatorial ocean surface temperature upon weather patterns in the Southwestern United States. Final report, United States Navy Contract N-0014-75-C-0126. University of California, Los Angeles, CA

Reed DW (2011) Letters in applied hydrology, DWR Consult, 88 pp

Reeves J, Chen J, Wang XL, Lund R, Lu QQ (2007) A review and comparison of change-point detection techniques for climate data. J Appl Meteorol Climatol 46(6):900–915. doi:10.1175/JAM2493.1

Rezacova D, Pesice P, Sokol Z (2005) An estimation of the probable maximum precipitation for river basins in the Czech Republic. Atmos Res 77(1–4):407–421. doi:10.1016/j.atmosres.2004.10.011

Risbey JS, Pook MJ, McIntosh PC, Wheeler MC, Hendon HH (2009) On the remote drivers of rainfall variability in Australia. Mon Weather Rev 137(10):3233–3253. doi:10.1175/2009MWR2861.1

Robson A, Reed D (1999) Statistical procedures for flood frequency estimation. In: Flood estimation handbook, vol 3. Institute of Hydrology, Crowmarsh Gifford/Wallingford

Schneider SH, Kuntz-Duriseti AC (2000) Costing non-linearities, surprises and irreversible events. Pacific Asian J Energy 10:81–106

Schwarz FK (1972) A proposal for estimating tropical storm Probable Maximum Precipitation (PMP) for sparse data regions. Floods and droughts proceedings second international symposium in hydrology, Fort Collins, Colorado, 11–13 Sept 1972

Seidou O, Ouarda TBMJ, Barbet M, Bruneau P, Bobée B (2006) A parametric Bayesian combination of local and regional information in flood frequency analysis. Water Resour Res 42:W11408. doi:10.1029/2005WR004397, 1-21

Sheffer NA, Enzel Y, Grodek T, Waldmann N, Benito G (2003) Claim of largest flood on record proves false. EOS Trans Am Geophys Union 84:109

Stainforth DA, Downing TE, Washington R, Lopez A, New M (2007) Issues in the interpretation of climate model ensembles to inform decisions. Philos Trans R Soc A-Math Phys Eng Sci 365:2163–2177

Stedinger JR, Lu L-H (1995) Appraisal of regional and index flood quantile estimators. Stoch Environ Res Risk 9(1):49–75. doi:10.1007/BF01581758

Strupczewski W, Singh V, Mitosek H (2001) Non-stationary approach to at-site flood frequency modelling III. Flood analysis of Polish rivers. J Hydrol 248(1–4):152–167. doi:10.1016/S0022-1694(01)00399-7

Sugahara S, Rocha RP, Silveira R (2009) Non-stationary frequency analysis of extreme daily rainfall in Sao Paulo, Brazil. Int J Climatol 29:1339–1349

Svensson S, Jones DA (2010) Review of rainfall frequency estimation methods. J Flood Risk Manag 3:296–313

Swain RE, Bowles D, Ostenaa D (1998) A framework for characterization of extreme floods for dam safety risk assessments. In: Proceedings of the 1998 USCOLD annual lecture, Buffalo, New York, August 1998

Thompson SM (2003) Duration of probable maximum precipitation on lake catchments: alternative analysis. J Hydrol Eng 8(4):190. doi:10.1061/(ASCE)1084-0699(2003) 8:4(190)

Thyer M, Renard B, Kavetski D, Kuczera G, Franks SW, Srikanthan S (2009) Critical evaluation of parameter consistency and predictive uncertainty in hydrological modeling: a case study using Bayesian total error analysis. Water Resour Res 45(3):1–22. doi:10.1029/2008WR006825

Toreti A, Kuglitsch FG, Xoplaki E, Della-Marta PM, Aguilar E, Prohom M et al (2011) A note on the use of the standard normal homogeneity test to detect inhomogeneities in climatic time series. Int J Climatol 31(4):630–632. doi:10.1002/joc.2088

van den Brink HW, Können GP (2008) The statistical distribution of meteorological outliers. Geophys Res Lett 35:L23702

Vaze J, Post DA, Chiew FHS, Perraud J-M, Viney NR, Teng J (2010) Climate non-stationarity – validity of calibrated rainfall-runoff models for use in climate change studies. J Hydrol 394 (3–4):447–457. doi:10.1016/j.jhydrol.2010.09.018

Villarini G, Smith JA, Serinaldi F, Bales J, Bates PD, Krajewski WF (2009) Flood frequency analysis for nonstationary annual peak records in an urban drainage basin. Adv Water Resour 32(8):1255–1266

Vogel RM (2010) Flood magnification factors in the United States. Workshop on nonstationarity, hydrologic frequency analysis, and water management, Boulder, Colorado, 13–15 Jan 2010. Colorado Water Institute Information Series No. 109

Vogel RM, Matalas NC, England JF, Castellarin A (2007) An assessment of exceedance probabilities of envelope curves. Water Resour Res 43(7):1–11. doi:10.1029/2006WR005586

Wang QJ (1997) LH moments for statistical analysis of extreme events. Water Resour Res 33(9):2841–2848

Wang QJ, Robertson DE, Chiew FHS (2009) A Bayesian joint probability modeling approach for seasonal forecasting of streamflows at multiple sites. Water Resour Res 45:W05407. doi:10.1029/2008WR007355

Water Utility Climate Alliance (2010) Decision support planning methods: incorporating climate change uncertainties into water planning, January 2010 http://www.wucaonline.org/html/actions_publications.html.

Water Utility Climate Alliance: Decision Support Planning Methods: Incorporating climate change uncertainties into water planning (2010) Retrieved 7 Apr 2011 from http://www.wucaonline.org/html/

Westra S, Sisson SA (2011) Detection of non-stationarity in precipitation extremes using a max-stable process model. J Hydrol 406(1–2):119–128. doi:10.1016/j.jhydrol.2011.06.014

Westra S, Varley I, Jordan P, Nathan R, Ladson A, Sharma A (2010) Addressing climatic nonstationarity in the assessment of flood risk. Aust J Water Resour 14(1):1–16
Wilby RL, Troni J, Biot Y, Tedd L, Hewitson BC, Smith DM, Sutton RT (2009) A review of climate risk information for adaptation and development planning. Int J Climatol 1215: 1193–215. doi:10.1002/joc.
Wilks DS (2006) On "field significance" and the false discovery rate. J Appl Meteorol Climatol 45:1181–1189. doi:10.1175/JAM2404.1
Willems P (2000) Compound intensity/duration/frequency-relationships of extreme precipitation for two seasons and two storm types. J Hydrol 233(1–4):189–205
WMO (1986) Manual for estimation of probable maximum precipitation. Second ed. Operational hydrology report no. 1, WMO – No. 332, Geneva.
WMO (2009) Manual on estimation of Probable Maximum Precipitation (PMP). WMO-No. 1045
Yip S, Ferro CAT, Stephenson DB, Hawkins E (2011) A simple, coherent framework for partitioning uncertainty in climate predictions. J Clim doi:10.1175/2011JCLI4085.1.
Zhu Y, Newell RE (1998) A proposed algorithm for moisture fluxes from atmospheric rivers. Mon Weather Rev 126:725–35.
Ziervogel G, Johnston P, Matthew M, Mukheibir P (2010) Using climate information for supporting climate change adaptation in water resource management in South Africa. Clim Chang 103:537–554. doi:10.1007/s10584-009-9771-3

Index

A. AghaKouchak et al. (eds.), *Extremes in a Changing Climate: Detection, Analysis and Uncertainty*, Water Science and Technology Library 65,
DOI 10.1007/978-94-007-4479-0,

MIX
Papier aus verantwortungsvollen Quellen
Paper from responsible sources
FSC® C105338

If you have any concerns about our products,
you can contact us on
ProductSafety@springernature.com

In case Publisher is established outside the EU,
the EU authorized representative is:
Springer Nature Customer Service Center GmbH
Europaplatz 3, 69115 Heidelberg, Germany

Printed by Libri Plureos GmbH
in Hamburg, Germany